Modelling Electroanalytical Experiments by the Integral Equation Method

Monographs in Electrochemistry

Series Editor: Fritz Scholz, University of Greifswald, Germany

Surprisingly, a large number of important topics in electrochemistry is not covered by Up-to-date monographs and series on the market, some topics are even not covered at all. The series Monographs in Electrochemistry fills this gap by publishing indepth monographs written by experienced and distinguished electrochemists, covering both theory and applications. The focus is set on existing as well as emerging methods for researchers, engineers, and practitioners active in the many and often interdisciplinary fields, where electrochemistry plays a key role. These fields will range – among others – from analytical and environmental sciences to sensors, materials sciences and biochemical research.

More information about the series at
http://www.springer.com/series/7386

Lesław K. Bieniasz

Modelling Electroanalytical Experiments by the Integral Equation Method

Springer

Lesław K. Bieniasz
Cracow University of Technology
Cracow
Poland

ISSN 1865-1836 ISSN 1865-1844 (electronic)
ISBN 978-3-662-44881-6 ISBN 978-3-662-44882-3 (eBook)
DOI 10.1007/978-3-662-44882-3
Springer Heidelberg New York Dordrecht London

Library of Congress Control Number: 2014957755

Printed on acid-free paper

Springer is part of Springer Science+Business Media (www.springer.com)

I dedicate this book to the memory of my mother, Maria

Preface

Theoretical modelling of transient electroanalytical experiments is commonly accomplished by formulating and solving initial or initial-boundary value problems that describe transport of reactants, homogeneous and heterogeneous reactions, and other phenomena taking place in electrochemical systems. The conversion of these problems into integral equations, subsequently solved analytically or numerically, is a classical modelling method which has contributed significantly to the development of electroanalytical techniques. Although this method has limitations (compared to direct solutions of differential equations by finite difference, finite element, or similar methods), it can be more rewarding (in cases when it is applicable) owing to additional theoretical insights that it offers. It can also bring a reduction of the costs of numerical simulations, albeit at the expense of extra human effort needed for the derivation of the integral equations. The method is therefore worthy of application and development. Unfortunately for anybody willing to learn and use the method, its theoretical principles and example applications are scattered in the electrochemical literature, and none of the existing textbooks provides its comprehensive description. Even the last edition of the monograph by Dieter Britz, "Digital Simulation in Electrochemistry" (Springer, Berlin, 2005), which is currently the most thorough compendium of numerical methods used for simulating transient experiments, devotes only two pages to the integral equation method.

At the same time, integral equations arising in electroanalytical chemistry remain almost unknown to mathematicians or computational scientists engaged in studies of integral equations, and are rarely used as examples, or as an inspiration for research.

The present book is intended to help electrochemists, mathematicians, and computational scientists, to get acquainted with the integral equation method in electroanalytical chemistry, by providing a (hopefully) comprehensive state-of-the-art overview of the method and its applications. It is hoped that the book will stimulate interdisciplinary research, resulting in new developments of the method, and enlargement of its application scope.

Cracow, Poland
July 2014

Lesław K. Bieniasz

Acknowledgements

I wish to thank all, the personal contacts or correspondence with whom have helped me, over the years, to prepare this book. They are (in alphabetical order): Dieter Britz, Davis Cope, Alexey Davydov, Milivoj Lovrić, Peter Mahon, Valentin Mirčeski, Michael Mirkin, Janice Myland, Keith Oldham, Waldemar Rachowicz, Palghat Ramachandran, Zdeněk Samec, Jörg Strutwolf, Ondřej Wein, Gunther Wittstock.

I am particularly grateful to Prof. Dieter Britz and Prof. Marian Jaskuła for their comments resulting from reading an early draft of this book.

I apologise for any possible omission of important citations or results in this book. Although I have devoted a lot of time to collecting literature, one can never be sure if such effort is fully successful. In a few cases I have not been able to access the source materials listed in databases or on the Internet. I will appreciate feedback from the Readers and indication of any remaining flaws in this book.

Acronyms

AC	Alternating current
AE	Algebraic equation
BEM	Boundary element method
BIM	Boundary integral method
BVP	Boundary value problem
CE	Counter electrode
DC	Direct current
DAE	Differential-algebraic equation
DME	Dropping mercury electrode
ES	Electrolyte solution
IBVP	Initial-boundary value problem
IE	Integral equation
IDE	Integro-differential equation
IVP	Initial value problem
ODE	Ordinary differential equation
PDE	Partial differential equation
RDE	Rotating disk electrode
RE	Reference electrode
SECM	Scanning electrochemical microscope
WE	Working electrode

Contents

Chapter 1
Introduction

This book is devoted to the theoretical modelling of transient experiments in electroanalytical chemistry, by means of a specific mathematical approach known under the name of the method of integral equations (IEs). Electroanalytical chemistry is a scientific discipline that studies physico-chemical processes occurring at (and in the neighbourhood of) the interfaces between electronic conductors (metals, semiconductors, conducting polymers, etc.) and/or ionic conductors (liquid electrolytic solutions, solid electrolytes, electrolytic gels, etc.). Liquid | liquid interfaces are also studied. Of particular interest are processes of electric charge transfer through the interfaces. The electric charge can be transferred either by electrons or ions. This gives rise to the flow of electric current, if a sequence of phases is constructed that forms a closed circuit. Electroanalytical studies are of considerable scientific and practical interest, since electrified interfaces and processes of the type studied in electroanalytical chemistry occur almost everywhere in nature, including both inorganic matter and living organisms. There are also many technical devices that make use of the electrochemical phenomena. Examples include galvanic cells, accumulators, electronic elements such as capacitors or certain kinds of electronic displays, electrochemical sensors, electro-filtration devices used e.g. in artificial kidneys or for the desalination of water, reactors serving for the galvanisation of metals, or for the electrochemical syntheses of various substances, and many other.

Figure 1.1 shows schematically experimental laboratory systems corresponding to the two most commonly studied interfaces. One of them (the situation depicted in Fig. 1.1a) is the interface between an electronic conductor and an ionic conductor. In this case the electroanalytical experiments are usually performed by means of the so-called three-electrode arrangement. One studies the interface between the working electrode (WE) and the electrolyte solution (ES). The counter electrode (CE), also called auxiliary electrode, is necessary to create a closed electric circuit, in which the electric current $I(t)$ may flow. The current generally depends on time t. An additional electrode, called reference electrode (RE) serves for measuring the electrode potential $E(t)$ of the WE under conditions of zero current flow. The

L.K. Bieniasz, *Modelling Electroanalytical Experiments by the Integral Equation Method*, Monographs in Electrochemistry, DOI 10.1007/978-3-662-44882-3_1

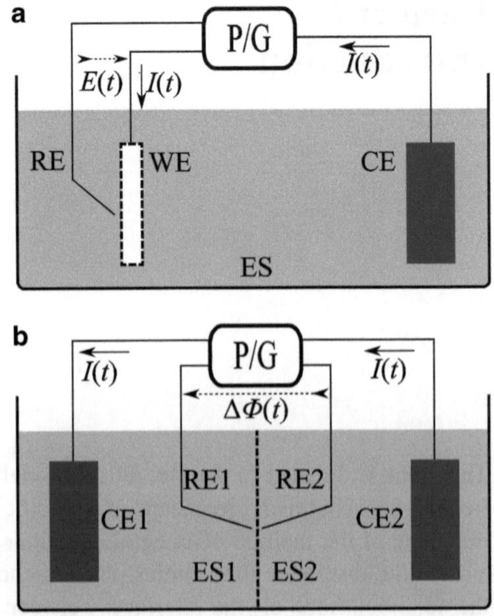

Fig. 1.1 Schematic views of the typical three-electrode (**a**) and four-electrode (**b**) experimental systems. The interfaces studied are marked by *dashed lines*. Symbol P/G denotes the potentiostat or galvanostat. Other symbols are explained in the text

second most frequently studied interface (depicted in Fig. 1.1b) is the interface between two ionic conductors (electrolytic solutions ES1 and ES2). In this case the electroanalytical experiments may require a four-electrode arrangement. Two counter electrodes (CE1 and CE2) ensure a closed circuit with the current $I(t)$. Two reference electrodes (RE1 and RE2) serve for measuring the Galvani potential difference $\Delta\Phi(t)$ across the interface, under conditions of zero current flowing through these electrodes.

Transient electroanalytical experiments can be divided into two essential categories: controlled potential experiments, and controlled current experiments. In the first case a specific known potential-time function $E(t)$ or $\Delta\Phi(t)$ is applied, and the current-time response $I(t)$ of the system to this perturbation is measured. In the second case a known $I(t)$ function is applied, and the $E(t)$ or $\Delta\Phi(t)$ response is measured. Suitable electronic devices for performing such experiments (known as potentiostats and galvanostats) are commercially available and widely used. There are also less frequently used methods in which other kinds of perturbations are employed, such as: temperature perturbation, light perturbation, concentration perturbation, etc.

The very schematic Fig. 1.1 should be perceived with a substantial dose of imagination. A large multiplicity of variations of the experimental laboratory systems may occur in reality. Electrodes may be of different shapes, and their sizes may vary between macroscopic dimensions (millimetre or centimetre sizes), microscopic (micrometre sizes), and even nanoscopic (nanometre sizes). Their dimensions may also vary in time. The electrodes can be embedded in insulators in various ways. There can be several working electrodes, each one polarised differently.

The electrolyte phases may be extensive enough to be considered as infinite or semi-infinite media, but they can also form thin layers or layers of large but finite thicknesses. Additional interfaces or membranes may separate the electrolyte phases into subdomains, possibly having different compositions. The electrolytes may be subject to flow or stirring, according to various methods and hydrodynamic regimes. Flowing electrodes (e.g. liquid mercury streaming electrodes) are also encountered. The discussion of all these nuances is outside the scope of the present book, since the IE method has only been applied thus far to a selection of experimental situations of interest to electroanalytical chemistry. However, Readers should be aware that the number of possible experimental situations is considerable. We direct interested Readers to standard textbooks about electroanalytical methods, such as, in particular [1, 3, 5, 7], and to the literature published in scientific electrochemical journals such as, in particular: the Journal of Electroanalytical Chemistry, Electrochimica Acta, the Journal of the Electrochemical Society, Electroanalysis, Electrochemistry Communications, and (especially older volumes of) Analytical Chemistry.

Just how the transient responses of the systems shown in Fig. 1.1 look further depends on the specific physico-chemical phenomena that occur in the systems. At the WE | ES and ES1 | ES2 interfaces we encounter electron and ion transfer reactions, and possibly also other heterogeneous reactions such as adsorption reactions. The charging of the electric double layer that is always present at the interfaces between any bulk phases is another phenomenon to deal with. In the electrolyte phases we encounter various kinds of transport processes. Most often the transport is by diffusion, but other kinds of transport, such as convection, electro-diffusion (electric migration), or combinations of these, are also conceivable. The transport can sometimes occur also in the working electrode phases, as is the case with amalgam electrodes (typically liquid mercury electrodes), some solid metal electrodes for gas electrode reactions (hydrogen diffusion in palladium electrodes), or electrodes made of conducting polymers. Apart from the transport processes, the electrolyte phase is the place where homogeneous chemical reactions may occur, between the neutral and/or ionic species present there. Another bulk phase phenomenon that may be necessary to be considered (both in the electrolyte and electrode phases) is the Ohmic drop, i.e. the electric potential drop due to a finite resistivity of the phase under flowing current conditions.

Theoretical description of the electroanalytical experiments involving the above phenomena is predominantly accomplished by means of continuum mathematical models, consisting of differential equations with appropriate initial and boundary conditions. The unknowns to be determined from these equations are usually concentrations of chemical species present in the electrochemical systems, and (in cases when the electric migration is considered) electric potential field within the systems. Due to the presence of the transport, apart from the time variable the models often depend on at least one spatial coordinate, so that the model equations are partial differential equations (PDEs). Probably the most numerous are models dependent on one spatial coordinate. The second, smaller, but growing group of models involves models dependent on two spatial coordinates, and the smallest (but also growing) group involves models dependent on three spatial coordinates. In cases when transport can be neglected or omitted, the models become independent of spatial coordinates, and the model equations turn into time-dependent ordinary differential equations (ODEs) or differential-algebraic equations (DAEs). The solution of the initial value problems (IVPs) or initial-boundary value problems (IBVPs) for the above differential equations allows one to calculate theoretical current-time $I(t)$ responses of the systems studied in the case of controlled potential experiments, or potential-time $E(t)$ or $\Delta\Phi(t)$ responses in the case of controlled current experiments.

In order to solve the IVPs or IBVPs, at least three distinct major approaches can be taken. The first approach is analytical. Although from the theoretical point of view this is the most desirable approach, analytical solutions are rarely possible. The second approach consists in a direct numerical simulation. The differential equations are discretised by means of finite-difference, finite element or other similar methods, and resulting algebraic equations (AEs) are solved numerically. The book "Digital Simulation in Electrochemistry" by Britz [2] provides an ample overview of the contemporary techniques used for such purposes. The IE method, discussed in the present book, represents the third approach. The IE method can be classified as semi-analytical, since it is, in some sense, intermediate between the first two approaches. In this method one attempts to solve the model equations partially in an analytical way. For example, in the case of models dependent on one spatial coordinate, the Laplace transform can be applied to the PDEs, and spatial dependencies are solved analytically. As a result, one obtains IEs that depend only on time. The IEs may be solved analytically (in rare cases), or numerically.

Owing to the partial or complete elimination of the spatial dependencies, the numerical solution of the IEs can be computationally less expensive than the direct numerical solution of the PDEs, which is a useful property, although the conversion of the PDEs into IEs usually means an additional cost of the human work. An appreciable feature of the IE method is the increased mathematical insight into the models, resulting from the need to perform the partial analytical solutions. For example, the analysis of the IEs may reveal some interesting regularities regarding the joint effect of various model parameters on the solution, that would be more difficult to recognise by employing the direct numerical PDE solution approach. The IEs may also facilitate obtaining analytical formulae for some limiting cases of the models (for example, steady state solutions). The extra intellectual effort associated with the use of the IE method may be perceived as a remedy against misinterpretations or mistakes accompanying blind uses of the modern black-box software for the direct numerical solution of the PDEs, indicated and criticised by Oldham [19].

Unfortunately, the contemporary IE method has also limitations. The main of them is the limitation to linear PDEs. This is a serious limitation, since many electroanalytical experiments involve second- (and higher-) order homogeneous reactions, or enzymatic homogeneous reactions subject to Michaelis–Menten kinetics, which give rise to nonlinear terms in the PDEs. Therefore, the contemporary IE method cannot be, in principle, applied to model such experiments, although we shall see that it is sometimes possible to consider second-order homogeneous reactions by the IE method, under additional assumptions. The limitation to the linear PDEs results from the involvement of the Laplace transformation in the derivation of the IEs. Other limitations of the IE method have been identified for complex networks of homogeneous reactions, for which the derivation of the IEs may become rather cumbersome, if not impossible. It should be noted, however, that some of these limitations may possibly be overcome in the future, if appropriate mathematical approaches are developed. Certainly, the electroanalytical IEs offer many interesting and challenging tasks for theoretical electrochemists, mathematicians, and computational scientists!

The IE method often makes use of the concentration-flux relationships, that result directly from analytical solutions of the PDEs. The relationships express interfacial concentrations as integrals of the interfacial fluxes, or of the related Faradaic current $I(t)$. This is probably the reason why the IE method has been used mostly for solving models of controlled potential experiments. In such models the fluxes are unknowns to be determined from the IEs. In models of controlled current experiments, for simple reaction schemes involving only one charge transfer reaction, the fluxes are entirely determined by the (known) Faradaic current. Therefore, the issue of solving the IEs does not arise. However, in models involving more than one charge transfer reaction or other complications, the fluxes are not easily expressed through the current. The use of the IE method in such cases is justified. Nevertheless, there have been only a few IE-based models published, of the controlled current experiments.

Apart from the aforementioned chapter in the last edition of the book by Britz [2], there have been very few reviews of the IE method and related developments. The seminal paper by Nicholson and Shain [15] (probably one of the most frequently cited papers in electrochemistry) provided a collection of IEs representing standard cyclic voltammetric experiments. It is often cited as a basic reference on the IE method. Nicholson and Olmstead [14] presented a frequently cited, but rather limited in scope review of some early numerical approaches to the IEs resulting from spatially one-dimensional electroanalytical models. Oldham and Mahon [10, 18] and Mahon et al. [9] presented selective reviews of the aforementioned interfacial concentration-flux relationships for spatially one-dimensional models, which play an essential role in the derivation of the IEs. Honeychurch [6] briefly described the method, in his book devoted to electroanalytical simulations using MATHEMATICA [13]. Fan et al. [4] reviewed fundamentals of the boundary integral method (BIM) with applications to electroanalysis. The BIM is one of the possible IE approaches to spatially two- and three-dimensional models.

The IE method for spatially one-dimensional models occurs in a few mutations, which in the opinion of the present author differ mostly by terminology, but not by principles. In the older literature it was usual to talk explicitly about the IEs and methods for solving them. However, authors of some of the recent approaches avoid using these terms. They prefer names like "convolutive modelling" [9, 11, 12], or "modelling without digital simulation" [20], and they emphasise connections of their approaches with the fractional calculus [21], rather than with the IEs.

Finally, it should be noted that the IE-based modelling route is closely related to the experimental data analysis technique developed for controlled potential electroanalytical experiments, and known under the broad name of "convolutive data analysis" (or less broadly "semi-integral data analysis"). The technique relies on the fact (mentioned above) that in the interfacial concentration-flux relationships the concentrations are often expressed as convolution integrals of the fluxes, or of the flux-dependent Faradaic current. This fact suggests that by analysing an appropriate convolution integral of the experimental Faradaic current, instead of the current itself, one may be able to extract an electrochemically relevant information from the experimental data in a more straightforward way. In particular, by convolving the current, interfacial concentrations may be considered as direct observables. A brief overview of the basic ideas of the convolutive data analysis can be found in the articles by Oldham [16, 17] and Mahon [8].

References

1. Bard AJ, Faulkner LR (2001) Electrochemical methods, fundamentals and applications. Wiley, New York
2. Britz D (2005) Digital simulation in electrochemistry. Springer, Berlin
3. Delahay P (1954) New instrumental methods in electrochemistry. Theory, instrumentation, and applications to analytical and physical chemistry. Interscience, New York
4. Fan TH, Mayle EJ, Kottke PA, Fedorov AG (2006) Simulation of electroanalysis using the boundary integral method. Trends Anal Chem 25:52–65
5. Galus Z (1994) Fundamentals of electrochemical analysis. Ellis Horwood, Chichester
6. Honeychurch MJ (2006) Simulating electrochemical reactions with MATHEMATICA. IBNH, St Lucia
7. Macdonald DD (1977) Transient techniques in electrochemistry. Plenum, New York
8. Mahon PJ (2009) Convolutive reshaping with applications for voltammetry. J Solid State Electrochem 13:573–582
9. Mahon PJ, Myland JC, Oldham KB (2002) A fresh approach to voltammetric modelling. J Electroanal Chem 537:1–5
10. Mahon PJ, Oldham KB (1998) Voltammetric modelling via extended semiintegrals. J Electroanal Chem 445:179–195
11. Mahon PJ, Oldham KB (1999) Convolutive modelling: A third route to predict voltammograms. Proc Electrochem Soc 99–5:90–97
12. Mahon PJ, Oldham KB (2001) Incorporating electrode kinetics into the convolutive modeling of reactions at planar, cylindrical and spherical electrodes. Electrochim Acta 46:953–965
13. MATHEMATICA (2014) Wolfram Res. Inc., Champaigne, IL. http://www.wolfram.com. Accessed 10 June 2014
14. Nicholson RS, Olmstead ML (1972) Numerical solution of integral equations. In: Mattson JS, Mark HB Jr, MacDonald HC Jr (eds) Electrochemistry, calculations, simulation, and instrumentation. Computers in chemistry and instrumentation, vol 2. Marcel Dekker, New York, pp 119–138
15. Nicholson RS, Shain I (1964) Theory of stationary electrode polarography. Single scan and cyclic methods applied to reversible, irreversible, and kinetic systems. Anal Chem 36:706–723
16. Oldham KB (1986) Convolution of voltammograms as a method of chemical analysis. J Chem Soc Faraday Trans I 82:1099–1104

17. Oldham KB (1986) Convolution: A general electrochemical procedure implemented by a universal algorithm. Anal Chem 58:2296–2300
18. Oldham KB (1991) Interrelation of current and concentration at electrodes. J Appl Electrochem 21:1068–1072
19. Oldham KB (2011) Trends in electrochemical instrumentation and modeling. J Solid State Electrochem 15:1697–1698
20. Oldham KB, Myland JC (2011) Modelling cyclic voltammetry without digital simulation. Electrochim Acta 56:10612–10625
21. Oldham KB, Spanier J (1974) The fractional calculus. Theory and applications of differentiation and integration to arbitrary order. Academic Press, New York

Chapter 2
Basic Assumptions and Equations of Electroanalytical Models

In this chapter we provide a summary of basic assumptions and equations representing physico-chemical laws, that are used in the modelling of electroanalytical experiments. This is not a comprehensive overview, but a selection dictated by the needs and up-to-date applications of the IE method, that will be discussed in subsequent chapters. Physico-chemical aspects, interpretations, and applicability issues regarding the equations listed here are not addressed. Readers are referred to specialised electrochemical literature for such information.

2.1 General Assumptions

Electrochemical systems are generally multiphase and multicomponent systems. The modelling of electroanalytical experiments in such systems is usually accomplished in terms of the continuum theory, which neglects the discrete nature of matter. It is assumed that the full description of such experiments is obtained by specifying the spatio-temporal evolution of the concentrations of chemical species present in the systems, and other relevant state variables, such as the electric potential, temperature distribution, magnetic field, etc. The chemical species occurring in the electrochemical systems can be neutral molecules, ions, radicals, ion radicals, complexes, as well as various interfacial structures or bound states. Electrons exchanged between reactants and products of electrochemical reactions can sometimes also be formally regarded as chemical species. The same refers to free states available for adsorption at interfaces. Yet another kind of formally chemical species can be solvated electrons. The chemical species (other than exchanged electrons) can be divided into *distributed species*, present in the spatially extended phases (such as, for example, the electrolyte solution phases ES, ES1, or ES2 in Fig. 1.1), and *localised species* (for example, adsorbed or deposited species), present only at the interfaces between the spatially extended phases. The

L.K. Bieniasz, *Modelling Electroanalytical Experiments by the Integral Equation Method*, Monographs in Electrochemistry, DOI 10.1007/978-3-662-44882-3_2

concentrations of the distributed species are usually expressed in moles per unit volume, and they will be denoted here by the letter c. Concentrations of the localised species are often expressed in moles per unit area, and they will be denoted here by the letter Γ. Alternatively, they can also be expressed as fractions of the interfacial area covered by the species, or fractions of the maximum possible concentration of the species at a given interface. For the purpose of the further discussions, it is also convenient to divide the chemical species (either distributed or localised) into *dynamic* and *static* (we partially adopt here the terminology from Bieniasz [12, 13]). Concentrations of the dynamic species are assumed to vary, whereas concentrations of the static species are assumed to be practically constant during the electroanalytical experiments. Static species are usually species occurring in excess.

The dynamic distributed species are subject to transport phenomena in the spatially extended phases. Basic equations for electrochemical and non-electrochemical transport are described in Sects. 2.2 and 2.3. The transport equations most generally take the mathematical form of partial differential equations (PDEs). The PDEs are defined over spatial domains coinciding with spatially extended phases. A basic classification of these domains is provided in Sect. 2.4. A surface transport of the dynamic localised species, along the interfaces, is also possible (see, for example, [63, 112, 117, 118]) but it is rather rarely taken into account in electroanalytical modelling, so that it will not be considered in this book.

Apart from the transport phenomena, an important role is played by chemical and electrochemical reactions. The reactions are transformations that occur at a molecular level between various species. Basic equations of the formal chemical and electrochemical kinetics are overviewed in Sect. 2.7. The reactions can be divided into *homogeneous* and *heterogeneous*. Homogeneous reactions take place in spatially extended phases, among distributed species. Usually these are electrolytic solution phases (such as phases ES, ES1, and ES2 in Fig. 1.1), or non-electrolytic solution phases. Homogeneous reactions between species distributed in the electrode phases (such as liquid mercury WEs) are hardly possible so that one cannot find electroanalytical models involving such reactions. Heterogeneous reactions occur only at interfaces, but they can involve both distributed and localised species. They can be divided into *heterogeneous electrochemical* reactions, characterised by a transfer of an electric charge across the interface, and *heterogeneous non-electrochemical* reactions, in which the transfer of the electric charge does not occur. Some homogeneous reactions may also involve a charge transfer, but in the models discussed in the present book this fact does not have to be taken into account, so that we shall not consider a separate class of such reactions. Homogeneous reactions have an effect on the transport equations. This issue is addressed in Sect. 2.8. Heterogeneous reactions, in turn, determine the boundary conditions (necessary for solving the transport equations), and possible additional equations governing the evolution of the concentrations of dynamic localised species at the interfaces. This issue is discussed in Sect. 2.9. All reactions may determine initial conditions for the concentrations of dynamic species. This aspect of the modelling is addressed in Sect. 2.10.

In Sect. 2.11 we provide a brief overview of the various electroanalytical techniques used to perturb the systems under study.

Anomalous diffusion phenomena, encountered in some systems, are briefly addressed in Sect. 2.12.

Some other physico-chemical phenomena that have to be taken into account in electroanalytical models are listed in Sect. 2.13.

2.2 Equations of Transport in Electrolytes

Consider an electrolyte solution phase such as phases ES, ES1 or ES2 in Fig. 1.1. Such phases typically contain a solvent and a supporting electrolyte (which are usually assumed not to participate in the processes occurring at an interface studied), and a number N_s^{distr} of distributed chemical species that take part in the processes studied, and therefore have to be accounted for in theoretical models. Let X_j be the jth of such distributed species. Assuming that the species is one of N_s^{dd} dynamic distributed species present in the phase, its movement in accordance with relevant transport laws has to be considered (the concentrations of the remaining $N_s^{\mathrm{sd}} = N_s^{\mathrm{distr}} - N_s^{\mathrm{dd}}$ static distributed species are assumed constant). The movement can be described quantitatively by using the vector variable $\boldsymbol{J}_j$ called *flux*. The vector indicates the direction in which the species is moving. The flux value, expressed in moles $\mathrm{m}^{-2}\mathrm{s}^{-1}$, gives the amount of the species passing per unit of time through a plane of unit area oriented perpendicularly to the vector of the average velocity of the species. The laws of electrochemical ionic transport have been discussed in numerous books and review papers (see, in particular, [3, 57, 82, 120]). For the purposes of the present book, we shall be concerned exclusively with the widely adopted theoretical transport model, known as the dilute solution model [3, 57, 82], according to which the flux consists of additive terms representing [in the order of their appearance in Eq. (2.1)]: electric migration, diffusion, and convection:

$$\boldsymbol{J}_j = -\mathfrak{z}_j\, \mathfrak{u}_j\, F\, c_j\, \mathbf{grad}\, \Phi - D_j\, \mathbf{grad}\, c_j + \boldsymbol{v}\, c_j \ . \tag{2.1}$$

In Eq. (2.1) $\mathfrak{z}_j$ is the number of proton charges carried by the species ($\mathfrak{z}_j \neq 0$ for ions, $\mathfrak{z}_j = 0$ for neutral species). The migration term vanishes for uncharged species. Symbol $\mathfrak{u}_j$ denotes the mobility of the species (the average velocity of the species resulting from a force of 1N per mole), c_j is the species concentration, Φ is the inner electric potential within the electrolyte phase (Galvani potential), D_j is the diffusion coefficient, and $\boldsymbol{v}$ is the convection velocity. Symbol F denotes the Faraday constant. The mobility and the diffusion coefficient are related by the Nernst–Einstein equation:

$$D_j = RT\, \mathfrak{u}_j \ , \tag{2.2}$$

where R is the gas constant, and T is the absolute temperature.

By summing up contributions from the fluxes of all ions in an electrolytic solution, the vector $\boldsymbol{i}_{\mathrm{F}}$ of the Faradaic current density flowing through the electrolyte can be expressed as

$$\boldsymbol{i}_{\mathrm{F}} = F \sum_j \mathfrak{z}_j \boldsymbol{J}_j \,. \tag{2.3}$$

Equations (2.1) and (2.2) originate from the early theories of Nernst and Planck [80, 81, 95, 96], and they are usually regarded valid for dilute electrolytic solutions. They ignore possible cross-effects, such as multicomponent diffusion (for which the flux $\boldsymbol{J}_j$ would also depend on concentration gradients of species other than X_j) and/or other complications specific for concentrated electrolytic solutions [3, 57, 82]. Variables $\boldsymbol{J}_j$, c_j, Φ, and $\boldsymbol{v}$ may, in general, depend on time t and on spatial coordinates (some possible choices for the spatial coordinates are indicated below). The presence of migration requires an additional equation, allowing one to determine the electric potential Φ. Most often this can be the electroneutrality equation or the Poisson equation.

In the case when the electrolytic solution contains an excess of the supporting electrolyte, it can be shown [82] that the migration component of the flux $\boldsymbol{J}_j$ becomes negligible in comparison with the diffusion term, because the conductivity of the solution is then increased, so that the electric field gradient is suppressed. In such a case Eq. (2.1) reduces to

$$\boldsymbol{J}_j = -D_j \,\mathbf{grad}\, c_j + \boldsymbol{v}\, c_j \,. \tag{2.4}$$

Looking from the perspective of the IE method, this is a convenient situation, because Eq. (2.1) is intrinsically nonlinear, owing to the product of c_j and $\mathbf{grad}\,\Phi$ (both variables are normally unknown and have to be determined simultaneously). The contemporary IE method cannot be applied (with the exception of a few special or simplified cases) to electroanalytical models involving this kind of nonlinearity. In contrast, there have been numerous applications of the method to models involving linear Eq. (2.4). Of course, Eq. (2.4) is linear if we assume that D_j does not depend on c_j, which is a typical assumption (although counterexamples are known, see, for example, Hernández et al. [53] and the references therein). In this book we further assume that D_j does not depend on spatial coordinates or time. Opposite situations are known (see, for example, Cattey et al. [22, 23]), but they are very rare. In the absence of convection, Eq. (2.4) further simplifies to the equation:

$$\boldsymbol{J}_j = -D_j \,\mathbf{grad}\, c_j \,, \tag{2.5}$$

known as the Fick first law of diffusion.

Equations (2.4) or (2.5) are usually not sufficient for the mathematical modelling, and have to be combined with other laws, especially the conservation laws.

In particular, the principle of the conservation of matter can be written in the differential form as

$$\frac{\partial c_j}{\partial t} = -\mathrm{div}\, \boldsymbol{J}_j + p_j^{\mathrm{hom}}\,, \tag{2.6}$$

where p_j^{hom} is the production rate (in moles per unit time and unit volume) of the species X_j, due to homogeneous reactions in the electrolyte, if there are any such reactions. Relevant equations for p_j^{hom} are given in Sect. 2.8. The production rate p_j^{hom} depends on concentrations c_j, and hence indirectly on time and spatial coordinates. Under the assumptions accepted, the combination of Eqs. (2.4) and (2.6) gives

$$\begin{aligned}\frac{\partial c_j}{\partial t} &= \mathrm{div}\left(D_j\, \mathbf{grad}\, c_j\right) - \mathrm{div}\left(\boldsymbol{v}\, c_j\right) + p_j^{\mathrm{hom}} \\ &= D_j\, \triangle\, c_j - c_j\, \mathrm{div}\, \boldsymbol{v} - \boldsymbol{v}\cdot \mathbf{grad}\, c_j + p_j^{\mathrm{hom}}\,. \end{aligned}\tag{2.7}$$

Assuming a low compressibility of the electrolyte solution phase, we have $\mathrm{div}\, \boldsymbol{v} \approx 0$, so that

$$\frac{\partial c_j}{\partial t} = D_j\, \triangle\, c_j - \boldsymbol{v}\cdot \mathbf{grad}\, c_j + p_j^{\mathrm{hom}}\,. \tag{2.8}$$

Equation (2.8) is a typical reaction-convection-diffusion PDE used in electroanalytical modelling. This equation, and its special subcases (diffusion, convection-diffusion, reaction-diffusion, etc.) will form a basis for a majority of the modelling considerations in the present book. Due to the aforementioned nonlinearity of Eq. (2.1), migration will not be considered, except for a few special cases or simplified models of it, mentioned in Sect. 10.2.

Most frequently $N_s^{\mathrm{dd}} > 1$, and the species considered may participate in more than one reaction. It is therefore convenient to write the model equations by using a vector–matrix notation, where the vector elements are associated with the various species or reactions. To avoid confusion, these vectors (and related matrices) have to be distinguished from the vectors already present in Eqs. (2.1)–(2.8), such as $\boldsymbol{J}_j$, $\boldsymbol{v}$, or **grad**, which are vectors in the ordinary space. For this reason, in this book the vectors in the spaces of variables associated with the various species or reactions are indicated by arrows over a vector symbol, and related matrices are indicated by bars over a matrix symbol. The vectors (and related matrices) in ordinary space are indicated by boldface characters. In particular, the vector of the concentrations of the dynamic distributed species is defined as $\vec{c} = \left[c_1, \ldots, c_{N_s^{\mathrm{dd}}}\right]^{\mathrm{T}}$, the vector of the production rates of these species, due to homogeneous reactions, is defined as $\vec{p}^{\,\mathrm{hom}} = \left[p_1^{\mathrm{hom}}, \ldots, p_{N_s^{\mathrm{dd}}}^{\mathrm{hom}}\right]^{\mathrm{T}}$, and the matrix of the (positive) diffusion coefficients is defined as a diagonal matrix $\overline{D} = \mathrm{diag}(D_1, \ldots, D_{N_s^{\mathrm{dd}}})$ (superscript T means a transposed vector). The vector of flux vectors can also be defined as $\vec{J} =$

$\left[\boldsymbol{J}_1, \ldots, \boldsymbol{J}_{N_s^{dd}}\right]^{T}$. This allows us to re-write systems of Eqs. (2.4) and (2.8) for $j = 1, \ldots, N_s^{dd}$ compactly as

$$\vec{\boldsymbol{J}} = -\overline{D}\,\mathbf{grad}\,\vec{c} + \boldsymbol{v}\,\vec{c}\,, \tag{2.9}$$

$$\frac{\partial \vec{c}}{\partial t} = \overline{D}\,\Delta\,\vec{c} - \boldsymbol{v}\cdot\mathbf{grad}\,\vec{c} + \vec{p}^{\,\mathrm{hom}}\,. \tag{2.10}$$

2.3 Equations of Transport in Non-electrolytes

The transport of chemical species can also occur in non-electrolytic spatially extended phases, for example in the working electrode phase WE in Fig. 1.1a, in the case of a liquid mercury electrode. The transport in non-electrolytes is usually modelled by reaction-convection-diffusion PDEs, expressed by Eqs. (2.8) or (2.10).

Further characterisation of the Eqs. (2.8) or (2.10) requires a specification of the spatial domains and related coordinate systems, a specification of the convection field $\boldsymbol{v}$, and a specification of the species production rates $\vec{p}^{\,\mathrm{hom}}$ in homogeneous reactions. These aspects of the transport equations are discussed in the following Sects. 2.4, 2.6, and 2.8.

2.4 Spatial Domains, Their Dimensionality, and Coordinate Systems

All real spatially extended phases are three-dimensional. Therefore, the spatial domains, on which the transport equations discussed in Sects. 2.2 and 2.3 are defined, are generally subdomains in the three-dimensional space. However, it often happens that, owing to a particular symmetry of the electrochemical system, the number of spatial coordinates necessary for the mathematical description is smaller than three. Our ability to consider a particular spatial domain as one- or two-dimensional depends on whether a suitable coordinate system exists that correctly reflects the above symmetry.

Electrochemical interfaces often have a planar, spherical, or cylindrical symmetry, and for this reason Cartesian, spherical, or cylindrical coordinate systems are in frequent use, often enabling the reduction of the dimensionality of the spatial domains. The spatial differential operators: **grad** (gradient), div (divergence), and Δ (Laplacian) in Eqs. (2.1) and (2.4)–(2.10) take different forms depending on the spatial coordinate system used. For the convenience of the Reader, Table 2.1 provides the formulae for these differential operators in the above three coordinate systems.

The spatial domains occurring in electroanalytical models can often be divided into *finite* and *semi-infinite*. Semi-infinite spatial (sub)domains represent a theoreti-

Table 2.1 Representation of differential operators **grad** (gradient), div (divergence), and Δ (Laplacian) in Cartesian, spherical, and cylindrical coordinates

Coordinate system	Operator[a]	Formula
Cartesian	$\mathbf{grad}\, f$	$\boldsymbol{e}_x \frac{\partial f}{\partial x} + \boldsymbol{e}_y \frac{\partial f}{\partial y} + \boldsymbol{e}_z \frac{\partial f}{\partial z}$
	$\mathrm{div}\, \boldsymbol{v}$	$\frac{\partial v_x}{\partial x} + \frac{\partial v_y}{\partial y} + \frac{\partial v_z}{\partial z}$
	Δf	$\frac{\partial^2 f}{\partial x^2} + \frac{\partial^2 f}{\partial y^2} + \frac{\partial^2 f}{\partial z^2}$
Spherical	$\mathbf{grad}\, f$	$\boldsymbol{e}_r \frac{\partial f}{\partial r} + \boldsymbol{e}_\varphi \frac{1}{r \sin\theta} \frac{\partial f}{\partial \varphi} + \boldsymbol{e}_\theta \frac{1}{r} \frac{\partial f}{\partial \theta}$
	$\mathrm{div}\, \boldsymbol{v}$	$\frac{1}{r^2} \frac{\partial (r^2 v_r)}{\partial r} + \frac{1}{r \sin\theta} \frac{\partial v_\varphi}{\partial \varphi} + \frac{1}{r \sin\theta} \frac{\partial (\sin\theta\, v_\theta)}{\partial \theta}$
	Δf	$\frac{\partial^2 f}{\partial r^2} + \frac{2}{r} \frac{\partial f}{\partial r} + \frac{1}{r^2 \sin^2\theta} \frac{\partial^2 f}{\partial \varphi^2} + \frac{1}{r^2} \frac{\partial^2 f}{\partial \theta^2} + \frac{\cot\theta}{r^2} \frac{\partial f}{\partial \theta}$
Cylindrical	$\mathbf{grad}\, f$	$\boldsymbol{e}_r \frac{\partial f}{\partial r} + \boldsymbol{e}_\varphi \frac{1}{r} \frac{\partial f}{\partial \varphi} + \boldsymbol{e}_z \frac{\partial f}{\partial z}$
	$\mathrm{div}\, \boldsymbol{v}$	$\frac{1}{r} \frac{\partial (r\, v_r)}{\partial r} + \frac{1}{r} \frac{\partial v_\varphi}{\partial \varphi} + \frac{\partial v_z}{\partial z}$
	Δf	$\frac{\partial^2 f}{\partial r^2} + \frac{1}{r} \frac{\partial f}{\partial r} + \frac{1}{r^2} \frac{\partial^2 f}{\partial \varphi^2} + \frac{\partial^2 f}{\partial z^2}$

[a]Operators **grad** and Δ act on scalar functions f. Operator div acts on vectors $\boldsymbol{v} = \left[v_x,\ v_y,\ v_z\right]^{\mathrm{T}}$, $\boldsymbol{v} = \left[v_r,\ v_\varphi,\ v_\theta\right]^{\mathrm{T}}$, or $\boldsymbol{v} = \left[v_r,\ v_\varphi,\ v_z\right]^{\mathrm{T}}$. Symbols $\boldsymbol{e}_x$, $\boldsymbol{e}_y$ and $\boldsymbol{e}_z$, or $\boldsymbol{e}_r$, $\boldsymbol{e}_\varphi$ and $\boldsymbol{e}_\theta$, or $\boldsymbol{e}_r$, $\boldsymbol{e}_\varphi$ and $\boldsymbol{e}_z$ denote unit vectors, correspondingly in the Cartesian, spherical, or cylindrical coordinate systems. At a given point in space, the unit vectors indicate directions in which the related coordinates increase

cal idealisation of situations when an interface studied is located at a large distance from other interfaces (such as the boundaries of a laboratory vessel or cell) in the electrochemical system. If the distance is sufficiently large so that a perturbation imposed at the interface studied is not likely to propagate in the system up to that distance during the experiment under consideration, the other interfaces can be formally considered as located at infinity. Such situations are typical, in particular, in the studies of solid macro- and microelectrodes [5, 7, 17, 21, 36, 37, 51, 99, 111], dropping mercury electrodes [126], and also (in some cases) liquid | liquid interfaces [42, 43, 103, 105]. In contrast, finite spatial domains must be assumed in situations when a perturbation imposed at the interface studied causes variations of the concentrations (or other state variables) within the entire volume of the adjacent phase(s). Finite spatial domains occur, for example, in the studies of amalgam electrodes [77, 121, 123], hydrogen dissolution in solid metals [97], thin layer cells [55], membranes [114, 115], modified electrodes [38], biosensors [11], and porous electrodes [50, 122]. Sometimes it is also convenient to artificially separate a finite subdomain out of a semi-infinite one. This happens in the modelling of hydrodynamic electrodes, when the region of a stationary diffusion layer is considered separately from a stirred part of the solution (see, for example, Sect. 2.6.2 below).

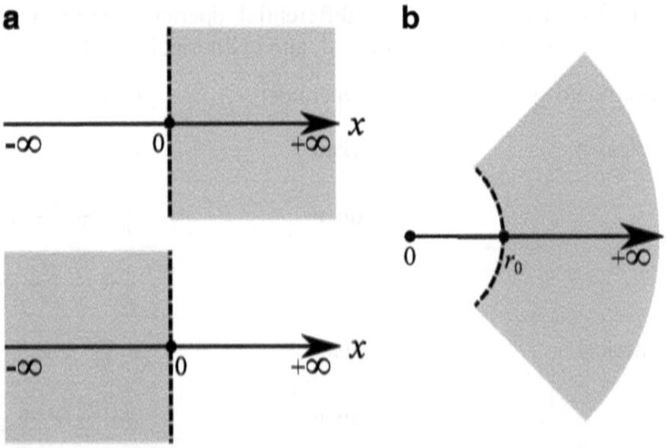

Fig. 2.1 One-dimensional semi-infinite spatial domains (*grey areas*) adjacent to a planar interface (**a**), and spherical or cylindrical interfaces (**b**) subject to electroanalytical studies. Boundaries corresponding to interfaces studied are indicated by *dashed lines*

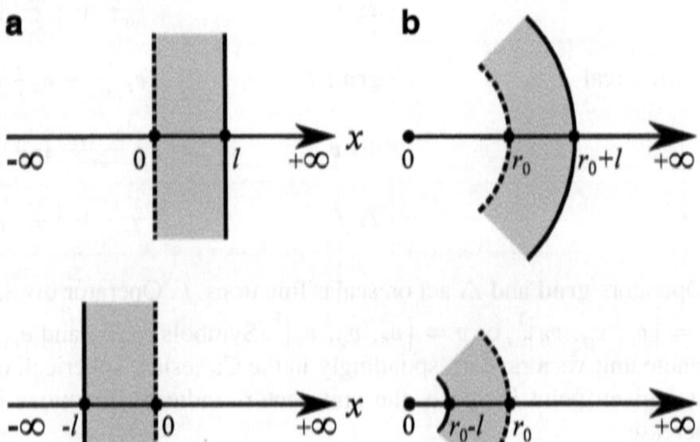

Fig. 2.2 One-dimensional finite spatial domains (*grey areas*) adjacent to a planar interface (**a**), and spherical or cylindrical interfaces (**b**) subject to electroanalytical studies. Boundaries corresponding to interfaces studied are indicated by *dashed lines*, and the remaining boundaries by *solid lines*

The IE method is particularly often applied to models defined over one-dimensional spatial domains, in which the single coordinate axis is perpendicular to the interface studied. Figures 2.1 and 2.2 present a collection of the typical one-dimensional spatial domains corresponding to such models. Figure 2.1 depicts schematically semi-infinite spatial domains that arise in the three often encountered cases of interfaces possessing a planar, spherical, or cylindrical geometry. Here, and later in this book we assume that a planar interface studied is located at $x = 0$, and a spherical or cylindrical interface studied is located at the radius $r = r_0$. As can be seen from Fig. 2.1a, the planar interface splits the space into two semi-infinite spatial subdomains adjacent to the interface: one for $x \in (-\infty, 0)$, and the second for $x \in (0, \infty)$. As these two subdomains are essentially mirror images of each other, it is sufficient to analyse only the second subdomain; any theory for the first subdomain can be obtained by replacing x by $-x$. The situation is different for spherical and cylindrical interfaces (cf. Fig. 2.1b). These interfaces also split the space into two subdomains, but only one of the subdomains is semi-infinite, namely the one for $r \in (r_0, \infty)$. The subdomains for $r \in (0, r_0)$ are finite.

By analogy with Fig. 2.1, Fig. 2.2 depicts schematically finite spatial domains that arise in the three often encountered cases of interfaces possessing a planar, spherical, or cylindrical geometry. The finite character of the spatial domains means that in addition to the interface studied, we have to pay attention to the second boundary located at a distance l from the interface studied. The second boundary may correspond to a real physical interface, or can be artificial, i.e. existing only in the mathematical model. As can be seen from Fig. 2.2a, the finite domains pertinent to the case of planar interface are similar to the semi-infinite domains from Fig. 2.1a, in that there can be two of them, symmetrical with respect to the interface studied. Hence, it is sufficient to analyse only the domain for $x \in (0, l)$, since any theory for the domain for $x \in (-l, 0)$ is obtained by replacing x by $-x$. A new situation arises for spherical and cylindrical interfaces, where there are now two different finite domains possible, one that we shall call "internal" (with respect to the interface studied), which corresponds to $r \in (r_0 - l, r_0)$ with $r_0 - l \geq 0$, and one that we shall call "external", which corresponds to $r \in (r_0, r_0 + l)$.

Two- and three-dimensional spatial domains, occurring in electroanalytical models, are more difficult to systematise and visualise, and a large diversity of such domains is conceivable.

2.5 Diffusion Fields

If diffusion is the sole mode of transport, and spatial domains can be regarded as one-dimensional, one can distinguish *planar diffusion*, *spherical diffusion*, and *cylindrical diffusion*, as diffusion fields rigorously described by one-dimensional PDEs, respectively, in Cartesian, spherical, and cylindrical coordinate systems. Planar diffusion is sometimes called *linear diffusion* in the literature. However, this is not a preferred terminology, because the latter name is also used to denote the independence of the diffusion coefficient on concentration(s).

If diffusion is the sole mode of transport, but spatial domains are two-dimensional, more diversified diffusion fields are conceivable. Taking, as an example, diffusion in phases adjacent to a planar interface consisting of an insulator with embedded electrodes, the spatial coordinate system, and the dimensionality of the spatial domain depend on the shape of the electrode(s). In particular, Figs. 2.3 and 2.4 present two popular examples of microelectrode arrangements, for which the spatial domains are two-dimensional: an infinite band electrode (or an array of parallel bands), and a disk electrode (or an array of concentric disk and/or rings). In the first example the Cartesian coordinate system is adequate, and the spatial domain is $x \in (-\infty, \infty)$ and $z \in [0, \infty)$, so that it is infinite for one coordinate, and semi-infinite for the second coordinate. It is assumed that any band is infinite in the direction of the omitted coordinate y, so that the diffusion field does not depend on y. In other words, edge effects resulting from a finite length of a real band are ignored. In the second example the cylindrical coordinate system is adequate, and

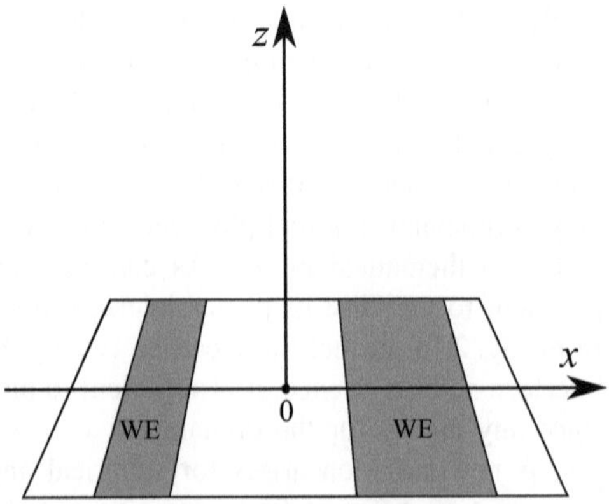

Fig. 2.3 Schematic view of an array of infinite parallel electrode bands (WE, *grey areas*) embedded in an insulating plane (*white areas*). Coordinate x is in the direction parallel to the plane, and perpendicular to the bands (which are parallel to the omitted coordinate y), whereas z is the coordinate in the direction perpendicular to the plane, located at $z = 0$

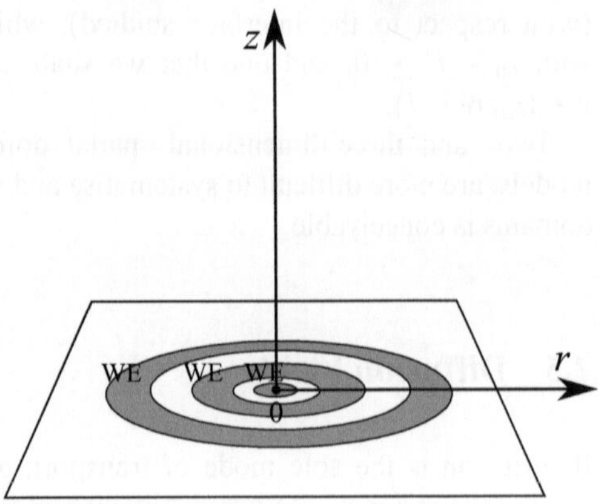

Fig. 2.4 Schematic view of an array of concentric electrode disk and ring(s) (WE, *grey areas*) embedded in an insulating plane (*white areas*). Coordinate r is in the radial direction parallel to the plane, whereas z is the coordinate in the direction perpendicular to the plane, located at $z = 0$

the spatial domain is $r \in [0, \infty)$ and $z \in [0, \infty)$, so that it is semi-infinite for both coordinates.

2.6 Convection-Diffusion Fields

There is a number of standard hydrodynamic arrangements resulting in forced convection-diffusion fields, used for electroanalytical experiments. Some of them are described in this section.

2.6.1 Dropping Mercury Electrode

The dropping mercury electrode (DME) is historically one of the first types of working electrodes WE of Fig. 1.1, where convection transport played an important role. This type of the electrode has been utilised in polarography for at least six decades.

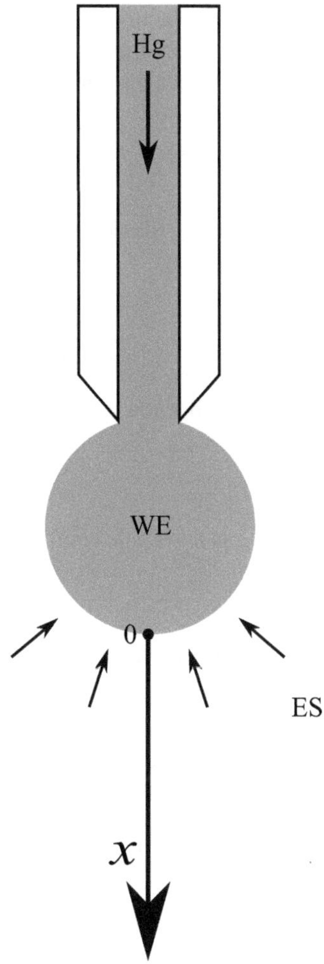

Fig. 2.5 Schematic view of the dropping mercury working electrode (WE, *grey area*), flowing out of a capillary (*white areas*) into an electrolyte solution (ES). Coordinate x is the distance from the surface of the drop ($x = 0$ corresponds to the surface). Consequently, there is convection of the electrolyte towards the electrode surface, indicated by *arrows*

Polarography has influenced very significantly the development of electroanalytical chemistry, and is indispensable for particular purposes, but currently electrodes of this type are gradually replaced by solid micro and nano electrodes or other devices.

The DME is depicted schematically in Fig. 2.5. Liquid mercury flows at a constant rate out of a capillary, and forms a drop. The size of the drop grows with time, until the drop falls down and another one begins to be formed. As a result, in a coordinate system associated with the surface of the drop, the electrolyte flows towards the surface of the drop. This electrolyte flow is most often described by the so-called expanding plane model, introduced by Ilkovič [58]. According to this model, the radius of the drop is assumed to be sufficiently large, so that during the experiment one can approximate the surface of the drop by a plane moving in the electrolyte in the direction of the x axis in Fig. 2.5. The spatial domain associated with this system is effectively one-dimensional and semi-infinite, with the distance

x from the electrode surface playing the role of the spatial coordinate. Detailed analysis [58] of the drop volume variations with time t (elapsed from the birth of the drop) yields the convection rate

$$v(x,t) = -\frac{2}{3}\frac{x}{t} . \tag{2.11}$$

Consequently, the convection-diffusion PDE for a jth dynamic distributed species is

$$\frac{\partial c_j(x,t)}{\partial t} = D_j \frac{\partial^2 c_j(x,t)}{\partial x^2} + \frac{2x}{3t}\frac{\partial c_j(x,t)}{\partial x} . \tag{2.12}$$

The electrode area in the expanding plane model is a growing function of time:

$$A = a\, t^{2/3} , \tag{2.13}$$

where a is a suitable coefficient.

2.6.2 Rotating Disk Electrode

Another quite popular experimental arrangement employing forced convection is the rotating disk electrode (RDE) system. Theoretical principles and experimental aspects of the RDE have been summarised in a number of books and review papers [68, 86, 101]. In this arrangement the working electrode WE of Fig. 1.1 is realised as a disk of a solid electronic conductor (metal or graphite, etc.) embedded in an insulator plate, and immersed in the electrolyte solution. The disk (together with the insulator) rotates, most often with a constant frequency, along the symmetry axis of the disk. As a result, a stationary field of the convection velocity is established throughout the electrolyte. The RDE arrangement is shown schematically in Fig. 2.6. The electrode can operate either under laminar or turbulent flow conditions, but we focus here on the laminar flow only. Close to the RDE surface the electrolyte fluid flows parallel to the surface, in the direction outwards the electrode symmetry axis. At a further distance from the electrode surface, the flow lines are perpendicular to the surface, and the flow is directed towards the surface. This perpendicular flow has the dominant effect on the results of electroanalytical experiments at the RDE. Therefore, in the theoretical modelling of such experiments the flow parallel to the surface is usually neglected. The spatial domain associated with this system is then considered as one-dimensional, and semi-infinite. As the single spatial coordinate one uses the Cartesian coordinate x along the symmetry axis of the RDE (see Fig. 2.6). The electrode surface is located at $x = 0$.

The modelling of electroanalytical experiments at the RDE is usually based on hydrodynamic calculations by Cochran [25]. By using the results of these

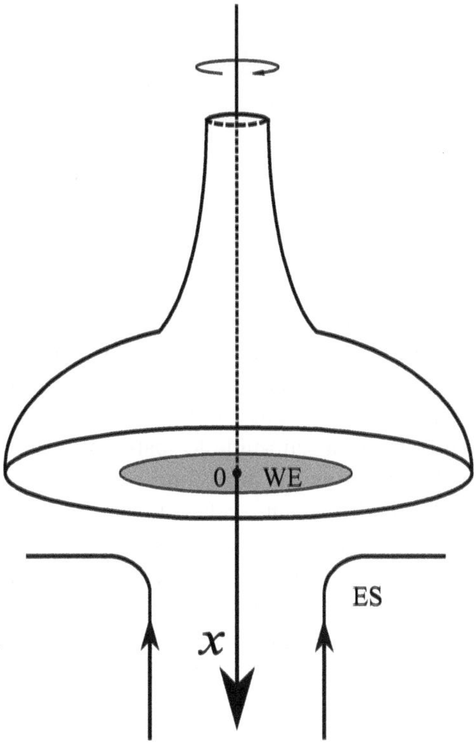

Fig. 2.6 Schematic view of the rotating disk working electrode (WE, *grey area*), embedded in an insulator (*white area*) and immersed into an electrolyte solution (ES). Coordinate x is the distance from the surface of the electrode ($x = 0$ corresponds to the surface). The insulated electrode rotates along the symmetry axis x. Consequently, there is convection of the electrolyte, indicated by *lines with arrows*

calculations it has been established (see, for example, [47, 119]) that under laminar flow conditions the convection velocity along the x axis equals

$$v(x) = (\omega\nu)^{1/2}\left[-0.51023\left(\frac{\omega}{\nu}\right)x^2 + 0.33333\left(\frac{\omega}{\nu}\right)^{3/2}x^3 - 0.10267\left(\frac{\omega}{\nu}\right)^2 x^4 + 0.0127\left(\frac{\omega}{\nu}\right)^{5/2}x^5 + 0.00283\left(\frac{\omega}{\nu}\right)^3 x^6 + \ldots\right], \tag{2.14}$$

where ω is the rotation frequency of the electrode (in rad s^{-1}), and ν is the kinematic viscosity of the solution (in m^2 s^{-1}). Often only the first term of Eq. (2.14) (corresponding to x^2) is taken into account for theoretical modelling purposes, so that the convection-diffusion PDE for a jth dynamic distributed species takes the form

$$\frac{\partial c_j(x,t)}{\partial t} = D_j\frac{\partial^2 c_j(x,t)}{\partial x^2} + 0.51023\,\omega^{3/2}\nu^{-1/2}x^2\frac{\partial c_j(x,t)}{\partial x}. \tag{2.15}$$

The fact that the convection velocity vanishes at $x = 0$, and increases rapidly with increasing x, implies that one may consider the spatial domain as composed of two subdomains. One of them is a thin layer close to the electrode, where the transport is practically by diffusion only. In the second, external, and semi-infinite subdomain, the concentration can be assumed practically constant and equal to the bulk concentration, owing to intensive mixing. The thickness l of the thin layer can be expressed [119] by the formula:

$$l \approx 1.6117\, D_j^{1/3}\, \omega^{-1/2} \nu^{1/6} \left[1 + 0.298 \left(\frac{D_j}{\nu} \right)^{1/3} + 0.14514 \left(\frac{D_j}{\nu} \right)^{2/3} \right]. \tag{2.16}$$

In view of this possibility, the applications of the IE method to Eq. (2.15) have followed two distinct routes. In one route (to be discussed in Sect. 5.3) Eq. (2.15) is replaced by a pure diffusion equation, and solved in a thin layer of thickness l. In the second route (to be discussed in Sect. 6.2.3) Eq. (2.15) is considered as defined over a semi-infinite spatial interval, and one attempts to take into account both convection and diffusion.

The hydrodynamic calculations of Cochran [25] have been recently questioned by Alexiadis et al. [2], who showed that they are inadequate except in the very vicinity of the electrode, if the finite size of the experimental vessel is taken into account. Therefore, it may be that the higher-order terms in Eqs. (2.14) and (2.16) are not very realistic.

2.6.3 Channel and Tubular Electrodes

Further examples of working electrodes, operating under conditions of forced convection, are channel and tubular electrodes. The simplest arrangements of such electrodes are schematically depicted in Fig. 2.7. In these arrangements an electrolytic solution moves past a conductive electrode plate embedded in a wall of a rectangular duct or tube. There is an abundant literature devoted to such electrodes. We refer the Reader to several available reviews [26, 27, 72, 79, 101].

Theoretical models of electroanalytical experiments involving channel and tubular electrodes are usually based on the assumption that the flow is laminar, and that an adequate lead-in length exists upstream of the electrode. It is further assumed that the flow velocity is significantly large, so that diffusion in the axial direction can be neglected, in comparison with axial convection. Under these assumptions, the convection-diffusion PDE (consistent with the coordinates and notation used in Fig. 2.7) for the concentration c_j of a jth dynamic distributed species is:

$$\frac{\partial c_j(y,z,t)}{\partial t} = D_j \frac{\partial^2 c_j(y,z,t)}{\partial y^2} - v_z(y)\, \frac{\partial c_j(y,z,t)}{\partial z} \tag{2.17}$$

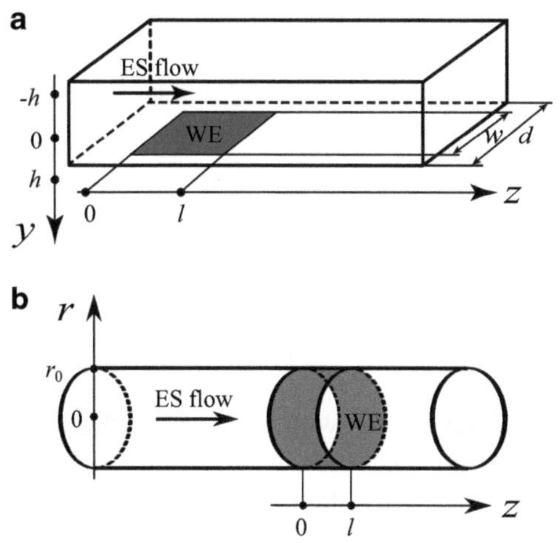

Fig. 2.7 Rectangular channel (**a**) and tubular (**b**) working electrode (WE) arrangements. In both arrangements an electrolyte solution (ES) flows past the WE (*grey areas*) embedded in an insulator (*white areas*). The channel has a width d and height $2h$. Normally $h \ll d$. The WE width w must be sufficiently smaller than d, to make edge effects negligible. Cartesian coordinates y and z are then sufficient for the description. The tube has a radius r_0, and the WE has a cylindrical symmetry. Therefore, cylindrical coordinates r and z are sufficient for the description

with

$$v_z(y) = v_0\left[1 - (y/h)^2\right] \tag{2.18}$$

for channel electrodes, and

$$\frac{\partial c_j(r,z,t)}{\partial t} = D_j\left[\frac{\partial^2 c_j(r,z,t)}{\partial r^2} + \frac{1}{r}\frac{\partial c_j(r,z,t)}{\partial r}\right] - v_z(r)\frac{\partial c_j(r,z,t)}{\partial z} \tag{2.19}$$

with

$$v_z(r) = v_0\left[1 - (r/r_0)^2\right] \tag{2.20}$$

for tubular electrodes. Parameter v_0 is the convection velocity in the middle of the channel or tube (along the symmetry axis z). We note that the PDEs (2.17) and (2.19) depend on two spatial coordinates (so that the spatial domain associated with this system is two-dimensional), and that the convection velocity variations (2.18) and (2.20) are parabolic in the directions perpendicular to the surface of the electrodes.

When the electrode length l is sufficiently small and/or the axial convection velocity v_0 large, then the concentration changes due to heterogeneous reactions at the electrode do not extend much into the solution, i.e. the thickness of the diffusion layer is much smaller than h or r_0 (that is h or r_0 can be formally regarded as infinitely large). Under such conditions it is usual to invoke the so-called Lévêque approximation [67], meaning a linearisation of the velocity profile within the diffusion layer at the electrode:

$$v_z(y) \approx 2v_0\,(1 - y/h) \tag{2.21}$$

for channel electrodes, and

$$v_z(r) \approx 2v_0\,(1 - r/r_0) \tag{2.22}$$

for tubular electrodes. In accord with this approximation one simultaneously assumes that the concentrations outside the diffusion layer remain practically equal to the (initial) concentrations at the inlet to the channel or tube.

2.7 Equations for Reaction Kinetics

There exists an ample literature devoted to formal chemical kinetics. For a general introduction, the classical book by Boudart [18] may be particularly helpful.

For a given set $X_1, X_2, \ldots, X_{N_s}$ of N_s species, any lth, out of N_r reactions, can be written as a pair of unidirectional processes that proceed in opposite directions:

$$\sum_{j=1}^{N_s} \nu_{j,l}^{f}\, X_j \rightleftarrows \sum_{j=1}^{N_s} \nu_{j,l}^{b}\, X_j \ , \tag{2.23}$$

where $\nu_{j,l}^{f}$ and $\nu_{j,l}^{b}$ are the so-called *stoichiometric coefficients* of the reactants and products, respectively. One should write the reactions in such a way that $\nu_{j,l}^{f}$ and $\nu_{j,l}^{b}$ are mutually prime non-negative integers [8]. The unidirectional transformation of the reactants into products is called *forward reaction*. The reverse transformation is called *backward reaction*. Basically all reactions are *reversible*, in the sense that if a forward (or backward) reaction is possible, then a reverse reaction must be possible as well. However, it sometimes happens that the rate of the reverse reaction is negligible, so that formally we can consider a given reaction as *irreversible*, proceeding only in one direction.

Electroanalytical models can involve *elementary* or *non-elementary* reactions. A reaction is elementary (represents an *elementary step*) if it takes place in a single irreducible act at the molecular level, just the way it is written in the *stoichiometric equation* (2.23) [18]. Non-elementary reactions occur in more than one elementary step. If such reactions are considered in a kinetic model, it is usually assumed that under the particular experimental conditions it is not possible to distinguish their elementary steps, and observe or identify possible intermediate species. Elementary reactions are usually unimolecular or bimolecular, which means that there is only one species with a stoichiometric coefficient 1 or 2, or two species with stoichiometric coefficients 1, at both sides of Eq. (2.23). Non-elementary reactions may involve more species, or have higher stoichiometric coefficients at one or both sides of Eq. (2.23).

The (specific) *rate of a reaction* (2.23) is defined as

$$r_l = r_l^{\mathrm{f}} - r_l^{\mathrm{b}} , \tag{2.24}$$

where r_l^{f} and r_l^{b} are (specific) rates of the forward and backward reactions. The latter can be most convincingly understood as average numbers of the (forward or backward) reaction acts (usually counted in moles) per unit time and unit volume of the solution (in the case of homogeneous reactions) or unit surface of an interface (in the case of heterogeneous reactions). An alternative definition involves the notion of the reaction extent [18]. A reaction is in *equilibrium*, when $r_l^{\mathrm{f}} = r_l^{\mathrm{b}}$. It is convenient to distinguish *positive equilibria* in which $r_l^{\mathrm{f}} > 0$ and $r_l^{\mathrm{b}} > 0$, and *zero equilibria*, in which $r_l^{\mathrm{f}} = r_l^{\mathrm{b}} = 0$, which implies that the reaction does not proceed at all, possibly owing to the absence of a particular reactant, or (in the case of electrochemical reactions) as a result of a particular electrode polarisation. There are reactions so fast that they are practically in a positive equilibrium all the time, even under conditions of transient experiments. In this book such reactions are termed *equilibrium reactions*. In contrast, reactions characterised by finite nonzero rates are termed *non-equilibrium reactions*. Such a terminology was postulated in Bieniasz [12].

The reaction rates may be subject to various rate laws defining how r_l^{f} and r_l^{b} depend on concentrations and other variables. In the case of homogeneous reactions in liquids, the most commonly followed rate law is the *power rate law*, according to which (see, for example, Boudart [18] or Ritchie [104]):

$$r_l^{\mathrm{f}} = k_l^{\mathrm{f}} \prod_{j=1}^{N_{\mathrm{s}}^{\mathrm{distr}}} \left(c_j\right)^{\lambda_{j,l}^{\mathrm{f}}} , \tag{2.25}$$

$$r_l^{\mathrm{b}} = k_l^{\mathrm{b}} \prod_{j=1}^{N_{\mathrm{s}}^{\mathrm{distr}}} \left(c_j\right)^{\lambda_{j,l}^{\mathrm{b}}} , \tag{2.26}$$

where $N_{\mathrm{s}}^{\mathrm{distr}}$ is the number of distributed species in a given phase, c_j is a local value of the concentration of a jth distributed species at some point in a spatially extended phase, k_l^{f} and k_l^{b} are coefficients known as the (forward and backward) *rate constants*, and $\lambda_{j,l}^{\mathrm{f}}$ and $\lambda_{j,l}^{\mathrm{b}}$ are called *reaction orders*. A nonzero reaction order means that the rate of the lth reaction depends on the concentration of the jth species. To be consistent with thermodynamics, the reaction orders generally have to obey the relationship ([16], [18, p. 91]):

$$\lambda_{j,l}^{\mathrm{f}} - \lambda_{j,l}^{\mathrm{b}} = \left(\nu_{j,l}^{\mathrm{f}} - \nu_{j,l}^{\mathrm{b}}\right) / \nu_l , \tag{2.27}$$

where ν_i is a positive constant. In the case of elementary reactions normally $\lambda^{\rm f}_{j,i} = \nu^{\rm f}_{j,i}$, $\lambda^{\rm b}_{j,i} = \nu^{\rm b}_{j,i}$, and $\nu_i = 1$. For non-elementary reactions composed of several elementary steps, if a reaction possesses a single rate determining step, then ν_i is the *stoichiometric number* of this step. This number is the number of times that the rate determining step must be repeated, in a closed sequence, in order to obtain by summation of all steps the overall stoichiometric equation for the reaction. In electroanalytical models non-elementary homogeneous reactions are rarely considered, hence the orders are usually equal to the stoichiometric coefficients, and orders greater than 2 are rare (examples of models involving third-order homogeneous reactions are in [28, 110]). We emphasise that second- or higher-order homogeneous reactions imply nonlinear dependencies of the reaction rates on concentrations, which presents a difficulty for the IE method, because the reaction-transport PDEs are then nonlinear (cf. Sect. 2.2). The rate constants $k^{\rm f}_i$ and $k^{\rm b}_i$ of homogeneous reactions in liquids depend on the temperature. Therefore, they can usually be regarded as constant parameters, since a majority of electroanalytical experiments are performed under isothermal conditions. Exceptions to this are temperature perturbation experiments (see, for example, Wildgoose et al. [124] and Gründler et al. [48]), but these are rarely modelled.

At a positive equilibrium, a homogeneous reaction subject to the power rate law (2.25) and (2.26) satisfies the equilibrium condition:

$$\prod_{j=1}^{N_{\rm s}^{\rm distr}} (c_j)^{\nu^{\rm b}_{j,i} - \nu^{\rm f}_{j,i}} = K_i \, , \tag{2.28}$$

where K_i is the classical *equilibrium constant* which should be distinguished from the thermodynamic equilibrium constant (see, for example, Everett [35, p. 112]). The equilibrium constant K_i is related to the rate constants by the equation:

$$k^{\rm f}_i / k^{\rm b}_i = (K_i)^{1/\nu_i} \, . \tag{2.29}$$

Some non-elementary homogeneous reactions occurring in electroanalytical models may be subject to other rate laws, for example to Michaelis–Menten kinetics [62] characteristic of enzymatic reactions (see, for example, Leskovac [66] or Marangoni [73]). The rates of enzymatic reactions are rational functions of the concentrations. Due to the nonlinearity of such functions, they have not been thus far handled by the IE method. Therefore, we do not write more about such rate laws here.

In the case of heterogeneous reactions a greater diversity of expressions for the reaction rates is encountered. References [1, 9, 31, 33, 40, 41] may serve as a source of basic information about such expressions, but some of the formulae used in the modelling may not be available in these references, and have to be searched in the literature. It is difficult to provide a complete list of possibilities.

In many modelling studies the power rate law analogous to Eqs. (2.25) and (2.26) is assumed:

$$r_i^{\mathrm{f}} = k_i^{\mathrm{f}} \prod_{j=1,\, j\neq e}^{N_{\mathrm{s}}^{\mathrm{distr}}+N_{\mathrm{s}}^{\mathrm{loc}}+1} \left(\Theta_j\right)^{\lambda_{j,i}^{\mathrm{f}}} , \tag{2.30}$$

$$r_i^{\mathrm{b}} = k_i^{\mathrm{b}} \prod_{j=1,\, j\neq e}^{N_{\mathrm{s}}^{\mathrm{distr}}+N_{\mathrm{s}}^{\mathrm{loc}}+1} \left(\Theta_j\right)^{\lambda_{j,i}^{\mathrm{b}}} , \tag{2.31}$$

where $N_{\mathrm{s}}^{\mathrm{distr}}$ is the number of distributed species in phases adjacent to a given interface, $N_{\mathrm{s}}^{\mathrm{loc}}$ is the number of localised species at the interface (including free sites, if relevant), Θ_j denotes either the concentration of a distributed species at the interface, further denoted by $c_j^{\dagger}$, or the concentration Γ_j of a localised species, and $\lambda_{j,i}^{\mathrm{f}}$ and $\lambda_{j,i}^{\mathrm{b}}$ are reaction orders. The orders satisfy the condition analogous to Eq. (2.27). Index $j = e$ refers to exchanged electrons, if they are included in the list of reactants. Rate constants k_i^{f} and k_i^{b} of heterogeneous non-electrochemical reactions are often assumed to depend only on the temperature, whereas those for electrochemical reactions depend not only on the temperature but also on the electrode potential E (in the case of electron transfers at electrode | electrolyte interfaces) or on the Galvani potential difference $\Delta\Phi$ across the interface (in the case of electron or ion transfers at liquid | liquid interfaces). According to the most popular Butler–Volmer kinetic model, for electron transfer reactions at electrodes the potential dependences are:

$$k_i^{\mathrm{f}} = k_i^{\mathrm{f}}(\varepsilon) = k_i^0 \exp\left(\frac{\nu_{e,i}\, \alpha_i^{\mathrm{f}}}{\nu_i}\, \varepsilon\right) , \tag{2.32}$$

$$k_i^{\mathrm{b}} = k_i^{\mathrm{b}}(\varepsilon) = k_i^0 \exp\left(-\frac{\nu_{e,i}\, \alpha_i^{\mathrm{b}}}{\nu_i}\, \varepsilon\right) , \tag{2.33}$$

with

$$\varepsilon = \frac{F}{RT}\left(E - E_i^0\right) , \tag{2.34}$$

where k_i^0 is the *conditional rate constant* of the reaction, $\nu_{e,i}$ is the stoichiometric coefficient of the electron exchanged in the reaction, α_i^{f} and $\alpha_i^{\mathrm{b}} = 1 - \alpha_i^{\mathrm{f}}$ are *charge transfer coefficients*, most often assumed to be constant, and E_i^0 is the *conditional (formal) potential* of the reaction. For elementary reactions α_i^{f} and α_i^{b} are called *symmetry factors*. Equations (2.32)–(2.34) are presented after IUPAC recommendations [92, 93], and we direct the Reader to those papers for further details. It must be noted, however, that alternative assumptions regarding the magnitude of the charge transfer coefficients are also encountered.

At a positive equilibrium, a heterogeneous non-electrochemical reaction subject to the power rate law satisfies the equilibrium condition analogous to Eq. (2.28):

$$\prod_{j=1,\, j\neq e}^{N_s^{\mathrm{distr}}+N_s^{\mathrm{loc}}+1} (\Theta_j)^{\nu_{j,l}^{\mathrm{b}}-\nu_{j,l}^{\mathrm{f}}} = K_l \; , \tag{2.35}$$

where K_l is a suitable equilibrium constant, whereas an electron transfer reaction at the electrode satisfies the Nernst equation:

$$\prod_{j=1,\, j\neq e}^{N_s^{\mathrm{distr}}+N_s^{\mathrm{loc}}+1} (\Theta_j)^{\nu_{j,l}^{\mathrm{b}}-\nu_{j,l}^{\mathrm{f}}} = \exp\left[\frac{\nu_{e,l} F}{RT}\left(E - E_l^0\right)\right] \; . \tag{2.36}$$

In electrochemistry, electron transfer reactions being in a permanent equilibrium are often called *Nernstian* or *reversible*, whereas those not in equilibrium, but having finite rates in both directions, are called *quasi-reversible*. This terminology is inconsistent with the previously mentioned general meaning of reversible reactions in chemistry. The inconsistency complicates theoretical discussions of the reaction kinetics in electrochemical systems, where electrochemical and non-electrochemical reactions simultaneously occur, because one has to use different names to denote analogous reaction properties, depending on the reaction type. This is a reason why it is more convenient to use the terminology adopted here after Bieniasz [12], according to which one simply distinguishes *equilibrium reactions* and *non-equilibrium reactions*, irrespective of whether electrochemical or non-electrochemical reactions are in mind. Any equilibrium reaction must be reversible, in a general sense. Any irreversible reaction must be a non-equilibrium reaction, and any non-equilibrium reversible reaction has finite rates in both directions.

The rate constants (2.32) and (2.33) are linked with the electrode potential by the equation

$$k_l^{\mathrm{f}}/k_l^{\mathrm{b}} = \exp\left[\frac{\nu_{e,l}}{\nu_l}\left(\frac{F}{RT}\right)\left(E - E_l^0\right)\right] \; . \tag{2.37}$$

In cases when free sites available for adsorption are considered as dynamic species, Eqs. (2.35) or (2.36) can be interpreted as Langmuir-type adsorption (or electrosorption) isotherms.

Among the various complications or alternative assumptions regarding the heterogeneous reaction rates, one can distinguish the following. Equations (2.30) and (2.31) may involve additional multiplicative factors exponentially dependent on the concentrations of localised species, and expressing the effect of the interactions between these species on the reaction rates. At a positive equilibrium, one then obtains Frumkin-type isotherms, instead of the Langmuir isotherms [40, 41]. Yet different dependences on concentrations may be needed to reproduce still other kinds of adsorption isotherms known in electrochemistry [31]. Rate constants of heterogeneous non-electrochemical adsorption reactions can sometimes be assumed exponentially dependent on the electrode potential (see, for example, the theory of Parsons [91], the IE-based model of cyclic voltammetry, reported by Wopschall and Shain [125], or the IE-based model of normal pulse polarography, reported by Puy et al. [98]). Charge transfer coefficients α_i^{f} and α_i^{b} of electrochemical reactions can sometimes be assumed potential-dependent (see, for example, the review by Sanecki and Skitał [107]). In recent years there is also an increasing interest in considering alternative models of the potential-dependent rate constants of the electrochemical reactions, originating from the Marcus [74, 75] and Hush [56] theory of electron transfer. In the Marcus–Hush theory the potential dependencies are more sophisticated, compared to Eqs. (2.32) and (2.33). The classical, "symmetric" variant of this theory, often identified by an additional reference to Chidsey [24], is usually termed the *Marcus–Hush–Chidsey* theory in the literature. According to this variant, for an elementary one-electron transfer $X_1 + e^- \rightleftarrows X_2$ the rate constants are given by

$$k_i^{\mathrm{f}} = k_i^{\mathrm{f}}(\varepsilon) = \begin{cases} k_i^0 \exp\left(\frac{\varepsilon}{2}\right) \frac{\mathrm{S}(\varepsilon,\Lambda)}{\mathrm{S}(0,\Lambda)} & \text{for } \varepsilon \geq 0 \\ \exp(\varepsilon)\, k_i^{\mathrm{f}}(-\varepsilon) & \text{for } \varepsilon < 0 \end{cases} , \tag{2.38}$$

$$k_i^{\mathrm{b}} = k_i^{\mathrm{b}}(\varepsilon) = \begin{cases} \exp(-\varepsilon)\, k_i^{\mathrm{f}}(\varepsilon) & \text{for } \varepsilon \geq 0 \\ k_i^{\mathrm{f}}(-\varepsilon) & \text{for } \varepsilon < 0 \end{cases} . \tag{2.39}$$

In Eq. (2.38),

$$\mathrm{S}(\varepsilon, \Lambda) = \pi^{-1/2} \Lambda^{-1/2} \exp\left(-\frac{\varepsilon^2}{4\Lambda}\right) \int\limits_0^\infty \exp\left(-\frac{x^2}{\Lambda}\right) \cosh\left(\frac{\varepsilon x}{\Lambda}\right) \operatorname{sech}(x)\, \mathrm{d}x \tag{2.40}$$

and $\Lambda = \lambda (RT)^{-1}$, with λ denoting the so-called reorganisation energy (in J mol^{-1}). Equations (2.38)–(2.40) are more difficult to handle than Eqs. (2.32) and (2.33), in mathematical models. Their use can be facilitated by the specially developed procedures for accurate and efficient computation of the above Marcus–Hush–Chidsey rate constants [15, 76, 85]. However, the physical adequacy of Eqs. (2.38)–(2.40) and also of other variants of the Marcus theory, for electroanalytical modelling, is still under debate (see, for example, the review by Henstridge et al. [52], and the references cited therein). The question seems to be open whether these equations ensure a better agreement between theoretical and experimental electrochemical responses, than the classical Butler–Volmer equations (2.32) and (2.33) do.

Other kinetic complications arise in the case of reactions involving solvated electrons (see, for example, Alpatova et al. [4]), but such reactions will not be of particular interest in this book.

It should be stressed that in contrast to homogeneous reactions, the nonlinearities of Eqs. (2.30) and (2.31) present no difficulty for the IE method, which will become clear in further chapters.

The above presentation is focused on single reactions. In practice, the number N_{r} of reactions is often greater than one. A set of reactions, presumably occurring in an electrochemical system studied (or a part of it), is usually called a *reaction scheme*. Alternatively (and sometimes interchangeably) the name *reaction mechanism* is also used. Strictly speaking, the latter name should be restricted to cases when a particular set of reactions represents elementary steps of a certain non-elementary reaction, or serves to explain how to obtain a certain product from particular substrates. The name *reaction network*, popular outside electrochemistry, is also sporadically used. There exist a number of standard reaction schemes, often encountered in electrochemical studies, for which a more or less widely adopted notation has been devised [6]. Elements of this notation are used in the present book. Thus, electron transfer reactions are noted by the letter E, other reactions by the letter C. For example, ECE denotes any reaction scheme comprising two electron transfer reactions separated by a homogeneous (or heterogeneous) non-electrochemical reaction, such as:

$$X_1 + e^- \rightleftarrows X_2 \,, \tag{2.41}$$

$$X_2 \rightleftarrows X_3 \,, \tag{2.42}$$

$$X_3 + e^- \rightleftarrows X_4 \,. \tag{2.43}$$

An important characteristic of the reaction schemes is the stoichiometric matrix $\overline{N}$, defined as $\overline{N} = \left\{\nu^{\mathrm{b}}_{j,i} - \nu^{\mathrm{f}}_{j,i}\right\}_{N_{\mathrm{s}} \times N_{\mathrm{r}}}$. Matrix $\overline{N}$ is usually non-square, as the number of the reactions is generally different from the number of species. Of particular interest is the submatrix $\overline{N}^{\,\mathrm{dyn}}$ of $\overline{N}$, corresponding to dynamic species (and electrons exchanged, if taken into account). A detailed analysis of the matrices $\overline{N}$ and $\overline{N}^{\,\mathrm{dyn}}$ allows one to derive many of the mathematical equations necessary for the description of electroanalytical experiments. Among other things, one can determine a set of linearly independent reactions, that might be called *base reactions*. The linear independence of the reactions is equivalent to the linear independence of the columns of the stoichiometrix matrix. A set of base reactions determines a set of independent model parameters, since in the case of linearly dependent reactions some of their parameters like equilibrium constants or conditional potentials are not arbitrary but depend on the relevant parameters of the base reactions (see Luo et al. [70] and Bieniasz [12]). The analysis of the stoichiometric matrix, and the derivation process of the mathematical equations can be automated and computerised, as was shown in [12–14, 69]. Given a vector $\overrightarrow{r} = [r_1, \ldots, r_{N_{\mathrm{r}}}]^{\mathrm{T}}$ of all reaction rates, the corresponding vector $\overrightarrow{p}$ of the production rates of all dynamic species involved in these reactions can be calculated as

$$\overrightarrow{p} = \overline{N}^{\,\mathrm{dyn}} \overrightarrow{r} \,. \tag{2.44}$$

The elements of the vector $\overrightarrow{p}$ are:

$$p_j = \frac{\partial c_j}{\partial t} \,, \tag{2.45}$$

in the case of dynamic distributed species subject to homogeneous reactions;

$$p_j = \pm J_j^{\perp} , \tag{2.46}$$

in the case of dynamic distributed species subject to heterogeneous reactions;

$$p_j = \frac{\mathrm{d}\Gamma_j}{\mathrm{d}t} , \tag{2.47}$$

in the case of dynamic localised species subject to heterogeneous reactions; and

$$p_j = \frac{i^{\perp}}{F} , \tag{2.48}$$

(with $j = e$) in the case of electrons exchanged in electron transfer reactions at electrodes. Symbol $J_j^{\perp}$ in Eq. (2.46) denotes the component perpendicular to the interface, of the flux of a given species. The plus or minus sign in Eq. (2.46) is chosen based on the following reasoning. Assume that ξ is a spatial coordinate along an axis perpendicular to the interface, with $\xi = 0$ corresponding to the location of the interface. Let $J_j^{\perp}$ be expressed in this coordinate system. Then plus is chosen for species present in the phase for which $\xi > 0$, and minus is chosen for species present in the phase for which $\xi < 0$. Symbol $i^{\perp}$ in Eq. (2.48) denotes the Faradaic current density at the electrode surface, in the direction perpendicular to the electrode. In cases when the current density $i^{\perp}$ is uniform along the electrode surface, Eq. (2.48) can be replaced by

$$p_j = \frac{I}{FA} , \tag{2.49}$$

where I is the total Faradaic current flowing through the electrode surface, and A is the electrode area.

In Eqs. (2.44)–(2.49), and in other related equations of this chapter, we omit lists of arguments of $\vec{p}$, $\vec{r}$, $i^{\perp}$ and I. However, one should keep in mind that under transient conditions these variables are functions of time. For some models they can also be functions of spatial coordinates.

2.8 The Effect of Homogeneous Reactions on Transport PDEs

Equation (2.44) implies, among other things, that the rates of the homogeneous reaction(s) are related to the production rates of the dynamic distributed species. Consider a spatially extended phase with $N_{\mathrm{r}}^{\mathrm{hom}}$ homogeneous reactions and $N_{\mathrm{s}}^{\mathrm{dd}}$

dynamic distributed species having concentrations $\vec{c}$. By rewriting Eq. (2.45) in the vector notation, we obtain:

$$\vec{p}^{\,\text{hom}} = \frac{\partial \vec{c}}{\partial t}, \tag{2.50}$$

where $\vec{p}^{\,\text{hom}} = \left[p_1^{\text{hom}}, \ldots, p_{N_s^{\text{dd}}}^{\text{hom}}\right]^{\text{T}}$ is the vector of the production rates of the dynamic distributed species, and the concentration changes are due solely to the homogeneous reactions. By introducing the vector $\vec{r}^{\,\text{hom}} = \left[r_1^{\text{hom}}, \ldots, r_{N_r^{\text{hom}}}^{\text{hom}}\right]^{\text{T}}$ of the homogeneous reaction rates, the relationship between $\vec{p}^{\,\text{hom}}$ and $\vec{r}^{\,\text{hom}}$, resulting from Eq. (2.44) is

$$\vec{p}^{\,\text{hom}} = \overline{N}^{\,\text{dd,hom}}\,\vec{r}^{\,\text{hom}}, \tag{2.51}$$

where $\overline{N}^{\,\text{dd,hom}}$ is the submatrix of the stoichiometric matrix $\overline{N}^{\,\text{dyn}}$, that takes into account only the above species and reactions. Expression (2.51) has to be included into Eq. (2.10) to obtain a complete reaction-convection-diffusion PDE system. Of special interest to this book is the situation when all homogeneous reactions are either of first order, or of pseudo first order. Pseudo first order means that each of the reaction rate expressions (2.25) and (2.26) may involve concentrations of at most one dynamic species with the reaction order equal 1. All remaining species participating in the reaction must be static species. In such a case it is possible to write Eq. (2.51) in the linear form

$$\vec{p}^{\,\text{hom}} = -\overline{K}\,\vec{c} + \vec{\sigma}, \tag{2.52}$$

where $\overline{K}$ is a real $N_s^{\text{dd}} \times N_s^{\text{dd}}$ matrix of coefficients dependent on the rate constants of the homogeneous reactions, and (in the case of pseudo first-order reactions) on the constant concentrations of the static distributed species; and $\vec{\sigma}$ is a constant source vector. Vector $\vec{\sigma}$ represents the homogeneous reaction rate terms dependent exclusively on the concentrations of the static distributed species.

2.9 The Effect of Heterogeneous Reactions on Boundary Conditions and Governing Equations for Localised Species

Consider an interface at which N_r^{het} heterogeneous reactions take place. Let N_s^{dd} be the number of dynamic distributed species in adjacent spatially extended phases, and N_s^{dl} be the number of dynamic localised species at this interface. By $\overline{N}^{\,\text{dyn,het}}$ let us denote the submatrix of the stoichiometric matrix $\overline{N}^{\,\text{dyn}}$ in Eq. (2.44), which corresponds to these species and reactions. If one of the adjacent spatially extended phases is an electrode, and electron transfer reactions are involved, electrons exchanged are considered as additional dynamic species, and are taken into account in the matrix $\overline{N}^{\text{dyn,het}}$. Hence, there are altogether $N_{\text{s}}^{\text{dyn}} = N_{\text{s}}^{\text{dd}} + N_{\text{s}}^{\text{dl}} + 1$ dynamic species to be considered at the interface. If the interface is a liquid | liquid interface, electrons exchanged are not taken into account in the matrix $\overline{N}^{\text{dyn,het}}$, even if there are electron transfer reactions, so that there are $N_{\text{s}}^{\text{dyn}} = N_{\text{s}}^{\text{dd}} + N_{\text{s}}^{\text{dl}}$ dynamic species to be considered. According to Eq. (2.44), the vector $\vec{p}^{\,\text{het}} = \left[p_1^{\text{het}}, \ldots, p_{N_{\text{s}}^{\text{dyn}}}^{\text{het}}\right]^{\text{T}}$ of the production rates of the dynamic species is linked with the vector $\vec{r}^{\,\text{het}} = \left[r_1^{\text{het}}, \ldots, r_{N_{\text{r}}^{\text{het}}}^{\text{het}}\right]^{\text{T}}$ of the heterogeneous reaction rates by the equation:

$$\vec{p}^{\,\text{het}} = \overline{N}^{\text{dyn,het}}\,\vec{r}^{\,\text{het}} \,. \tag{2.53}$$

In cases when all heterogeneous reactions are non-equilibrium reactions, Eq. (2.53) can sometimes be used directly as a coupled set of equations containing boundary conditions for dynamic distributed species, and equations governing the evolution of the concentrations of dynamic localised species. If electrons exchanged are taken into account in the stoichiometric matrix $\overline{N}^{\text{dyn,het}}$, Eq. (2.53) contains one more equation that can be used for calculating the Faradaic current (in the case of controlled potential experiments) or electrode potential (in the case of controlled current experiments). If electrons exchanged are not taken into account in $\overline{N}^{\text{dyn,het}}$, an additional equation defining the current is needed, based on the balance of electric charge in the charge transfer reactions involved.

In practice, however, Eq. (2.53) is rarely used in the direct form. Analysis of many electroanalytical models published in the literature reveals that usually Eq. (2.53) is presented in a modified form that can be obtained by partially inverting Eq. (2.53), so that some of the reaction rates are expressed as linear combinations of the other reaction rates, and/or species production rates. Such a modification of Eq. (2.53) is necessary in cases when the reaction scheme involves heterogeneous equilibrium reactions. The detailed procedure for performing the inversion is described in Bieniasz [13]. The procedure transforms Eq. (2.53) into the equation:

$$\overline{V}\,\vec{r}^{\,\text{het}} = \overline{Z}\,\vec{p}^{\,\text{het}} \,, \tag{2.54}$$

where $\overline{V}_{N_{\text{s}}^{\text{dyn}} \times N_{\text{r}}^{\text{het}}}$ and $\overline{Z}_{N_{\text{s}}^{\text{dyn}} \times N_{\text{s}}^{\text{dyn}}}$ are matrices of rational coefficients. Matrix $\overline{V}$ has in the general case the following block structure:

$$\begin{array}{cc} & \begin{array}{cccc} C1 & C2 & C3 & C4 \end{array} \\ \overline{V} = & \left[\begin{array}{cccc} \overline{1}_1 & \overline{0} & \overline{M}_1 & \overline{M}_2 \\ \overline{0} & \overline{1}_2 & \overline{0} & \overline{M}_3 \\ \overline{0} & \overline{0} & \overline{0} & \overline{0} \end{array}\right] \begin{array}{c} R1 \\ R2 \\ R3 \end{array} , \end{array} \tag{2.55}$$

where $\overline{1}_1$ and $\overline{1}_2$ are unit submatrices (not necessarily of the same size), $\overline{M}_1$, $\overline{M}_2$, and $\overline{M}_3$ are some generally non-square submatrices, the elements of which can be different from zero, and $\overline{0}$ denotes null submatrices. Sets $C1$ and $C3$ of columns of $\overline{V}$ correspond to equilibrium reactions, sets $C2$ and $C4$ to non-equilibrium reactions. The procedure of Bieniasz [13] ensures that all columns within sets $C1$ and $C2$ are linearly independent, columns $C3$ are dependent on columns $C1$, and columns $C4$ are dependent on columns $C1$ and $C2$. Each equation corresponding to set $R1$ of rows in matrix $\overline{V}$ allows one to express the rate of some equilibrium reaction from set $C1$ as a linear combination of the following variables: rates of equilibrium reactions from set $C3$, rates of non-equilibrium reactions from set $C4$, and species production rates. One can then neglect all finite reaction rates from set $C4$, and (finite) species production rates in equations from the rows $R1$. Terms resulting from dependent equilibrium reactions in set $C3$ can also be removed from equations corresponding to $R1$, because dependent reactions do not alter equilibria of independent reactions. In this way, all equations from the set $R1$ are replaced by equilibrium equations for reactions from set $C1$. Each equation corresponding to set $R2$ of rows allows one to express a finite rate of one of the independent non-equilibrium reactions from set $C2$, as a linear combination of the rates of dependent non-equilibrium reactions from set $C4$, and species production rates. Finally, equations corresponding to set $R3$ of rows do not involve reaction rates. They only express relationships between species production rates.

In addition to replacing Eq. (2.53) by Eq. (2.54), one can also apply steady state conditions to some of the intermediates, by zeroing some of the elements of $\overline{Z}$, thereby neglecting respective species production rates.

The above transformation of Eq. (2.53) into Eq. (2.54) is similar to the process of looking for the so-called first linear invariants of a dynamical system, performed in kinetic studies of complex non-electrochemical reaction networks (see, for example, Reder [100], Corio [29], and Sauro and Ingalls [108]). By following this analogy, equations belonging to the group $R3$ should be called *structural conservation relationships* [100], because they express invariants resulting from the structure of the reaction network, independently of the rate laws.

If a particular dynamic species does not take part in any of the heterogeneous reactions at the interface, then Eq. (2.53) implies that the production rate of this species is simply zero. In view of Eqs. (2.46) and (2.47), for a distributed species this implies a zero flux boundary condition at the interface, and for a localised species this implies a time-invariant concentration.

For models not including dynamic distributed species, Eq. (2.53) yields generally an ordinary differential equation (ODE) system, and Eq. (2.54) yields generally a differential-algebraic equation (DAE) system, for the temporal evolution of the concentrations of the dynamic localised species (plus one more equation for calculating the Faradaic current or electrode potential, if electrons exchanged are taken into account in the stoichiometric matrix $\overline{N}^{\,\mathrm{dyn,het}}$). In the case of models not involving localised species Eqs. (2.53) or (2.54) yield boundary conditions for the transport PDEs or steady state ODEs (plus optionally one more equation, as above). If both dynamic distributed and dynamic localised species occur, all these equations may be coupled in the way which makes it impossible to distinguish among them boundary conditions for transport PDEs, and ODEs or DAEs for the localised species; all equations must be considered and solved jointly.

In the case of controlled current experiments it is not always possible to obtain an explicit formula for the electrode potential (or interfacial potential difference). Implicit expressions, requiring numerical potential determinations, may occur.

There is also a special situation when dynamic distributed species can be formally considered as localised species, with all consequences regarding their mathematical representation in Eqs. (2.53) or (2.54). This is the situation of dynamic distributed species confined to a very thin spatially extended phase (for example, a thin electrolyte layer) adjacent to the interface studied. The distributed species must not be allowed to leave such a thin layer through the other interfaces. In such a case it can be shown (see Sect. 5.3) that despite diffusion, there is almost no concentration gradient inside the layer, in the direction perpendicular to the interface studied. Consequently, we can treat a distributed species as a localised one, having concentration $\Gamma_j = c_j\, l$, where c_j is the uniform concentration of the distributed species in the layer, and l is the layer thickness. The production rate of such a quasi-localised species must be expressed as $p_j^{\text{het}} = \mathrm{d}\Gamma_j/\mathrm{d}t$, not as $p_j^{\text{het}} = \pm J_j^{\perp}$. Examples of electroanalytical models involving such quasi-localised species can be found in [54, 65], and in the references cited therein.

2.10 The Effect of Reactions on Initial Conditions

In order for the transport Eq. (2.10), and equations governing the evolution of the concentrations of localised species, contained in Eqs. (2.53) or (2.54), to be solvable, electroanalytical models have to provide initial conditions. Most often it is assumed that the system studied is initially in equilibrium, so that all reactions in the reaction scheme considered are in equilibrium, and there are also no concentration gradients in spatially extended phases. Consequently, all initial concentrations are assumed to be time-invariant prior to the electroanalytical experiment, and in the case of distributed species they are also spatially uniform. Alternative initial conditions are conceivable, but they occur rarely. For example, when flash photolysis is used to produce an electroactive species to be studied in an electroanalytical experiment, the initial concentration of this species is not in equilibrium, and it decays in time, prior to the experiment [20, 102, 116]. Such situations will not be considered in this book. Another example, of non-equilibrium initial conditions, is encountered in the studies of adsorption at DMEs, where it is often assumed that every newly created drop is initially free of the adsorbates, even if the adsorbates are present in the electrolyte. Such situations will be addressed in Chap. 9. The initial concentrations will be denoted by $c_j^{\star}$ in the case of dynamic distributed species, and by $\Gamma_j^{\star}$ in the case of dynamic localised species.

In the initial equilibrium the various reactions can be either at positive or zero equilibria. Depending on how many electrochemical reactions are initially at positive equilibria, the following three situations are conceivable: (A) All electrochemical reactions are at zero equilibria, so that the rest potential E_{rest} (in the case when the interface studied is an electrode | electrolyte interface) or the rest value of the Galvani potential difference $\Delta\Phi_{rest}$ (in the case when the interface studied is a liquid | liquid interface) is not governed by the given reaction scheme. (B) There is only one linearly independent electrochemical reaction at positive equilibrium, and E_{rest} or $\Delta\Phi_{rest}$ correspond to this equilibrium (E_{rest} is an equilibrium potential or $\Delta\Phi_{rest}$ is an equilibrium Galvani potential difference). The reaction can be accompanied by any number of other electrochemical reactions that are also in positive equilibrium, but only if they are linearly dependent on the first reaction. (C) There is more than one linearly independent electrochemical reaction initially at positive equilibrium, but such a case is acceptable only for a special combination of kinetic parameters and initial concentrations, which may not always be possible to precisely achieve or maintain experimentally. In general, if there is more than one linearly independent electrochemical reaction not at zero equilibrium, a non-equilibrium, mixed-potential situation is expected (see, for example, Kortüm [64, p. 510]), so that equilibrium initial conditions cannot be assumed in such a case.

In the case of the power rate law (2.25) and (2.26) or (2.30) and (2.31), any relationships between nonzero initial concentrations of the dynamic species and concentrations of static species can always be expressed in a closed, explicit form. A procedure for deriving these relationships was described in Bieniasz [12]. It relies on solving analytically equilibrium equations (2.28), (2.35), and (2.36), transformed into linear AEs by a logarithmic transformation.

2.11 Electroanalytical Methods

Boundary conditions depend not only on the kinetics of heterogeneous reactions but also on the electroanalytical method used in an experiment of interest. Experimental electroanalytical methods can be divided into two major categories: controlled potential methods, and controlled current methods. In the controlled potential methods a specific potential-time perturbation is applied to the WE(s), and the current-time response is measured. In the controlled current methods a current-time perturbation is applied and the potential-time response is measured. The various methods differ mostly by the perturbation functions. Instead of providing detailed equations of the various perturbation functions used, Figs. 2.8 and 2.9 compare the most common perturbations visually. It is assumed in the figures that the WE is polarised to study anodic processes. Replacing $E(t)$ by $-E(t)$, or $I(t)$ by $-I(t)$ would give the functions for studying cathodic processes. Apart from the perturbation functions, the various methods may also differ by the ways in which the responses are sampled and analysed. The figures are based on IUPAC recommendations [59] and on the Bard and Faulkner book [10], to which the Reader

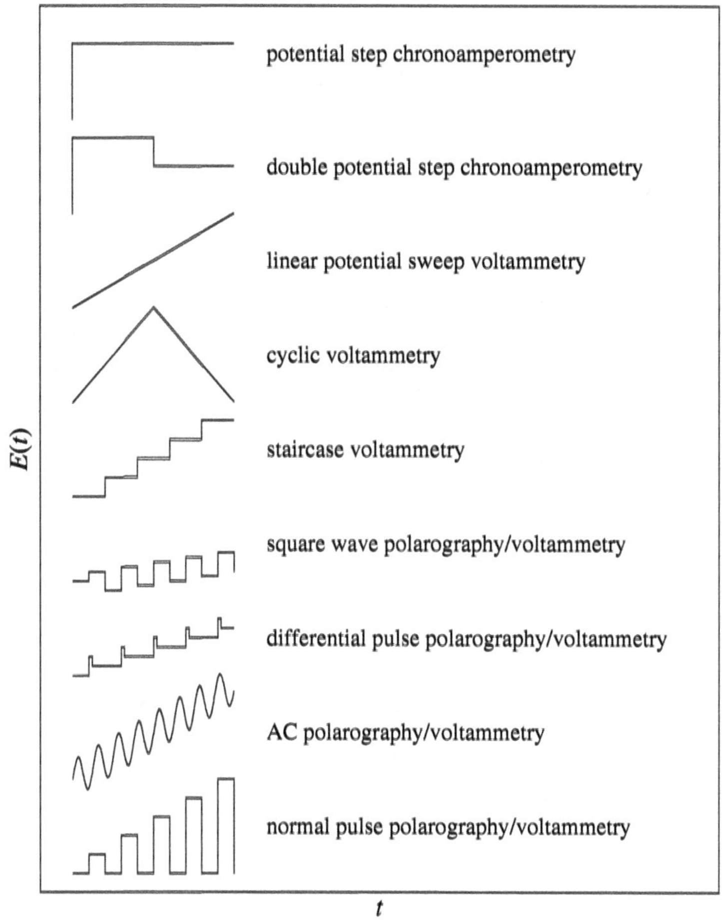

Fig. 2.8 Comparison of common potential-time perturbation functions used in controlled potential electroanalytical methods

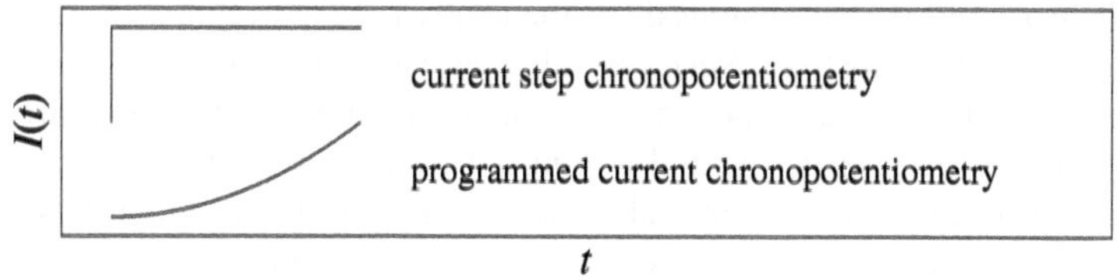

Fig. 2.9 Comparison of common current-time perturbation functions used in controlled current electroanalytical methods. The programmed current chronopotentiometry may involve any sort of current-time functions, often defined by power or exponential functions [61]

is referred for the details of the methods. Further information can be found in other similar books [32, 39, 71], and in reviews or monographs dedicated to particular methods, such as [34, 46, 49, 61, 78, 113, 126].

2.12 Anomalous Diffusion

There exist electrochemical systems, in which electrochemical observables exhibit specific deviations from the predictions of the Fick law (2.5), even if diffusion is the sole mode of transport. Such unusual transport is known under the name of *anomalous diffusion*. We refer the Reader to the reviews by Pajkossy et al. [87, 90], Go and Pyun [45], and Gmucová [44], and the references cited therein, for a detailed discussion of the anomalous diffusion phenomena in electrochemistry. Below we present a few basic equations.

Ordinary diffusion is generally interpreted [45], at a molecular level, as a random walk process, in which the mean squared displacement $< r^2(t) >$ of a molecule is proportional to the time t:

$$< r^2(t) > \sim t \; . \tag{2.56}$$

As a consequence, the various electrochemical variables and observables, recorded in the presence of ordinary diffusion, often depend on $t^{1/2}$ rather than on t explicitly. In particular, the thickness of the diffusion layer, arising under transient conditions at electrochemical interfaces, is often proportional to $t^{1/2}$, the Faradaic current often decays as $t^{-1/2}$, etc. In the case of anomalous diffusion, the same variables or observables evolve differently in time. They behave as if Eq. (2.56) was replaced by

$$< r^2(t) > \sim t^{\alpha} \; , \tag{2.57}$$

where $\alpha \neq 1$ [45]. Symbol α in Eq. (2.57) should not be confused with the charge transfer coefficients in Eqs. (2.32) and (2.33).

In electrochemistry, anomalous diffusion is observed at irregular (rough, porous, or partially active) electrode surfaces. Theoretical description of the anomalous diffusion at such interfaces appears to be far from complete, but there are models that explain the basic experimental findings. The models interpret the anomalous diffusion as an effect of the fractal nature of the surface irregularities. For the purpose of the present book it is sufficient to limit the presentation to the semi-empirical continuum model of Pajkossy and Nyikos [83, 84, 88, 89]. The model assumes that the structure of the surface has a fractal character at distances within the interval $(\lambda_{\rm inn}, \lambda_{\rm out})$, where $\lambda_{\rm inn}$ is the size of the smallest structural feature (termed *inner cut-off*), and $\lambda_{\rm out}$ is the size of the largest feature (termed *outer cut-off*). For any jth dynamic distributed species possessing a diffusion coefficient

D_j one can define characteristic times $t_{j,\mathrm{inn}}$ and $t_{j,\mathrm{out}}$, corresponding to λ_{inn} and λ_{out}:

$$t_{j,\mathrm{inn}} = \gamma\, D_j^{-1}\, \lambda_{\mathrm{inn}}^2\,, \tag{2.58}$$

$$t_{j,\mathrm{out}} = \gamma\, D_j^{-1}\, \lambda_{\mathrm{out}}^2\,, \tag{2.59}$$

where γ is a dimensionless empirical geometrical factor dependent on the shape of the surface irregularities ($\gamma = \pi^{-1}$ is an orientational value suggested based on the solution of a one-dimensional planar diffusion equation). In a transient experiment, electrochemical responses consistent with ordinary diffusion are observed at $t \ll t_{j,\mathrm{inn}}$ and $t \gg t_{j,\mathrm{out}}$, whereas anomalous diffusion effects are seen at $t \in \left(t_{j,\mathrm{inn}}, t_{j,\mathrm{out}}\right)$. In particular, if one performs a cathodic potential step chronoamperometric experiment for the electron transfer reaction $\mathrm{X}_j + ne^- \rightleftarrows \mathrm{X}_{j+1}$ taking place at a (macroscopically) planar electrode in a semi-infinite spatial domain under conditions of one-dimensional planar diffusion, then the limiting Faradaic current obeys the classical Cottrell equation [30]:

$$I(t) = -nFA_{\mathrm{micr}}\, c_j^{\star}\, D_j^{1/2}\, \pi^{-1/2}\, t^{-1/2} \tag{2.60}$$

for $t \ll t_{j,\mathrm{inn}}$, and

$$I(t) = -nFA\, c_j^{\star}\, D_j^{1/2}\, \pi^{-1/2}\, t^{-1/2} \tag{2.61}$$

for $t \gg t_{j,\mathrm{out}}$ (The Reader may find one possible derivation of the Cottrell equation in Sect. 11.1.1). In Eqs. (2.60) and (2.61) $c_j^{\star}$ is the initial/bulk concentration of species X_j, A is the macroscopic area of the electrode, and A_{micr} is the microscopic area. The two areas are related by the equation:

$$\frac{A}{A_{\mathrm{micr}}} = \left(\frac{\lambda_{\mathrm{inn}}}{\lambda_{\mathrm{out}}}\right)^{d_{\mathrm{F}}-2}\,, \tag{2.62}$$

where d_{F} is the fractal dimension of the irregular interface. In the case of rough surfaces $d_{\mathrm{F}} > 2$ was assumed, whereas in the case of partially active ones $d_{\mathrm{F}} < 2$ [83, 84, 88, 89]. For $t \in \left(t_{j,\mathrm{inn}}, t_{j,\mathrm{out}}\right)$ the anomalous current response occurs, satisfying

$$I(t) = -\sigma_{j,\mathrm{F}}\, t^{-\alpha}\,, \tag{2.63}$$

with

$$\alpha = \frac{d_{\mathrm{F}} - 1}{2}\,. \tag{2.64}$$

Coefficient $\sigma_{j,\mathrm{F}}$ results from the assumed continuity of the formulae (2.60), (2.61), and (2.63) at $t = t_{j,\mathrm{inn}}$ and $t = t_{j,\mathrm{out}}$, which gives

$$\sigma_{j,\mathrm{F}} = nFA_{\mathrm{micr}}\, c_j^\star\, D_j^{1/2}\, \pi^{-1/2} \left(\gamma\, D_j^{-1}\, \lambda_{\mathrm{inn}}^2\right)^{\alpha-1/2} \tag{2.65}$$

and

$$\sigma_{j,\mathrm{F}} = nFA\, c_j^\star\, D_j^{1/2}\, \pi^{-1/2} \left(\gamma\, D_j^{-1}\, \lambda_{\mathrm{out}}^2\right)^{\alpha-1/2} . \tag{2.66}$$

Equations (2.65) and (2.66) are equivalent, in view of the relationship (2.62). Similar derivations can be performed for other types of transient experiments [83,84,88,89].

2.13 Uncompensated Ohmic Drop and Double Layer Charging

In considering the kinetics of the electrochemical reactions in Sects. 2.7, 2.9, and 2.10 it was tacitly assumed that the controlled or measured quantity $E(t)$ represents the true electrode potential of the WE in Fig. 1.1a, or that $\Delta\Phi(t)$ represents the true Galvani potential difference across the liquid | liquid interface in Fig. 1.1b. It was also assumed that the entire current $I(t)$ is associated with electrochemical reactions taking place at the interfaces studied. In practice, none of these assumptions is obeyed strictly, even though the modern experimental equipment is designed to approach them as closely as possible. The two basic departures from the above assumptions are: the uncompensated Ohmic potential drop, and the double layer charging. The role played by these two phenomena is best illustrated by standard electric circuits corresponding to the experimental systems from Fig. 1.1. These equivalent circuits are depicted in Fig. 2.10. The review by Britz [19] provides a comprehensive discussion of the problem of the uncompensated Ohmic potential drop. The reviews by Parsons [94] and Schmickler [109] provide an introduction to the subject of the double layer.

The uncompensated Ohmic potential drop is the potential drop across the uncompensated resistance R_U in Fig. 2.10a, or across the uncompensated resistances R_{U1} and R_{U2} in Fig. 2.10b, which result from the finite conductivity of the electrolyte, and the technical difficulty in placing the reference electrodes at the outer Helmholtz planes of the electric double layers of interfaces studied. For a particular experimental setup, the parameters R_U, R_{U1}, and R_{U2} are usually considered to be constant resistances. However, in view of Eqs. (2.1)–(2.3), for the ionic conductivity this is only an approximation valid in the absence of concentration gradients. As a consequence of the Ohmic drop, the controlled, or measured potential $E(t)$ in Fig. 2.10a is related with the true electrode potential $E'(t)$ by the formula:

$$E(t) = E'(t) + I(t)\, R_\mathrm{U} . \tag{2.67}$$

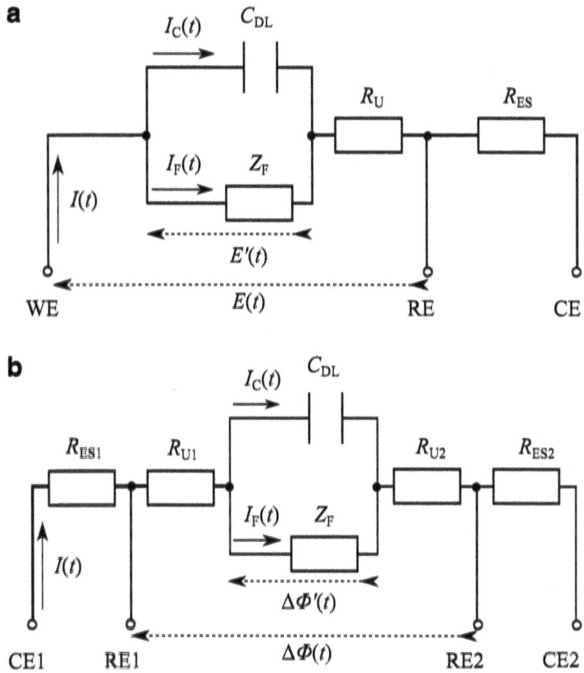

Fig. 2.10 Simplified equivalent circuits corresponding to the typical three-electrode (**a**) and four-electrode (**b**) experimental systems from Fig. 1.1. Notation: Z_F—Faradaic impedance; C_{DL}—differential double layer capacitance; R_U, R_{U1}, R_{U2}—uncompensated resistances; R_{ES}, R_{ES1}, R_{ES2}—solution resistances; $I_F(t)$—Faradaic current; $I_C(t)$—double layer charging (i.e. capacitive) current; $I(t)$—total current; $E(t)$—controlled or measured potential of the WE; $\Delta\Phi(t)$—controlled or measured Galvani potential difference between RE1 and RE2; $E'(t)$—true potential of the WE; $\Delta\Phi'(t)$—true Galvani potential difference across the liquid | liquid interface. The circuit **a** results from the well-known Randles equivalent circuit [10]. The circuit **b** results from the equivalent circuit presented by Samec et al. [106]

Analogous relationship for the ES1 | ES2 interface, referring to the controlled or measured Galvani potential difference $\Delta\Phi(t)$, and the true potential difference $\Delta\Phi'(t)$ in Fig. 2.10b is:

$$\Delta\Phi(t) = \Delta\Phi'(t) + I(t)(R_{U1} + R_{U2}) . \tag{2.68}$$

In Eqs. (2.67) and (2.68) $I(t)$ denotes the total current, composed of the Faradaic and capacitive components:

$$I(t) = I_F(t) + I_C(t) . \tag{2.69}$$

These currents obey the IUPAC convention [60], according to which anodic currents are regarded positive, and cathodic negative.

The double layer charging is caused by the presence of the electric double layer at all kinds of electrochemical interfaces. The double layer can be thought of as a capacitor, having differential capacitance C_{DL}, and charged with charge $Q_{\mathrm{DL}}(t)$. The non-direct current flowing through the double layer is given by the formula

$$I_{\mathrm{C}}(t) = \frac{\mathrm{d}\,Q_{\mathrm{DL}}(t)}{\mathrm{d}\,t} = C_{\mathrm{DL}} \frac{\mathrm{d}\,E'(t)}{\mathrm{d}\,t} \tag{2.70}$$

in the case of Fig. 2.10a, or

$$I_{\mathrm{C}}(t) = \frac{\mathrm{d}\,Q_{\mathrm{DL}}(t)}{\mathrm{d}\,t} = C_{\mathrm{DL}} \frac{\mathrm{d}\,\Delta\Phi'(t)}{\mathrm{d}\,t} \tag{2.71}$$

in the case of Fig. 2.10b. The differential capacitance C_{DL} is generally not constant, although in the modelling of electroanalytical experiments it is often assumed constant. It depends among other things on $E'(t)$ or $\Delta\Phi'(t)$, and on the history of the processes at the interface (for example on the coverages by localised species).

It should be stressed that only $E(t)$ or $\Delta\Phi(t)$, and $I(t)$ can be controlled or measured. There is no way to access directly $E'(t)$ or $\Delta\Phi'(t)$, or control or measure separately the components $I_{\mathrm{F}}(t)$ or $I_{\mathrm{C}}(t)$ of $I(t)$. If the uncompensated Ohmic drop and double layer charging are taken into account, all occurrences of $I(t)$, $E(t)$ and $\Delta\Phi(t)$ in Sects. 2.7, 2.9, and 2.10 must be replaced by $I_{\mathrm{F}}(t)$, $E'(t)$, and $\Delta\Phi'(t)$, respectively.

References

1. Albery J (1975) Electrode kinetics. Clarendon Press, Oxford
2. Alexiadis A, Cornell A, Dudukovic MP (2012) Comparison between CFD calculations of the flow in a rotating disk cell and the Cochran/Levich equations. J Electroanal Chem 669:55–66
3. Alkire RC (1985) Transport processes in electrochemical systems. Chem Eng Commun 38:401–413
4. Alpatova NM, Krishtalik LI, Pleskov YuV (1987) Electrochemistry of solvated electrons. Top Curr Chem 138:149–219
5. Amatore C (1995) Electrochemistry at ultramicroelectrodes. In: Rubinstein I (ed) Physical electrochemistry, principles, methods, and applications. Dekker, New York, pp 131–208
6. Andrieux CP (1994) Terminology and notations for multistep electrochemical reaction mechanisms (IUPAC Recommendations 1994). Pure Appl Chem 66:2445–2450
7. Aoki K (1993) Theory of ultramicroelectrodes. Electroanalysis 5:627–639
8. Aris R (1965) Prolegomena to the rational analysis of systems of chemical reactions. Arch Ration Mech Anal 19:81–99
9. Bagotsky VS (2006) Fundamentals of electrochemistry. Wiley-Interscience, Hoboken
10. Bard AJ, Faulkner LR (2001) Electrochemical methods, fundamentals and applications. Wiley, New York
11. Baronas R, Ivanauskas F, Kulys J (2010) Mathematical modeling of biosensors. An introduction for chemists and mathematicians. Springer, Dordrecht

12. Bieniasz LK (1996) Automatic derivation of the governing equations that describe a transient electrochemical experiment, given a reaction mechanism of arbitrary complexity. Part 1. Problem parameters and initial conditions. J Electroanal Chem 406:33–43
13. Bieniasz LK (1996) Automatic derivation of the governing equations that describe a transient electrochemical experiment, given a reaction mechanism of arbitrary complexity. Part 2. Governing equations in one-dimensional geometry. J Electroanal Chem 406:45–52
14. Bieniasz LK (1996) A reaction compiler for electrochemical kinetics. Comput Chem 20:403–418
15. Bieniasz LK (2012) A procedure for rapid and highly accurate computation of Marcus–Hush–Chidsey rate constants. J Electroanal Chem 683:112–118
16. Blum EH, Luus R (1964) Thermodynamic consistency of reaction rate expressions. Chem Eng Sci 19:322–323
17. Bond AM (1994) Past, present and future contributions of microelectrodes to analytical studies employing voltammetric detection. A review. Analyst 119:R1–R21
18. Boudart M (1968) Kinetics of chemical processes. Prentice–Hall, Englewood Cliffs
19. Britz D (1978) iR elimination in electrochemical cells. J Electroanal Chem 88:309–352
20. Britz D, Kastening B (1974) On the electrochemical observation of a second-order decay of radicals generated by flash photolysis or pulse radiolysis. J Electroanal Chem 56:73–90
21. Britz D, Strutwolf J (2014) Digital simulation of electrochemistry at microelectrodes. In: Lei KF (ed) Microelectrodes: techniques, structures for biosensing and potential applications. Nova Science Publishers, Hauppauge, pp 1–85
22. Cattey H, Audebert P, Sanchez C, Hapiot P (1997) Electrochemical investigations on the sol–gel polymerization of transition-metal alkoxides. J Mater Chem 7:1461–1466
23. Cattey H, Audebert P, Sanchez C, Hapiot P (1998) Electrochemical investigations on liquid-state polymerizing systems: case of sol–gel polymerization of transition-metal alkoxides. J Phys Chem B 102:1193–1202
24. Chidsey CED (1991) Free energy and temperature dependence of electron transfer at the metal–electrolyte interface. Science 251:919–922
25. Cochran WG (1934) The flow due to a rotating disc. Proc Camb Philos Soc 30:365–375
26. Compton RG, Unwin PR (1986) Channel and tubular electrodes. J Electroanal Chem 205:1–20
27. Cooper JA, Compton RG (1998) Channel electrodes—A review. Electroanalysis 10:141–155
28. Cooper JA, Alden JA, Oyama M, Compton RG, Okazaki S (1998) Channel electrode voltammetry: the kinetics of the complexation of the chloranil radical anion with M^{2+} ions by waveshape analysis. J Electroanal Chem 442:201–206
29. Corio PL (1989) Theory of reaction mechanisms. Top Curr Chem 150:249–283
30. Cottrell FG (1903) Der Reststrom bei galvanischer Polarisation, betrachtet als ein Diffusionsproblem. Z Phys Chem 42:385–431
31. Damaskin BB, Petrii OA, Batrakov W (1975) Adsorption organischer Verbindungen an Elektroden. Akademie, Berlin
32. Delahay P (1954) New instrumental methods in electrochemistry. Theory, instrumentation, and applications to analytical and physical chemistry. Interscience, New York
33. Delahay P (1965) Double layer and electrode kinetics. Interscience, New York
34. Eklund JC, Bond AM, Alden JA, Compton RG (1999) Perspectives in modern voltammetry: basic concepts and mechanistic analysis. Adv Phys Org Chem 32:1–120
35. Everett DH (1959) An introduction to the study of chemical thermodynamics. Longmans, London
36. Fleischmann M, Pons S, Rolison DR, Schmidt PP (1987) Ultramicroelectrodes. Datatech Systems, Morganton
37. Forster RJ (1994) Microelectrodes: new dimensions in electrochemistry. Chem Soc Rev 23:289–297
38. Fujihira M, Rubinstein I, Rusling JF (eds) (2007) Modified electrodes. In: Bard AJ, Stratmann M (eds) Encyclopedia of electrochemistry, vol 10. Wiley-VCH, Weinheim
39. Galus Z (1994) Fundamentals of electrochemical analysis. Ellis Horwood, Chichester

40. Gileadi E (1993) Electrode kinetics for chemists, chemical engineers and materials' scientists. Wiley-VCH, New York
41. Gileadi E, Conway BAE (1964) The behaviour of intermediates in electrochemical catalysis. Mod Asp Electrochem 3:347–443
42. Girault HH (1993) Charge transfer across liquid–liquid interfaces. Mod Asp Electrochem 25:1–62
43. Girault HH (2010) Electrochemistry at liquid–liquid interfaces. Electroanal Chem 23:1–104
44. Gmucová K (2012) A review of non-Cottrellian diffusion towards micro- and nano-structured electrodes. In: Shao Y (ed) Electrochemical cells—new advances in fundamental researches and applications. http://www.intechopen.com. Accessed 10 June 2014
45. Go JY, Pyun SI (2007) A review of anomalous diffusion phenomena at fractal interface for diffusion-controlled and non-diffusion-controlled transfer processes. J Solid State Electrochem 11:323–334
46. Gosser DK Jr (1993) Cyclic voltammetry. Simulation and analysis of reaction mechanisms. Wiley-VCH, New York
47. Gregory DP, Riddiford AC (1956) Transport to the surface of a rotating disc. J Chem Soc Lond 3:3756–3764
48. Gründler P, Kirbs A, Dunsch L (2009) Modern thermoelectrochemistry. Chem Phys Chem 10:1722–1746
49. Hammerich O (2000) Methods for studies of electrochemical reactions. In: Lund H, Hammerich O (eds) Organic electrochemistry. Marcel Dekker, New York, pp 95–182
50. Hampson NA, McNeil AJS (1984) The electrochemistry of porous electrodes: flow-through and three-phase electrodes. In: Pletcher D (ed) Specialist periodical reports. Electrochemistry, vol. 9. RSC, London, pp 1–65
51. Heinze J (1993) Ultramicroelectrodes in electrochemistry. Angew Chem Int Ed Engl 22:1268–1288
52. Henstridge MC, Laborda E, Rees NV, Compton RG (2012) Marcus–Hush–Chidsey theory of electron transfer applied to voltammetry: A review. Electrochim Acta 84:12–20
53. Hernández-Labrado GR, Collazos-Castro JE, Polo JL (2008) Digital simulations to solve electrochemical processes involving a diffusion coefficient varying linearly with the concentration. J Electroanal Chem 615:62–68
54. Hubbard AT (1969) Study of the kinetics of electrochemical reactions by thin-layer voltammetry. J Electroanal Chem 22:165–174
55. Hubbard AT, Anson FC (1970) The theory and practice of electrochemistry with thin layer cells. Electroanal Chem 4:129–214
56. Hush NS (1968) Homogeneous and heterogeneous optical and thermal electron transfer. Electrochim Acta 13:1005–1023
57. Ibl N (1983) Fundamentals of transport phenomena in electrolytic systems. In: Yeager E, Bockris J O'M, Conway BAE, Sarangapani S (eds) Comprehensive Treatise of Electrochemistry, vol. 6, Electrodics: Transport. Plenum Press, New York, pp 1–63
58. Ilkovič D (1938) Sur la valeur des courants de diffusion observés dans l'électrolyse a l'aide de l'électrode a gouttes de mercure. Étude polarographique. J Chim Phys 35:129–135
59. International Union of Pure and Applied Chemistry, Analytical Chemistry Division, Commission on Electroanalytical Chemistry (1976) Classification and nomenclature of electroanalytical techniques. Pure Appl Chem 45:81–97
60. International Union of Pure and Applied Chemistry, Analytical Chemistry Division, Commission on Electroanalytical Chemistry (1976) Recommendations for sign conventions and plotting of electrochemical data. Pure Appl Chem 45:131–134
61. Jain RK, Gaur HC, Welch BJ (1977) Chronopotentiometry: a review of theoretical principles. J Electroanal Chem 79:211–236
62. Johnson KA, Goody RS (2011) The original Michaelis constant: translation of the 1913 Michaelis–Menten paper. Biochem 50:8264–8269

63. Koper MTM, Lebedeva NP, Hermse CGM (2002) Dynamics of CO at the solid/liquid interface studied by modeling and simulation of CO electro-oxidation on Pt and PtRu electrodes. Faraday Discuss 121:301–311
64. Kortüm G (1965) Treatise on electrochemistry. Elsevier, Amsterdam
65. Laviron E (1978) Theory of regeneration mechanims in thin layer potential sweep voltammetry. J Organomet Chem 87:31–37
66. Leskovac V (2003) Comprehensive enzyme kinetics. Kluwer/Plenum, New York
67. Lévêque MA (1928) Les lois de la transmission de chaleur par convection, Chapitre V. Résolution théorique de quelques problèmes simples. Ann Mines Mem Ser 12, 13: 237–304
68. Levich VG (1962) Physicochemical Hydrodynamics. Prentice–Hall, Englewood Cliffs
69. Ludwig K, Speiser B (2004) EChem++—An object-oriented problem solving environment for electrochemistry. 2. The kinetic facilities of Ecco—a compiler for (electro-)chemistry. J Chem Inf Comput Sci 44:2051–2060
70. Luo W, Feldberg SW, Rudolph M (1994) Ensuring self-consistent assignment of thermodynamic parameters in simulations of electrochemical-chemical systems. J Electroanal Chem 368:109–113
71. Macdonald DD (1977) Transient techniques in electrochemistry. Plenum, New York
72. Macpherson JV, Simjee N, Unwin PR (2001) Hydrodynamic ultramicroelectrodes: kinetic and analytical applications. Electrochim Acta 47:29–45
73. Marangoni AG (2003) Enzyme kinetics. A modern approach. Wiley, Hoboken
74. Marcus RA (1964) Chemical and electrochemical electron-transfer theory. Ann Rev Phys Chem 15:155–196
75. Marcus RA (1993) Electron transfer reactions in chemistry: theory and experiment (Nobel lecture). Angew Chem Int Ed Engl 32:1111–1121
76. Migliore A, Nitzan A (2012) On the evaluation of the Marcus–Hush–Chidsey integral. J Electroanal Chem 671:99–101
77. Mikkelsen Ø, Schrøder KH (2003) Amalgam electrodes for electroanalysis. Electroanalysis 15:679–687
78. Mirčeski V, Komorsky-Lovrić Š, Lovrić M (2007) Square-wave voltammetry, theory and application. Springer, Berlin
79. Mount AR (2003) Hydrodynamic electrodes. In: Bard AJ, Stratmann M (eds) Encyclopedia of electrochemistry, vol. 3, Instrumentation and electroanalytical chemistry. Wiley-VCH, Weinheim, pp 134–159
80. Nernst W (1888) Zur Kinetik der in Lösung befindlichen Körper. I. Theorie der Diffusion. Z Phys Chem 2:613–637
81. Nernst W (1889) Die elektromotorische Wirksamkeit der Jonen. Z Phys Chem 4:129–181
82. Newman JS (1973) Electrochemical systems. Prentice–Hall, Englewood Cliffs
83. Nyikos L, Pajkossy T (1986) Diffusion to fractal surfaces. Electrochim Acta 31:1347–1350
84. Nyikos L, Pajkossy T, Borosy AP, Martemyanov SA (1990) Diffusion to fractal surfaces—IV. The case of the rotating disc electrode of fractal surface. Electrochim Acta 35:1423–1424
85. Oldham KB, Myland JC (2011) On the evaluation and analysis of the Marcus–Hush–Chidsey integral. J Electroanal Chem 655:65–72
86. Opekar F, Beran P (1976) Rotating disk electrodes. J Electroanal Chem 69:1–105
87. Pajkossy T (1991) Electrochemistry at fractal surfaces. J Electroanal Chem 300:1–11
88. Pajkossy T, Nyikos L (1989) Diffusion to fractal surfaces—II. Verification of theory. Electrochim Acta 34:171–179
89. Pajkossy T, Nyikos L (1989) Diffusion to fractal surfaces—III. Linear sweep and cyclic voltammograms. Electrochim Acta 34:181–186
90. Pajkossy T, Borosy AP, Imre A, Martemyanov SA, Nagy G, Schiller R, Nyikos L (1994) Diffusion kinetics at fractal electrodes. J Electroanal Chem 366:69–73
91. Parsons R (1964) The description of adsorption at electrodes. J Electroanal Chem 7:136–152
92. Parsons R (1974) Manual of symbols and terminology for physicochemical quantities and units. Appendix III Electrochemical nomenclature. Pure Appl Chem 37:499–516

93. Parsons R (1979) Electrode reaction orders, transfer coefficients and rate constants. Amplification of definitions and recommendations for publication of parameters. Pure Appl Chem 52:233–240
94. Parsons R (1990) Electrical double layer: recent experimental and theoretical developments. Chem Rev 90:813–826
95. Planck M (1890) Ueber die Erregung von Electricität und Wärme in Electrolyten. Ann Phys Chem NF 39:161–186
96. Planck M (1890) Ueber die Potentialdifferenz zwischen zwei verdünnten Lösungen binärer Electrolyte. Ann Phys Chem NF 40:561–576
97. Pound BC (1993) Electrochemical techniques to study hydrogen ingress in metals. Mod Asp Electrochem, 25:63–133
98. Puy J, Mas F, Díaz-Cruz JM, Esteban M, Casassas E (1992) Induced reactant adsorption in normal pulse polarography of labile metal+polyelectrolyte systems. Part 2. Study of the current–potential relationship assuming potential-dependent adsorption parameters. J Electroanal Chem 328:271–285
99. Rajendran L (2013) Mathematical modeling in microelectrodes. In: Brennan CR (ed) Mathematical modelling. Nova Science Publishers, Hauppauge, pp 1–110
100. Reder C (1988) Metabolic control theory: a structural approach. J Theor Biol 135:175–201
101. Rees NV, Compton RG (2008) Hydrodynamic microelectrode voltammetry. Russ J Electrochem 44:368–389
102. Reinert KE, Berg H (1962) Theorie der polarographischen Verfolgung schneller chemischer Reaktionen in Lösung mittels reaktionsbedingter Diffusionsstrom–Zeit–Kurven
103. Reymond F, Fermin D, Lee HJ, Girault HH (2000) Electrochemistry at liquid | liquid interfaces: methodology and potential applications. Electrochim Acta 45:2647–2662
104. Ritchie M (1966) Chemical kinetics in homogeneous systems. Oliver & Boyd, Edinburgh
105. Samec Z (2004) Electrochemistry at the interface between two immiscible electrolyte solutions (IUPAC technical report). Pure Appl Chem 76:2147–2180
106. Samec Z, Langmaier J, Trojánek A (1996) Evaluation of parasitic elements contributing to experimental cell impedance: Impedance measurements at interfaces between two immiscible electrolyte solutions. J Chem Soc Faraday Trans 92:3843–3849
107. Sanecki PT, Skitał PM (2012) Mathematical modeling of electrode processes—potential dependent transfer coefficient in electrochemical kinetics. http://www.intechopen.com. Accessed 10 June 2014
108. Sauro HM, Ingalls B (2004) Conservation analysis in biochemical networks: computational issues for software writers. Biophys Chem 109:1–15
109. Schmickler W (1996) Electronic effects in the electric double layer. Chem Rev 96:3177–3200
110. Scott K, Sun YP (1995) Simulation and reaction engineering of electrochemical gas/liquid reactions. Electrochim Acta 40:423–431
111. Sequeira CAC, Santos DMF (2010) Mass transfer to microelectrodes and arrays. Z Phys Chem 224:1297–1336
112. Shi Y, Cai N, Li C (2007) Numerical modeling of an anode-supported SOFC button cell considering anodic surface diffusion. J Power Sources 164:639–648
113. Speiser B (1999) From cyclic voltammetry to scanning electrochemical microscopy: modern electroanalytical methods to study organic compounds, materials, and reactions. Curr Org Chem 3:171–191
114. Strathmann H (2011) Introduction to membrane science and technology. Wiley-VCH, Weinheim
115. Strathmann H, Giorno L, Drioli E (2006) An introduction to membrane science and technology. Institute on Membrane Technology, Rende
116. Stutter E (1974) Zur photopolarographischen Untersuchung schneller, der Durchtrittsreaktion parallelgelagerter Reaktionen zweiter Ordnung. J Electroanal Chem 50:315–321
117. Su L, Wu BL (2004) Investigation of surface diffusion and recombination reaction kinetics of H-adatoms in the process of the hydrogen evolution reaction (her) at Au electrodes. J Electroanal Chem 565:1–6

118. Tansel T, Magnussen OM (2006) Video STM studies of adsorbate diffusion at electrochemical interfaces. Phys Rev Lett 96:026101:1–4
119. Tokuda K, Matsuda H (1977) Theory of A.C. voltammetry at a rotating disk electrode. Part I. A Reversible electrode process. J Electroanal Chem 82:157–171
120. Volgin VM, Davydov AD (2012) Mass-transfer problems in the electrochemical systems. Russ J Electrochem 48:565–569
121. Vydra F, Štulik K, Juláková (1976) Electrochemical stripping analysis. Ellis Horwood, Coll House
122. Walcarius A (2010) Template-directed porous electrodes in electroanalysis. Anal Bioanal Chem 396:261–272
123. Wang J (1985) Stripping analysis, principles, instrumentation, and applications. Wiley-VCH, Deerfield Beach
124. Wildgoose GG, Giovanelli D, Lawrence NS, Compton RG (2004) High-temperature electrochemistry: A review. Electroanalysis 16:421–433
125. Wopschall RH, Shain I (1967) Effects of adsorption of electroactive species in stationary electrode polarography. Anal Chem 39:1514–1527
126. Zutshi K (2006) Introduction to polarography and allied techniques. New Age International Publishers, New Delhi

Chapter 3
Mathematical Preliminaries

In this chapter we briefly summarise mathematical definitions and theorems necessary for the understanding of the IE method. The summary is not intended to be a comprehensive outline of the mathematics of the IEs and related subjects. Similarly to Chap. 2, it is limited to the issues most important for the practice of the IE method in electroanalytical chemistry.

3.1 Integral Equations: Basic Concepts and Definitions

There is an extensive mathematical literature devoted to the subject of the IEs. For the practitioners of the IE method in electrochemistry, basic textbooks, and textbooks combining the elementary mathematics of the IEs, with the discussion of analytical and numerical solution techniques, are probably the most useful. References [2–5, 10, 13, 15, 17, 19, 20, 33, 34, 38, 39, 41] represent a selection of such books.

Even a superficial look at the mathematical literature reveals that the terminology used by various authors is not always consistent and universally accepted. It is therefore important to define the terms to be used in the present book, so that possible misinterpretations are avoided. The definitions listed below have been compiled with the help of the book by Hackbusch [15], the Springer on-line Encyclopedia of Mathematics [12], and a few other references.

There seems to be a fairly general agreement that by an *integral equation* (IE) one should generally understand a *functional equation*, in which the unknown function occurs under the sign of some integral. By functional equations we understand here equations in which the unknowns are elements of some functional space. Hence, the solution of a functional equation (and hence of an IE as well) is a certain function Ψ (but see the critique of such a definition of the functional equations in the Encyclopedia of Mathematics [12]). The above definition of the IE clearly

L.K. Bieniasz, *Modelling Electroanalytical Experiments by the Integral Equation Method*, Monographs in Electrochemistry, DOI 10.1007/978-3-662-44882-3_3

implies a diversity of the IEs, since various kinds of integrals are conceivable (e.g. "ordinary" integrals over intervals in $\mathbb{R}^1$, n-dimensional integrals in $\mathbb{R}^n$, contour and surface integrals in $\mathbb{R}^n$, etc.). The authors of many textbooks often narrow down the definition of the IE to a particular type of the integrals, according to their area of interests. Such narrow definitions are not sufficient for the purpose of the present book, since in electrochemistry various types of integrals may be encountered in the IEs.

The integrals occurring in the IEs can be viewed as results of applying certain *integral operators* to the (unknown) function Ψ. By an *integral operator* (or *integral transformation*) we understand a mapping $\Psi \to \mathfrak{A}\Psi$, in which the law of the correspondence $\mathfrak{A}$ is given by an integral. If Ψ is a function of a scalar or vector variable t, the result of applying $\mathfrak{A}$ to Ψ can be quite generally expressed in the form

$$\mathfrak{A}\Psi(t) = \int_{\mathbb{D}} \mathscr{P}(t, \tau, \Psi(\tau))\, \mathrm{d}\tau \,, \tag{3.1}$$

where $\mathbb{D}$ is a given measurable set of finite Lebesgue measure in a finite-dimensional space, and $\mathscr{P}(t, \tau, \Psi(\tau))$ (for $t,\ \tau \in \mathbb{D}$ and $+\infty < \Psi < \infty$) is a given measurable function. Symbol t does not have to denote time, although in electrochemistry it is often so. It is assumed that Ψ and $\mathscr{P}$ satisfy conditions that ensure the existence of the integral (3.1) in the sense of Lebesgue. If the function $\mathscr{P}$ is linear with respect to Ψ, so that

$$\mathscr{P}(t, \tau, \Psi(\tau)) = \mathscr{K}(t, \tau)\, \Psi(\tau) \tag{3.2}$$

and

$$\mathfrak{A}\Psi(t) = \int_{\mathbb{D}} \mathscr{K}(t, \tau)\, \Psi(\tau)\, \mathrm{d}\tau \,, \tag{3.3}$$

then the integral operator $\mathfrak{A}$ is called *linear*, and the function $\mathscr{K}(t, \tau)$ is called a *kernel* of the operator $\mathfrak{A}$. Otherwise, the operator is called *nonlinear*. Some authors (see, for example, [5, 19, 33]) use the term *kernel* also to denote the function $\mathscr{P}$ of the nonlinear integral operator (3.1). Operators with function $\mathscr{P}$ given in the form

$$\mathscr{P}(t, \tau, \Psi(\tau)) = \mathscr{K}(t, \tau)\, \mathscr{R}(\Psi(\tau)) \,, \tag{3.4}$$

where $\mathscr{R}$ is a nonlinear function of Ψ, are special cases of the nonlinear operator (3.1), and are also termed nonlinear integral operators in the mathematical literature. In the present book we deal almost exclusively with the IEs involving integrals corresponding to linear integral operators (3.3) with kernels $\mathscr{K}(t, \tau)$ (but see exclusions in Sect. 9.1).

Some of the integral operators may have associated inverse integral operators, allowing one to determine a function $\Psi(t)$ from its image $\mathfrak{A}\Psi(t)$. Kernels of such inverse operators are called *inverse kernels*.

In the electrochemical context the kernels may depend on certain additional variables or parameters (apart from variables t and τ). In such cases we shall write the additional variables or parameters in the formulae for the kernels, but for simplicity we shall not include them into the list of the variables of the kernel functions.

When an integral such as (3.3) occurs in an IE, function $\mathscr{K}(t,\tau)$ is often called a *kernel* of the IE. In the opinion of the present author, such a terminology is not very satisfactory, because it makes sense only if there is exclusively one integral in the given IE. Although many of the mathematical textbooks indeed discuss only such IEs, there is a priori no reason why an IE should not contain several integrals with different kernels. In fact, we shall see that IEs with several different integrals and kernels are typical in the electroanalytical modelling. The actual number of integrals in an IE may sometimes depend on how we write the IE. For example, if a complicated kernel function $\mathscr{K}(t,\tau)$ can be expressed as a combination of simpler kernel functions $\mathscr{K}_1(t,\tau)$ and $\mathscr{K}_2(t,\tau)$:

$$\mathscr{K}(t,\tau) = f_1(t)\,\mathscr{K}_1(t,\tau) + f_2(t)\,\mathscr{K}_2(t,\tau)\,, \tag{3.5}$$

then the (one) integral involving $\mathscr{K}(t,\tau)$ is equivalent to an expression involving two integrals with kernels $\mathscr{K}_1(t,\tau)$ and $\mathscr{K}_2(t,\tau)$:

$$\int\limits_{\mathbb{D}} \mathscr{K}(t,\tau)\,\Psi(\tau)\,\mathrm{d}\tau = f_1(t)\int\limits_{\mathbb{D}} \mathscr{K}_1(t,\tau)\,\Psi(\tau)\,\mathrm{d}\tau + f_2(t)\int\limits_{\mathbb{D}} \mathscr{K}_2(t,\tau)\,\Psi(\tau)\,\mathrm{d}\tau\,. \tag{3.6}$$

However, there are also cases when two different integrals in an IE cannot be easily replaced by a single integral.

Among various types of integral transforms, the so-called *convolution transform* represents an important category. According to the Encyclopedia of Mathematics [12], the convolution transform is defined by the following integral taken over $\mathbb{R}^1$:

$$\mathfrak{A}\Psi(t) = \int\limits_{-\infty}^{\infty} \mathscr{K}(t,\tau)\,\Psi(\tau)\,\mathrm{d}\tau\,, \tag{3.7}$$

where the kernel takes the form

$$\mathscr{K}(t,\tau) = \varphi(t-\tau)\,, \tag{3.8}$$

with φ denoting a function of one variable. This means that the kernel is a function of the difference $t-\tau$ only. Such a kernel is called a *convolution kernel* or a *difference kernel* [33]. In the present book this definition of the convolution kernel is adopted, but the above definition of the convolution transform needs to be modified, because in electrochemistry somewhat different convolution integrals are in use:

$$\mathfrak{A}\Psi(t) = \int_0^t \mathcal{K}(t,\tau)\,\Psi(\tau)\,\mathrm{d}\tau\ , \tag{3.9}$$

where $\mathcal{K}(t,\tau)$ is a convolution kernel.

In a particular IE, the unknown function Ψ may appear exclusively under the integral sign (of the integrals involved in the IE). In such a case, following Hackbusch [15] we can say that the IE is of the *first kind*. If the unknown function appears also outside the integral(s), we can say the IE is of the *second kind*. It should be noted that these definitions are usually expressed exclusively for IEs with the unknown function dependent on a single independent variable $t \in \mathbb{R}^1$, which opens a way to distinguishing also IEs of the third kind [17], that are some mutations of the second kind IEs according to the present definition. However, the more general definition given by Hackbusch [15] seems logical and most useful.

Of crucial importance is the concept of the linearity/nonlinearity of the IEs. We adopt here the general definition from Hackbusch [15], also seemingly shared by Baker [3] and a few other authors, but apparently not shared by the Encyclopedia of Mathematics [12] and quite many mathematicians. By a *linear/nonlinear* IE we shall understand an IE that is linear/nonlinear with respect to the unknown function Ψ. This means that a linear IE must be linear with respect to all occurrences of Ψ outside the integral(s), it must be linear with respect to all the integral(s), and that all integral operators involved must be linear. Such a definition may seem natural and obvious, but it should be stressed that the mathematical literature regarding the nonlinear IEs is almost exclusively restricted to the IEs which are linear with respect to the occurrences of Ψ outside the integral(s), and linear with respect to the integral(s); the only presence of the nonlinearities is through nonlinear integral operators. In other words, nonlinear expressions involving Ψ occur only in the integrands. Such a narrow view of the nonlinear IEs is represented both in the classical old papers [16, 36] and in the most recent textbooks [41]. As we shall see, that narrow concept of the nonlinear IEs is inadequate for electrochemical IEs. Most of the nonlinear IEs in electrochemistry exhibit nonlinearities with respect to the occurrences of Ψ outside the integral(s), and with respect to the integral(s), but the integral operators are linear. It may be that, as a consequence of the neglect of this matter, the existing body of the mathematical theory concerning the solvability and various properties of the nonlinear IEs is insufficient for the type of nonlinearities present in the electrochemical IEs. Mathematical studies aimed at the clarification of this issue are desirable, and should be of interest to mathematicians.

If the integral in an IE is taken over a domain $\mathbb{D}$ that is a fixed interval in $\mathbb{R}^1$, or a more general fixed subset of $\mathbb{R}^n$, including curves and surfaces, etc., then the IE is called the *Fredholm* IE. If, in turn, $\mathbb{D}$ is not fixed, but varies with the variable t, then the IE is called the *Volterra* IE. These fairly general definitions are taken from Hackbusch [15], but one should be aware that in many textbooks the notions of the Fredholm and Volterra IEs are limited to $\mathbb{D}$ being an interval in $\mathbb{R}^1$, possibly for historical reasons. But even the present definition is not entirely satisfactory, because it tacitly assumes that the IE involves only one integral, or, if there are more integrals in the IE, all of them share the domain $\mathbb{D}$ of integration. In general, mixed (Fredholm–Volterra) IEs should be conceivable, if the different integrals in the IE are taken over different domains. Luckily, in electrochemical IEs such mixed situations do not occur, and we shall see that most of the IEs relevant to the electrochemical modelling fall into the category of the Volterra IEs.

Depending on the dimension n of the integration domain $\mathbb{D}$, the corresponding IEs are sometimes called *one-dimensional*, *two-dimensional*, *multidimensional*, etc. [33]. In electroanalytical chemistry, the *dimensionality* of the IEs should be clearly distinguished from the dimensionality of the spatial domains over which the differential equations representing electroanalytical models (see Sect. 2.4) are defined.

The IEs can be *regular*, if the integral(s) exist as proper integrals. They can be *weakly singular*, if the integral(s) exist as improper integrals, or *strongly singular*, if the integral(s) must be defined by special regularisation (such as the Cauchy principal value). According to some sources [15, 20, 41], weakly singular IEs together with strongly singular IEs form the class of *singular* IEs. However, some other sources (like the Encyclopedia of Mathematics [12]) apparently restrict the name singular IEs to the strongly singular IEs. These definitions again seem to be limited to IEs involving one integral, because in the case of several integrals, mixed cases (regular–singular, etc.) might occur. It seems more reasonable to talk about *regular*, *weakly singular*, and *strongly singular kernels* of integral operators, and such a terminology is also in use. In the case of integrals taken over $\mathbb{D} \subset \mathbb{R}^n$, a typical example of a weakly singular kernel is [12]:

$$\mathscr{K}(t,\tau) = \frac{\mathscr{M}(t,\tau)}{|t-\tau|^{\alpha}}\,, \tag{3.10}$$

where $\mathscr{M}(t,\tau)$ is bounded for $(t,\tau) \in \mathbb{D} \times \mathbb{D}$, and $0 < \alpha < n$. Kernel (3.10) occurs in the so-called *Abel* IE [13], in which $\mathbb{D} \subset \mathbb{R}^1$, $\alpha = 1/2$ and $f(t)$ is a given function:

$$\int_0^t (t-\tau)^{-1/2}\,\Psi(\tau)\,\mathrm{d}\tau = f(t)\,, \tag{3.11}$$

and in generalisations of this equation onto $\alpha \neq 1/2$. In electroanalytical chemistry the most typical are weakly singular kernels of the type (3.10), although regular ones occur as well. Strongly singular kernels are rare (see Chap. 7 and Appendix D).

If in a functional equation the unknown function occurs not only inside and (possibly) outside some integral(s), but also in a derivative form, then the equation is called *integro-differential equation* (IDE). IEs and differential equations (ODEs or PDEs) are special cases of the IDEs. Examples of IDEs, pertinent to electrochemistry, will be indicated in further Chaps. 8, 9, and 10. Such examples are relatively rare, and many of them can be replaced by equivalent IEs.

In the above list of definitions we had in mind single IEs. However, systems of the IEs, from which several unknown functions are to be determined, are also considered in the mathematical literature, albeit less frequently than single IEs. The above concepts and definitions can be extended to such IE systems, possibly giving rise to an even wider spectrum of mixed cases. Both single IEs and IE systems occur in electrochemistry.

3.2 The Laplace Transformation

The Laplace transformation is one of the basic and most powerful mathematical tools utilised in the derivation and discussion of the IEs of electroanalytical chemistry. There are numerous textbooks and book chapters devoted to the Laplace transformation. References [6,7,11,18,24,32,37] may be particularly lucid or useful from the point of view of an electrochemist, but there are other choices possible. Hence, there is no need to repeat the contents of the books here, and we list only the basic definitions and theorems, to facilitate further discussion.

Consider a real function $f(t)$ of a real variable t. Most often t denotes time or a variable derived from time, but it does not have to be so. The direct Laplace transformation of this function is defined by the integral:

$$\hat{f}(s) = \mathscr{L}\{f(t)\} = \int_0^\infty \exp(-st)\, f(t)\, \mathrm{d}t \;, \tag{3.12}$$

where s is a complex variable called Laplace variable. Symbol $\mathscr{L}\{\cdot\}$ denotes the transformation. The transformed function is often called the image function, or just the transform. In this book we also adopt the usual notation according to which the transformed function is denoted by adding a hat over the function name, and replacing t by s. Hence, $\hat{f}(s)$ is the transformed function $f(t)$. This notation will be used whenever it is necessary to distinguish the original function from its transform. Not every function $f(t)$ possesses a Laplace transform. The function must be such that the integral (3.12) converges. It must be defined for all $t \in [0, \infty)$. It may be singular at $t = 0$, provided that $\lim_{t \to 0+} t\, f(t) = 0$. The Laplace transformation (3.12) can be applied to functions dependent on more than one variable, but the integration is then restricted to one variable only. For example, for a function $f(x,t)$ of two variables x and t we can have

$$\hat{f}(x,s) = \mathscr{L}\{f(x,t)\} = \int_0^\infty \exp(-st)\, f(x,t)\, \mathrm{d}t \;, \tag{3.13}$$

where the t variable is chosen for integration. It is always desirable (if not necessary) to specify which variable is to be used as the integration variable, unless there is no doubt about it.

The inverse Laplace transformation allows one to obtain the original function $f(t)$, given its transform $\hat{f}(s)$. The inverse transformation $\mathscr{L}^{-1}\{\cdot\}$ is defined by the Bromwich integral:

$$f(t) = \mathscr{L}^{-1}\left\{\hat{f}(s)\right\} = \frac{1}{2\pi \mathrm{i}} \int\limits_{a-\mathrm{i}\infty}^{a+\mathrm{i}\infty} \exp(ts)\,\hat{f}(s)\,\mathrm{d}s \tag{3.14}$$

in the complex plane of variable s, where i is the imaginary unit, and a is a real constant such that all possible singularities of $\hat{f}(s)$ are located in the subdomain $\Re\{s\} < a$. It is rarely realised that apart from this requirement a can actually be arbitrary. This property is important for the electrochemical IEs, since there are situations when a right choice of a can facilitate obtaining the inverse transform. Not every function of the complex variable s possesses an inverse Laplace transform. One obvious requirement for a function $\hat{f}(s)$ to be invertible is that $\lim\limits_{s\to\infty} \hat{f}(s) = 0$, otherwise the integral (3.14) would not be convergent.

Using Eq. (3.14) directly is usually a difficult task. Therefore, in practice one often looks for the original functions in tables providing numerous functions and their transforms. Many such tables have been published; tables contained in [1, 7, 11, 35] may be particularly useful for electrochemists. Contemporary programs for symbolic computing, such as MATHEMATICA [22], MAPLE [21] (and other, including the freely available MAXIMA [23]) have capabilities for deriving the formulae for the direct and inverse Laplace transforms of user-defined functions. In particularly complicated cases, when the tables or symbolic software fail, the integral (3.14) has to be calculated analytically by special mathematical techniques. One such technique is the method of residues [24].

If any analytical methods fail, the last resort may be to compute the inverse transform numerically, and if the need arises, to interpolate or approximate the obtained data by some simple formulae (e.g. by polynomials) to have analytically manageable formulae for particular applications. There is a number of methods available for the numerical inversion of the Laplace transforms (see, for example, Davies and Martin [9] or Cohen [8], and the literature therein). Some of the methods have been tested, in the electrochemical context, by Montella and co-workers [25–31]. Out of these techniques, the Gaver–Wynn–rho method of Valkó and Abate [40] has been found by the present author convenient and useful. Their GWR procedure [14] is available freely for MATHEMATICA [22], and it seems to work infallibly.

The Laplace transformation is a linear transformation, implying that the direct/inverse transform of a linear combination of given functions is a linear combination of their transforms. There are also theorems referring to the inverse transforms of functions with linearly transformed variable s:

$$\mathscr{L}^{-1}\left\{\hat{f}(a\,s)\right\} = \frac{1}{a} f\left(\frac{t}{a}\right) , \tag{3.15}$$

where a is any positive real constant, and

$$\mathscr{L}^{-1}\left\{\hat{f}(s-a)\right\} = \exp(at)\, f(t) , \tag{3.16}$$

where a is any real constant. Both theorems are often utilised in the derivations of the electrochemical IEs. But the most important theorem, of fundamental consequences for the electrochemical IEs, is the so-called convolution theorem (also referred to as the Duhamel theorem), which states that the inverse Laplace transform of the product of two Laplace transforms is a convolution integral of the original functions:

$$\mathscr{L}^{-1}\left\{\hat{f}(s)\,\hat{g}(s)\right\} = \int_0^t f(t-\tau)\,g(\tau)\,\mathrm{d}\tau = \int_0^t f(\tau)\,g(t-\tau)\,\mathrm{d}\tau . \tag{3.17}$$

Further useful theorems refer to the transforms of function derivative(s) for $n = 1, 2, \ldots$:

$$\mathscr{L}\left\{\frac{\mathrm{d}^n f(t)}{\mathrm{d}t^n}\right\} = s^n \hat{f}(s) - s^{n-1} f(0^+) - s^{n-2} \left.\frac{\mathrm{d}f(t)}{\mathrm{d}t}\right|_{t=0+} - \ldots - \left.\frac{\mathrm{d}^{n-1} f(t)}{\mathrm{d}t^{n-1}}\right|_{t=0+} \tag{3.18}$$

[where $f(0^+) = \lim_{t\to 0+} f(t), \ldots, \mathrm{d}^{n-1} f(t)/\mathrm{d}t^{n-1}|_{t=0+} = \lim_{t\to 0+} \mathrm{d}^{n-1} f(t)/\mathrm{d}t^{n-1}$], and to the transforms of the integral(s):

$$\mathscr{L}\left\{\int_0^t f(\tau)\,\mathrm{d}\tau\right\} = \frac{\hat{f}(s)}{s} . \tag{3.19}$$

3.3 The Fourier Transformation

Another integral transformation that has found some application in the derivation of electrochemical IEs is the Fourier transformation. According to one definition [24], given a function $f(x)$ of a real variable x defined over the interval $x \in (-\infty, \infty)$, the direct Fourier transform of $f(x)$ is given by the integral:

$$\breve{f}(k) = \mathscr{F}\{f(x)\} = \int_{-\infty}^{\infty} f(x)\exp(-\mathrm{i}kx)\,\mathrm{d}x\,, \tag{3.20}$$

where i is the imaginary unit. Symbol $\mathscr{F}\{\cdot\}$ denotes the transformation. We also adopt a notation according to which the transformed function is denoted by adding the symbol $\smile$ over the function name, and by replacing the variable x by the Fourier variable k. Similarly to the Laplace transformation, the Fourier transformation can be applied to functions of several variables, but the integration is restricted to a single variable. For example,

$$\breve{f}(k,t) = \mathscr{F}\{f(x,t)\} = \int_{-\infty}^{\infty} f(x,t)\exp(-\mathrm{i}kx)\,\mathrm{d}x\,. \tag{3.21}$$

The inverse Fourier transformation $\mathscr{F}^{-1}\{\cdot\}$ is defined by the integral:

$$f(x) = \mathscr{F}^{-1}\left\{\breve{f}(k)\right\} = \frac{1}{2\pi}\int_{-\infty}^{\infty} \breve{f}(k)\exp(\mathrm{i}kx)\,\mathrm{d}k\,. \tag{3.22}$$

The Fourier transformation satisfies the convolution theorem, which for the transform defined by Eq. (3.20) takes the form:

$$\mathscr{F}^{-1}\left\{\breve{f}(k)\,\breve{g}(k)\right\} = \int_{-\infty}^{\infty} f(\xi)\,g(x-\xi)\,\mathrm{d}\xi = \int_{-\infty}^{\infty} f(x-\xi)\,g(\xi)\,\mathrm{d}\xi\,. \tag{3.23}$$

Important for the present book is the formula for the transform of the nth derivative of a function $f(x)$:

$$\mathscr{F}\left\{\frac{\mathrm{d}^n f(x)}{\mathrm{d}x^n}\right\} = (\mathrm{i}k)^n\,\breve{f}(k)\,, \tag{3.24}$$

which holds provided that $f(x)$ and its first $n-1$ derivatives vanish at $x \to \pm\infty$.

3.4 The Hankel Transformation

A yet another integral transformation, employed in the derivations of the IEs of electroanalytical chemistry, is the Hankel transformation. The direct Hankel transform of order ν (such that $\nu \geq -1/2$) of a function $f(r)$ of a real variable

r, defined over the interval $r \in [0, \infty)$, is given by the integral [24]:

$$\check{f}(p) = \mathscr{H}_\nu \{ f(r) \} = \int_0^\infty f(r) \, \mathrm{J}_\nu (pr) \, r \, \mathrm{d}r \, , \tag{3.25}$$

where $\mathrm{J}_\nu (\cdot)$ is the Bessel function of the first kind and order ν. Symbol $\mathscr{H}_\nu \{\cdot\}$ denotes the transformation, and p denotes the Hankel variable. In this book we also denote the transformed function by adding the symbol $\vee$ over the function name. Similarly to the Laplace and Fourier transformations, the effect of applying the Hankel transformation to functions of two variables is:

$$\check{f}(p,t) = \mathscr{H}_\nu \{ f(r,t) \} = \int_0^\infty f(r,t) \, \mathrm{J}_\nu (pr) \, r \, \mathrm{d}r \, . \tag{3.26}$$

The inverse Hankel transformation $\mathscr{H}_\nu^{-1} \{\cdot\}$ is defined by the integral:

$$f(r) = \mathscr{H}_\nu^{-1} \left\{ \check{f}(p) \right\} = \int_0^\infty \check{f}(p) \, \mathrm{J}_\nu (pr) \, p \, \mathrm{d}p \, . \tag{3.27}$$

A property of the Hankel transformation, important for this book, is the formula for the Hankel transform of the Bessel operator applied to a function $f(r)$:

$$\mathscr{H}_\nu \left\{ \frac{\mathrm{d}^2 f(r)}{\mathrm{d}r^2} + \frac{1}{r} \frac{\mathrm{d}f(r)}{\mathrm{d}r} - \left(\frac{\nu}{r} \right)^2 f(r) \right\} = -p^2 \check{f}(p) \, . \tag{3.28}$$

Equation (3.28) holds provided that $\lim_{r \to \infty} f(r) = 0$.

References

1. Abramowitz M, Stegun IA (1972) Handbook of mathematical functions. Dover Publications, New York
2. Atkinson KE (1997) The numerical solution of integral equations of the second kind. Cambridge University Press, Cambridge
3. Baker CTH (1978) The numerical treatment of integral equations. Clarendon Press, Oxford
4. Brunner H (2004) Collocation methods for Volterra integral and related functional differential equations. Cambridge University Press, Cambridge
5. Brunner H, Van der Houwen PJ (1986) The numerical solution of Volterra equations. North-Holland, Amsterdam
6. Carslaw HS, Jaeger JC (1949) Operational methods in applied mathematics. Oxford University Press, London
7. Churchill RV (1972) Operational mathematics. McGraw-Hill, New York
8. Cohen AM (2007) Numerical methods for Laplace transform inversion. Springer, New York

9. Davies B, Martin B (1979) Numerical inversion of the Laplace transform: a survey and comparison of methods. J Comput Phys 33:1–32
10. Delves LM, Mohamed JL (1985) Computational methods for integral equations. Cambridge University Press, Cambridge
11. Doetsch G (1974) Introduction to the theory and application of the Laplace transformation. Springer, Berlin
12. Encyclopedia of Mathematics (2014). http://www.encyclopediaofmath.org. Accessed 10 June 2014
13. Gorenflo R, Vessella S (1991) Abel integral equations, analysis and applications. Springer, Berlin
14. GWR (2014). http://library.wolfram.com/infocenter/MathSource/4738. Accessed 10 June 2014
15. Hackbusch W (1995) Integral equations, theory and numerical treatment. Birkhäuser, Basel
16. Hammerstein A (1930) Nichtlineare Integralgleichungen nebst Anwendungen. Acta Math 54:117–176
17. Kythe PK, Puri P (2002) Computational methods for linear integral equations. Birkhäuser, Boston
18. LePage WR (1980) Complex variables and the Laplace transform for engineers. Dover Publications, New York
19. Linz P (1985) Analytical and numerical methods for Volterra equations. SIAM, Philadelphia
20. Mandal BN, Chakrabarti A (2011) Applied singular integral equations. CRC Press, Boca Raton, and Science Publishers, Enfield
21. MAPLE (2014). http://www.maplesoft.com/products/maple. Accessed 10 June 2014
22. MATHEMATICA (2014) Wolfram Res. Inc., Champaigne, IL. http://www.wolfram.com. Accessed 10 June 2014
23. MAXIMA (2014). http://maxima.sourceforge.net. Accessed 10 June 2014
24. Miles JW (1971) Integral transforms in applied mathematics. Cambridge University Press, Cambridge
25. Montella C (2008) LSV modelling of electrochemical systems through numerical inversion of Laplace transforms. I—The GS–LSV algorithm. J Electroanal Chem 614:121–130
26. Montella C (2009) Re-examination of the potential-step chronoamperometry method through numerical inversion of Laplace transforms. I. General formulation and numerical solution. J Electroanal Chem 633:35–44
27. Montella C (2009) Re-examination of the potential-step chronoamperometry method through numerical inversion of Laplace transforms. II. Application examples. J Electroanal Chem 633:45–56
28. Montella C, Diard JP (2008) New approach of electrochemical systems dynamics in the time-domain under small-signal conditions. I. A family of algorithms based on numerical inversion of Laplace transforms. J Electroanal Chem 623:29–40
29. Montella C, Diard JP (2009) New approach of electrochemical systems dynamics in the time-domain under small-signal conditions. II. Modelling the responses of electrochemical systems by numerical inversion of Laplace transforms. J Electroanal Chem 625:156–164
30. Montella C, Michel R (2009) New approach of electrochemical systems dynamics in the time-domain under small-signal conditions. III. Discrimination between nine candidate models for analysis of PITT experimental data from $Li_xCo\,O_2$ film electrodes. J Electroanal Chem 628:97–112
31. Montella C, Michel R, Diard JP (2007) Numerical inversion of Laplace transforms. A useful tool for evaluation of chemical diffusion coefficients in ion-insertion electrodes investigated by PITT. J Electroanal Chem 608:37–46
32. Oldham K, Myland J, Spanier J (2009) An atlas of functions. Springer, New York
33. Polyanin AD, Manzhirov AV (2008) Handbook of integral equations, 2nd edn. Chapman & Hall, Boca Raton
34. Rahman M (2007) Integral equations and their applications. WIT Press, Southampton
35. Roberts GE, Kaufman H (1966) Table of Laplace transforms. Saunders, Philadelphia

36. Schmidt E (1908) Zur Theorie der Linearen und Nichtlinearen Integralgleichungen. III Teil. Über die Auflösung der nichtlinearen Integralgleichung und die Verzweigung ihrer Lösungen. Math Ann 65:370–399
37. Spiegel MR (1965) Schaum's outline of the theory and problems of Laplace Transforms. McGraw-Hill, New York
38. Tricomi FG (1957) Integral equations. Interscience, New York
39. Vainikko G (1993) Multidimensional weakly singular integral equations. Springer, Berlin
40. Valkó PP, Abate J (2004) Comparison of sequence accelerators for the Gaver method of numerical Laplace transform inversion. Comput Math Appl 48:629–636
41. Wazwaz AM (2011) Linear and nonlinear integral equations. Higher Education Press, Beijing, and Springer, Berlin

Chapter 4
Models Independent of Spatial Coordinates

Electroanalytical models independent of spatial coordinates arise in situations when transport phenomena, and related spatial nonuniformities in an electrochemical system considered, can be neglected or omitted. Most often these are models describing heterogeneous reactions taking place uniformly along the interface studied; homogeneous reaction systems without transport are almost never considered in electroanalytical chemistry. Models independent of spatial coordinates are typically represented by IVPs for systems of ODEs or DAEs, for the concentrations of species localised at the interface studied. The presence of species distributed in adjacent spatially extended phases is not forbidden in such models, but only if they are static species, having practically constant concentrations, or when they are dynamic species confined to a very thin extended phase and can be formally treated as quasi-localised species (see Sect. 2.9).

Although actually every ODE or DAE system can be rewritten in the form of an IE system [21], such representations have been rarely practiced in electrochemistry. Most often the ODEs or DAEs are solved numerically by direct discretisations, such as finite-difference techniques. The conversion to IEs was used only sporadically in the past (see, for example, [11, 22–24]), and this fact was not much emphasised by the authors. Hence, it is not entirely clear whether the IEs seen in [11, 22–24] were used as final model representations, or only as intermediate ones. It is also not clear how they were solved.

However, in recent years there have been several publications [1–10, 12–20, 25] describing in much detail the IE representations and solutions of such models. We therefore provide a relevant presentation for completeness. In Sect. 4.1 we describe systematic procedures of obtaining the IEs representing models independent of spatial coordinates. The procedures are a generalisation of the approaches taken for particular models in [1–10, 12–20]. A brief review of the models considered in those references is given in Sect. 4.2. Finally, in Sect. 4.3 we address the issue of the pros and cons of the IE formulation of models independent of spatial coordinates.

L.K. Bieniasz, *Modelling Electroanalytical Experiments by the Integral Equation Method*, Monographs in Electrochemistry, DOI 10.1007/978-3-662-44882-3_4

4.1 Derivation of the IEs

Consider an electroanalytical experiment, in which a set of $N_\mathrm{r}^\mathrm{het}$ heterogeneous reactions takes place among N_s^dl dynamic localised species $\mathrm{X}_{1,\mathrm{ad}}, \ldots, \mathrm{X}_{N_\mathrm{s}^\mathrm{dl},\mathrm{ad}}$ at an interface studied. We restrict the discussion to the case of the controlled potential experiment, further assuming that all reactions are non-equilibrium reactions, so that the experiment is described by an ODE system rather than by a more general DAE system. We also assume that the interface studied is between an electrolytic phase and an electronic conductor (or semi-conductor), so that electrons exchanged are formally considered as reactants, and taken into account in the stoichiometric matrix $\overline{N}^{\,\mathrm{dyn,het}}$ for the dynamic species at the interface (to simplify the presentation we assume that the last row of matrix $\overline{N}^{\,\mathrm{dyn,het}}$ corresponds to electrons exchanged). Hence, in view of Eq. (2.53) the experiment is completely described by the equation:

$$\vec{p}^{\,\mathrm{het}}(t) = \overline{N}^{\,\mathrm{dyn,het}}\,\vec{r}^{\,\mathrm{het}}(t)\,, \tag{4.1}$$

where $\vec{p}^{\,\mathrm{het}}(t) = \left[\mathrm{d}\Gamma_1(t)/\mathrm{d}t, \ldots, \mathrm{d}\Gamma_{N_\mathrm{s}^\mathrm{dl}}(t)/\mathrm{d}t, I(t)(FA)^{-1}\right]^\mathrm{T}$ is the vector of production rates of the dynamic localised species and electrons exchanged, and vector $\vec{r}^{\,\mathrm{het}}(t) = \left[r_1^\mathrm{het}(t), \ldots, r_{N_\mathrm{r}^\mathrm{het}}^\mathrm{het}(t)\right]^\mathrm{T}$ contains relevant expressions for the rates of heterogeneous reactions. In general, the expressions may be nonlinear with respect to the concentrations $\vec{\Gamma}(t) = \left[\Gamma_1(t), \ldots, \Gamma_{N_\mathrm{s}^\mathrm{dl}}(t)\right]^\mathrm{T}$ of the localised species. In the case of models independent of spatial coordinates, variables $\vec{p}^{\,\mathrm{het}}(t)$, $\vec{r}^{\,\mathrm{het}}(t)$, and $\vec{\Gamma}(t)$ may depend on time t only, but not on any spatial coordinate, which we explicitly indicate in every occurrence of the symbols of these variables. Let $\overline{N}^{\,\mathrm{dl,het}}$ be the submatrix of $\overline{N}^{\,\mathrm{dyn,het}}$, corresponding to the dynamic localised species (that is the matrix resulting from the omission of the last row of $\overline{N}^{\,\mathrm{dyn,het}}$, corresponding to electrons exchanged). By using this submatrix, the ODE system for the surface concentrations can be written as

$$\frac{\mathrm{d}\vec{\Gamma}(t)}{\mathrm{d}t} = \overline{N}^{\,\mathrm{dl,het}}\,\vec{r}^{\,\mathrm{het}}(t)\,. \tag{4.2}$$

For solving Eq. (4.2) an initial condition is required. This condition can be written as

$$\vec{\Gamma}(0) = \vec{\Gamma}^{\,\star}\,, \tag{4.3}$$

where $\vec{\Gamma}^{\,\star} = \left[\Gamma_1^\star, \ldots, \Gamma_{N_\mathrm{s}^\mathrm{dl}}^\star\right]^\mathrm{T}$ is the vector of initial concentrations. We adopt the usual assumption (see Sect. 2.10) that the initial concentrations correspond to the equilibrium state of the reactions considered. Once the IVP (4.2) and (4.3) is solved, the last equation in the system (4.1) allows one to calculate the unknown Faradaic current as an appropriate linear combination of the reaction rates.

The conversion of the IVP (4.2) and (4.3) into an IE system can be performed according to at least three alternative procedures described below.

Procedure 1 This procedure is always applicable. It gives IEs in which the reaction rates $\overrightarrow{r}^{\,\mathrm{het}}(t)$ play the role of the unknowns. By integrating both sides of Eq. (4.2) over the interval $[0, t]$, one obtains:

$$\int_0^t \frac{\mathrm{d}\overrightarrow{\Gamma}(\tau)}{\mathrm{d}\tau}\,\mathrm{d}\tau = \overrightarrow{\Gamma}(t) - \overrightarrow{\Gamma}^{\star} = \overline{N}^{\,\mathrm{dl,het}} \int_0^t \overrightarrow{r}^{\,\mathrm{het}}(\tau)\,\mathrm{d}\tau\,. \tag{4.4}$$

This gives

$$\overrightarrow{\Gamma}(t) = \overrightarrow{\Gamma}^{\star} + \overline{N}^{\,\mathrm{dl,het}}\,\overrightarrow{Y}(t)\,, \tag{4.5}$$

where $\overrightarrow{Y}(t) = \left[Y_1(t), \ldots, Y_{N_\mathrm{r}^\mathrm{het}}(t)\right]^\mathrm{T}$ with

$$Y_i(t) = \int_0^t r_i^{\mathrm{het}}(\tau)\,\mathrm{d}\tau = \int_0^t \mathscr{K}_i(t,\tau)\,r_i^{\mathrm{het}}(\tau)\,\mathrm{d}\tau \tag{4.6}$$

and

$$\mathscr{K}_i(t,\tau) = 1\,, \tag{4.7}$$

for $i = 1, \ldots, N_\mathrm{r}^\mathrm{het}$. Equation (4.5) allows one to express the unknown surface concentrations as linear combinations of integrals of the reaction rates. But the model must also provide expressions for the reaction rates as functions of the concentrations. Therefore, after setting the concentrations (4.5) into these reaction rate expressions, we obtain a set of $N_\mathrm{r}^\mathrm{het}$ second kind Volterra IEs for the unknown reaction rates. From Eq. (4.6) we see that all integrals in the IEs are characterised by simple integral transformation kernels (4.7). Once we solve the IEs for the reaction rates $\overrightarrow{r}^{\,\mathrm{het}}(t)$, the last equation in the system (4.1) yields the Faradaic current.

Procedure 2 The second procedure is less general. Similarly to Procedure 1, it gives IEs in which (some of) the reaction rates $\overrightarrow{r}^{\,\mathrm{het}}(t)$ play the role of the unknowns. It requires that some of the reaction rates be linear expressions of the concentrations, with coefficients (rate constants) independent of time. Assume that $N_\mathrm{r}^\mathrm{het,lin}$ reactions obey this condition, and the remaining $N_\mathrm{r}^\mathrm{het,ess} = N_\mathrm{r}^\mathrm{het} - N_\mathrm{r}^\mathrm{het,lin}$ reactions do not obey it. Let us call the latter reactions *essential reactions*. The essential reactions should not be confused with *base reactions* (see Sect. 2.7). In contrast to the latter reactions, the essential reactions need not be linearly independent. Electrochemical reactions have rate constants dependent on the electrode potential, and the potential often depends on time, but even if this is not the case, it is preferable to always include the electrochemical reactions into the set of essential reactions. One applies the Laplace transformation (3.12) to Eq. (4.2). This gives

$$\mathscr{L}\left\{\frac{\mathrm{d}\vec{\Gamma}(t)}{\mathrm{d}t}\right\} = \mathscr{L}\left\{\overline{N}^{\,\mathrm{dl,het}}\vec{r}^{\,\mathrm{het}}(t)\right\} , \tag{4.8}$$

so that in view of Eqs. (3.18) and (4.3):

$$s\hat{\vec{\Gamma}}(s) - \vec{\Gamma}^{\,\star} = \overline{N}^{\,\mathrm{dl,het}}\hat{\vec{r}}^{\,\mathrm{het}}(s) , \tag{4.9}$$

where $N_{\mathrm{r}}^{\mathrm{het,lin}}$ elements of $\hat{\vec{r}}^{\,\mathrm{het}}(s)$ are now linear combinations of some of the elements of $\hat{\vec{\Gamma}}(s)$. We solve Eq. (4.9) for $\hat{\vec{\Gamma}}(s)$. As this is a linear system of equations, the solutions are linear combinations of the Laplace transforms of the rates of the essential reactions, with linear coefficients that are simple fractions involving the Laplace variable s. The simplest simple fractions are all in the form $(s + k_l)^{-1}$, where k_l are constants dependent on the rate constants of the reactions, the rates of which were assumed linear with respect to the concentrations (that is reactions other than the essential reactions). In view of Eq. (3.16), this implies that after inversion of the transforms by means of the convolution theorem (3.17) we obtain concentrations $\vec{\Gamma}(t)$ as linear combinations of the integrals

$$Y_l(t) = \int_0^t \mathscr{K}_l(t,\tau)\, r_l^{\mathrm{het}}(\tau)\, \mathrm{d}\tau , \tag{4.10}$$

with the convolution-type kernels given by exponential functions:

$$\mathscr{K}_l(t,\tau) = \exp\left[-k_l(t-\tau)\right] . \tag{4.11}$$

If more complicated simple fractions occur, different kernels are expected, but in all models considered thus far [1–6, 8, 9, 12–16, 18, 19] only exponential kernels (4.11) were observed. Substitution of the concentrations $\vec{\Gamma}(t)$ into the expressions for the rates of essential reactions gives a system of $N_{\mathrm{r}}^{\mathrm{het,ess}}$ second kind Volterra IEs for these rates. As the electrochemical reactions were assumed to be among the essential reactions, the solution of the IEs obtained gives immediately the rates of the electrochemical reactions, and permits the calculation of the Faradaic current from the last equation in the system (4.1).

Procedure 3 This procedure is generally applicable. It yields IEs in which the species production rates $\mathrm{d}\vec{\Gamma}(t)/\mathrm{d}t$ play the role of the unknowns. Integration of $\mathrm{d}\vec{\Gamma}(t)/\mathrm{d}t$ gives

$$\vec{\Gamma}(t) = \vec{\Gamma}^{\,\star} + \int_0^t \frac{\mathrm{d}\vec{\Gamma}(\tau)}{\mathrm{d}\tau}\, \mathrm{d}\tau . \tag{4.12}$$

Substitution of the elements of $\vec{\Gamma}(t)$, expressed by Eq. (4.12), into the formulae for the reaction rates in Eq. (4.1), gives a system of N_s^{dl} IEs for the elements of $d\vec{\Gamma}(t)/dt$. All integrals in this system have simple kernels (4.7). The last equation in the system (4.1) then gives the Faradaic current, most generally as an expression involving elements of the integrals of $d\vec{\Gamma}(t)/dt$, rather than the elements of the solution vector $d\vec{\Gamma}(t)/dt$. This might seem to be a complication, but numerical methods for solving IEs usually calculate the integrals together with the solutions (see Chap. 12). It often happens that the number of IEs obtainable by this procedure can be reduced by taking into account structural conservation relationships resulting from the system (4.1), which express connections between the various elements of $d\vec{\Gamma}(t)/dt$. Such relationships may also allow one to express the current directly by using the elements of $d\vec{\Gamma}(t)/dt$.

We illustrate the above procedures by two examples. As a first example, let us consider a controlled potential experiment for the surface EC catalytic reaction scheme:

$$X_{1,\mathrm{ad}} + n_1\, e^- \rightleftarrows X_{2,\mathrm{ad}} \,, \tag{4.13}$$

$$X_{2,\mathrm{ad}} \underset{k_b}{\overset{k_f}{\rightleftarrows}} X_{1,\mathrm{ad}} \,, \tag{4.14}$$

with the electrochemical reaction (4.13) subject to Butler–Volmer kinetics with the rate $r_1^{het}(t)$, and the heterogeneous non-electrochemical reaction (4.14) subject to first-order kinetics with the rate $r_2^{het}(t)$. When $k_f = k_b = 0$ the reaction scheme reduces to the electron transfer reaction (4.13), so that the model equations apply also to this case. For this reaction scheme we have:

$$\overline{N}^{\,\mathrm{dyn,het}} = \begin{bmatrix} -1 & 1 \\ 1 & -1 \\ -n_1 & 0 \end{bmatrix} , \tag{4.15}$$

$$\overline{N}^{\,\mathrm{dl,het}} = \begin{bmatrix} -1 & 1 \\ 1 & -1 \end{bmatrix} , \tag{4.16}$$

$$r_1^{het}(t) = k_1^0 \left[f_{1,f}(t)\Gamma_1(t) - f_{1,b}(t)\Gamma_2(t) \right] \,, \tag{4.17}$$

$$r_2^{het}(t) = k_f \Gamma_2(t) - k_b \Gamma_1(t) \,, \tag{4.18}$$

with

$$f_{1,f}(t) = \exp\left\{ -\alpha_1^f \frac{n_1 F}{RT} \left[E(t) - E_1^0 \right] \right\} \tag{4.19}$$

and

$$f_{1,b}(t) = \exp\left\{ \alpha_1^b \frac{n_1 F}{RT} \left[E(t) - E_1^0 \right] \right\} \,. \tag{4.20}$$

In Eqs. (4.17), (4.19), and (4.20), k_1^0, α_1^{f}, α_1^{b} and E_1^0 denote the conditional rate constant, charge transfer coefficients, and the conditional potential of reaction (4.13), respectively. By applying Procedure 1, from Eq. (4.5) with matrix (4.16) we obtain

$$\Gamma_1(t) = \Gamma_1^\star - \int_0^t r_1^{\mathrm{het}}(\tau)\,\mathrm{d}\tau + \int_0^t r_2^{\mathrm{het}}(\tau)\,\mathrm{d}\tau\,, \tag{4.21}$$

$$\Gamma_2(t) = \Gamma_2^\star + \int_0^t r_1^{\mathrm{het}}(\tau)\,\mathrm{d}\tau - \int_0^t r_2^{\mathrm{het}}(\tau)\,\mathrm{d}\tau\,, \tag{4.22}$$

where $\Gamma_2^\star = (k_{\mathrm{b}}/k_{\mathrm{f}})\,\Gamma_1^\star$. Substitution of Eqs. (4.21) and (4.22) into Eqs. (4.17) and (4.18) gives the IE system:

$$\Psi_1(t) - k_1^0\left\{ f_{1,\mathrm{f}}(t)\left[\Gamma_1^\star - \int_0^t \Psi_1(\tau)\,\mathrm{d}\tau + \int_0^t \Psi_2(\tau)\,\mathrm{d}\tau\right] - f_{1,\mathrm{b}}(t)\left[\Gamma_2^\star + \int_0^t \Psi_1(\tau)\,\mathrm{d}\tau - \int_0^t \Psi_2(\tau)\,\mathrm{d}\tau\right]\right\} = 0\,, \tag{4.23}$$

$$\Psi_2(t) - k_{\mathrm{f}}\left[\Gamma_2^\star + \int_0^t \Psi_1(\tau)\,\mathrm{d}\tau - \int_0^t \Psi_2(\tau)\,\mathrm{d}\tau\right] + k_{\mathrm{b}}\left[\Gamma_1^\star - \int_0^t \Psi_1(\tau)\,\mathrm{d}\tau + \int_0^t \Psi_2(\tau)\,\mathrm{d}\tau\right] = 0\,. \tag{4.24}$$

In Eqs. (4.23) and (4.24) the vector of unknowns is equivalent to the vector of reaction rates, i.e. $\vec{\Psi}(t) = \vec{r}^{\,\mathrm{het}}(t)$. After solving this system for $\Psi_1(t)$ and $\Psi_2(t)$, the last equation in the system (4.1) with matrix (4.15) gives

$$\frac{I(t)}{FA} = -n_1\, r_1^{\mathrm{het}}(t) = -n_1\,\Psi_1(t)\,. \tag{4.25}$$

Alternatively, by applying Procedure 2, Eq. (4.9) gives:

$$s\hat{\Gamma}_1(s) - \Gamma_1^\star = -\hat{r}_1^{\mathrm{het}}(s) + k_{\mathrm{f}}\hat{\Gamma}_2(s) - k_{\mathrm{b}}\hat{\Gamma}_1(s)\,, \tag{4.26}$$

$$s\hat{\Gamma}_2(s) - \Gamma_2^\star = \hat{r}_1^{\mathrm{het}}(s) - k_{\mathrm{f}}\hat{\Gamma}_2(s) + k_{\mathrm{b}}\hat{\Gamma}_1(s)\,, \tag{4.27}$$

with reaction (4.13) taken as an essential reaction. Solving Eqs. (4.26) and (4.27) for $\hat{\Gamma}_1(s)$ and $\hat{\Gamma}_2(s)$ gives

$$\hat{\Gamma}_1(s) = \frac{\Gamma_1^\star}{s} - \frac{\hat{r}_1^{\text{het}}(s)}{s + (k_\text{f} + k_\text{b})} , \tag{4.28}$$

$$\hat{\Gamma}_2(s) = \frac{\Gamma_2^\star}{s} + \frac{\hat{r}_1^{\text{het}}(s)}{s + (k_\text{f} + k_\text{b})} , \tag{4.29}$$

so that the convolution theorem (3.17) applied to Eqs. (4.28) and (4.29) yields

$$\Gamma_1(t) = \Gamma_1^\star - \int_0^t \exp\left[-(k_\text{f} + k_\text{b})(t - \tau)\right] r_1^{\text{het}}(\tau) \, \mathrm{d}\tau , \tag{4.30}$$

$$\Gamma_2(t) = \Gamma_2^\star + \int_0^t \exp\left[-(k_\text{f} + k_\text{b})(t - \tau)\right] r_1^{\text{het}}(\tau) \, \mathrm{d}\tau . \tag{4.31}$$

Substitution of Eqs. (4.30) and (4.31) into Eq. (4.17) gives the single IE:

$$\Psi(t) - k_1^0 \left\langle f_{1,\text{f}}(t) \left\{ \Gamma_1^\star - \int_0^t \exp\left[-(k_\text{f} + k_\text{b})(t - \tau)\right] \Psi(\tau) \, \mathrm{d}\tau \right\} \right.$$
$$\left. - f_{1,\text{b}}(t) \left\{ \Gamma_2^\star + \int_0^t \exp\left[-(k_\text{f} + k_\text{b})(t - \tau)\right] \Psi(\tau) \, \mathrm{d}\tau \right\} \right\rangle = 0 , \tag{4.32}$$

where the unknown function $\Psi(t)$ is the reaction rate $r_1^{\text{het}}(t)$. Solving Eq. (4.32) gives $\Psi(t)$, and subsequently $I(t)$:

$$\frac{I(t)}{FA} = -n_1 r_1^{\text{het}}(t) = -n_1 \Psi(t) . \tag{4.33}$$

Finally, by applying Procedure 3, we write the two ODEs of Eq. (4.2) with matrix (4.16) in the form:

$$\frac{\mathrm{d}\Gamma_1(t)}{\mathrm{d}t} = \left[k_\text{f} - k_1^0 f_{1,\text{f}}(t)\right] \left[\Gamma_1^\star + \int_0^t \frac{\mathrm{d}\Gamma_1(\tau)}{\mathrm{d}\tau} \mathrm{d}\tau\right]$$
$$- \left[k_\text{b} - k_1^0 f_{1,\text{b}}(t)\right] \left[\Gamma_2^\star + \int_0^t \frac{\mathrm{d}\Gamma_2(\tau)}{\mathrm{d}\tau} \mathrm{d}\tau\right] , \tag{4.34}$$

$$\frac{\mathrm{d}\Gamma_2(t)}{\mathrm{d}t} = -\left[k_\mathrm{f} - k_1^0\, f_{1,\mathrm{f}}(t)\right]\left[\Gamma_1^\star + \int_0^t \frac{\mathrm{d}\Gamma_1(\tau)}{\mathrm{d}\tau}\,\mathrm{d}\tau\right]$$

$$+\left[k_\mathrm{b} - k_1^0\, f_{1,\mathrm{b}}(t)\right]\left[\Gamma_2^\star + \int_0^t \frac{\mathrm{d}\Gamma_2(\tau)}{\mathrm{d}\tau}\,\mathrm{d}\tau\right] . \quad (4.35)$$

One of the unknowns can be eliminated, because of the structural conservation relationship $\mathrm{d}\Gamma_1(t)/\mathrm{d}t + \mathrm{d}\Gamma_2(t)/\mathrm{d}t = 0$, resulting from adding Eqs. (4.34) and (4.35). We thus obtain the single IE:

$$\Psi(t) - \left[k_\mathrm{f} - k_1^0\, f_{1,\mathrm{f}}(t)\right]\left[\Gamma_1^\star + \int_0^t \Psi(\tau)\,\mathrm{d}\tau\right]$$

$$+\left[k_\mathrm{b} - k_1^0\, f_{1,\mathrm{b}}(t)\right]\left[\Gamma_2^\star - \int_0^t \Psi(\tau)\,\mathrm{d}\tau\right] = 0 . \quad (4.36)$$

where the unknown function is $\Psi(t) = \mathrm{d}\Gamma_1(t)/\mathrm{d}t$. After determining $\Psi(t)$ and $\int_0^t \Psi(\tau)\,\mathrm{d}\tau$, the Faradaic current is calculated as

$$\frac{I(t)}{FA} = -n_1 r_1^\mathrm{het}(t)$$

$$= -n_1\, k_1^0 \left\{ f_{1,\mathrm{f}}(t)\left[\Gamma_1^\star + \int_0^t \Psi(\tau)\,\mathrm{d}\tau\right] - f_{1,\mathrm{b}}(t)\left[\Gamma_2^\star - \int_0^t \Psi(\tau)\,\mathrm{d}\tau\right]\right\} . \quad (4.37)$$

As a second example, we consider a controlled potential experiment for the surface ECE reaction scheme:

$$\mathrm{X}_{1,\mathrm{ad}} + n_1\,\mathrm{e}^- \rightleftarrows \mathrm{X}_{2,\mathrm{ad}} , \quad (4.38)$$

$$\mathrm{X}_{2,\mathrm{ad}} \underset{k_\mathrm{b}}{\overset{k_\mathrm{f}}{\rightleftarrows}} \mathrm{X}_{3,\mathrm{ad}} , \quad (4.39)$$

$$\mathrm{X}_{3,\mathrm{ad}} + n_3\,\mathrm{e}^- \rightleftarrows \mathrm{X}_{4,\mathrm{ad}} , \quad (4.40)$$

with the electrochemical reactions (4.38) and (4.40) subject to Butler–Volmer kinetics with the rates $r_1^\mathrm{het}(t)$ and $r_3^\mathrm{het}(t)$, and the heterogeneous non-electrochemical reaction (4.39) subject to first-order kinetics with the rate $r_2^{\text{het}}(t)$. Only species $X_{1,\text{ad}}$ is assumed to be initially present. For this reaction scheme we have

$$\overline{N}^{\text{dyn,het}} = \begin{bmatrix} -1 & 0 & 0 \\ 1 & -1 & 0 \\ 0 & 1 & -1 \\ 0 & 0 & 1 \\ -n_1 & 0 & -n_3 \end{bmatrix}, \tag{4.41}$$

$$\overline{N}^{\text{dl,het}} = \begin{bmatrix} -1 & 0 & 0 \\ 1 & -1 & 0 \\ 0 & 1 & -1 \\ 0 & 0 & 1 \end{bmatrix}, \tag{4.42}$$

$$r_1^{\text{het}}(t) = k_1^0 \left[f_{1,\text{f}}(t)\Gamma_1(t) - f_{1,\text{b}}(t)\Gamma_2(t) \right] , \tag{4.43}$$

$$r_2^{\text{het}}(t) = k_\text{f}\Gamma_2(t) - k_\text{b}\Gamma_3(t) , \tag{4.44}$$

$$r_3^{\text{het}}(t) = k_3^0 \left[f_{3,\text{f}}(t)\Gamma_3(t) - f_{3,\text{b}}(t)\Gamma_4(t) \right] , \tag{4.45}$$

with

$$f_{i,\text{f}}(t) = \exp\left\{ -\alpha_i^\text{f} \frac{n_i F}{RT} \left[E(t) - E_i^0 \right] \right\} \tag{4.46}$$

and

$$f_{i,\text{b}}(t) = \exp\left\{ \alpha_i^\text{b} \frac{n_i F}{RT} \left[E(t) - E_i^0 \right] \right\} \tag{4.47}$$

for $i = 1, 3$. By applying Procedure 1, from Eq. (4.5) with matrix (4.42) we obtain

$$\Gamma_1(t) = \Gamma_1^\star - \int_0^t r_1^{\text{het}}(\tau)\, d\tau , \tag{4.48}$$

$$\Gamma_2(t) = \int_0^t r_1^{\text{het}}(\tau)\, d\tau - \int_0^t r_2^{\text{het}}(\tau)\, d\tau , \tag{4.49}$$

$$\Gamma_3(t) = \int_0^t r_2^{\text{het}}(\tau)\, d\tau - \int_0^t r_3^{\text{het}}(\tau)\, d\tau , \tag{4.50}$$

$$\Gamma_4(t) = \int_0^t r_3^{\text{het}}(\tau)\, d\tau . \tag{4.51}$$

Substitution of Eqs. (4.48)–(4.51) into the reaction rates (4.43)–(4.45) gives the system of three IEs:

$$\Psi_1(t) - k_1^0 \left\{ f_{1,\mathrm{f}}(t) \left[\Gamma_1^\star - \int_0^t \Psi_1(\tau)\,\mathrm{d}\tau \right] - f_{1,\mathrm{b}}(t) \left[\int_0^t \Psi_1(\tau)\,\mathrm{d}\tau - \int_0^t \Psi_2(\tau)\,\mathrm{d}\tau \right] \right\} = 0\,, \tag{4.52}$$

$$\Psi_2(t) - k_\mathrm{f} \left[\int_0^t \Psi_1(\tau)\mathrm{d}\tau - \int_0^t \Psi_2(\tau)\mathrm{d}\tau \right] + k_\mathrm{b} \left[\int_0^t \Psi_2(\tau)\mathrm{d}\tau - \int_0^t \Psi_3(\tau)\mathrm{d}\tau \right] = 0\,, \tag{4.53}$$

$$\Psi_3(t) - k_3^0 \left\{ f_{3,\mathrm{f}}(t) \left[\int_0^t \Psi_2(\tau)\,\mathrm{d}\tau - \int_0^t \Psi_3(\tau)\,\mathrm{d}\tau \right] - f_{3,\mathrm{b}}(t) \int_0^t \Psi_3(\tau)\,\mathrm{d}\tau \right\} = 0\,. \tag{4.54}$$

In Eqs. (4.52)–(4.54) the vector of unknowns is equivalent to the vector of reaction rates: $\overrightarrow{\Psi}(t) = \overrightarrow{r}^{\,\mathrm{het}}(t)$. After solving this system, the last equation in the system (4.1) with matrix (4.41) gives the Faradaic current:

$$\frac{I(t)}{FA} = -n_1 r_1^{\mathrm{het}}(t) - n_3 r_3^{\mathrm{het}}(t) = -n_1 \Psi_1(t) - n_3 \Psi_3(t)\,. \tag{4.55}$$

Alternatively, by applying Procedure 2, Eq. (4.9) becomes

$$s\hat{\Gamma}_1(s) - \Gamma_1^\star = -\hat{r}_1^{\mathrm{het}}(s)\,, \tag{4.56}$$

$$s\hat{\Gamma}_2(s) - 0 = \hat{r}_1^{\mathrm{het}}(s) - k_\mathrm{f}\hat{\Gamma}_2(s) + k_\mathrm{b}\hat{\Gamma}_3(s)\,, \tag{4.57}$$

$$s\hat{\Gamma}_3(s) - 0 = -\hat{r}_3^{\mathrm{het}}(s) + k_\mathrm{f}\hat{\Gamma}_2(s) - k_\mathrm{b}\hat{\Gamma}_3(s)\,, \tag{4.58}$$

$$s\hat{\Gamma}_4(s) - 0 = \hat{r}_3^{\mathrm{het}}(s)\,, \tag{4.59}$$

with reactions (4.38) and (4.40) taken as essential reactions. Solving Eqs. (4.56)–(4.59) for $\hat{\Gamma}_1(s)$, $\hat{\Gamma}_2(s)$, $\hat{\Gamma}_3(s)$, and $\hat{\Gamma}_4(s)$ gives

$$\hat{\Gamma}_1(s) = \frac{\Gamma_1^\star}{s} - \frac{\hat{r}_1^{\mathrm{het}}(s)}{s}\,, \tag{4.60}$$

$$\hat{\Gamma}_2(s) = \frac{k_\mathrm{f}}{k_\mathrm{f}+k_\mathrm{b}} \frac{\hat{r}_1^{\mathrm{het}}(s)}{s+(k_\mathrm{f}+k_\mathrm{b})} + \frac{k_\mathrm{b}}{k_\mathrm{f}+k_\mathrm{b}} \left[\frac{\hat{r}_1^{\mathrm{het}}(s)}{s} - \frac{\hat{r}_3^{\mathrm{het}}(s)}{s} + \frac{\hat{r}_3^{\mathrm{het}}(s)}{s+(k_\mathrm{f}+k_\mathrm{b})} \right]\,, \tag{4.61}$$

$$\hat{\Gamma}_3(s) = \frac{k_\mathrm{f}}{k_\mathrm{f}+k_\mathrm{b}}\left[\frac{\hat{r}_1^\mathrm{het}(s)}{s} - \frac{\hat{r}_1^\mathrm{het}(s)}{s+(k_\mathrm{f}+k_\mathrm{b})} - \frac{\hat{r}_3^\mathrm{het}(s)}{s}\right] - \frac{k_\mathrm{b}}{k_\mathrm{f}+k_\mathrm{b}}\frac{\hat{r}_3^\mathrm{het}(s)}{s+(k_\mathrm{f}+k_\mathrm{b})}\,, \tag{4.62}$$

$$\hat{\Gamma}_4(s) = \frac{\hat{r}_3^\mathrm{het}(s)}{s}\,. \tag{4.63}$$

Consequently, the convolution theorem (3.17) applied to Eqs. (4.60)–(4.63) yields:

$$\Gamma_1(t) = \Gamma_1^\star - \int_0^t r_1^\mathrm{het}(\tau)\,\mathrm{d}\tau\,, \tag{4.64}$$

$$\Gamma_2(t) = \frac{k_\mathrm{f}}{k_\mathrm{f}+k_\mathrm{b}}\int_0^t \exp\left[-(k_\mathrm{f}+k_\mathrm{b})(t-\tau)\right]\,r_1^\mathrm{het}(\tau)\,\mathrm{d}\tau + \frac{k_\mathrm{b}}{k_\mathrm{f}+k_\mathrm{b}}\int_0^t r_1^\mathrm{het}(\tau)\,\mathrm{d}\tau$$
$$-\frac{k_\mathrm{b}}{k_\mathrm{f}+k_\mathrm{b}}\left\{\int_0^t r_3^\mathrm{het}(\tau)\,\mathrm{d}\tau - \int_0^t \exp\left[-(k_\mathrm{f}+k_\mathrm{b})(t-\tau)\right]\,r_3^\mathrm{het}(\tau)\,\mathrm{d}\tau\right\}\,, \tag{4.65}$$

$$\Gamma_3(t) = \frac{k_\mathrm{f}}{k_\mathrm{f}+k_\mathrm{b}}\left\{\int_0^t r_1^\mathrm{het}(\tau)\,\mathrm{d}\tau - \int_0^t \exp\left[-(k_\mathrm{f}+k_\mathrm{b})(t-\tau)\right]\,r_1^\mathrm{het}(\tau)\,\mathrm{d}\tau\right\}$$
$$-\frac{k_\mathrm{f}}{k_\mathrm{f}+k_\mathrm{b}}\int_0^t r_3^\mathrm{het}(\tau)\,\mathrm{d}\tau - \frac{k_\mathrm{b}}{k_\mathrm{f}+k_\mathrm{b}}\int_0^t \exp\left[-(k_\mathrm{f}+k_\mathrm{b})(t-\tau)\right]\,r_3^\mathrm{het}(\tau)\,\mathrm{d}\tau\,, \tag{4.66}$$

$$\Gamma_4(t) = \int_0^t r_3^\mathrm{het}(\tau)\,\mathrm{d}\tau\,. \tag{4.67}$$

Substitution of Eqs. (4.64)–(4.67) into Eqs. (4.43) and (4.45) gives the system of two IEs:

$$\Psi_1(t) - k_1^0\Bigg\langle f_{1,\mathrm{f}}(t)\left\{\Gamma_1^\star - \int_0^t \Psi_1(\tau)\,\mathrm{d}\tau\right\}$$
$$-\frac{f_{1,\mathrm{b}}(t)}{k_\mathrm{f}+k_\mathrm{b}}\Bigg\{k_\mathrm{b}\int_0^t \Psi_1(\tau)\,\mathrm{d}\tau + k_\mathrm{f}\int_0^t \exp\left[-(k_\mathrm{f}+k_\mathrm{b})(t-\tau)\right]\,\Psi_1(\tau)\,\mathrm{d}\tau$$

$$-k_{\mathrm{b}} \int_0^t \Psi_2(\tau)\,\mathrm{d}\tau + k_{\mathrm{b}} \int_0^t \exp[-(k_{\mathrm{f}}+k_{\mathrm{b}})(t-\tau)]\,\Psi_2(\tau)\,\mathrm{d}\tau \Bigg\}\Bigg) = 0\,, \tag{4.68}$$

$$\Psi_2(t) - k_3^0 \Bigg\langle \frac{f_{3,\mathrm{f}}(t)}{k_{\mathrm{f}}+k_{\mathrm{b}}} \Bigg\{ k_{\mathrm{f}} \int_0^t \Psi_1(\tau)\,\mathrm{d}\tau - k_{\mathrm{f}} \int_0^t \exp[-(k_{\mathrm{f}}+k_{\mathrm{b}})(t-\tau)]\,\Psi_1(\tau)\,\mathrm{d}\tau$$

$$-k_{\mathrm{f}} \int_0^t \Psi_2(\tau)\,\mathrm{d}\tau - k_{\mathrm{b}} \int_0^t \exp[-(k_{\mathrm{f}}+k_{\mathrm{b}})(t-\tau)]\,\Psi_2(\tau)\,\mathrm{d}\tau \Bigg\}$$

$$-f_{3,\mathrm{b}}(t) \int_0^t \Psi_2(\tau)\,\mathrm{d}\tau \Bigg\rangle = 0\,. \tag{4.69}$$

In Eqs. (4.68) and (4.69) $\overrightarrow{\Psi}(t) = [\Psi_1(t), \Psi_2(t)]^{\mathrm{T}} = \left[r_1^{\mathrm{het}}(t), r_3^{\mathrm{het}}(t)\right]^{\mathrm{T}}$ denotes the vector of unknown functions. After solving Eqs. (4.68) and (4.69) one calculates the Faradaic current as

$$\frac{I(t)}{FA} = -n_1 r_1^{\mathrm{het}}(t) - n_3 r_3^{\mathrm{het}}(t) = -n_1 \Psi_1(t) - n_3 \Psi_2(t)\,. \tag{4.70}$$

Finally, application of Procedure 3 gives four equations:

$$\frac{\mathrm{d}\Gamma_1(t)}{\mathrm{d}t} = -k_1^0 \left\{ f_{1,\mathrm{f}}(t) \left[\Gamma_1^\star + \int_0^t \frac{\mathrm{d}\Gamma_1(\tau)}{\mathrm{d}\tau}\,\mathrm{d}\tau \right] - f_{1,\mathrm{b}}(t) \int_0^t \frac{\mathrm{d}\Gamma_2(\tau)}{\mathrm{d}\tau}\,\mathrm{d}\tau \right\}\,, \tag{4.71}$$

$$\frac{\mathrm{d}\Gamma_2(t)}{\mathrm{d}t} = k_1^0 \left\{ f_{1,\mathrm{f}}(t) \left[\Gamma_1^\star + \int_0^t \frac{\mathrm{d}\Gamma_1(\tau)}{\mathrm{d}\tau}\,\mathrm{d}\tau \right] - f_{1,\mathrm{b}}(t) \int_0^t \frac{\mathrm{d}\Gamma_2(\tau)}{\mathrm{d}\tau}\,\mathrm{d}\tau \right\}$$

$$-k_{\mathrm{f}} \int_0^t \frac{\mathrm{d}\Gamma_2(\tau)}{\mathrm{d}\tau}\,\mathrm{d}\tau + k_{\mathrm{b}} \int_0^t \frac{\mathrm{d}\Gamma_3(\tau)}{\mathrm{d}\tau}\,\mathrm{d}\tau\,, \tag{4.72}$$

$$\frac{\mathrm{d}\Gamma_3(t)}{\mathrm{d}t} = -k_3^0 \left\{ f_{3,\mathrm{f}}(t) \int_0^t \frac{\mathrm{d}\Gamma_3(\tau)}{\mathrm{d}\tau}\,\mathrm{d}\tau - f_{3,\mathrm{b}}(t) \int_0^t \frac{\mathrm{d}\Gamma_4(\tau)}{\mathrm{d}\tau}\,\mathrm{d}\tau \right\}$$

$$+k_{\mathrm{f}} \int_0^t \frac{\mathrm{d}\Gamma_2(\tau)}{\mathrm{d}\tau}\,\mathrm{d}\tau - k_{\mathrm{b}} \int_0^t \frac{\mathrm{d}\Gamma_3(\tau)}{\mathrm{d}\tau}\,\mathrm{d}\tau\,, \tag{4.73}$$

$$\frac{\mathrm{d}\Gamma_4(t)}{\mathrm{d}t} = k_3^0 \left\{ f_{3,\mathrm{f}}(t) \int_0^t \frac{\mathrm{d}\Gamma_3(\tau)}{\mathrm{d}\tau}\,\mathrm{d}\tau - f_{3,\mathrm{b}}(t) \int_0^t \frac{\mathrm{d}\Gamma_4(\tau)}{\mathrm{d}\tau}\,\mathrm{d}\tau \right\} . \tag{4.74}$$

System (4.71)–(4.74) can be reduced to three equations, by making use of the structural conservation relationship $\mathrm{d}\Gamma_1(t)/\mathrm{d}t + \mathrm{d}\Gamma_2(t)/\mathrm{d}t + \mathrm{d}\Gamma_3(t)/\mathrm{d}t + \mathrm{d}\Gamma_4(t)/\mathrm{d}t = 0$, resulting from adding all Eqs. (4.71)–(4.74). By eliminating, e.g., the unknown $\mathrm{d}\Gamma_4(t)/\mathrm{d}t$, we thus obtain three IEs:

$$\Psi_1(t) + k_1^0 \left\{ f_{1,\mathrm{f}}(t) \left[\Gamma_1^\star + \int_0^t \Psi_1(\tau)\,\mathrm{d}\tau \right] - f_{1,\mathrm{b}}(t) \int_0^t \Psi_2(\tau)\,\mathrm{d}\tau \right\} = 0 , \tag{4.75}$$

$$\Psi_2(t) - k_1^0 \left\{ f_{1,\mathrm{f}}(t) \left[\Gamma_1^\star + \int_0^t \Psi_1(\tau)\,\mathrm{d}\tau \right] - f_{1,\mathrm{b}}(t) \int_0^t \Psi_2(\tau)\,\mathrm{d}\tau \right\}$$
$$+ k_\mathrm{f} \int_0^t \Psi_2(\tau)\,\mathrm{d}\tau - k_\mathrm{b} \int_0^t \Psi_3(\tau)\,\mathrm{d}\tau = 0 , \tag{4.76}$$

$$\Psi_3(t) + k_3^0 \left\{ f_{3,\mathrm{f}}(t) \int_0^t \Psi_3(\tau)\,\mathrm{d}\tau \right.$$
$$\left. + f_{3,\mathrm{b}}(t) \left[\int_0^t \Psi_1(\tau)\,\mathrm{d}\tau + \int_0^t \Psi_2(\tau)\,\mathrm{d}\tau + \int_0^t \Psi_3(\tau)\,\mathrm{d}\tau \right] \right\}$$
$$- k_\mathrm{f} \int_0^t \Psi_2(\tau)\,\mathrm{d}\tau + k_\mathrm{b} \int_0^t \Psi_3(\tau)\,\mathrm{d}\tau = 0 . \tag{4.77}$$

In Eqs. (4.75)–(4.77) $\vec{\Psi}(t) = [\Psi_1(t), \Psi_2(t), \Psi_3(t)]^\mathrm{T} = [\mathrm{d}\Gamma_1(t)/\mathrm{d}t, \mathrm{d}\Gamma_2(t)/\mathrm{d}t, \mathrm{d}\Gamma_3(t)/\mathrm{d}t]^\mathrm{T}$ denotes the vector of unknown functions. In view of the equivalences $\mathrm{d}\Gamma_1(t)/\mathrm{d}t = -r_1^\mathrm{het}(t)$ and $\mathrm{d}\Gamma_4(t)/\mathrm{d}t = r_3^\mathrm{het}(t)$, the Faradaic current is calculated as

$$\frac{I(t)}{FA} = -n_1 r_1^\mathrm{het}(t) - n_3 r_3^\mathrm{het}(t) = n_1 \Psi_1(t) + n_3 [\Psi_1(t) + \Psi_2(t) + \Psi_3(t)] . \tag{4.78}$$

4.2 Literature Examples

All currently available literature examples of IEs refer to controlled potential experiments. Most of them were obtained by means of Procedures 1 and 2. A number of surface reaction schemes and transient methods were considered. In particular, IEs for the single non-equilibrium reversible surface electron transfer reaction

$$X_{1,\mathrm{ad}} + n_1\,\mathrm{e}^- \rightleftarrows X_{2,\mathrm{ad}} \tag{4.79}$$

subject to Butler–Volmer kinetics, were formulated and solved assuming AC voltammetry [12], staircase cyclic voltammetry [9], and square wave voltammetry [4, 5, 8, 18, 20]. The limiting case of an irreversible reaction (4.79) was also considered by Gulaboski et al. [4]. An additional effect of interactions between adsorbed reactants $X_{1,\mathrm{ad}}$ and $X_{2,\mathrm{ad}}$ was included into the IEs for cyclic staircase and square wave voltammetry in [14, 19]. IEs for the surface EE reaction scheme involving two consecutive non-equilibrium reversible electron transfer reactions:

$$X_{1,\mathrm{ad}} + n_1\,\mathrm{e}^- \rightleftarrows X_{2,\mathrm{ad}}\ , \tag{4.80}$$

$$X_{2,\mathrm{ad}} + n_2\,\mathrm{e}^- \rightleftarrows X_{3,\mathrm{ad}}\ , \tag{4.81}$$

subject to Butler–Volmer kinetics were formulated and solved in [15], assuming square wave voltammetry, and in [6] assuming cyclic staircase voltammetry. IEs for a non-equilibrium reversible surface electron transfer, subject to Butler–Volmer kinetics, and preceded by a reversible heterogeneous chemical reaction (CE reaction scheme):

$$X_{1,\mathrm{ad}} \underset{k_\mathrm{b}}{\overset{k_\mathrm{f}}{\rightleftarrows}} X_{2,\mathrm{ad}}\ , \tag{4.82}$$

$$X_{2,\mathrm{ad}} + n\,\mathrm{e}^- \rightleftarrows X_{3,\mathrm{ad}}\ , \tag{4.83}$$

were formulated and solved by Gulaboski et al. [3], assuming square wave voltammetry. Probably due to misspelling, the heterogeneous reaction (4.82) was termed "homogeneous chemical reaction" in that reference, and the IEs provided therein contain errors: exponential terms in the kernel functions should have differences $t - \tau$ in place of the variable τ. IEs for the surface EC reaction scheme:

$$X_{1,\mathrm{ad}} + n\,\mathrm{e}^- \rightleftarrows X_{2,\mathrm{ad}}\ , \tag{4.84}$$

$$X_{2,\mathrm{ad}} \overset{k_\mathrm{f}}{\rightarrow} X_{3,\mathrm{ad}}\ , \tag{4.85}$$

involving a non-equilibrium reversible reaction (4.84) (subject to Butler–Volmer kinetics), and irreversible reaction (4.85) were formulated and solved in [17], assuming square wave voltammetry. IEs for the surface catalytic EC reaction scheme:

$$X_{1,\mathrm{ad}} + n\,\mathrm{e}^- \rightleftarrows X_{2,\mathrm{ad}}\ , \tag{4.86}$$

$$X_{2,\mathrm{ad}} \stackrel{k_f}{\rightarrow} X_{1,\mathrm{ad}}\ , \tag{4.87}$$

involving a non-equilibrium reversible reaction (4.86) (subject to Butler–Volmer kinetics), and an irreversible reaction (4.87) were formulated and solved in [13, 16], assuming cyclic staircase voltammetry and square wave voltammetry. A more complicated surface EEC catalytic reaction scheme:

$$X_{1,\mathrm{ad}} + n\,\mathrm{e}^- \rightleftarrows X_{2,\mathrm{ad}}\ , \tag{4.88}$$

$$X_{2,\mathrm{ad}} + n\,\mathrm{e}^- \rightleftarrows X_{3,\mathrm{ad}}\ , \tag{4.89}$$

$$X_{3,\mathrm{ad}} \stackrel{k_f}{\rightarrow} X_{2,\mathrm{ad}}\ , \tag{4.90}$$

was also discussed in an analogous way in [2, 6]. Finally, IEs for the surface ECE reaction scheme:

$$X_{1,\mathrm{ad}} + n\,\mathrm{e}^- \rightleftarrows X_{2,\mathrm{ad}}\ , \tag{4.91}$$

$$X_{2,\mathrm{ad}} \stackrel{k_f}{\rightarrow} X_{3,\mathrm{ad}}\ , \tag{4.92}$$

$$X_{3,\mathrm{ad}} + n\,\mathrm{e}^- \rightleftarrows X_{4,\mathrm{ad}}\ , \tag{4.93}$$

involving non-equilibrium reversible electron transfer reactions (4.91) and (4.93), and an irreversible heterogeneous reaction (4.92) were given in [1, 6], assuming cyclic staircase voltammetry and square wave voltammetry. A collection of the numerical IE solutions for square wave voltammetry, corresponding to surface E, CE, EC, EE, and ECE reaction schemes, was provided by Gulaboski et al. [7].

Vvedenskii et al. [25] presented a theory of linear potential sweep voltammetry, for a few reaction schemes in which reactants, products (or both reactants and products) of electron transfer reactions were adsorbed, and all remaining species were static distributed species. Several adsorption isotherms (Langmuir, Temkin, Frumkin) were considered. The IEs were obtained by a procedure different from Procedures 1–3 of Sect. 4.1, by making use of the particular forms of the expressions for reaction rates or reaction equilibria.

All the above literature examples of IEs refer to electrochemical systems involving adsorbed species. But there is also one published example of IEs obtained for a system where dynamic distributed species confined to thin solution layers were involved. It is the model of cyclic voltammetry for an electrode reaction coupled with ion transfer across the liquid | liquid interface, described by Komorsky–Lovrić and Lovrić [10].

4.3 The Relative Merits: ODEs vs. IEs

There are usually two main general arguments in favour of using the IE method. One argument is theoretical: we expect to improve our understanding of the model considered, by performing the necessary IE derivations. The second argument is computational: we expect that the numerical solution of the IEs is computationally less expensive than the numerical solution of the differential equations. It should be noted that the both expectations are somewhat debatable in the case of models independent of spatial coordinates. The mathematical and computational complexity of the ODE and IE systems is rather comparable in this case. If all reactions in the reaction scheme are linearly independent, then the conversion of the ODEs into IEs by Procedure 1 of Sect. 4.1 reduces the number of equations to be solved at least by 1, because the number of species is greater than the number of reactions at least by 1 (each reaction involves at least two species). However, if the reactions are linearly dependent, the reduction of the number of equations is not certain, as the number of reactions may then be greater than the number of species. For example, the following hypothetical reaction scheme:

$$\mathrm{X}_{1,\mathrm{ad}} \rightleftarrows \mathrm{X}_{2,\mathrm{ad}} \,, \tag{4.94}$$

$$\mathrm{X}_{2,\mathrm{ad}} \rightleftarrows \mathrm{X}_{3,\mathrm{ad}} \,, \tag{4.95}$$

$$\mathrm{X}_{3,\mathrm{ad}} \rightleftarrows \mathrm{X}_{4,\mathrm{ad}} \,, \tag{4.96}$$

$$\mathrm{X}_{1,\mathrm{ad}} \rightleftarrows \mathrm{X}_{3,\mathrm{ad}} \,, \tag{4.97}$$

$$\mathrm{X}_{2,\mathrm{ad}} \rightleftarrows \mathrm{X}_{4,\mathrm{ad}} \,, \tag{4.98}$$

$$\mathrm{X}_{1,\mathrm{ad}} \rightleftarrows \mathrm{X}_{4,\mathrm{ad}} \,, \tag{4.99}$$

involves 4 species, but 6 reactions, out of which only three are linearly independent. A greater reduction of the number of the equations is achieved by Procedure 2 (if the reactions are linearly independent), because the number of essential reactions is smaller than the total number of the reactions. However, in the case of Procedure 2 one obtains IEs with convolution kernels (4.11), whereas Procedure 1 yields constant kernels (4.7) only. The numerical solution of the IEs with kernels (4.7) can be usually accomplished in a computational time that is proportional to the length of the t interval, similarly to the numerical solution of ODEs. In the case of the kernels (4.11) the computational time is usually proportional to the square of the t interval (see Chap. 12). Therefore, in the case of the IEs obtained by Procedure 2, any speedup resulting from the reduction of the number of equations is likely to be lost by the increased computational effort. In the case of Procedure 3 the reduction of the number of equations is possible if suitable structural conservation relationships exist. The number of IEs cannot exceed the number of dynamic localised species. The computational time increases linearly with t.

Perhaps the only benefit from using the IE formalism for models independent of spatial coordinates occurs in situations when one considers a series of alternative models, some independent of spatial coordinates, and some involving a space-dependent transport. Examples of such situations are found in [5,8,17]. If the models of the space-dependent transport are solved by the IE method, it then becomes natural and convenient to apply the analogous IE formalism and solution techniques also to the remaining, space-independent models. This facilitates comparisons and saves some intellectual effort.

References

1. Gulaboski R (2009) Surface ECE mechanism in protein film voltammetry—a theoretical study under conditions of square-wave voltammetry. J Solid State Electrochem 13:1015–1024
2. Gulaboski R, Mihajlov L (2011) Catalytic mechanism in successive two-step protein-film voltammetry—theoretical study in square-wave voltammetry. Biophys Chem 155:1–9
3. Gulaboski R, Mirčeski V, Lovrić M, Bogeski I (2005) Theoretical study of a surface electrode reaction preceded by a homogeneous chemical reaction under conditions of square-wave voltammetry. Electrochem Commun 7:515–522
4. Gulaboski R, Lovrić M, Mirčeski V, Bogeski I, Hoth M (2008) Protein-film voltammetry: a theoretical study of the temperature effect using square-wave voltammetry. Biophys Chem 137:49–55
5. Gulaboski R, Lovrić M, Mirčeski V, Bogeski I, Hoth M (2008) A new rapid and simple method to determine the kinetics of electrode reactions of biologically relevant compounds from the half-peak width of the square-wave voltammograms. Biophys Chem 138:130–137
6. Gulaboski R, Kokoškarova P, Mitrev S (2012) Theoretical aspects of several successive two-step redox mechanisms in protein-film cyclic staircase voltammetry. Electrochim Acta 69:86–96
7. Gulaboski R, Mirčeski V, Bogeski I, Hoth M (2012) Protein film voltammetry: electrochemical enzymatic spectroscopy. A review on recent progress. J Solid State Electrochem 16:2315–2328
8. Komorsky-Lovrić Š, Lovrić M (1995) Kinetic measurements of a surface confined redox reaction. Anal Chim Acta 305:248–255
9. Komorsky-Lovrić Š, Lovrić M (1996) A peak current—scan rate relationship in staircase voltammetry of a surface redox reaction. Electroanalysis 8:959–962
10. Komorsky-Lovrić Š, Lovrić M (2005) Kinetics of electrode reaction coupled to ion transfer across the liquid | liquid interface. Cent Eur J Chem 3:216–229
11. Laviron E (1966) Etude de phenomenes d'autoinhibition en voltammetrie a variation lineaire de tension. I. Le film inhibiteur est forme par deposition reversible d'un compose insoluble. Application aux chlorures. J Electroanal Chem 12:516–523
12. Lovrić M, Komorsky-Lovrić Š (1990) Faradaic alternating current response of the adsorbed redox couple. Mikrochim Acta 1:321–325
13. Mirčeski V, Gulaboski R (2001) Surface catalytic mechanism in square-wave voltammetry. Electroanalysis 13:1326–1334
14. Mirčeski V, Gulaboski R (2002) Adsorptive stripping voltammetric behavior of probucole. Experimental and theoretical treatment. Mikrochim Acta 138:33–42
15. Mirčeski V, Gulaboski R (2003) A theoretical and experimental study of a two-step quasireversible surface redox reaction by square-wave voltammetry. Croat Chem Acta 76:37–48
16. Mirčeski V, Gulaboski R (2003) The surface catalytic mechanism: a comparative study with square-wave and staircase cyclic voltammetry. J Solid State Electrochem 7:157–165
17. Mirčeski V, Lovrić M (2000) Adsorption effects in square-wave voltammetry of an EC mechanism. Croat Chem Acta 73:305–329

18. Mirčeski V, Gulaboski R, Petrovska-Jovanović S, Stojanova K (2001) Characterization of the redox reaction of V(V) in ammonia buffers with square-wave voltammetry. Port Electrochim Acta 19:25–41
19. Mirčeski V, Lovrić M, Gulaboski R (2001) Theoretical and experimental study of the surface redox reaction involving interactions between the adsorbed particles under conditions of square-wave voltammetry. J Electroanal Chem 515:91–100
20. Quentel F, Mirčeski V, Laouenan A, Elleouet C, Madec CL (2003) Square-wave voltammetry of the molybdenum-1,10 phenanthroline-fulvic acids complex: redox kinetics measurements. Electroanalysis 15:270–277
21. Rahman M (2007) Integral equations and their applications. WIT Press, Southampton
22. Stonehart P (1968) Potentiodynamic determination of electrode kinetics for chemisorbed reactants: the $Ag/Ag_2O/OH^-$ system. Electrochim Acta 13:1789–1803
23. Stonehart P, Portante FP (1968) Potentiodynamic examination of surface processes and kinetics for the $Ag_2O/AgO/OH^-$ system. Electrochim Acta 13:1805–1814
24. Stonehart P, Kozlowska HA, Conway BE (1969) Potentiodynamic examination of electrode kinetics for electroactive adsorbed species: applications to the reduction of noble metal surface oxides. Proc R Soc A 310:541–563
25. Vvedenskii AV, Kartashova TV, Bobrinskaya EV (2008) Voltammetry of redox reactions using the general form of the oxidized or reduced species adsorption isotherm. Russ J Electrochem 44:1373–1381

Chapter 5
Models Involving One-Dimensional Diffusion

In this chapter we consider electroanalytical models described by systems of diffusion PDEs defined over spatial domains that can be regarded effectively as one-dimensional. In such models, owing to a particular symmetry, just one spatial coordinate is sufficient for the mathematical description (see Sect. 2.4). We assume also that the diffusion is in the direction perpendicular to the interface studied, and that the models do not involve homogeneous reactions, dynamic localised species, nor other complications. We begin with the general description in Sect. 5.1, of how the IEs can be obtained for this kind of models. The description reveals the need to derive concentration–production rate relationships by analytically solving relevant PDEs. The relationships arising in the simplest case of diffusion in semi-infinite spatial domains, and a literature overview of the published IEs corresponding to this case, are provided in Sect. 5.2. Then, in Sect. 5.3 we pass to the concentration–production rate relationships and literature IEs for the somewhat more difficult to handle diffusion in finite spatial domains. Anomalous diffusion is considered in Sect. 5.4.2. Finally, Sect. 5.5 presents benefits resulting from using the IE method.

5.1 Derivation of the IEs

Consider an electroanalytical experiment, in which a set of $N_{\mathrm{r}}^{\mathrm{het}}$ heterogeneous reactions takes place at an interface studied, among $N_{\mathrm{s}}^{\mathrm{distr}}$ distributed species $\mathrm{X}_1, \ldots, \mathrm{X}_{N_{\mathrm{s}}^{\mathrm{distr}}}$ present in spatially extended phase(s) adjacent to this interface. Let $N_{\mathrm{s}}^{\mathrm{dd}}$ of the species be dynamic distributed species. The dynamic distributed species are subject to one-dimensional diffusion PDEs, accompanied by appropriate initial and boundary conditions. There are no dynamic localised species at the interface studied.

For any jth dynamic distributed species, the one-dimensional diffusion equation results from Eq. (2.8) written in a suitable system of coordinates. In particular,

L.K. Bieniasz, *Modelling Electroanalytical Experiments by the Integral Equation Method*, Monographs in Electrochemistry, DOI 10.1007/978-3-662-44882-3_5

taking an appropriate operator from Table 2.1, the one-dimensional PDE for planar diffusion is:

$$\frac{\partial c_j(x,t)}{\partial t} = D_j \frac{\partial^2 c_j(x,t)}{\partial x^2} . \tag{5.1}$$

For spherical diffusion:

$$\frac{\partial c_j(r,t)}{\partial t} = D_j \left[\frac{\partial^2 c_j(r,t)}{\partial r^2} + \frac{2}{r} \frac{\partial c_j(r,t)}{\partial r} \right] . \tag{5.2}$$

For cylindrical diffusion:

$$\frac{\partial c_j(r,t)}{\partial t} = D_j \left[\frac{\partial^2 c_j(r,t)}{\partial r^2} + \frac{1}{r} \frac{\partial c_j(r,t)}{\partial r} \right] . \tag{5.3}$$

Equations (5.1)–(5.3) are assumed to hold in every spatial subdomain adjacent to an interface studied, but to avoid an overly complicated notation we do not introduce here any additional index or symbol to distinguish the subdomains. Equations (5.1), (5.2), or (5.3) must be accompanied by initial conditions. The typical initial conditions corresponding to the initial equilibrium state (see Sect. 2.10) take the form:

$$c_j(x,0) = c_j^\star \tag{5.4}$$

for planar diffusion, or

$$c_j(r,0) = c_j^\star , \tag{5.5}$$

for spherical or cylindrical diffusion, with the uniform initial concentration $c_j^\star$ that can be positive or zero. The boundary conditions must also be provided at the interface studied, as well as at other boundaries.

In the case of controlled potential experiments, the conversion of the above IBVPs into IEs is usually performed according to the following procedure. At the interface studied, the boundary conditions are usually given by equation system (2.54), with matrices $\overline{V}$ and $\overline{Z}$ resulting from the partial inversion of Eq. (2.53). Equation system (2.54) may contain one additional equation for the Faradaic current, if electrons exchanged are formally considered as dynamic species. As the reaction rates depend on the concentrations $c_j^\dagger$ of the dynamic distributed species at the interface studied [see Eqs. (2.30) and (2.31)], the boundary conditions in Eq. (2.54) contain, in general, $2N_s^{dd}$ unknown variables: N_s^{dd} concentrations $c_j^\dagger$, and N_s^{dd} production rates of the same species at this interface. We have to determine all these unknowns, because they are needed for calculating the Faradaic current (according to the relevant one of the equations in Eq. (2.54), or another relevant equation). However, there are only N_s^{dd} boundary conditions in

Eq. (2.54). Therefore, we need another independent set of N_s^{dd} relationships between concentrations $c_j^\dagger$ and species production rates. Such a set can be obtained by finding general solutions (integrals) of the diffusion PDEs, satisfying boundary conditions at the remaining interfaces. We strive to obtain this set analytically, and the work needed to accomplish this task represents the main theoretical effort associated with the use of the IE method for spatially one-dimensional models. Luckily, once derived relationships can often be re-used in a variety of different models or electrochemical systems, provided that the models share the boundary conditions at interface(s) other than the one we study. If this is not the case, one has to make appropriate derivations and related analyses. Sections 5.2, 5.3, and Appendix A can serve as a guide how to proceed in such yet unresolved cases. Since in the case of the distributed species the production rates are related to the fluxes by Eq. (2.46), the task amounts to determining analytical concentration–flux relationships at the interface studied. The task is relatively easy under assumptions accepted in this chapter, because when the transport of the distributed species is governed solely by diffusion, without any interference of other modes of transport or homogeneous reactions, the PDE for each distributed species is independent of the concentrations of other distributed species, and can be analysed separately. Consequently, a separate concentration–flux relationship can be obtained for every dynamic species. As is discussed in Sects. 5.2 and 5.3 below, these relationships take the general form:

$$c_j^\dagger = c_j^\dagger(t) = c_j^\star + \int_0^t \mathscr{K}_j(t,\tau)\; p_j^{\mathrm{het}}(\tau)\; \mathrm{d}\tau\,, \tag{5.6}$$

where $\mathscr{K}_j(t,\tau)$ is a suitable integral transformation kernel, consistent with the definition (3.3) of linear integral transformations. The kernel depends on the diffusion coefficient of the jth species, geometry of the spatial domains, the system of coordinates used, and boundary conditions at other interfaces. For models discussed in this chapter, $c_j^\dagger$ and p_j^{het} are functions of time only, which is indicated in Eq. (5.6). The occurrence of the integral in Eq. (5.6) is the reason why, after combining the relationships (5.6) with the boundary conditions in Eq. (2.54) for the interface studied, we obtain IEs. The IEs may result from the boundary conditions belonging to groups $R1$ and $R2$ of equations in Eq. (2.54), where the unknown concentrations $c_j^\dagger(t)$ occur that are replaced by Eq. (5.6). Structural conservation relationships belonging to the group $R3$ of equations in Eq. (2.54) do not involve $c_j^\dagger(t)$, but only species production rates (equal to fluxes). As the IEs obtained in the above way have the species production rates as unknown functions, the structural conservation relationships can be used to diminish the number of unknowns in the IEs. As a result, the number of final IEs to be solved is usually smaller than N_s^{dd}.

In Sects. 5.2 and 5.3 we shall see that there are also special situations or limiting cases when Eq. (5.6) can be replaced by algebraic equations linking $c_j^\dagger(t)$ and $p_j^{\mathrm{het}}(t)$. In such special situations we obtain integro-algebraic equation systems or even purely algebraic ones, for the unknown species production rates.

Once the species production rates and interfacial concentrations are determined by solving the IEs, the Faradaic current can be calculated from the relevant equation in Eq. (2.53) or (2.54), or from another appropriate equation.

The above procedure is somewhat analogous to Procedure 3 applicable to models independent of spatial coordinates (see Sect. 4.1), since in both procedures we obtain IEs for the species production rates. It is difficult to formulate counterparts of Procedures 1 and 2 from Sect. 4.1, applicable to models dependent on one spatial coordinate.

Let us pass now to controlled current experiments. For such experiments, equation system (2.54) contains N_s^{dd} equations linking reaction rates and production rates of dynamic distributed species. If electrons exchanged are considered as dynamic species, then equation system (2.54) contains one more equation involving Faradaic current. Otherwise, equation system (2.54) must be supplemented with an additional equation defining the current. Consequently, one has $N_s^{dd}+1$ equations which one tries to separate into N_s^{dd} boundary conditions for the concentrations of dynamic distributed species, and one equation for determining the unknown electrode potential $E(t)$ or Galvani potential difference $\Delta\Phi(t)$. This may be impossible to achieve for sufficiently complicated reaction schemes, but we shall not consider such complicated cases here. If the reaction scheme involves only one charge transfer reaction, then the production rates of dynamic distributed species are immediately obtained as functions of the (known) Faradaic current $I(t)$, and there is no need to formulate and solve IEs. However, if there are more charge transfer reactions, the production rates are not so easily predictable. In such cases, we can make use of the integral concentration–production rate relationships (5.6), to derive the IEs in much the same way as was shown above for the controlled potential experiments. Solving the IEs yields the production rates, and Eq. (5.6) gives also $c_j^{\dagger}(t)$. Consequently, $E(t)$ or $\Delta\Phi(t)$ can be calculated, as these quantities generally depend on $c_j^{\dagger}(t)$ and the species production rates.

We illustrate the above procedures by four examples. As a first example, let us consider a controlled potential experiment for the electron transfer reaction:

$$X_1 + n\,e^- \rightleftarrows X_2 \tag{5.7}$$

taking place between dynamic distributed species X_1 and X_2 at some electrode. Hence, the electrons exchanged are regarded as one of the dynamic species. If we take these electrons as the third dynamic species, after X_1 and X_2, then in accordance with Eqs. (2.46) and (2.49) the vector of the species production rates is $\overrightarrow{p}^{\,\mathrm{het}}(t) = \left[p_1^{\mathrm{het}}(t),\ p_2^{\mathrm{het}}(t),\ p_3^{\mathrm{het}}(t)\right]^{\mathrm{T}} = \left[J_1^{\perp}(t),\ J_2^{\perp}(t),\ I(t)(FA)^{-1}\right]^{\mathrm{T}}$, and the corresponding stoichiometric matrix $\overline{N}^{\,\mathrm{dyn,het}}$ is:

$$\overline{N}^{\,\mathrm{dyn,het}} = \begin{bmatrix} -1 \\ 1 \\ -n \end{bmatrix}. \tag{5.8}$$

Let us assume initially that reaction (5.7) is a non-equilibrium reversible reaction subject to Butler–Volmer kinetics (see Sect. 2.7), so that its rate $r^{\text{het}}(t)$ is given by

$$r^{\text{het}}(t) = k^0 \left[f_{\text{f}}(t)\, c_1^{\dagger}(t) - f_{\text{b}}(t)\, c_2^{\dagger}(t) \right] , \tag{5.9}$$

with

$$f_{\text{f}}(t) = \exp\left\{ -\alpha^{\text{f}} \frac{nF}{RT} \left[E(t) - E^0 \right] \right\} \tag{5.10}$$

and

$$f_{\text{b}}(t) = \exp\left\{ \alpha^{\text{b}} \frac{nF}{RT} \left[E(t) - E^0 \right] \right\} . \tag{5.11}$$

Partial inversion of Eq. (2.53) with matrix (5.8) gives Eq. (2.54) with the following matrices $\overline{V}$ and $\overline{Z}$:

$$\overline{V} = \begin{bmatrix} 1 \\ 0 \\ 0 \end{bmatrix} , \tag{5.12}$$

$$\overline{Z} = \begin{bmatrix} -1 & 0 & 0 \\ -1 & -1 & 0 \\ -n & 0 & 1 \end{bmatrix} . \tag{5.13}$$

Equation system (2.54) with matrices (5.12) and (5.13) involves one kinetic equation relating the rate of reaction (5.7) with the production rate of species X_1:

$$r^{\text{het}}(t) = -p_1^{\text{het}}(t) , \tag{5.14}$$

and two structural conservation relationships:

$$0 = -p_1^{\text{het}}(t) - p_2^{\text{het}}(t) , \tag{5.15}$$

$$0 = -n\, p_1^{\text{het}}(t) + \frac{I(t)}{FA} . \tag{5.16}$$

Equations (5.14) and (5.15) serve as boundary conditions at the electrode, whereas Eq. (5.16) can be used for calculating the Faradaic current:

$$\frac{I(t)}{FA} = n\, p_1^{\text{het}}(t) . \tag{5.17}$$

Although there are two unknown production rates $p_1^{\text{het}}(t)$ and $p_2^{\text{het}}(t)$ in vector $\vec{p}^{\,\text{het}}(t)$, they are connected by the structural conservation relationship (5.15), so that in fact there is only one unknown to be determined. Let us denote this unknown function by

$$\Psi(t) = -p_1^{\text{het}}(t) = p_2^{\text{het}}(t) . \tag{5.18}$$

By replacing the boundary concentrations $c_1^\dagger(t)$ and $c_2^\dagger(t)$ in Eq. (5.9) by the relevant integral concentration–production rate relationships (5.6), and setting the results into Eq. (5.14), we therefore obtain the single second kind Volterra IE:

$$\Psi(t) - k^0 \left\{ f_\mathrm{f}(t) \left[c_1^\star - \int_0^t \mathcal{K}_1(t,\tau)\, \Psi(\tau)\, \mathrm{d}\tau \right] - f_\mathrm{b}(t) \left[c_2^\star + \int_0^t \mathcal{K}_2(t,\tau)\, \Psi(\tau)\, \mathrm{d}\tau \right] \right\} = 0 . \tag{5.19}$$

If reaction (5.7) were irreversible, then instead of Eq. (5.9) we would have

$$r^{\text{het}}(t) = k^0\, f_\mathrm{f}(t)\, c_1^\dagger(t) , \tag{5.20}$$

and Eq. (5.19) would reduce to

$$\Psi(t) - k^0\, f_\mathrm{f}(t) \left[c_1^\star - \int_0^t \mathcal{K}_1(t,\tau)\, \Psi(\tau)\, \mathrm{d}\tau \right] = 0 . \tag{5.21}$$

Alternatively, if reaction (5.7) was an equilibrium (Nernstian) reaction, then the boundary condition (5.9) would be replaced by the Nernst equation:

$$c_1^\dagger(t) - c_2^\dagger(t)\, \exp\left\{ \frac{nF}{RT} [E(t) - E^0] \right\} = 0 , \tag{5.22}$$

so that instead of Eq. (5.19) we would obtain the first kind Volterra IE:

$$\left[c_1^\star - \int_0^t \mathcal{K}_1(t,\tau)\, \Psi(\tau)\, \mathrm{d}\tau \right] - \left[c_2^\star + \int_0^t \mathcal{K}_2(t,\tau)\, \Psi(\tau)\, \mathrm{d}\tau \right] \exp\left\{ \frac{nF}{RT} [E(t) - E^0] \right\} = 0 . \tag{5.23}$$

As a second example, let us consider a controlled potential experiment for the EE reaction scheme:

$$X_1 + n_1\,e^- \rightleftarrows X_2\,, \tag{5.24}$$

$$X_2 + n_2\,e^- \rightleftarrows X_3\,, \tag{5.25}$$

taking place between dynamic distributed species X_1, X_2, and X_3 at a WE. If we take the electrons exchanged as the fourth dynamic species, after X_1, X_2, and X_3, then in accordance with Eqs. (2.46) and (2.49) the vector of the species production rates is $\vec{p}^{\,\text{het}}(t) = [p_1^{\text{het}}(t), p_2^{\text{het}}(t), p_3^{\text{het}}(t), p_4^{\text{het}}(t)]^{\text{T}} = [J_1^{\perp}(t), J_2^{\perp}(t), J_3^{\perp}(t), I(t)(FA)^{-1}]^{\text{T}}$, and the corresponding stoichiometric matrix $\overline{N}^{\,\text{dyn,het}}$ is:

$$\overline{N}^{\,\text{dyn,het}} = \begin{bmatrix} -1 & 0 \\ 1 & -1 \\ 0 & 1 \\ -n_1 & -n_2 \end{bmatrix}. \tag{5.26}$$

Assuming that the reactions (5.24) and (5.25) are non-equilibrium reversible and subject to Butler–Volmer kinetics, their rates are:

$$r_1^{\text{het}}(t) = k_1^0\left[f_{1,\text{f}}(t)\,c_1^{\dagger}(t) - f_{1,\text{b}}(t)\,c_2^{\dagger}(t)\right], \tag{5.27}$$

$$r_2^{\text{het}}(t) = k_2^0\left[f_{2,\text{f}}(t)\,c_2^{\dagger}(t) - f_{2,\text{b}}(t)\,c_3^{\dagger}(t)\right], \tag{5.28}$$

with

$$f_{i,\text{f}}(t) = \exp\left\{-\alpha_i^{\text{f}}\frac{n_i F}{RT}\left[E(t) - E_i^0\right]\right\} \tag{5.29}$$

and

$$f_{i,\text{b}}(t) = \exp\left\{\alpha_i^{\text{b}}\frac{n_i F}{RT}\left[E(t) - E_i^0\right]\right\} \tag{5.30}$$

for $i = 1, 2$. Partial inversion of Eq. (2.53) with matrix (5.26) gives Eq. (2.54) with the following matrices $\overline{V}$ and $\overline{Z}$:

$$\overline{V} = \begin{bmatrix} 1 & 0 \\ 0 & 1 \\ 0 & 0 \\ 0 & 0 \end{bmatrix}, \tag{5.31}$$

$$\overline{Z} = \begin{bmatrix} -1 & 0 & 0 & 0 \\ -1 & -1 & 0 & 0 \\ -1 & -1 & -1 & 0 \\ -(n_1+n_2) & -n_2 & 0 & 1 \end{bmatrix} . \tag{5.32}$$

Equation system (2.54) with matrices (5.31) and (5.32) involves two kinetic equations relating the rates of reactions (5.24) and (5.25) with the species production rates:

$$r_1^{\mathrm{het}}(t) = -p_1^{\mathrm{het}}(t) , \tag{5.33}$$

$$r_2^{\mathrm{het}}(t) = -p_1^{\mathrm{het}}(t) - p_2^{\mathrm{het}}(t) , \tag{5.34}$$

and two structural conservation relationships:

$$0 = -p_1^{\mathrm{het}}(t) - p_2^{\mathrm{het}}(t) - p_3^{\mathrm{het}}(t) , \tag{5.35}$$

$$0 = -(n_1+n_2)\, p_1^{\mathrm{het}}(t) - n_2\, p_2^{\mathrm{het}}(t) + \frac{I(t)}{FA} . \tag{5.36}$$

Equations (5.33)–(5.35) serve as boundary conditions at the electrode, whereas Eq. (5.36) can be used for calculating the Faradaic current:

$$\frac{I(t)}{FA} = (n_1+n_2)\, p_1^{\mathrm{het}}(t) + n_2\, p_2^{\mathrm{het}}(t) . \tag{5.37}$$

The structural conservation relationship (5.35) allows one to reduce the number of unknown species production rates, so that for the unknown functions we can take, for example:

$$\Psi_1(t) = -p_1^{\mathrm{het}}(t) , \tag{5.38}$$

$$\Psi_2(t) = -p_2^{\mathrm{het}}(t) , \tag{5.39}$$

resulting in

$$p_3^{\mathrm{het}}(t) = \Psi_1(t) + \Psi_2(t) . \tag{5.40}$$

By replacing the boundary concentrations $c_1^{\dagger}(t)$, $c_2^{\dagger}(t)$, and $c_3^{\dagger}(t)$ in Eqs. (5.27) and (5.28), by the relevant integral concentration–production rate relationships (5.6), and setting the results into Eqs. (5.33) and (5.34), we therefore obtain the following system of two second kind Volterra IEs:

$$\Psi_1(t) - k_1^0 \left\{ f_{1,\mathrm{f}}(t) \left[c_1^\star - \int_0^t \mathscr{K}_1(t,\tau)\, \Psi_1(\tau)\, \mathrm{d}\tau \right] - f_{1,\mathrm{b}}(t) \left[c_2^\star - \int_0^t \mathscr{K}_2(t,\tau)\, \Psi_2(\tau)\, \mathrm{d}\tau \right] \right\} = 0\,, \tag{5.41}$$

$$\Psi_1(t) + \Psi_2(t) - k_2^0 \left\{ f_{2,\mathrm{f}}(t) \left[c_2^\star - \int_0^t \mathscr{K}_2(t,\tau)\, \Psi_2(\tau)\, \mathrm{d}\tau \right] - f_{2,\mathrm{b}}(t) \left[c_3^\star + \int_0^t \mathscr{K}_3(t,\tau)\, \Psi_1(\tau)\, \mathrm{d}\tau + \int_0^t \mathscr{K}_3(t,\tau)\, \Psi_2(\tau)\, \mathrm{d}\tau \right] \right\} = 0\,. \tag{5.42}$$

Similarly to the first example, if any of reactions (5.24) and (5.25) were an irreversible or an equilibrium reaction, then a corresponding IE (5.41) or (5.42) would have to be replaced by an alternative IE. For example, if reaction (5.25) were an equilibrium reaction, then the boundary condition (5.34) would be replaced by the Nernst equation, so that instead of Eq. (5.42) we would obtain the first kind Volterra IE:

$$\left[c_2^\star - \int_0^t \mathscr{K}_2(t,\tau)\, \Psi_2(\tau)\, \mathrm{d}\tau \right] - \left[c_3^\star + \int_0^t \mathscr{K}_3(t,\tau)\, \Psi_1(\tau)\, \mathrm{d}\tau + \int_0^t \mathscr{K}_3(t,\tau)\, \Psi_2(\tau)\, \mathrm{d}\tau \right] \times \exp\left\{ \frac{n_2 F}{RT} \left[E(t) - E_2^0 \right] \right\} = 0\,. \tag{5.43}$$

Most often $c_2^\star = c_3^\star = 0$ in Eqs. (5.41) and (5.42) or (5.43), which simplifies the IEs.

As a third example, we consider a controlled current experiment for the single electron transfer (5.7). Similarly to the first example, the experiment is described by Eqs. (5.14)–(5.16), but the equations are used differently. Equations (5.15) and (5.16) serve as boundary conditions at the electrode, and Eq. (5.14) is used for calculating the electrode potential response $E(t)$. From Eqs. (5.15) and (5.16) we see that both conceivable unknowns, $p_1^{\mathrm{het}}(t)$ and $p_2^{\mathrm{het}}(t)$, are uniquely determined by the known $I(t)$:

$$p_1^{\mathrm{het}}(t) = -p_2^{\mathrm{het}}(t) = \frac{I(t)}{nFA}\,. \tag{5.44}$$

Therefore, the interfacial concentrations are obtainable immediately as

$$c_1^\dagger(t) = c_1^\star + \int_0^t \mathscr{K}_1(t,\tau)\,\frac{I(\tau)}{nFA}\,\mathrm{d}\tau\,, \tag{5.45}$$

$$c_2^\dagger(t) = c_2^\star - \int_0^t \mathscr{K}_2(t,\tau)\,\frac{I(\tau)}{nFA}\,\mathrm{d}\tau\,. \tag{5.46}$$

The same can be said about the electrode potential. There is no closed-form expression for $E(t)$ in the case of a non-equilibrium reversible reaction (5.7), but for an irreversible reaction (5.7) Eq. (5.14) can be re-arranged into the formula

$$E(t) = E^0 + \frac{RT}{nF\alpha^{\mathrm{f}}}\ln\frac{-\frac{I(t)}{nFA}}{k^0\left[c_1^\star + \int_0^t \mathscr{K}_1(t,\tau)\,\frac{I(\tau)}{nFA}\,\mathrm{d}\tau\right]}\,. \tag{5.47}$$

Alternatively, assuming equilibrium reaction (5.7), Eq. (5.14) is replaced by the Nernst equation:

$$E(t) = E^0 + \frac{RT}{nF}\ln\frac{c_1^\star + \int_0^t \mathscr{K}_1(t,\tau)\,\frac{I(\tau)}{nFA}\,\mathrm{d}\tau}{c_2^\star - \int_0^t \mathscr{K}_2(t,\tau)\,\frac{I(\tau)}{nFA}\,\mathrm{d}\tau}\,. \tag{5.48}$$

As was expected, no IEs arise in this example. However, we shall see in Chap. 10 that this is no longer true in the presence of double layer charging.

As a fourth example, we consider a controlled current experiment for the EE reaction scheme (5.24) and (5.25), assuming equilibrium reactions. One possible re-arrangement of the Eqs. (5.33)–(5.36) gives the explicit formula for the unknown electrode potential:

$$E(t) = E_1^0 + \frac{RT}{n_1F}\ln\frac{c_1^\dagger(t)}{c_2^\dagger(t)}\,, \tag{5.49}$$

and three boundary conditions not involving $E(t)$:

$$E_1^0 + \frac{RT}{n_1F}\ln\frac{c_1^\dagger(t)}{c_2^\dagger(t)} = E_2^0 + \frac{RT}{n_2F}\ln\frac{c_2^\dagger(t)}{c_3^\dagger(t)}\,, \tag{5.50}$$

$$\frac{I(t)}{FA} - (n_1+n_2)\,p_1^{\mathrm{het}}(t) - n_2\,p_2^{\mathrm{het}}(t) = 0\,, \tag{5.51}$$

$$\frac{I(t)}{FA} - n_1\,p_1^{\mathrm{het}}(t) + n_2\,p_3^{\mathrm{het}}(t) = 0\,. \tag{5.52}$$

A more convenient form of Eq. (5.50) is

$$c_3^\dagger(t)\, c_1^\dagger(t)^{n_2/n_1} - K\, c_2^\dagger(t)^{(n_1+n_2)/n_1} = 0\,, \tag{5.53}$$

where $K = \exp\left[n_2 F (RT)^{-1} \left(E_2^0 - E_1^0\right)\right]$, and the interfacial concentrations are expressed by the following formulae. By defining the unknown function as

$$\Psi(t) = -p_1^{\mathrm{het}}(t)\,, \tag{5.54}$$

and making use of Eq. (5.6), we obtain:

$$c_1^\dagger(t) = c_1^\star - \int_0^t \mathcal{K}_1(t,\tau)\, \Psi(\tau)\, \mathrm{d}\tau\,, \tag{5.55}$$

and [in view of Eqs. (5.51) and (5.52)]:

$$c_2^\dagger(t) = c_2^\star + \int_0^t \mathcal{K}_2(t,\tau)\, \frac{I(\tau)}{n_2 FA}\, \mathrm{d}\tau + \frac{n_1+n_2}{n_2} \int_0^t \mathcal{K}_2(t,\tau)\, \Psi(\tau)\, \mathrm{d}\tau\,, \tag{5.56}$$

$$c_3^\dagger(t) = c_3^\star - \int_0^t \mathcal{K}_3(t,\tau)\, \frac{I(\tau)}{n_2 FA}\, \mathrm{d}\tau - \frac{n_1}{n_2} \int_0^t \mathcal{K}_3(t,\tau)\, \Psi(\tau)\, \mathrm{d}\tau\,. \tag{5.57}$$

Since $I(t)$ is known, the integrals of $I(t)$, occurring in Eqs. (5.56) and (5.57) are determined functions of time. Setting Eqs. (5.55)–(5.57) into Eq. (5.53) then gives a single first kind Volterra IE with the sole unknown $\Psi(t)$. Having determined $\Psi(t)$ and all integrals, one can use Eq. (5.49) to calculate $E(t)$. Interestingly, the IE arising for this example is nonlinear, even though the reaction scheme involves only first-order reactions.

5.2 Diffusion in Semi-Infinite Spatial Domains

In this section we assume that diffusion takes place in one of the semi-infinite spatial domains depicted in Fig. 2.1. The assumption of a semi-infinite spatial domain simplifies the formulation of the boundary conditions at the second boundary, and consequently results also in relatively uncomplicated integral transformation kernels and IEs. This is because we can assume that the concentrations of the distributed species at infinity (usually termed bulk concentrations) are simply equal to the initial, equilibrium concentrations. Hence, the boundary conditions at $x \to \infty$ (in the case of planar diffusion) or at $r \to \infty$ (in the case of spherical or cylindrical diffusion) are:

$$c_j(\infty, t) = c_j^\star\,. \tag{5.58}$$

5.2.1 Concentration–Production Rate Relationships

In order to derive the concentration–production rate relationship for a semi-infinite domain, one needs to solve analytically the (incomplete) IBVP comprising one of the Eqs. (5.1)–(5.3), with an appropriate one of the initial conditions (5.4) or (5.5), and boundary condition (5.58).

Such concentration–flux relationships for planar diffusion were derived in connection with the modelling of the linear potential sweep voltammetry by the IE method. In particular, Ševčik [139] obtained such relationships, although his interest was rather in deriving an integral expression for the flux (or Faradaic current). Delahay [41], Matsuda and Ayabe [94], and Senda [135] derived and presented the relationships in the explicit form. All these authors used the Laplace transform method. Gokhshtein [62] obtained the same relationships in an alternative way, by using fundamental solutions of the diffusion PDE, assuming additionally nonuniform initial concentrations. Fujioka [59] used a similar approach. The relationships for spherical diffusion were derived by Hurwitz [71] in connection with the theory of current step chronopotentiometry. The relationships for cylindrical diffusion were also obtained by Hurwitz [71], but the integral transformation kernel was given only in the form of a series expansion. A complete presentation of the kernel was given by Aoki et al. [15] in connection with the theory of the linear potential sweep voltammetry.

For the purpose of the kernel derivations, it is convenient to introduce an auxiliary unknown variable u_j (or several such variables), and consider the IBVP(s) for the auxiliary variable(s), rather than for the concentrations c_j. In the case of the uncomplicated diffusion considered here, for the auxiliary variable we can take the departure of the concentration from its initial value, i.e.

$$u_j = c_j - c_j^\star \,. \tag{5.59}$$

As u_j is a linear function of c_j, it obeys the diffusion equations analogous to Eqs. (5.1), (5.2), or (5.3), but with simpler initial and boundary conditions. Thus, we have to solve the following incomplete IBVPs. For planar diffusion:

$$\frac{\partial u_j(x,t)}{\partial t} = D_j \frac{\partial^2 u_j(x,t)}{\partial x^2} \,, \tag{5.60}$$

$$u_j(x,0) = 0 \,, \tag{5.61}$$

$$u_j(\infty,t) = 0 \,. \tag{5.62}$$

For spherical diffusion:

$$\frac{\partial u_j(r,t)}{\partial t} = D_j \left[\frac{\partial^2 u_j(r,t)}{\partial r^2} + \frac{2}{r} \frac{\partial u_j(r,t)}{\partial r} \right] , \tag{5.63}$$

$$u_j(r,0) = 0 \,, \tag{5.64}$$

$$u_j(\infty, t) = 0 \, . \tag{5.65}$$

For cylindrical diffusion:

$$\frac{\partial u_j(r,t)}{\partial t} = D_j \left[\frac{\partial^2 u_j(r,t)}{\partial r^2} + \frac{1}{r} \frac{\partial u_j(r,t)}{\partial r} \right] , \tag{5.66}$$

$$u_j(r, 0) = 0 \, , \tag{5.67}$$

$$u_j(\infty, t) = 0 \, . \tag{5.68}$$

An important characteristic of pure diffusion transport, considered in this chapter, is that the fluxes of variables c_j and u_j are identical, as long as the initial concentrations $c_j^\star$ are uniform. From Eqs. (2.5) and (5.59) we obtain:

$$J_j(x,t) = -D_j \frac{\partial c_j(x,t)}{\partial x} = -D_j \frac{\partial u_j(x,t)}{\partial x} \tag{5.69}$$

for planar, and

$$J_j(r,t) = -D_j \frac{\partial c_j(r,t)}{\partial r} = -D_j \frac{\partial u_j(r,t)}{\partial r} \tag{5.70}$$

for spherical and cylindrical diffusion.

In Appendix A we follow [15, 41, 62, 71, 94, 139], and present detailed analytical solutions of the problems (5.60)–(5.68), by employing the Laplace transform method. As a result, we obtain the following relationships between the solution u_j and its flux J_j at the interface studied [cf. Eqs. (A.24), (A.25), (A.85), (A.86), (A.152), and (A.154)].

For planar diffusion:

$$u_j(0,t) = \int_0^t \mathcal{K}_j^{\mathrm{p}}(t,\tau) \; J_j(0,\tau) \; \mathrm{d}\tau \, , \tag{5.71}$$

with

$$\mathcal{K}_j^{\mathrm{p}}(t,\tau) = D_j^{-1/2} \left[\pi(t-\tau)\right]^{-1/2} \, . \tag{5.72}$$

For spherical diffusion:

$$u_j(r_0,t) = \int_0^t \mathcal{K}_j^{\mathrm{s}}(t,\tau) \; J_j(r_0,\tau) \; \mathrm{d}\tau \, , \tag{5.73}$$

with

$$\mathscr{K}_j^{s}(t,\tau) = D_j^{-1/2}\left\{[\pi(t-\tau)]^{-1/2} - \varrho_j \,\mathrm{erex}\left[\varrho_j (t-\tau)^{1/2}\right]\right\} , \tag{5.74}$$

where

$$\varrho_j = D_j^{1/2}/r_0 , \tag{5.75}$$

and

$$\mathrm{erex}(z) = \exp(z^2)\,\mathrm{erfc}(z) . \tag{5.76}$$

In Eq. (5.76) erfc(·) is the complementary error function:

$$\mathrm{erfc}(z) = 1 - \mathrm{erf}(z) , \tag{5.77}$$

and erf(·) is the error function [1]:

$$\mathrm{erf}(z) = 2\pi^{-1/2}\int_0^z \exp(-\zeta^2)\,\mathrm{d}\zeta . \tag{5.78}$$

The issue how to compute erf(·), erfc(·), and erex(·) numerically will be addressed in Sect. 12.1.1.4.

For cylindrical diffusion we obtain

$$u_j(r_0,t) = \int_0^t \mathscr{K}_j^{c}(t,\tau)\; J_j(r_0,\tau)\,\mathrm{d}\tau , \tag{5.79}$$

with

$$\mathscr{K}_j^{c}(t,\tau) = D_j^{-1/2}\varrho_j\,\varphi\left[\varrho_j^2(t-\tau)\right] , \tag{5.80}$$

where the function $\varphi(\vartheta)$ is defined as

$$\varphi(\vartheta) = \mathscr{L}^{-1}\left\{s^{-1/2}\,\frac{\mathrm{K}_0(s^{1/2})}{\mathrm{K}_1(s^{1/2})}\right\} , \tag{5.81}$$

the Laplace transformation is from the ϑ domain to the s domain, and $\mathrm{K}_0(\cdot)$ and $\mathrm{K}_1(\cdot)$ are modified Bessel functions of the second kind and orders 0 and 1 [1]. Function $\varphi(\vartheta)$ can also be expressed as

$$\varphi(\vartheta) = 4\,\pi^{-2}\,\mathrm{G}(1,0;\vartheta) , \tag{5.82}$$

where $G(1,0;\vartheta)$ is one of the special integrals discussed by Carslaw and Jaeger [35] in the theory of heat conduction in solids (we use here the notation introduced by Phillips and Mahon [123], who analysed these integrals mathematically):

$$G(1,0;\vartheta) = \int_0^\infty \frac{\exp(-\vartheta\xi^2)}{\left[J_1^2(\xi) + Y_1^2(\xi)\right]} \frac{d\xi}{\xi} . \tag{5.83}$$

In Eq. (5.83) $J_1(\cdot)$ and $Y_1(\cdot)$ are Bessel functions of the first and second kind, and order 1 [1].

The form (5.80) of the kernel was originally presented in the literature [15]. However, it is more convenient to introduce a special function defined by the formula

$$\mathrm{kcylw}(z) = \mathrm{kcylw}(\vartheta^{1/2}) = \mathscr{L}^{-1}\left\{\frac{K_1(s^{1/2}) - K_0(s^{1/2})}{s^{1/2}\,K_1(s^{1/2})}\right\} , \tag{5.84}$$

where the Laplace transformation is from the domain of the variable $\vartheta = z^2$ to the s domain. By using this function, the kernel $\mathscr{K}_j^c(t,\tau)$ can be expressed in the form analogous to Eq. (5.74) for spherical diffusion:

$$\mathscr{K}_j^c(t,\tau) = D_j^{-1/2}\left\{[\pi(t-\tau)]^{-1/2} - \varrho_j\,\mathrm{kcylw}\left[\varrho_j(t-\tau)^{1/2}\right]\right\} . \tag{5.85}$$

The name "kcylw" has been chosen by the present author [22] as an acronym for "kernel for cylindrical wire". The issue how to compute $\mathrm{kcylw}(\cdot)$ numerically will be addressed in Sect. 12.1.1.4.

In view of Eq. (2.46) the fluxes J_j at the interface studied are equal to the production rates $p_j^{\mathrm{het}}(t)$ of the jth species at the interface. Therefore, the concentration–production rate relationships resulting from Eqs. (5.71), (5.73), and (5.79) are as follows. For planar diffusion:

$$c_j^\dagger(t) \equiv c_j(0,t) = c_j^\star + \int_0^t \mathscr{K}_j^p(t,\tau)\; p_j^{\mathrm{het}}(\tau)\; d\tau . \tag{5.86}$$

For spherical diffusion:

$$c_j^\dagger(t) \equiv c_j(r_0,t) = c_j^\star + \int_0^t \mathscr{K}_j^s(t,\tau)\; p_j^{\mathrm{het}}(\tau)\; d\tau . \tag{5.87}$$

For cylindrical diffusion:

$$c_j^\dagger(t) \equiv c_j(r_0,t) = c_j^\star + \int_0^t \mathscr{K}_j^c(t,\tau)\; p_j^{\mathrm{het}}(\tau)\; d\tau . \tag{5.88}$$

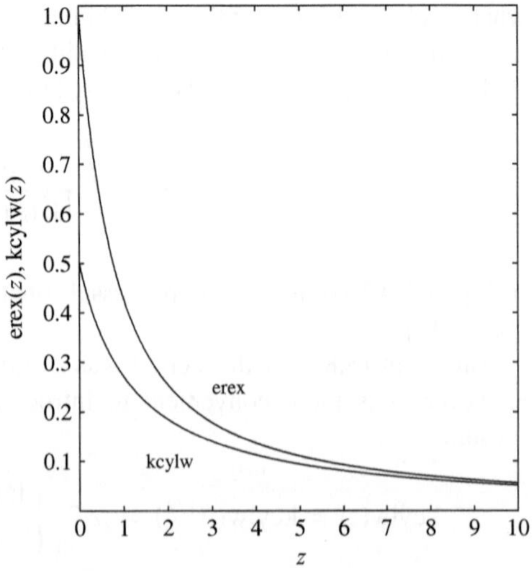

Fig. 5.1 Comparison of functions $\mathrm{erex}(z)$ and $\mathrm{kcylw}(z)$, representing the effect of the spherical and cylindrical geometry, respectively, on the kernels arising for diffusion in semi-infinite spatial domains

We note that $\mathscr{K}_j^{\mathrm{p}}(t,\tau)$, $\mathscr{K}_j^{\mathrm{s}}(t,\tau)$, and $\mathscr{K}_j^{\mathrm{c}}(t,\tau)$ are all convolution kernels (see Sect. 3.1). The kernels $\mathscr{K}_j^{\mathrm{s}}(t,\tau)$ and $\mathscr{K}_j^{\mathrm{c}}(t,\tau)$ are composed of two terms, the first being identical to the kernel $\mathscr{K}_j^{\mathrm{p}}(t,\tau)$ for the planar diffusion case. The second term, proportional to ϱ_j, represents a "spherical" or "cylindrical" contribution, which vanishes when $r_0 \to \infty$, that is when the interface studied becomes effectively planar, because in this limit $\varrho_j \to 0$, whereas $\mathrm{erex}\left[\varrho_j (t-\tau)^{1/2}\right] \to 1$ or $\mathrm{kcylw}\left[\varrho_j (t-\tau)^{1/2}\right] \to 1/2$, which are finite values. The terms representing "spherical" or "cylindrical" contributions look similar; in the cylindrical case function $\mathrm{kcylw}(z)$ replaces the function $\mathrm{erex}(z)$ pertinent to the spherical case. Figure 5.1 demonstrates the difference between the functions $\mathrm{erex}(z)$ and $\mathrm{kcylw}(z)$.

A special feature of semi-infinite spherical diffusion (not shared by planar and cylindrical diffusion) is the possibility of the occurrence of a steady state, in which there is a nonzero solution of Eq. (5.63) satisfying $\partial u_j(r,t)/\partial t \approx 0$. In Appendix A, Sect. A.2 the analytical solution of such a steady state diffusion equation is obtained, and we conclude that at steady state Eq. (5.73) takes a special limiting form [cf. Eq. (A.89)]:

$$u_j(r_0,t) = D_j^{-1/2} \varrho_j^{-1} J_j(r_0,t) \,, \tag{5.89}$$

so that in view of Eqs. (2.46) and (5.59)

$$c_j^{\dagger}(t) = c_j(r_0,t) = c_j^{\star} + D_j^{-1/2} \varrho_j^{-1} p_j^{\mathrm{het}}(t) \,. \tag{5.90}$$

Equation (5.90) is the concentration–production rate relationship at steady state. We note that no convolution integral is involved in Eq. (5.90), because the steady state assumption holds independently at every time moment, so that the connections between different time moments are lost. Although the dependences of the transport PDEs on time t are formally removed under steady state conditions, u_j or c_j may still generally depend on t, owing to the possibly time-dependent boundary conditions at $x = 0$ (not explicitly written here). Therefore, throughout this book we retain the t variable in the list of arguments of the PDE solutions under steady state conditions. Relationship (5.90) can be obtained either by considering the steady state diffusion equation, or it can be deduced from the general formula (5.73), as was demonstrated in Appendix A in Bieniasz et al. [25]. Both approaches are described in Sect. A.2 of Appendix A of the present book. The possibility to deduce the steady state concentration–production rate relationship from the transient formula (5.73) is one of the advantages of the IE method.

In the case of controlled potential experiments under spherical diffusion conditions, as an alternative to using the integral relationship (5.87) one can use approximate corrections to the current, accounting for sphericity. The corrections were obtained by Goodisman [65], by means of a perturbative approach, in which the inverse r_0^{-1} of the electrode radius played the role of the perturbation parameter. Consequently, the corrections are applicable when r_0 is large, so that the departures from planar diffusion are small. It seems that the direct use of Eq. (5.87) should be preferred as a more general and viable approach.

5.2.2 *Literature Examples*

The following list of examples is limited to IE-based models in which diffusion uncomplicated by convection, homogeneous reactions or other phenomena was assumed, together with one-dimensional semi-infinite spatial domains. Models involving such complications are listed in further Chaps. 6, 8, 9, and 10.

5.2.2.1 Planar Diffusion

Theory of linear potential sweep and cyclic voltammetry for the single electron transfer reaction

$$\mathrm{X_1} + n\,\mathrm{e^-} \rightleftarrows \mathrm{X_2} \tag{5.91}$$

involving dynamic distributed species X_1 and X_2 was considered by Ševčik [139], Delahay [41], Matsuda and Ayabe [94], Gokhshtein [62], Gokhshtein and Gokhshtein [61], Nicholson and Shain [112], Nicholson [111], and Mareček and Honz [92]. Seelig and Blount [134] used related IE solutions in their approach to data analysis based on the Kalman filter. Savéant [132] presented a theory of cyclic voltammetry with asymmetrical potential sweeps. The effect of the nonzero initial concentration of the reaction product, and/or of starting the scan from the equilibrium potential was addressed by Farsang et al. [55], and Weidner and Fedkiw

[158]. Bonciocat [28] analysed long-time behaviour of cyclic voltammograms. The effect of reaction order higher than one, on linear potential sweep voltammetry for irreversible reaction (5.91) was studied by Kohler et al. [76]. The resulting theory involved a nonlinear IE. A somewhat similar nonlinear IE was obtained by Mendelson and Tenno [98], while modelling the effect of concentration-dependent activity coefficients on linear potential sweep voltammetry for irreversible reaction (5.91). Bonciocat et al. [30] studied connections between the IE corresponding to irreversible reaction (5.91), and some ODE. Seralathan et al. [138] discussed the equivalence of linear potential sweep and staircase voltammetry. Theory of cyclic staircase voltammetry was presented by Murphy et al. [106].

The above authors assumed equilibrium, non-equilibrium reversible, or irreversible reaction (5.91) subject to Butler–Volmer kinetics with constant charge transfer coefficients. Andrieux et al. [6–8] solved IEs describing cyclic voltammetry for irreversible reaction (5.91), subject to a Butler–Volmer kinetic model modified to incorporate a quadratic activation-driving force relationship. Such a model leads to potential-dependent charge transfer coefficients. Linear potential sweep voltammetry for reaction (5.91) subject to Marcus–Hush–Chidsey kinetics was considered by Henstridge et al. [68].

Delahay, Matsuda, and their co-workers [42, 95] used IEs for the modelling of the Faradaic impedance for reaction (5.91). Smith [141] discussed the relationships between Faradaic impedances obtained by four techniques, in which small amplitude controlled potential or controlled current AC perturbations were combined with controlled potential or controlled current DC perturbations. The theory of square wave polarography/voltammetry for reaction (5.91) was presented by Tachi and Kambara [146], Okamoto [116, 117], O'Dea et al. [115], Nuwer et al. [114], and Fatouros and Krulic [56], who utilised the former results of Mauduit et al. [96] for normal and differential pulse voltammetries. Senda and Tachi [137] considered AC polarography with an arbitrary periodic electrode potential perturbation, assuming non-equilibrium reversible and equilibrium reaction (5.91). They obtained IDEs rather than IEs. The theory of large amplitude AC voltammetry and sinusoidal voltammetry for equilibrium reaction (5.91) was presented by Bell et al. [19, 20]. The theory of differential pulse voltammetry was discussed by Aoki et al. [14]. A high frequency AC technique for investigating fast non-Volmerian reaction (5.91) was analysed by Garreau et al. [60]. Bonciocat and Cotârţă [29] presented a derivation of the IE for reaction (5.91) and arbitrary potential perturbation, by considering a non-stationary process as a continuous succession of stationary states. De Vries [46] used IEs to describe potential step chronoamperometry followed by linear potential sweep voltammetry, for reaction (5.91). Wein and Tovchigrechko [159, 160] discussed voltage step transients in a two-electrode cell, assuming non-equilibrium reaction (5.91). Camacho et al. [34] utilised IE solutions for verifying their theory of reverse pulse voltammetry and polarography for reaction (5.91). Chen et al. [36] discussed an electroanalytical technique, in which the electrode potential was an exponential function of time, assuming equilibrium reaction (5.91).

Gueshi et al. [67] analysed the effect of partially covered electrodes on linear potential sweep and cyclic voltammetry for non-equilibrium reversible reaction (5.91). Their theory leads to IEs with kernels analogous to those arising in the presence of homogeneous reactions, that will be considered in Chap. 8 (see Sects. 8.2 and 8.3). A similar theory, focused on microscopic inhomogeneities of the electrode surface, was presented by Amatore et al. [3]. Kokhanovskii et al. [77] considered the effect of a limited solubility of the product X_2 of equilibrium reaction (5.91), leading to the passivation of the electrode, on the peak potential in linear potential sweep voltammetry.

IEs describing linear potential sweep and cyclic voltammetry for the equilibrium or non-equilibrium reversible reaction of metal electrodeposition:

$$X_1 + n\,e^- \rightleftarrows X_{2,s} \tag{5.92}$$

were reported by White and Lawson [161], and Pnev et al. [125]. In reaction (5.92) X_1 is a dynamic distributed species (metal ion), and $X_{2,s}$ is a static localised species (solid metal). Affoune et al. [2] modelled cyclic voltammetry for the reverse reaction (5.92). Both cases were also studied by Avaca et al. [17]. An IE formalism similar to that arising for reaction (5.92) was utilised by Vogel et al. [156], and Ross and Stonehart [131], in the modelling of cyclic voltammetry for the hydrogen electrode reaction on platinum.

A rare example of controlled current oscillographic polarography for a reaction scheme involving two independent electron transfer reactions of the type (5.91), and one more independent electron transfer reaction of the type (5.92), was studied by Matsuda [93].

Shuman [140] presented an IE describing cyclic voltammetry for the reaction:

$$m\,X_1 + n\,e^- \rightleftarrows q\,X_2\,, \tag{5.93}$$

with non-unity stoichiometric coefficients m and q for dynamic distributed species X_1 and X_2. His calculations have been redone by Ito et al. [72]. Whereas in [72, 140] the initial absence of X_2 was assumed, Bieniasz [21] studied an alternative case, of potential scan started from the equilibrium potential, assuming $m = 1$ and $q = 2$, and $m = q = 1$.

Bieniasz and Suski [24] reported the theory of potential step chronoamperometry and linear potential sweep voltammetry for the equilibrium reaction

$$X_1 + X_{2,s} \rightleftarrows X_3 + X_4 + n\,e^-\,, \tag{5.94}$$

where X_1, X_3, and X_4 are dynamic distributed species, and $X_{2,s}$ is a static distributed species. The theory of linear potential sweep voltammetry for the equilibrium reaction (5.94) was also considered by Tkalenko et al. [148].

The EE reaction scheme:

$$X_1 + n_1\,e^- \rightleftarrows X_2\,, \tag{5.95}$$

$$X_2 + n_2\,e^- \rightleftarrows X_3\,, \tag{5.96}$$

involving dynamic distributed species X_1, X_2, and X_3, was a subject of several studies. Gokhshtein and Gokhshtein [63,64], and Polcyn and Shain [127] presented the theory of cyclic voltammetry under various assumptions concerning the rates of reactions (5.95) and (5.96). Lovrić and Komorsky-Lovrić [78–80,87,88] elaborated the theory of square wave voltammetry. Hung and Smith [69, 70] presented the theory of AC polarography. Aoki [10] analysed linear potential sweep voltammetry for an extension of the scheme (5.95) and (5.96), comprising several consecutive electron transfers characterised by the same standard potential.

There are also examples of IEs describing charge transfers at liquid | liquid interfaces. Stewart et al. [142] modelled cyclic voltammetry for an electron transfer at such interfaces. Mirčeski and Scholz [101] considered square wave voltammetry for reaction schemes involving electron and ion transfers. Senda [136] presented a model of voltammetry for an electron transfer, taking into account the bimolecular character of the transfer. The model resulted in a simplified IE with a kernel characteristic of planar diffusion. Katano et al. [74] described a model of cyclic voltammetry for an ion transfer reaction. Tatsumi et al. [147] discussed a theory of normal pulse voltammetry for an electron transfer. In this kind of models one usually assumes that the two adjacent liquid phases are semi-infinite in the direction perpendicular to the liquid | liquid interface studied.

A similar situation arises at the interface between a liquid electrolyte and a solid electrolyte, characteristic of the so-called insertion electrodes. An interesting example model of cyclic voltammetry at such interfaces was presented by Lovrić et al. [91], based on the former model in [89].

5.2.2.2 Spherical Diffusion

Spherical diffusion was assumed in models of transient experiments for the single reaction

$$X_1 + n\,e^- \rightleftarrows X_2 \tag{5.97}$$

involving dynamic distributed species X_1 and X_2, with only X_1 being initially present. Delmastro and Smith [43] presented a model of the first harmonic of the current in AC and DC polarography, assuming equal diffusion coefficients and non-equilibrium reversible reaction (5.97). The model was extended by Delmastro and Smith [44] to the first harmonic of the current in AC polarography, corresponding to the case of different diffusion coefficients. McCord et al. [97] discussed second and third harmonic of the AC polarographic current under the same conditions. Bond et al. [31] analysed the second harmonic of the current in AC cyclic voltammetry, assuming different diffusion coefficients, and equilibrium, non-equilibrium reversible, or irreversible reaction (5.97). Komorsky-Lovrić et al. [81] considered a theory of square wave voltammetry, and Lovrić [85] considered a theory of differential staircase voltammetry, for a non-equilibrium reversible reaction (5.97). Lovrić [86] has also used IE solutions to verify his theory of differential pulse polarography for equilibrium reaction (5.97). Diao and Zhang [50] presented a theory of cyclic voltammetry at a (hemi)spherical electrode, assuming non-equilibrium reversible reaction (5.97), different diffusion coefficients, and both species X_1 and X_2 initially present at equilibrium of reaction (5.97).

Stewart et al. [143] presented a model of linear potential sweep voltammetry for an equilibrium ion transfer reaction at a liquid | liquid interface supported at the tip of a micropipette. The model approximated the transport of the ions for ingress transfer by spherical diffusion, and for egress transfer by planar diffusion.

5.2.2.3 Cylindrical Diffusion

Cylindrical diffusion was assumed in models of linear potential sweep and cyclic voltammetry [11, 15, 108], normal pulse voltammetry [16], and AC voltammetry [154] for the single reaction:

$$X_1 + n\,e^- \rightleftarrows X_2 \tag{5.98}$$

involving dynamic distributed species X_1 and X_2. Initial presence of X_2 only was assumed in [11, 15, 16]. Initial presence of X_1 only was assumed by Neudeck and Dittrich [108], and simultaneous initial presence of X_1 and X_2 was assumed by Tokuda et al. [154]. All these authors assumed equal diffusion coefficients of X_1 and X_2. Equilibrium reaction (5.98) was considered in [15, 154], whereas non-equilibrium reversible and irreversible reaction (5.98) was considered in [11, 16, 108].

5.3 Diffusion in Finite Spatial Domains

In this section we assume that diffusion takes place in one of the finite spatial domains depicted in Fig. 2.2. The major complication for the IE method, resulting from such finite spatial domains, is the need to take into account boundary conditions at the second boundary, while deriving analytical concentration–production rate relationships. Potentially, an unlimited number of different boundary conditions at the second boundary (and different corresponding relationships) is conceivable, which may seem discouraging. Fortunately, in the majority of known models solved by the IE method, only two types of boundary conditions have been of interest.

The first type is analogous to the boundary condition (5.58) for semi-infinite domains. For planar diffusion

$$c_j(l,t) = c_j^\star \,, \tag{5.99}$$

and for spherical or cylindrical diffusion

$$c_j(r_0 \pm l,t) = c_j^\star \,. \tag{5.100}$$

Typically, Eqs. (5.99) or (5.100) hold when the concentrations at the second boundary are maintained constant owing to a physical or chemical equilibrium with the phase (or artificially separated part of it) located outside the second boundary. This means that the jth distributed species is allowed to cross the second boundary. Therefore, the second boundary satisfying Eqs. (5.99) or (5.100) is called "permeable" in this book.

In the second type of boundary conditions one assumes that the flux of the jth distributed species is zero at the second boundary, or:

$$\left.\frac{\partial c_j(x,t)}{\partial x}\right|_{x=l} = 0 \tag{5.101}$$

for planar diffusion, or

$$\left.\frac{\partial c_j(r,t)}{\partial r}\right|_{r=r_0 \pm l} = 0 \tag{5.102}$$

for spherical or cylindrical diffusion. Equations (5.101) or (5.102) mean that the jth distributed species cannot cross the second boundary. Therefore, the second boundary satisfying Eqs. (5.101) or (5.102) is called "impermeable" in this book. In the special case when $l = r_0$ and the second boundary is located at $r = r_0 - l = 0$, there is of course no physical interface associated with this boundary, but Eq. (5.102) expresses the fact that there is no transfer of matter into the interval of (physically nonsensical) negative r.

By combining the finite spatial domains of Fig. 2.2 with the boundary conditions (5.99) and (5.101), or (5.100) and (5.102), we obtain two distinct situations to be considered for planar diffusion, four distinct situations for spherical diffusion, and analogously four situations for cylindrical diffusion. The four situations pertinent to spherical or cylindrical diffusion are: internal domain with a permeable second boundary (I–P); internal domain with an impermeable second boundary (I–I); external domain with a permeable second boundary (E–P); and external domain with an impermeable second boundary (E–I). Concentration–production rate relationships corresponding to all these situations are discussed below.

Alternative boundary conditions at the second boundary are possible, but they are rarely modelled by the IE method. One such alternative was investigated by Diard et al. [51], in connection with the modelling of linear potential sweep voltammetry for rotating metal electrodes covered by polymer films. In [51] the spatial domain was composed of a thin finite interval corresponding to the polymer film, and an adjacent finite or semi-infinite interval corresponding to the diffusion layer in the stirred electrolyte solution. The boundary condition at the polymer | electrolyte interface assumed a linear relationship between the concentrations at both sides of the interface. This particular situation will not be analysed further in the present book.

5.3.1 Concentration–Production Rate Relationships

In order to derive the concentration–production rate relationship for a finite domain, one needs to solve analytically the (incomplete) IBVP comprising one of the Eqs. (5.1)–(5.3), with one of the initial conditions (5.4) or (5.5) (whichever is appropriate), and one of the boundary conditions (5.99)–(5.102).

For planar diffusion and a permeable second boundary relevant relationships were obtained by Fried and Elving [58], in connection with the modelling of the RDE. Several years later they were rediscovered by Tokuda and Matsuda [150]. Analogous relationships for an impermeable second boundary were obtained by Rangarajan and Doss [129, 130], in connection with the modelling of Faradaic admittance experiments. However, in the electrochemical literature this result is often attributed to the later works of De Vries and Van Dalen [45,49], who modelled stripping and linear potential sweep voltammetry at a plane mercury film electrode. Birke [26] reconsidered the concentration–flux relationships for planar diffusion (and both permeable and impermeable second boundaries), by using the formalism of semi-integration. Laplace transforms of the concentration–flux relationships for planar diffusion in a finite spatial domain with an impermeable second boundary were also derived by Danielsson et al. [38], in connection with the theory of differential pulse polarography.

For spherical diffusion some of the concentration–flux relationships were obtained by Nigmatullin [113], who considered the case I–I with $l < r_0$ and $l = r_0$. This was motivated by the modelling of electrolytic cells with stationary spherical mercury electrodes and solid spherical electrodes covered by mercury layers. However, the relationships were not given explicitly, but were embedded in some final model equations, which might be the reason why the Nigmatullin work [113] remained almost unnoticed by the IE method users. Keller and Reinmuth [75] reported concentration–flux relationships applicable to all cases I–I, I–P, E–I, and E–P, although not all of these cases were given equal attention. The study was motivated by the modelling of linear potential sweep voltammetry. The relationships for the I–I case with $l = r_0$ were analysed in much detail by Tokuda et al. [153], in connection with the modelling of cyclic voltammetry at hanging mercury drop electrodes, for reversible amalgam formation and dissolution reactions. Further discussion of this case was given by Myland et al. [107], with the aim to study concentrations at the surface of a stationary mercury drop electrode with amalgam formation. Laplace transforms of the concentration–flux relationships for spherical diffusion in a finite spatial domain with an impermeable second boundary were also derived by Danielsson et al. [38].

The concentration–flux relationships for cylindrical diffusion were obtained by Keller and Reinmuth [75], who considered the cases I–I, I–P, E–I, and E–P. The case I–I with $l = r_0$ was elaborated by Weidner [157], in connection with the modelling of linear potential sweep voltammetry of a soluble redox couple in a cylindrical pore electrode. The kernel function corresponding to the case I–I with $l = r_0$ was also published by Montella [105], in connection with his modelling approach based on the numerical inversion of the Laplace transforms.

For the purpose of the derivations, as for the semi-infinite diffusion case it is convenient to introduce an auxiliary unknown variable u_j defined by Eq. (5.59), which leads to the following incomplete IBVPs. For planar diffusion:

$$\frac{\partial u_j(x,t)}{\partial t} = D_j \frac{\partial^2 u_j(x,t)}{\partial x^2} , \tag{5.103}$$

$$u_j(x,0) = 0 , \tag{5.104}$$

and

$$u_j(l,t) = 0 \tag{5.105}$$

or

$$\left. \frac{\partial u_j(x,t)}{\partial x} \right|_{x=l} = 0 . \tag{5.106}$$

For spherical diffusion:

$$\frac{\partial u_j(r,t)}{\partial t} = D_j \left[\frac{\partial^2 u_j(r,t)}{\partial r^2} + \frac{2}{r} \frac{\partial u_j(r,t)}{\partial r} \right] , \tag{5.107}$$

$$u_j(r,0) = 0 , \tag{5.108}$$

and

$$u_j(r_0 \pm l, t) = 0 \tag{5.109}$$

or

$$\left. \frac{\partial u_j(r,t)}{\partial r} \right|_{r=r_0 \pm l} = 0 . \tag{5.110}$$

For cylindrical diffusion:

$$\frac{\partial u_j(r,t)}{\partial t} = D_j \left[\frac{\partial^2 u_j(r,t)}{\partial r^2} + \frac{1}{r} \frac{\partial u_j(r,t)}{\partial r} \right] , \tag{5.111}$$

$$u_j(r,0) = 0 , \tag{5.112}$$

and

$$u_j(r_0 \pm l, t) = 0 \tag{5.113}$$

or

$$\left.\frac{\partial u_j(r,t)}{\partial r}\right|_{r=r_0\pm l} = 0\,. \tag{5.114}$$

The flux of c_j (and u_j) is again given by Eqs. (5.69) or (5.70).

In Appendix A we follow [45, 49, 58, 75, 107, 113, 129, 130, 150, 153, 157], and present detailed analytical solutions of the problems (5.103)–(5.114), by employing the Laplace transform method. As a result, we obtain the following relationships between the solution u_j and its flux J_j at the interface studied.

For planar diffusion and a permeable second boundary we get [cf. Eqs. (A.35) and (A.36)]:

$$u_j(0,t) = \int_0^t \mathscr{K}_j^{\mathrm{plp}}(t,\tau)\; J_j(0,\tau)\; \mathrm{d}\tau\,, \tag{5.115}$$

with

$$\mathscr{K}_j^{\mathrm{plp}}(t,\tau) = l^{-1}\,\Theta_2\left[0\left|\mu_j^2\;(t-\tau)\right.\right], \tag{5.116}$$

and for planar diffusion and an impermeable second boundary we get [cf. Eqs. (A.53) and (A.54)]:

$$u_j(0,t) = \int_0^t \mathscr{K}_j^{\mathrm{pli}}(t,\tau)\; J_j(0,\tau)\; \mathrm{d}\tau\,, \tag{5.117}$$

with

$$\mathscr{K}_j^{\mathrm{pli}}(t,\tau) = l^{-1}\,\Theta_3\left[0\left|\mu_j^2\;(t-\tau)\right.\right]. \tag{5.118}$$

In Eqs. (5.116) and (5.118)

$$\mu_j = D_j^{1/2}/l\,, \tag{5.119}$$

and $\Theta_2(\nu\,|\,z)$ and $\Theta_3(\nu\,|\,z)$ are two of the so-called exponential theta functions (see, for example, Oldham et al. [119, p. 261]) with parameter $\nu = 0$. These forms of the kernels were originally adopted in the literature [45, 49, 129, 130, 150], but in Sects. A.1.2 and A.1.3 of Appendix A we derive more convenient equivalent forms, similar to the kernels previously discussed in Sect. 5.2:

$$\mathscr{K}_j^{\mathrm{plp}}(t,\tau) = D_j^{-1/2}\left\{[\pi(t-\tau)]^{-1/2} - \mu_j\,\mathrm{kplp}\left[\mu_j(t-\tau)^{1/2}\right]\right\}\,, \tag{5.120}$$

$$\mathscr{K}_j^{\mathrm{pli}}(t,\tau) = D_j^{-1/2}\left\{[\pi(t-\tau)]^{-1/2} + \mu_j\,\mathrm{kpli}\left[\mu_j(t-\tau)^{1/2}\right]\right\}\,. \tag{5.121}$$

In Eqs. (5.120) and (5.121) functions kplp(·) and kpli(·) are special functions defined by the formulae:

$$\mathrm{kplp}(z) = \mathrm{kplp}(\vartheta^{1/2}) = \mathscr{L}^{-1}\left\{s^{-1/2}\left[1 - \tanh(s^{1/2})\right]\right\} , \tag{5.122}$$

$$\mathrm{kpli}(z) = \mathrm{kpli}(\vartheta^{1/2}) = \mathscr{L}^{-1}\left\{s^{-1/2}\left[\coth(s^{1/2}) - 1\right]\right\} , \tag{5.123}$$

where the Laplace transformation is from the domain of the variable $\vartheta = z^2$ to the s domain. The names "kplp" and "kpli" have been chosen by the present author [23] as acronyms for "kernel for a planar layer with a permeable boundary", and "kernel for a planar layer with an impermeable boundary", respectively. The issue of how to compute kplp(·) and kpli(·) numerically will be addressed in Sect. 12.1.1.4.

In view of Eq. (2.46) the fluxes J_j at the interface studied are equal to the production rates $p_j^{\mathrm{het}}(t)$ of the jth species at the interface. Therefore, the concentration–production rate relationships resulting from Eqs. (5.115) and (5.117) are as follows. For a planar layer with a permeable second boundary

$$c_j^{\dagger}(t) = c_j(0,t) = c_j^{\star} + \int_0^t \mathscr{K}_j^{\mathrm{plp}}(t,\tau)\; p_j^{\mathrm{het}}(\tau)\; \mathrm{d}\tau . \tag{5.124}$$

For a planar layer with an impermeable second boundary

$$c_j^{\dagger}(t) = c_j(0,t) = c_j^{\star} + \int_0^t \mathscr{K}_j^{\mathrm{pli}}(t,\tau)\; p_j^{\mathrm{het}}(\tau)\; \mathrm{d}\tau . \tag{5.125}$$

We note that in view of Eqs. (5.120) and (5.121) the kernels $\mathscr{K}_j^{\mathrm{plp}}(t,\tau)$ and $\mathscr{K}_j^{\mathrm{pli}}(t,\tau)$ are of the convolution type, and they are composed of two terms, the first weakly singular term being identical to the kernel $\mathscr{K}_j^{\mathrm{p}}(t,\tau)$ for planar diffusion in a semi-infinite domain. The second, nonsingular term, proportional to μ_j, represents a contribution resulting from the finite length of the spatial interval, dependent on the boundary condition at the second boundary. The contribution vanishes when $l \to \infty$, that is when the spatial domain becomes semi-infinite, because in this limit $\mu_j \to 0$, whereas $\mathrm{kplp}\left[\mu_j(t-\tau)^{1/2}\right]$ and $\mathrm{kpli}\left[\mu_j(t-\tau)^{1/2}\right]$ are bounded. Figure 5.2 demonstrates the difference between the functions kplp(z) and kpli(z).

At low z an approximate equality holds: $\mathrm{kplp}(z) \approx \mathrm{kpli}(z)$, and both functions tend to zero when $z \to 0$. When $z \to \infty$, then $\mathrm{kplp}(z) \to 0$, whereas $\mathrm{kpli}(z) \to 1$. Function kplp(z) has a maximum of about 0.414 at $z \approx 1.148$, and kpli(z) is an increasing function of z.

A non-trivial steady state solution, satisfying $\partial u_j(x,t)/\partial t \approx 0$ is possible for a permeable second boundary. In Appendix A, Sect. A.1.2 the analytical solution of such a steady state diffusion equation is obtained, and we show that at steady state

Fig. 5.2 Comparison of functions kplp(·) and kpli(·), representing the effect of the finite spatial domain for planar diffusion with a permeable or impermeable second boundary, respectively

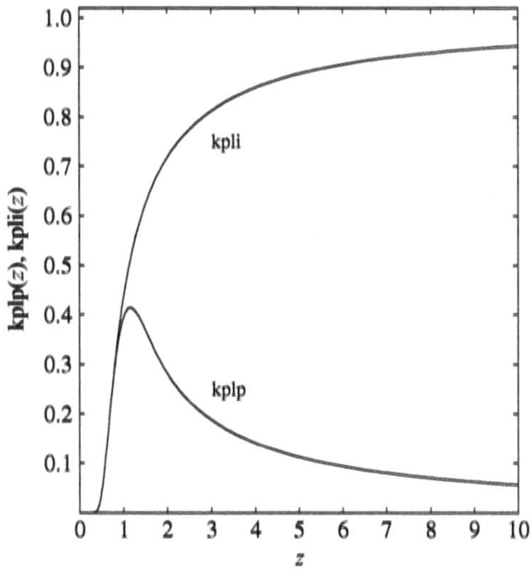

Eq. (5.115) takes a special limiting form [cf. Eq. (A.43)]:

$$u_j(0,t) = D_j^{-1/2} \mu_j^{-1} J_j(0,t) \; , \tag{5.126}$$

so that in view of Eq. (5.59)

$$c_j^{\dagger}(t) = c_j(0,t) = c_j^{\star} + D_j^{-1/2} \mu_j^{-1} p_j^{\text{het}}(t) \; . \tag{5.127}$$

Equation (5.127) is the concentration–production rate relationship for the steady state. The steady state will prevail when parameter μ_j is large, that is when the layer is very thin. No such steady state is possible in the case of an impermeable second boundary and large μ_j. However, in this limiting case we can neglect the integral corresponding to the term $[\pi(t-\tau)]^{-1/2}$ of the kernel (5.121) in Eq. (5.125), and replace the function kpli $\left[\mu_j(t-\tau)^{1/2}\right]$ by unity, which gives

$$c_j^{\dagger}(t) = c_j(0,t) \approx c_j^{\star} + D_j^{-1/2} \mu_j \int\limits_0^t p_j^{\text{het}}(\tau)\, \mathrm{d}\tau \; . \tag{5.128}$$

By multiplying Eq. (5.128) by l, defining $\Gamma_j(t) = c_j(0,t)\, l$ and $\Gamma_j^{\star} = c_j^{\star}\, l$, and taking into account Eq. (5.119), we obtain

$$\Gamma_j(t) \approx \Gamma_j^{\star} + \int\limits_0^t p_j^{\text{het}}(\tau)\, \mathrm{d}\tau \; . \tag{5.129}$$

This relationship is equivalent to Eq. (4.12) characteristic of localised (adsorbed) species, if we define $p_j^{\text{het}}(t) = \mathrm{d}\Gamma_j(t)/\mathrm{d}t$. Therefore, distributed species present in very thin planar layers with an impermeable second boundary can be formally considered as quasi-localised species. This possibility was indicated in Sect. 2.9.

The transition between Eqs. (5.125) and (5.129) was discussed by Leddy [84], with emphasis on its consequences for convolutive data analysis.

For spherical diffusion, when the second boundary is located at $r = r_0 \pm l > 0$, the solution–flux relationships are as follows. In the case of a permeable second boundary we get [cf. Eqs. (A.101) and (A.102)]:

$$u_j(r_0, t) = \int_0^t \mathscr{K}_j^{\text{exsp/insp}}(t, \tau) \left[\pm J_j(r_0, \tau)\right] \mathrm{d}\tau \;, \tag{5.130}$$

where

$$\mathscr{K}_j^{\text{exsp/insp}}(t, \tau) = \varphi_j^{\text{exsp/insp}}(t - \tau) \tag{5.131}$$

with

$$\varphi_j^{\text{exsp/insp}}(\vartheta) = \pm D_j^{-1/2}\, \mathscr{L}^{-1} \left\{ \frac{\tanh\left(\pm \frac{s^{1/2}}{\mu_j}\right)}{s^{1/2} + \varrho_j \tanh\left(\pm \frac{s^{1/2}}{\mu_j}\right)} \right\} , \tag{5.132}$$

and ϱ_j and μ_j are defined by Eqs. (5.75) and (5.119). In the case of an impermeable second boundary we get [cf. Eqs. (A.117) and (A.118)]:

$$u_j(r_0, t) = \int_0^t \mathscr{K}_j^{\text{exsi/insi}}(t, \tau) \left[\pm J_j(r_0, \tau)\right] \mathrm{d}\tau \;, \tag{5.133}$$

where

$$\mathscr{K}_j^{\text{exsi/insi}}(t, \tau) = \varphi_j^{\text{exsi/insi}}(t - \tau) \tag{5.134}$$

with

$$\varphi_j^{\text{exsi/insi}}(\vartheta) = D_j^{-1/2}\, \mathscr{L}^{-1} \left\{ \frac{s^{1/2}\left(\frac{1}{\varrho_j} \pm \frac{1}{\mu_j}\right) - \tanh\left(\pm \frac{s^{1/2}}{\mu_j}\right)}{s^{1/2}\left(\frac{\varrho_j}{\mu_j}\right) \pm \left[s\left(\frac{1}{\varrho_j} \pm \frac{1}{\mu_j}\right) - \varrho_j\right] \tanh\left(\pm \frac{s^{1/2}}{\mu_j}\right)} \right\} . \tag{5.135}$$

In Eqs. (5.132) and (5.135) the Laplace transformation is between the ϑ and s domains, and the upper/lower signs and superscipts at the kernel symbol refer to external/internal domains, respectively. It is difficult to find more transparent expressions for the kernels $\mathscr{K}_j^{\mathrm{exsp/insp}}(t,\tau)$ and $\mathscr{K}_j^{\mathrm{exsi/insi}}(t,\tau)$, as they generally depend on the two parameters ϱ_j and μ_j in a complicated, nonlinear way, so that no alternative expressions have been reported thus far.

Equations (5.131) and (5.134) are derived for the second boundary located at $r = r_0 \pm l > 0$, but the case when this boundary is located at $r = r_0 - l = 0$ is also of interest. Equation (5.131) cannot be extended to this limiting case, but the kernel $\mathscr{K}_j^{\mathrm{exsi/insi}}(t,\tau)$ possesses a limiting form, denoted here by $\mathscr{K}_j^{\mathrm{ins}}(t,\tau)$ [cf. Eq. (A.120)]:

$$\mathscr{K}_j^{\mathrm{ins}}(t,\tau) = \varphi_j^{\mathrm{ins}}(t-\tau) \ , \tag{5.136}$$

where

$$\varphi_j^{\mathrm{ins}}(\vartheta) = D_j^{-1/2}\,\mathscr{L}^{-1}\left\{\frac{\tanh\left(\frac{s^{1/2}}{\varrho_j}\right)}{s^{1/2}-\varrho_j\tanh\left(\frac{s^{1/2}}{\varrho_j}\right)}\right\} \ . \tag{5.137}$$

This form of the kernel describing internal spherical diffusion (within the entire sphere) was originally adopted in the literature [107, 153], but by defining a nonsingular special function

$$\mathrm{kins}(z) = \mathrm{kins}(\vartheta^{1/2}) = \mathscr{L}^{-1}\left\{\frac{\tanh(s^{1/2})}{s^{1/2}-\tanh(s^{1/2})} - s^{-1/2}\right\} \ , \tag{5.138}$$

where the Laplace transformation is from the domain of the variable $\vartheta = z^2$ to the s domain, it is more convenient to express the kernel $\mathscr{K}_j^{\mathrm{ins}}(t,\tau)$ by the formula analogous to the number of previously discussed kernels:

$$\mathscr{K}_j^{\mathrm{ins}}(t,\tau) = D_j^{-1/2}\left\{[\pi(t-\tau)]^{-1/2} + \varrho_j\,\mathrm{kins}\left[\varrho_j(t-\tau)^{1/2}\right]\right\} \ . \tag{5.139}$$

The name "kins" is an acronym for "<u>k</u>ernel for <u>in</u>ternal <u>s</u>pherical" diffusion (or domain). Function kins(·) is a growing function of its argument. As can be seen in Fig. 5.3, it increases from 1 (for $z = 0$) to 3 (for $z \to \infty$). This function is in some sense complementary to the function erex(·) that occurs in the kernel for external spherical diffusion [cf. Eq. (5.74)].

In view of Eq. (5.59), the concentration–production rate relationships for spherical diffusion, resulting from Eqs. (5.130) and (5.133), are as follows. For a permeable second boundary:

$$c_j^{\dagger}(t) = c_j(r_0,t) = c_j^{\star} + \int\limits_0^t \mathscr{K}_j^{\mathrm{exsp/insp}}(t,\tau)\; p_j^{\mathrm{het}}(\tau)\,\mathrm{d}\tau \ . \tag{5.140}$$

Fig. 5.3 Comparison of functions kpli(·), kinc(·), and kins(·), representing the effect of finite spatial domains on the kernels, in the case of planar, cylindrical, and spherical diffusion, respectively

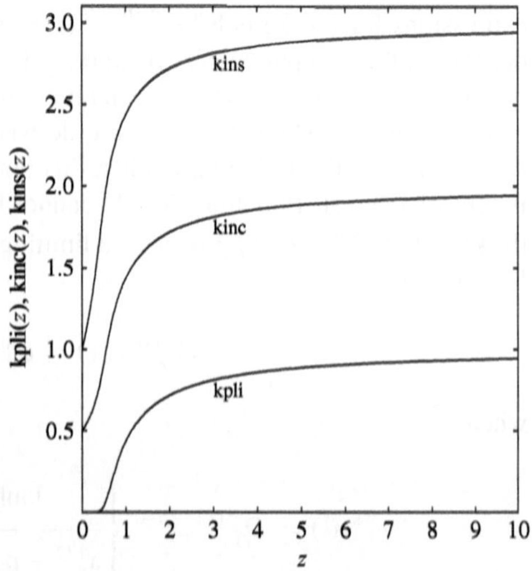

For an impermeable second boundary:

$$c_j^\dagger(t) = c_j(r_0, t) = c_j^\star + \int_0^t \mathscr{K}_j^{\text{exsi/insi}}(t, \tau)\; p_j^{\text{het}}(\tau)\, \mathrm{d}\tau\,, \tag{5.141}$$

and for its limiting case I–I with $r_0 - l \to 0$:

$$c_j^\dagger(t) = c_j(r_0, t) = c_j^\star + \int_0^t \mathscr{K}_j^{\text{ins}}(t, \tau)\; p_j^{\text{het}}(\tau)\, \mathrm{d}\tau\,. \tag{5.142}$$

A non-trivial steady state solution, satisfying $\partial u_j(r,t)/\partial t \approx 0$ is possible for a permeable second boundary located at $r = r_0 \pm l > 0$ (cases E–P and I–P). In Appendix A, Sect. A.2.2 the analytical solution of such a steady state diffusion equation is obtained, and we show that at steady state Eq. (5.130) takes a special limiting form [cf. Eq. (A.108)]:

$$u_j(r_0, t) = D_j^{-1/2}\, \frac{\varrho_j^{-1}\, \mu_j^{-1}}{\varrho_j^{-1} \pm \mu_j^{-1}}\, \left[\pm J_j(r_0, t)\right]\,, \tag{5.143}$$

so that in view of Eqs. (2.46) and (5.59)

$$c_j^\dagger(t) = c_j(r_0, t) = c_j^\star + D_j^{-1/2}\, \frac{\varrho_j^{-1}\, \mu_j^{-1}}{\varrho_j^{-1} \pm \mu_j^{-1}}\, p_j^{\text{het}}(t)\,. \tag{5.144}$$

No such steady state is possible for an impermeable second boundary, in the E–I and I–I cases. However, in the I–I case with $r_0 = l$ and large ϱ_j we can neglect the integral corresponding to the term $[\pi(t-\tau)]^{-1/2}$ of the kernel (5.139) in Eq. (5.142), and replace the term kins$\left[\varrho_j(t-\tau)^{1/2}\right]$ by 3, which gives

$$c_j^{\dagger}(t) = c_j(r_0,t) \approx c_j^{\star} + D_j^{-1/2}\, 3\mu_j \int\limits_0^t p_j^{\text{het}}(\tau)\, d\tau\,, \tag{5.145}$$

since now $\varrho_j = \mu_j$.

For cylindrical diffusion, when the second boundary is located at $r = r_0 \pm l > 0$, the solution–flux relationships are as follows. In the case of a permeable second boundary we get [cf. Eqs. (A.163) and (A.164)]:

$$u_j(r_0,t) = \int\limits_0^t \mathcal{K}_j^{\text{excp/incp}}(t,\tau)\, \left[\pm J_j(r_0,\tau)\right]\, d\tau\,, \tag{5.146}$$

with

$$\mathcal{K}_j^{\text{excp/incp}}(t,\tau) = \varphi_j^{\text{excp/incp}}(t-\tau)\,, \tag{5.147}$$

where

$$\varphi_j^{\text{excp/incp}}(\vartheta) = \pm D_j^{-1/2}\, \mathcal{L}^{-1}\left\{ \frac{\mathrm{K}_0\left(\frac{s^{1/2}}{\varrho_j}\right) - \frac{\mathrm{K}_0\left[s^{1/2}\left(\frac{1}{\varrho_j} \pm \frac{1}{\mu_j}\right)\right]}{\mathrm{I}_0\left[s^{1/2}\left(\frac{1}{\varrho_j} \pm \frac{1}{\mu_j}\right)\right]}\, \mathrm{I}_0\left(\frac{s^{1/2}}{\varrho_j}\right)}{s^{1/2}\left\{\mathrm{K}_1\left(\frac{s^{1/2}}{\varrho_j}\right) + \frac{\mathrm{K}_0\left[s^{1/2}\left(\frac{1}{\varrho_j} \pm \frac{1}{\mu_j}\right)\right]}{\mathrm{I}_0\left[s^{1/2}\left(\frac{1}{\varrho_j} \pm \frac{1}{\mu_j}\right)\right]}\, \mathrm{I}_1\left(\frac{s^{1/2}}{\varrho_j}\right)\right\}} \right\}, \tag{5.148}$$

and ϱ_j and μ_j are defined by Eqs. (5.75) and (5.119). Symbols $\mathrm{I}_0(\cdot)$ and $\mathrm{I}_1(\cdot)$ denote the modified Bessel functions of the first kind, and of orders zero and one. Symbols $\mathrm{K}_0(\cdot)$ and $\mathrm{K}_1(\cdot)$ denote the modified Bessel functions of the second kind, and of orders zero and one. In the case of an impermeable second boundary we get [cf. Eqs. (A.181) and (A.182)]:

$$u_j(r_0,t) = \int\limits_0^t \mathcal{K}_j^{\text{exci/inci}}(t,\tau)\, \left[\pm J_j(r_0,\tau)\right]\, d\tau\,, \tag{5.149}$$

with

$$\mathcal{K}_j^{\text{exci/inci}}(t,\tau) = \varphi_j^{\text{exci/inci}}(t-\tau)\,, \tag{5.150}$$

where

$$\varphi_j^{\mathrm{exci/inci}}(\vartheta) = \pm D_j^{-1/2}\,\mathscr{L}^{-1}\left\{\frac{\mathrm{K}_0\left(\frac{s^{1/2}}{\varrho_j}\right) + \frac{\mathrm{K}_1\left[s^{1/2}\left(\frac{1}{\varrho_j}\pm\frac{1}{\mu_j}\right)\right]}{\mathrm{I}_1\left[s^{1/2}\left(\frac{1}{\varrho_j}\pm\frac{1}{\mu_j}\right)\right]}\mathrm{I}_0\left(\frac{s^{1/2}}{\varrho_j}\right)}{s^{1/2}\left\{\mathrm{K}_1\left(\frac{s^{1/2}}{\varrho_j}\right) - \frac{\mathrm{K}_1\left[s^{1/2}\left(\frac{1}{\varrho_j}\pm\frac{1}{\mu_j}\right)\right]}{\mathrm{I}_1\left[s^{1/2}\left(\frac{1}{\varrho_j}\pm\frac{1}{\mu_j}\right)\right]}\mathrm{I}_1\left(\frac{s^{1/2}}{\varrho_j}\right)\right\}}\right\}. \tag{5.151}$$

In Eqs. (5.148) and (5.151) the Laplace transformation is between the ϑ and s domains, and the upper/lower signs and superscripts at the kernel symbol refer to external/internal domains, respectively. It is difficult to find more transparent expressions for the kernels $\mathscr{K}_j^{\mathrm{excp/incp}}(t,\tau)$ and $\mathscr{K}_j^{\mathrm{exci/inci}}(t,\tau)$, as they generally depend on the two parameters ϱ_j and μ_j in a complicated, nonlinear way, so that no alternative expressions have been reported thus far.

Equations (5.147) and (5.150) are derived for the second boundary located at $r = r_0 \pm l > 0$, but the case when this boundary is located at $r = r_0 - l = 0$ is also of interest. Equation (5.147) cannot be extended to this limiting case, but the kernel $\mathscr{K}_j^{\mathrm{exci/inci}}(t,\tau)$ possesses a limiting form, denoted here by $\mathscr{K}_j^{\mathrm{inc}}(t,\tau)$ [cf. Eq. (A.186)]:

$$\mathscr{K}_j^{\mathrm{inc}}(t,\tau) = \varphi_j^{\mathrm{inc}}(t-\tau)\,, \tag{5.152}$$

where

$$\varphi_j^{\mathrm{inc}}(\vartheta) = D_j^{-1/2}\,\mathscr{L}^{-1}\left\{\frac{\mathrm{I}_0\left(\frac{s^{1/2}}{\varrho_j}\right)}{s^{1/2}\,\mathrm{I}_1\left(\frac{s^{1/2}}{\varrho_j}\right)}\right\}. \tag{5.153}$$

This form of the kernel describing internal cylindrical diffusion (within the entire cylinder) was originally adopted in the literature [157], but by defining a nonsingular special function

$$\mathrm{kinc}(z) = \mathrm{kinc}(\vartheta^{1/2}) = \mathscr{L}^{-1}\left\{\frac{\mathrm{I}_0\left(s^{1/2}\right) - \mathrm{I}_1\left(s^{1/2}\right)}{s^{1/2}\,\mathrm{I}_1\left(s^{1/2}\right)}\right\}, \tag{5.154}$$

where the Laplace transformation is from the domain of the variable $\vartheta = z^2$ to the s domain, it is more convenient to express the kernel $\mathscr{K}_j^{\mathrm{inc}}(t,\tau)$ by the formula analogous to the number of previously discussed kernels:

$$\mathscr{K}_j^{\mathrm{inc}}(t,\tau) = D_j^{-1/2}\left\{[\pi(t-\tau)]^{-1/2} + \varrho_j\,\mathrm{kinc}\left[\varrho_j (t-\tau)^{1/2}\right]\right\}. \tag{5.155}$$

The name "kinc" is an acronym for "kernel for internal cylindrical" diffusion (or domain). Function kinc(z) is a growing function of its argument z. As can be seen in Fig. 5.3, it increases from 1/2 (for $z = 0$) to 2 (for $z \to \infty$). This function is in some sense complementary to the function kcylw($\cdot$) that occurs in the kernel for external cylindrical diffusion [cf. Eq. (5.85)].

In view of Eq. (5.59), the concentration–production rate relationships for cylindrical diffusion, resulting from Eqs. (5.146) and (5.149), are as follows. For a permeable second boundary:

$$c_j^\dagger(t) = c_j(r_0, t) = c_j^\star + \int_0^t \mathcal{K}_j^{\mathrm{excp/incp}}(t, \tau)\ p_j^{\mathrm{het}}(\tau)\ \mathrm{d}\tau\ . \tag{5.156}$$

For an impermeable second boundary:

$$c_j^\dagger(t) = c_j(r_0, t) = c_j^\star + \int_0^t \mathcal{K}_j^{\mathrm{exci/inci}}(t, \tau)\ p_j^{\mathrm{het}}(\tau)\ \mathrm{d}\tau\ , \tag{5.157}$$

and for its limiting case I–I with $r_0 - l \to 0$:

$$c_j^\dagger(t) = c_j(r_0, t) = c_j^\star + \int_0^t \mathcal{K}_j^{\mathrm{inc}}(t, \tau)\ p_j^{\mathrm{het}}(\tau)\ \mathrm{d}\tau\ . \tag{5.158}$$

A non-trivial steady state solution, satisfying $\partial u_j(r,t)/\partial t \approx 0$ is possible for a permeable second boundary, located at $r = r_0 \pm l > 0$ (cases E–P and I–P). In Appendix A, Sect. A.3.2 the analytical solution of such a steady state diffusion equation is obtained, and we show that at steady state Eq. (5.146) takes a special limiting form [cf. Eq. (A.172)]:

$$u_j(r_0, t) = \mp D_j^{-1/2}\, \varrho_j^{-1}\, \ln\left(\frac{\varrho_j^{-1}}{\varrho_j^{-1} \pm \mu_j^{-1}}\right) \left[\pm J_j(r_0, t)\right]\ , \tag{5.159}$$

so that in view of Eqs. (2.46) and (5.59)

$$c_j^\dagger(t) = c_j(r_0, t) = c_j^\star \mp D_j^{-1/2}\, \varrho_j^{-1}\, \ln\left(\frac{\varrho_j^{-1}}{\varrho_j^{-1} \pm \mu_j^{-1}}\right) p_j^{\mathrm{het}}(t)\ . \tag{5.160}$$

No such steady state is possible for an impermeable second boundary, in the E–I and I–I cases. However, in the I–I case with $r_0 = l$ and large ϱ_j we can neglect the integral corresponding to the term $[\pi(t - \tau)]^{-1/2}$ of the kernel (5.155) in Eq. (5.158), and replace the term $\mathrm{kinc}\left[\varrho_j (t-\tau)^{1/2}\right]$ by 2, which gives

$$c_j^{\dagger}(t) = c_j(r_0,t) \approx c_j^{\star} + D_j^{-1/2}\, 2\mu_j \int\limits_0^t p_j^{\mathrm{het}}(\tau)\, \mathrm{d}\tau\,, \tag{5.161}$$

since now $\varrho_j = \mu_j$.

It is interesting to compare contributions to the kernels, resulting from finite spatial domains with an impermeable second boundary in the planar, cylindrical, and spherical geometry (assuming internal diffusion within the entire cylinder or sphere in the latter two cases, which means $\varrho_j = \mu_j$). This can be done by comparing functions kpli(z), kinc(z) and kins(z). Figure 5.3 reveals that for a particular μ_j the smallest effect of the finite domain occurs in the planar case, the largest in the spherical case, and the cylindrical case is intermediate. At large μ_j the effect differs only by the multiplicative factor in Eqs. (5.128), (5.145), and (5.161), which equals 1 for planar diffusion, 2 for cylindrical, and 3 for spherical.

5.3.2 *Literature Examples*

The following list of examples is limited to IE-based models in which diffusion uncomplicated by convection, homogeneous reactions, or other phenomena was assumed, together with one-dimensional finite spatial domains. Models involving such complications are listed in further Chaps. 6, 8, 9, and 10.

5.3.2.1 Planar Diffusion

For the diffusion in a finite layer with a permeable second boundary there are several examples of IEs, or closely related equations, motivated by the theory of transient experiments at the RDE, in which a semi-infinite spatial domain is divided into a thin diffusion layer and a stirred outer domain. Fried and Elving [58] derived a general current–potential relationship at the RDE, for the reaction

$$X_1 + n\,e^- \rightleftarrows X_2 \tag{5.162}$$

involving dynamic distributed species X_1 and X_2. The theory of AC voltammetry was presented by Tokuda and Matsuda [150–152], assuming equilibrium, non-equilibrium reversible, and irreversible reaction (5.162). The theory of linear potential sweep voltammetry was provided by Andricacos and Cheh [4, 5] and Quintana et al. [128], at analogous assumptions. Apart from the above studies devoted to the RDE, Andrieux et al. [9] considered repetitive cyclic voltammetry for irreversible reaction (5.162) in a finite layer resulting from arbitrary natural or forced convection. Miles and Compton [99] modelled square wave voltammetry for equilibrium, non-equilibrium reversible, and irreversible reaction (5.162), assuming similar transport conditions. Daruházi et al. [39] presented a model of cyclic voltammetry for equilibrium reaction (5.162) occurring in a thin layer cell with closely separated working and auxiliary electrodes, which is mathematically equivalent to a thin electrolyte layer with a permeable second boundary.

There are also several examples of IEs for the diffusion in a finite layer with an impermeable second boundary. One group of such examples is connected with the modelling of the reaction

$$X_1 + n\,e^- \rightleftarrows X_2 \tag{5.163}$$

involving dynamic species X_1 distributed in an electrolyte, and dynamic species X_2 distributed in a mercury film electrode. IEs describing linear potential sweep voltammetry for reaction (5.163) were presented and solved by De Vries and Van Dalen [45, 48, 49], Bakanov et al. [18], Seelig and Blount [134], Donten et al. [52, 53], and Schiewe et al. [133]. De Vries [47] considered a potential step chronoamperometry followed by linear potential sweep voltammetry. Osteryoung and Christie discussed pulsed voltammetric [120] and staircase voltammetric [37] stripping. Goto et al. [66] analysed semidifferentiated voltammetric current. Kounaves et al. [82, 83] considered square wave voltammetry. Bos [32] presented a rare example of IEs describing current step chronopotentiometry for two parallel reactions:

$$X_1 + n_1\,e^- \rightleftarrows X_2\,, \tag{5.164}$$

$$X_3 + n_2\,e^- \rightleftarrows X_4\,, \tag{5.165}$$

analogous to reaction (5.163).

A finite layer with an impermeable second boundary occurred also in the work by Bucur et al. [33], who analysed linear potential sweep voltammetry for the reaction

$$X_1 \rightleftarrows X_{2,s} + n\,e^-\,. \tag{5.166}$$

Here X_1 is a dynamic species distributed in a thin metal membrane (hydrogen in palladium), and $X_{2,s}$ is a static species (hydrogen ion in an electrolyte).

Further examples involving an impermeable second boundary refer to the reaction

$$X_1 + n\,e^- \rightleftarrows X_2\,, \tag{5.167}$$

in which both dynamic distributed species X_1 and X_2 belong to a thin layer. Aoki et al. [12, 13] presented the theory of linear potential sweep voltammetry for equilibrium, non-equilibrium reversible and irreversible reaction (5.167). Zamponi et al. [163] extended their results to derivative linear sweep voltabsorptometry.

Lovrić et al. [90] discussed staircase voltammetry. Mirčeski et al. [100, 104] considered square wave voltammetry.

Pnev et al. [124, 126] discussed linear potential sweep voltammetry for the non-equilibrium reversible and irreversible electrodeposition reaction

$$X_1 + n\,e^- \rightleftarrows X_{2,s}\,, \tag{5.168}$$

where X_1 is a dynamic species distributed in a thin electrolyte layer, and $X_{2,s}$ is a static species (deposit). The potential sweep was started from the equilibrium potential of the reaction (5.168).

The IE and IDE formalism for diffusion in a finite layer with an impermeable second boundary was also utilised by Jones et al. [73] in a biosensor modelling study.

5.3.2.2 Spherical Diffusion

Nigmatullin [113] modelled linear potential sweep voltammetry for the reaction:

$$X_1 + n\,e^- \rightleftarrows X_2 \tag{5.169}$$

involving dynamic distributed species X_1 (in the electrolyte) and X_2 (in the electrode), assuming the case I–I with $l < r_0$ and $l = r_0$. Keller and Reinmuth [75] derived corrections, to the linear potential sweep voltammetric current for reaction (5.169), resulting from spherical diffusion in finite layers. Reaction (5.169) involved dynamic species X_1 and X_2, distributed in the same phase, or in two different phases adjacent to the interface studied. Tokuda et al. [149, 153] discussed cyclic voltammetry for reaction (5.169) with species X_1 distributed in the electrolyte and X_2 in the electrode, assuming the case I–I with $l = r_0$. Equilibrium reaction (5.169) was assumed. Analogous equations of linear potential sweep voltammetry for a non-equilibrium reversible reaction (5.169) were derived by Zakharov and Pneva [162], whereas Park et al. [122] reported equations for double potential step chronoamperometry under limiting current conditions. Engblom and Oldham [54] discussed the problem of the reshaping of the staircase voltammograms into classical wave shapes characteristic of polarograms. Fatouros et al. [57] presented approximate IEs describing square wave and normal pulse voltammetries.

5.3.2.3 Cylindrical Diffusion

Keller and Reinmuth [75] derived corrections, to the linear sweep voltammetric current for the single reaction:

$$X_1 + n\,e^- \rightleftarrows X_2\,, \tag{5.170}$$

resulting from cylindrical diffusion in finite layers. Reaction (5.170) involved dynamic species X_1 and X_2, distributed in the same phase, or in two different phases adjacent to the interface studied. Weidner [157] presented the theory of linear potential sweep voltammetry for reaction (5.170), applicable to a cylindrical pore (tube) electrode, that is for the case I–I with $r_0 = l$, and both species belonging to the electrolyte phase. Only species X_1 was initially present. Equilibrium, non-equilibrium reversible, and irreversible reactions (5.170) were considered. Equal diffusion coefficients of X_1 and X_2 were assumed. An analogous theory was obtained by Neudeck and Dunsch [109, 110], but for cyclic voltammetry [109], and potential step chronoamperometric and UV-visible spectroelectrochemical experiments [110]. The theory served as an approximation to the more complicated problem of diffusion in honeycombed LIGA-structured electrodes.

5.4 Anomalous Diffusion

Equations presented in Sects. 5.2 and 5.3 are not applicable in the presence of anomalous diffusion effects. In particular, the concentration–production rate relationships have to be modified.

5.4.1 Concentration–Production Rate Relationship

For the (macroscopically) one-dimensional, planar diffusion in a semi-infinite domain, Pajkossy and Nyikos [121] proposed a formal concentration–production rate relationship representing the anomalous diffusion effects due to the irregular electrode surface. The equation proposed was based on their semi-empirical model of potential step chronoamperometry at fractal electrodes, described in Sect. 2.12. Pajkossy and Nyikos assumed that the relevant concentration–production rate relationship should be analogous to that for ordinary planar diffusion in a semi-infinite domain, given by Eq. (5.86), that is:

$$c_j^{\dagger}(t) = c_j(0,t) = c_j^{\star} + \int_0^t \mathscr{K}_j^{\mathrm{ap}}(t,\tau)\; p_j^{\mathrm{het}}(\tau)\; \mathrm{d}\tau\,, \tag{5.171}$$

where $\mathscr{K}_j^{\mathrm{ap}}(t,\tau)$ is a suitable convolution kernel to be determined. As it is a convolution kernel, we can write $\mathscr{K}_j^{\mathrm{ap}}(t,\tau) = \varphi_j(t-\tau)$, where $\varphi_j(\vartheta)$ with $\vartheta = t - \tau$ is a function of one variable. After Laplace transformation (3.12), Eq. (5.171) becomes

$$\hat{c}_j(0,s) = \frac{c_j^{\star}}{s} + \hat{\varphi}_j(s)\; \hat{p}_j^{\mathrm{het}}(s)\,. \tag{5.172}$$

Pajkossy and Nyikos postulated that the function $\varphi_j(\vartheta)$ should be such that Eq. (5.171) correctly predicts the model formula (2.63) for the chronoamperometric limiting current for the reaction $X_j + ne^- \rightleftarrows X_{j+1}$. In such a case, the production rate $p_j^{het}(t)$ (equal to the flux of the species X_j at the electrode) is related to the Faradaic current by

$$p_j^{het}(t) = J_j(0,t) = \frac{I(t)}{nFA} . \tag{5.173}$$

In addition, under limiting current conditions $\hat{c}_j(0,s)$ is zero, so that Eq. (5.172) combined with the Laplace transform of Eq. (5.173) gives

$$\hat{\varphi}_j(s) = \frac{-nFAc_j^\star}{s\,\hat{I}(s)} . \tag{5.174}$$

The Laplace transform of Eq. (2.63) is

$$\hat{I}(s) = -\sigma_{j,F}\,\Gamma(1-\alpha)\,s^{\alpha-1} , \tag{5.175}$$

where parameter α is defined by Eq. (2.64), parameter $\sigma_{j,F}$ is given by Eqs. (2.65) and (2.66), and $\Gamma(\cdot)$ is the Euler gamma function [1]. Setting Eq. (5.175) into Eq. (5.174) gives

$$\hat{\varphi}_j(s) = \frac{nFAc_j^\star}{\sigma_{j,F}\,\Gamma(1-\alpha)}\,s^{-\alpha} , \tag{5.176}$$

so that after inverse transformation

$$\varphi_j(\vartheta) = \frac{nFAc_j^\star}{\sigma_{j,F}\,\Gamma(\alpha)\Gamma(1-\alpha)}\,\vartheta^{\alpha-1} . \tag{5.177}$$

Consequently, taking into account Eq. (2.66) for $\sigma_{j,F}$ we get

$$\mathscr{K}_j^{ap}(t,\tau) = D_j^{\alpha-1}\left(\gamma\,\lambda_{out}^2\right)^{1/2-\alpha}\frac{\Gamma(1/2)}{\Gamma(\alpha)\Gamma(1-\alpha)}\,(t-\tau)^{\alpha-1} . \tag{5.178}$$

It is easy to check that for $\alpha = 1/2$ (which corresponds to the ordinary diffusion at the planar surface of dimension $d_F = 2$) Eq. (5.178) reduces to Eq. (5.72) for the kernel $\mathscr{K}_j^p(t,\tau)$.

It seems important to point out that Eq. (5.178) must be considered as an approximation. The equation is tacitly assumed to hold for any $t > 0$, whereas strictly speaking anomalous diffusion occurs only for $t \in (t_{j,inn}, t_{j,out})$ (cf. Sect. 2.12).

Apart from the above derivation, there have been interesting attempts by Dassas and Duby [40], and Mirčeski and Tomovski [102] to deduce the relationship (5.171) with the kernel proportional to $(t-\tau)^{\alpha-1}$, by solving the so-called time-fractional diffusion equations. The time-fractional diffusion equations are formally similar to the ordinary diffusion PDEs, but they have the temporal derivative replaced by a fractional derivative, understood in the sense of the fractional calculus [118]. Mirčeski and Tomovski also followed this route to obtain the concentration–production rate relationship for planar anomalous diffusion in a finite spatial domain [103]. However, it seems that this approach raises a number of yet unresolved questions regarding the physical interpretation of the fractional diffusion equations. There are also some mathematical difficulties regarding the handling and understanding of initial conditions in such equations (see, for example, the discussion by Bisquert and Compte [27]). For these reasons we do not expose this approach further in the present book.

5.4.2 Literature Examples

Strømme et al. [144] used Eqs. (5.171) and (5.178) for obtaining IEs describing cyclic voltammetric experiments at fractal electrodes. Theoretical predictions obtained by solving these IEs were applied to determine the fractal dimension of the electrode surfaces [145]. Mirčeski and Tomovski [102] also obtained detailed solutions of the IEs for cyclic voltammetry assuming kernels analogous to Eq. (5.178), and a semi-infinite spatial domain. In a later work [103], they extended these results onto a finite spatial domain (but as was noted above, there are some unresolved questions regarding their derivations, related to the assumed formalism of the equations of fractional diffusion).

Apart from the above examples, IEs involving kernels having the form (5.178), but strongly singular, were used in the modelling of impedance spectroscopy for equivalent circuits involving constant phase elements [155].

5.5 Benefits from Using the IE Method

In the case of models dependent on one spatial coordinate, such as the models involving one-dimensional diffusion, discussed in this chapter, the use of the IE method may bring important benefits, compared to the direct PDE solving. Firstly, the replacement of the PDEs by the IEs, as described in Sect. 5.1, eliminates the need to determine the spatial concentration profiles. The IEs depend on one independent variable only (variable t), typically representing time. Consequently, the numerical solution of the IEs is likely to be computationally less expensive than the direct numerical solution of the PDEs, since in the latter case the numerical calculation of the spatial concentration profiles is necessary. The actual reduction of the computational cost depends on the numerical techniques, accuracy requirements, and the length of the interval of t. The latter factor is important, because as we shall see in Chap. 12, typical numerical methods for solving IEs are characterised by the computational time proportional to t^2, whereas the computing time in discrete PDE solving techniques is typically proportional to t. Therefore, for very large t intervals the computational cost advantage of the IE-based simulations may be lost.

Secondly, the partially analytical solution of the PDEs, inherent in the IE method, allows one to gain more insights into the solutions obtained, compared to the direct numerical PDE solutions. To illustrate this aspect of the IE method by an example, let us have a closer look at the standard Eq. (5.23) for the equilibrium electron transfer (5.7). Assume, for simplicity, planar diffusion in a semi-infinite domain, and let only the species X_1 be initially present. In such a case $\mathscr{K}_j(t,\tau) = \mathscr{K}_j^{\mathrm{p}}(t,\tau) = D_j^{-1/2}[\pi(t-\tau)]^{-1/2}$ (for $j = 1, 2$), and Eq. (5.23) becomes, after some rearrangement, the following Abel IE:

$$\int_0^t [\pi(t-\tau)]^{-1/2}\,\Psi(\tau)\,\mathrm{d}\tau = \frac{c_1^\star D_1^{1/2}}{1+\exp\left\{\frac{nF}{RT}\left[E(t)-E_{1/2}\right]\right\}}, \tag{5.179}$$

where

$$E_{1/2} = E^0 - \frac{RT}{nF}\ln\left(\frac{D_1}{D_2}\right)^{1/2} \tag{5.180}$$

is known as the half-wave potential (for the equilibrium electron transfer). As can be seen, Eqs. (5.179) and (5.180) predict that the effect of the diffusion coefficient ratio on the Faradaic current (proportional to $\Psi(t)$), amounts to a shift by $RT(nF)^{-1}\ln(D_1/D_2)^{1/2}$, along the electrode potential axis, of the current–potential curve. This exact analytical result would be rather difficult to reveal by means of the direct numerical PDE solving.

Other examples of additional insights, offered by the IE method, can be the various analytical predictions for steady state or other limiting cases, easily obtainable by the method (see, in particular, Sects. 11.1.5 and A.2.1). Direct numerical PDE solving does not provide such predictions.

References

1. Abramowitz M, Stegun IA (1972) Handbook of mathematical functions. Dover Publications, New York
2. Affoune AM, Abedelkader S, Bouteillon J, Poignet JC (2007) Electrochemical behaviour of silver and rhenium electrodes in molten alkali fluorides. J Appl Electrochem 37:155–160
3. Amatore C, Savéant JM, Tessier D (1983) Charge transfer at partially blocked surfaces. A model for the case of microscopic active and inactive sites. J Electroanal Chem 147:39–51
4. Andricacos PC, Cheh HY (1980) The application of linear sweep voltammetry to a rotating disk electrode for a reversible reaction with soluble product. J Electrochem Soc 127:2385–2388

5. Andricacos PC, Cheh HY (1981) The application of linear sweep voltammetry to a rotating disk electrode for a first-order irreversible reaction. J Electroanal Chem 124:95–101
6. Andrieux CP, Merz A, Savéant JM (1985) Dissociative electron transfer. Autocatalysis and kinetic characteristics of the electrochemical reductive cleavage of the carbon-halogen bond in alkyl halides giving rise to stable radicals and carbanions. J Am Chem Soc 107:6097–6103
7. Andrieux CP, Savéant JM, Su KB (1986) Kinetics of dissociative electron transfer. Direct and mediated electrochemical reductive cleavage of the carbon-halogen bond. J Phys Chem 90:3815–3823
8. Andrieux CP, Le Gorande A, Savéant JM (1992) Electron transfer and bond breaking. Examples of passage from a sequential to a concerted mechanism in the electrochemical reductive cleavage of arylmethyl halides. J Am Chem Soc 114:6892–6904
9. Andrieux CP, Hapiot P, Savéant JM (1993) Repetitive cyclic voltammetry of irreversible systems. J Electroanal Chem 349:299–309
10. Aoki K (2005) Is voltammetric current proportional to the number of transferred electrons for multi-charged ions or to 3/2 power of the number? Electroanalysis 17:1379–1383
11. Aoki K, Kaneko H (1988) Theory of irreversible cyclic voltammograms at microcylinder electrodes. J Electroanal Chem 247:17–27
12. Aoki K, Tokuda K, Matsuda H (1983) Theory of linear sweep voltammetry with finite diffusion space. J Electroanal Chem 146:417–424
13. Aoki K, Tokuda K, Matsuda H (1984) Theory of linear sweep voltammetry with finite diffusion space. Part II. Totally irreversible and quasi-reversible cases. J Electroanal Chem 160:33–45
14. Aoki K, Tokuda K, Matsuda H (1984) Theory of differential pulse voltammetry at stationary planar electrodes. J Electroanal Chem 175:1–13
15. Aoki K, Honda K, Tokuda K, Matsuda H (1985) Voltammetry at microcylinder electrodes. Part I. Linear sweep voltammetry. J Electroanal Chem 182:267–279
16. Aoki K, Tokuda K, Matsuda H (1986) Voltammetry at microcylinder electrodes. Part V. Pulse voltammetric current–potential curves for quasi-reversible and totally irreversible electrode reactions. J Electroanal Chem 206:47–56
17. Avaca LA, Kaufmann S, Kontturi K, Murtomäki L, Schiffrin DJ (1993) Theory of cyclic voltammetry for quasi-reversible electrodeposition reactions with insoluble products. Ber Bunsenges Phys Chem 97:70–76
18. Bakanov VI, Zakharov MS, Antipyeva VA (1975) Theory of irreversible electrode processes in inversion polarography on a mercury film electrode. Elektrohim 11:689–693 (in Russian)
19. Bell CG, Anastassiou CA, O'Hare D, Parker KH, Siggers JH (2011) Large-amplitude AC voltammetry: theory for reversible redox reactions in the "slow scan limit approximation". Electrochim Acta 56:6131–6141
20. Bell CG, Anastassiou CA, O'Hare D, Parker KH, Siggers JH (2011) Theory of large-amplitude sinusoidal voltammetry for reversible redox reactions. Electrochim Acta 56:8492–8508
21. Bieniasz LK (1991) Kinetics of the oxygen electrode reaction in molten (Li / Na) carbonate eutectic. Part 2. Theory of linear scan voltammetry and potential-step chronoamperometry for the reaction $m\mathrm{O} + ne^- \rightleftharpoons q\mathrm{R}$ initially at equilibrium. J Electroanal Chem 304:101–109
22. Bieniasz LK (2011) A highly accurate, inexpensive procedure for computing integral transformation kernel and its moment integrals for cylindrical wire electrodes. J Electroanal Chem 661:280–286
23. Bieniasz LK (2012) Automatic simulation of electrochemical transients assuming finite diffusion space at planar interfaces, by the adaptive Huber method for Volterra integral equations. J Electroanal Chem 684:20–31
24. Bieniasz L, Suski L (1983) Kinetics of anodic processes in molten carbonate fuel cell. Part II. Equations for non-stationary diffusion controlled second order electrochemical process: $\mathrm{Red} + \mathrm{A} \rightleftharpoons \mathrm{Ox}_1 + \mathrm{Ox}_2 + ne$. Pol J Chem 57:161–168

25. Bieniasz LK, González J, Molina Á, Laborda E (2010) Theory of linear sweep/cyclic voltammetry for the electrochemical reaction mechanism involving a redox catalyst couple attached to a spherical electrode. Electrochim Acta 56:543–552
26. Birke RL (1973) Generalization of the semioperator approach for electrochemical problems involving diffusion. Anal Chem 45:2292–2294
27. Bisquert J, Compte A (2001) Theory of the electrochemical impedance of anomalous diffusion. J Electroanal Chem 499:112–120
28. Bonciocat N (1993) Application of integral Volterra-equations in the study of new nonstationary DC and AC methods for investigating electrochemical reactions. Sov Electrochem 29:128–136
29. Bonciocat N, Cotârţă A (1998) Electrode redox processes occurring with nonstationary interfacial generation of the electrochemical active species. I. The kinetic model and the general functional equation of the current density. Rev Roum Chim 43:925–933
30. Bonciocat N, Papadopol E, Borca S, Marian IO (2000) A new theoretical approach to the voltammetry with linear scanning of the potential. II. Comparison with the standard approach given in the literature. Rev Roum Chim 45:1047–1053
31. Bond AM, O'Halloran RJ, Ruzic I, Smith DE (1976) Fundamental and second harmonic alternating current cyclic voltammetric theory and experimental results for simple electrode reactions involving solution-soluble redox couples. Anal Chem 48:872–883
32. Bos P (1971) Chronopotentiometry of mixtures. 1. Reduction at a mercury pool and at a mercury-film electrode. J Electroanal Chem 33:379–391
33. Bucur RV, Covaci I, Miron C (1967) Voltametrie a tension lineairement variable, en couche mince rigide. I. L'oxydation de l'hydrogene dissous dans une couche mince de palladium, en solution agitee. J Electroanal Chem 13:263–274
34. Camacho L, Ruiz JJ, Serna C, Molina A, Martínez-Ortiz F (1994) Reverse pulse voltammetry and polarography: a general analytical solution. Can J Chem 72:2369–2377
35. Carslaw HS, Jaeger JC (1959) Conduction of heat in solids. Clarendon Press, Oxford
36. Chen XH, Liu SL, Chen JH, Du M, Lin XH (2012) Numerical simulation based accurate models of improved chronoamperometry and its experimental validation. Acta Phys Chim Sin 28:877–884 (in Chinese)
37. Christie JH, Osteryoung RA (1976) Theoretical treatment of staircase voltammetric stripping from the thin film mercury electrode. Anal Chem 48:869–872
38. Danielsson R, Baecklund P, Källström A (1984) The applicability of the superposition principle in differential pulse polarography. Anal Chim Acta 158:305–322
39. Daruházi L, Tokuda K, Farsang G (1989) Cyclic voltammetry for reversible redox-electrode reactions in thin-layer cells with closely separated working and auxiliary electrodes of the same size. J Electroanal Chem 264:77–89
40. Dassas Y, Duby P (1995) Diffusion toward fractal interfaces. Potentiostatic, galvanostatic, and linear sweep voltammetric techniques. J Electrochem Soc 142:4175–4180
41. Delahay P (1953) Theory of irreversible waves in oscillographic polarography. J Am Chem Soc 75:1190–1196
42. Delahay P, Senda M, Weis CH (1961) Faradaic rectification and electrode processes. J Am Chem Soc 83:312–322
43. Delmastro JR, Smith DE (1965) Influence of mercury drop growth and geometry on the A.C. polarographic wave. J Electroanal Chem 9:192–217
44. Delmastro JR, Smith DE (1966) Influence of spherical diffusion on the alternating current polarographic wave. Anal Chem 38:169–179
45. De Vries WT (1965) Exact treatment of anodic stripping voltammetry with a plane mercury-film electrode. J Electroanal Chem 9:448–456
46. De Vries WT (1967) Potential-step electrolysis followed by linear-sweep voltammetry. J Electroanal Chem 14:75–81
47. De Vries WT (1968) Potential-step electrolysis followed by linear-sweep voltammetry at a plane mercury-film electrode. J Electroanal Chem 16:295–312

48. De Vries WT, Van Dalen E (1964) Theory of anodic stripping voltammetry with a plane, thin mercury-film electrode. J Electroanal Chem 8:366–377
49. De Vries WT, Van Dalen E (1967) Linear potential-sweep voltammetry at a plane mercury-film electrode. J Electroanal Chem 14:315–327
50. Diao G, Zhang Z (1996) Theory and application of cyclic voltammetry at a hemispherical microelectrode for a quasi-reversible reaction. J Electroanal Chem 410:155–162
51. Diard JP, Le Gorrec B, Montella C (2000) O_2 mass-transfer function for polymer-filmed metal electrodes. Application to the linear sweep voltammetry method. J Electrochem Soc 147:296–307
52. Donten M, Kublik Z (1985) Theory of cyclic voltammetry at the mercury film electrode. Reversible case with substrate initially present in the mercury phase. J Electroanal Chem 195:251–261
53. Donten M, Stojek Z, Kublik Z (1984) Theory of cyclic voltammetry at the mercury film electrode. Reversible case with substrate present initially in aqueous solution. J Electroanal Chem 163:11–21
54. Engblom SO, Oldham KB (1990) Reshaping of staircase voltammograms by discrete spherical convolution. Anal Chem 62:625–630
55. Farsang G, Rozsondai B, Tomcsányi L (1970) A voltametriás redoxilépcső nyugvó elektródon, I. A voltametriás redoxilépcső elméleti áram–potenciál–egyenlete. Magy Kém Foly 76:233–236 (in Hungarian)
56. Fatouros N, Krulic D (1998) Contribution to the study of square wave voltammetry in the case of simple irreversible reactions. J Electroanal Chem 443:262–265
57. Fatouros N, Krulic D, López-Tenés M, El Belamachi MM (1996) Single and multiple pulse voltammetries at the static mercury drop electrode for reversible reactions with and without amalgam formation. J Electroanal Chem 405:197–204
58. Fried I, Elving PJ (1965) General equation for current–potential relationships at rotating disk electrode. Anal Chem 37:464
59. Fujioka J (1987) Integral equations in linear sweep voltammetry of irreversible reactions with adsorption at stationary and rotating disk electrodes. J Electrochem Soc 134:1446–1450
60. Garreau D, Savéant JM, Tessier D (1979) A computer coupled high frequency A.C. technique for investigating fast non-Volmerian electron transfers. J Electroanal Chem 103:321–333
61. Gokhshtein AYa, Gokhshtein YaP (1960) Kinetic equation for irreversible reactions in oscillographic polarography. Dokl Akad Nauk SSSR 131:601–604 (in Russian)
62. Gokhshtein YaP (1959) General equations of oscillographic polarography, reversible processes at the cathodic and anodic polarisation. Dokl Akad Nauk SSSR 126:598–601 (in Russian)
63. Gokhshtein YaP, Gokhshtein AYa (1959) Consecutive electrochemical reactions in oscillographic polarography. Dokl Akad Nauk SSSR 128:985–987 (in Russian)
64. Gokhshtein YaP, Gokhshtein AYa (1960) Multistage electrochemical reactions in oscillographic polarography. In: Longmuir IS (ed) Advances in polarography, vol 2, Pergamon Press, Oxford, pp 465–481
65. Goodisman J (1983) General correction for electrode sphericity in voltammetry of Nernstian systems. J Electroanal Chem 144:33–43
66. Goto M, Ikenoya K, Ishii D (1979) Anodic stripping semidifferential electroanalysis with thin mercury film electrode formed in situ. Anal Chem 51:110–115
67. Gueshi T, Tokuda K, Matsuda H (1979) Voltammetry at partially covered electrodes. Part II. Linear potential sweep and cyclic voltammetry. J Electroanal Chem 101:29–38
68. Henstridge MC, Laborda E, Dickinson EJF, Compton RG (2012) Redox systems obeying Marcus-Hush-Chidsey electrode kinetics do not obey the Randles-Ševčik equation for linear sweep voltammetry. J Electroanal Chem 664:73–79
69. Hung HL, Smith DE (1966) Alternating current polarography with multi-step charge transfer. I. Theory for systems with reversible two-step charge transfer. J Electroanal Chem 11:237–254

70. Hung HL, Smith DE (1966) Alternating current polarography with multi-step charge transfer. II. Theory for systems with quasi-reversible two-step charge transfer. J Electroanal Chem 11:425–461
71. Hurwitz HD (1964) Propriétés générales des méthodes chronopotentiométriques en régime de diffusion cylindrique et sphérique. Détermination du courant compensant les effets liés au caractère non linéaire du mode de diffusion. J Electroanal Chem 7:368–381
72. Ito H, Hasegawa Y, Ito Y (2011) Numerical simulation of cyclic voltammetry for reversible systems with complex stoichiometry. Russ J Electrochem 47:1006–1015
73. Jones S, Jumarhon B, McKee S, Scott JA (1996) A mathematical model of a biosensor. J Eng Math 30:321–337
74. Katano H, Tatsumi H, Senda M (2004) Ion-transfer voltammetry at 1,6-dichlorohexane | water and 1,4-dichlorobutane | water interfaces. Talanta 63:185–193
75. Keller HE, Reinmuth WH (1972) Potential scan voltammetry with finite diffusion—a unified theory. Anal Chem 44:434–442
76. Kohler H, Piron DL, Bélanger G (1987) A linear sweep voltammetry theory for irreversible electrode reactions with an order of one or higher. I. Mathematical formulation. J Electrochem Soc 134:120–125
77. Kokhanovskii IN, Tkalenko DA, Bezdetnyi BP, Gorodyskii AV (1983) Chronovoltamperometry of complex electrode processes in molten salts. Peak potential in the case of the formation of weakly soluble products of an electrode reaction. Elektrohim 19:1649–1651 (in Russian)
78. Komorsky-Lovrić Š, Lovrić M (2011) Theory of square-wave voltammetry of two-step electrode reaction with kinetically stabilized intermediate. J Electroanal Chem 660:22–25
79. Komorsky-Lovrić Š, Lovrić M (2011) Simulation of square-wave voltammograms of three-electron redox reaction. Electrochim Acta 56:7189–7193
80. Komorsky-Lovrić Š, Lovrić M (2014) Square-wave voltammetry of two-step electrode reaction. Int J Electrochem Sci 9:435–444
81. Komorsky-Lovrić Š, Lovrić M, Bond AM (1993) Square-wave voltammetry at spherical and disk microelectrodes as a function of electrode radius and frequency. Electroanalysis 5:29–40
82. Kounaves SP, O'Dea JJ, Chandresekhar P, Osteryoung J (1986) Square wave voltammetry at the mercury film electrode: theoretical treatment. Anal Chem 58:3199–3202, with correction in Anal Chem 59:1888
83. Kounaves SP, O'Dea JJ, Chandresekhar P, Osteryoung J (1987) Square wave anodic stripping voltammetry at the mercury film electrode: theoretical treatment. Anal Chem 59:386–389
84. Leddy J (1991) Ultrathin layer convolution. J Electroanal Chem 300:295–307
85. Lovrić M (1995) Theory of differential staircase voltammetry. Croat Chem Acta 68:335–341
86. Lovrić M (1999) Differential pulse voltammetry on spherical microelectrodes. Electroanalysis 11:1089–1093
87. Lovrić M, Komorsky-Lovrić Š (2011) Theory of square-wave voltammetry of two-step electrode reaction using an inverse scan direction. Int J Electrochem ID538341
88. Lovrić M, Komorsky-Lovrić Š (2012) Theory of square-wave voltammetry of kinetically controlled two-step electrode reactions. Croat Chem Acta 85:569–575
89. Lovrić M, Scholz F (1997) A model for the propagation of a redox reaction through microcrystals. J Solid State Electrochem 1:108–113
90. Lovrić M, Komorsky-Lovrić Š, Scholz F (1997) Staircase voltammetry with finite diffusion space. Electroanalysis 9:575–577
91. Lovrić M, Hermes M, Scholz F (1998) The effect of the electrolyte concentration in the solution on the voltammetric response of insertion electrodes. J Solid State Electrochem 2:401–404
92. Mareček V, Honz J (1973) Polarization of an electrode with a potentiostatic pulse in the form of an equilateral triangle. Collect Cecoslov Chem Commun 38:965–975
93. Matsuda H (1956) Zur Theorie der Heyrovský-Forejtschen oszillographischen Polarographie. Z Elektrochem 60:617–626
94. Matsuda H, Ayabe Y (1955) Zur Theorie der Randles-Sevčikschen Kathodenstrahl-Polarographie. Z Elektrochem 59:494–503

95. Matsuda H, Delahay P (1960) Faradaic rectification with control of alternating potential variations—application to electrode kinetics for fast processes. J Am Chem Soc 82:1547–1550
96. Mauduit C, Fatouros N, Chevalet J, Reeves RM (1985) Studies of the polarographic wave with superimposed potential perturbations. Part II. Generalised theoretical treatment of potentiostatic techniques for simple redox processes at a static electrode for all rate constants. J Electroanal Chem 184:221–238
97. McCord TG, Brown ER, Smith DE (1966) Influence of spherical diffusion in second harmonic A.C. polarography and related techniques. Anal Chem 38:1615–1620
98. Mendelson A, Tenno R (2008) Modelling and parameter identification of electrochemical Cu–Cu cell. In: Proceedings of the UKACC International Conference on Control 2008, University of Manchester, 2–4 Sept 2008
99. Miles AB, Compton RG (2000) The theory of square wave voltammetry at uniformly accessible hydrodynamic electrodes. J Electroanal Chem 487:75–89
100. Mirčeski V (2004) Charge transfer kinetics in thin-film voltammetry. Theoretical study under conditions of square-wave voltammetry. J Phys Chem B 108:13719–13725
101. Mirčeski V, Scholz F (2002) Reduction of iodine at the organic liquid | aqueous solution | graphite electrode three-phase arrangement. J Electroanal Chem 522:189–198
102. Mirčeski V, Tomovski Ž (2009) Voltammetry based on fractional diffusion. J Phys Chem B 113:2794–2799
103. Mirčeski V, Tomovski Ž (2011) Modelling of a voltammetric experiment in a limiting diffusion space. J Solid State Electrochem 15:197–204
104. Mirčeski V, Gulaboski R, Scholz F (2004) Square-wave thin-film voltammetry: influence of uncompensated resistance and charge transfer kinetics. J Electroanal Chem 566:351–360
105. Montella C (2008) LSV modelling of electrochemical systems through numerical inversion of Laplace transforms. I—The GS–LSV algorithm. J Electroanal Chem 614:121–130
106. Murphy MM, O'Dea JJ, Arn D, Osteryoung JG (1989) Theory of cyclic staircase voltammetry for electrode kinetics. Anal Chem 61:2249–2254
107. Myland JC, Oldham KB, Zoski CG (1985) Electroactive concentrations at the surface of a spherical electrode. J Electroanal Chem 193:3–22
108. Neudeck A, Dittrich J (1991) The determination of diffusion coefficients and rate constants from the dependence of the peak separation and peak current on the scan rate of cyclic voltammograms at micro-cylindrical electrodes. J Electroanal Chem 313:37–59
109. Neudeck A, Dunsch L (1994) Cyclic voltammetry at microstructured electrodes. J Electroanal Chem 370:17–32
110. Neudeck A, Dunsch L (1995) Microstructured electrode materials in UV-visible spectroelectrochemistry. J Electroanal Chem 386:135–148
111. Nicholson RS (1965) Theory and application of cyclic voltammetry for measurement of electrode reaction kinetics. Anal Chem 37:1351–1355
112. Nicholson RS, Shain I (1964) Theory of stationary electrode polarography. Single scan and cyclic methods applied to reversible, irreversible, and kinetic systems. Anal Chem 36:706–723
113. Nigmatullin RSh (1963) General equation and electrical analogue of an electrolytic cell with spherical stationary microelectrode. Dokl Akad Nauk SSSR 151:1383–1386 (in Russian)
114. Nuwer MJ, O'Dea JJ, Osteryoung J (1991) Analytical and kinetic investigations of totally irreversible electron transfer reactions by square-wave voltammetry. Anal Chim Acta 251:13–25
115. O'Dea JJ, Osteryoung J, Osteryoung RA (1983) Square wave voltammetry and other pulse techniques for the determination of kinetic parameters. The reduction of Zinc(II) at mercury electrodes. J Phys Chem 87:3911–3918
116. Okamoto K (1963) Some theoretical considerations of quasi-reversible and irreversible electrode reactions in oscillographic square-wave polarography. Bull Chem Soc Jpn 36:1381–1389

117. Okamoto K (1964) Theoretical considerations on reversible electrode reaction in oscillographic square wave polarography. Rev Polarogr 12:40–49
118. Oldham KB, Spanier J (1974) The fractional calculus. Theory and applications of differentiation and integration to arbitrary order. Academic Press, New York
119. Oldham K, Myland J, Spanier J (2009) An atlas of functions. Springer, New York
120. Osteryoung RA, Christie JH (1974) Theoretical treatment of pulsed voltammetric stripping at the thin film mercury electrode. Anal Chem 46:351–355
121. Pajkossy T, Nyikos L (1989) Diffusion to fractal surfaces—III. Linear sweep and cyclic voltammograms. Electrochim Acta 34:181–186
122. Park SY, Aoki K, Tokuda K, Matsuda H (1983) Determination of diffusion coefficients of metals dissolved in mercury by double potential step chronoamperometry at hanging mercury drop electrodes. Bull Chem Soc Jpn 56:2133–2137
123. Phillips WRC, Mahon PJ (2011) On approximations to a class of Jaeger integrals. Proc R Soc A 467:3570–3589
124. Pnev VV, Moskovskikh LA (1974) Equations of polarographic curves in thin solution layers. II. Irreversible processes. Elektrohim 10:1765–1768 (in Russian)
125. Pnev VV, Popov GN, Zakharov MS (1974) Equations for polarograms on smooth solid electrodes. Elektrohim 10:1768–1772 (in Russian)
126. Pnev VV, Moskovskikh LA, Zakharov MS (1977) Equations of polarographic curves in thin solution layers. Deposition of insoluble material. Sov Electrochem 13:872–875
127. Polcyn DS, Shain I (1966) Multistep charge transfers in stationary electrode polarography. Anal Chem 38:370–375
128. Quintana GC, Andricacos PC, Cheh HY (1983) Linear sweep voltammetry at a rotating disk electrode for first-order quasi-reversible reactions. J Electroanal Chem 144:77–85
129. Rangarajan SK, Doss KSG (1962) Faradaic admittance and adsorption: a diffusion model. J Electroanal Chem 4:237–241
130. Rangarajan SK, Doss KSG (1963) Faradaic admittance and adsorption: a diffusion model. II. J Electroanal Chem 5:114–123
131. Ross PN, Stonehart P (1974) Correlations between electrochemical activity and heterogeneous catalysis for hydrogen dissociation on platinum. J Res Inst Catal Hokkaido Univ 22:22–41
132. Saveant JM (1967) Cyclic voltammetry with asymmetrical potential scan: a simple approach to mechanisms involving moderately fast chemical reactions. Electrochim Acta 12:999–1030
133. Schiewe J, Oldham KB, Myland JC, Bond AM, Vicente-Beckett VA, Fletcher S (1997) Linear-scan anodic stripping voltammetry with thin-film electrodes: theory of the stripping stage and experimental tests. Anal Chem 69:2673–2681
134. Seelig PF, Blount HN (1976) Kalman filter applied to anodic stripping voltammetry: theory. Anal Chem 48:252–258
135. Senda M (1956) Application of Laplace-transformation to polarography. III Convolution and some fundamental expression in polarography and voltammetry. Rev Polarogr 4:89–96 (in Japanese)
136. Senda M (2003) Theory of voltammetric current of electron transfer at organic(O)/aqueous(W) solution interface. Rev Polarogr 49:31–35
137. Senda M, Tachi I (1955) Studies on A.C. polarography. V. Theory of reversible wave. Bull Chem Soc Jpn 28:632–636
138. Seralathan M, Osteryoung RA, Osteryoung J (1987) General equivalence of linear scan and staircase voltammetry. J Electroanal Chem 222:69–100
139. Ševčik A (1948) Oscillographic polarography with periodical triangular voltage. Collect Cecoslov Chem Commun 13:349–377
140. Shuman MS (1969) Nonunity electrode reaction orders and stationary electrode polarography. Anal Chem 41:142–146
141. Smith DE (1964) Theory of the Faradaic impedance. Relationship between Faradaic impedances for various small amplitude alternating current techniques. Anal Chem 36:962–970

142. Stewart AA, Campbell JA, Girault HH, Eddowes M (1990) Cyclic voltammetry for electron transfer reactions at liquid/liquid interfaces. Ber Bunsenges Phys Chem 94:83–87
143. Stewart AA, Taylor G, Girault HH, McAleer J (1990) Voltammetry at microITIES supported at the tip of a micropipette. Part I. Linear sweep voltammetry. J Electroanal Chem 296:491–515
144. Strømme M, Niklasson GA, Granqvist CG (1995) Voltammetry on fractals. Solid State Commun 96:151–154
145. Strømme M, Niklasson GA, Granqvist CG (1995) Determination of fractal dimension by cyclic I–V studies: the Laplace-transform method. Phys Rev B 52:14192–14197
146. Tachi I, Kambara T (1955) Studies on A.C. polarography. I. Theoretical treatment. Bull Chem Soc Jpn 28:25–31
147. Tatsumi H, Katano H, Senda M (2006) Theory of normal-pulse voltammetric current of electron transfer reaction at liquid/liquid interface. J Electroanal Chem 590:82–89
148. Tkalenko DA, Kokhanovskii IN, Tchmilenko NA, Bezdetnyi BP, Fedorov NV (1986) Chronovoltamperometry of complex electrode processes. Peak potential and current in the case of the formation of two electrode reaction products. Elektrohim 22:1589–1592 (in Russian)
149. Tokuda K, Koizumi N (1980) Cyclic voltammetry at spherical electrodes for reversible amalgam formation and dissolution reactions. Rev Polarogr 26:16
150. Tokuda K, Matsuda H (1977) Theory of A.C. voltammetry at a rotating disk electrode. Part I. A reversible electrode process. J Electroanal Chem 82:157–171
151. Tokuda K, Matsuda H (1977) Theory of AC voltammetry at a rotating disk electrode. Redox-electrode reactions. Rev Polarogr 23:10–11
152. Tokuda K, Matsuda H (1978) Theory of A.C. voltammetry at a rotating disk electrode. Part II. Quasi-reversible and irreversible redox-electrode reactions. J Electroanal Chem 90:149–163
153. Tokuda K, Enomoto N, Matsuda H, Koizumi N (1983) Cyclic voltammetry of hanging mercury drop electrodes for reversible amalgam formation and dissolution reactions. J Electroanal Chem 159:23–32
154. Tokuda K, Kitamura F, Kumagai A, Ohsaka T (1995) AC voltammetry at microcylinder electrodes. J Electroanal Chem 396:365–370
155. Van Heuveln FH (1994) Analysis of nonexponential transient response due to a constant-phase element. J Electrochem Soc 141:3423–3428
156. Vogel W, Lundquist J, Ross P, Stonehart P (1975) Reaction pathways and poisons—II. The rate controlling step for electrochemical oxidation of hydrogen on Pt in acid and poisoning of the reaction by CO. Electrochim Acta 20:79–93
157. Weidner JW (1991) Linear-sweep voltammetry of a soluble redox couple in a cylindrical electrode. J Electrochem Soc 138:258C–264C
158. Weidner JW, Fedkiw PS (1990) Reversible, linear-sweep voltammetry of a soluble redox couple: effect of initial concentrations. Anal Chem 62:875–877
159. Wein O (2010) Voltage-step transient in redox systems II. 1D approximation (revisited). Research Report ICPF No. 2010/6, Institute of Chemical Process Fundamentals ASCR, Prague
160. Wein O, Tovchigrechko VV (2011) Voltage-step transient on circular electrodes. J Appl Electrochem 41:1065–1075
161. White N, Lawson F (1970) Potential sweep voltammetry of metal deposition and dissolution. Part I. Theoretical analysis. J Electroanal Chem 25:409–419
162. Zakharov MS, Pneva AP (1979) On the theory of inversion voltammetry on a mercury–graphite electrode. Ž Anal Him 34:1908–1913
163. Zamponi S, Czerwinski A, Gambini G, Marassi R (1992) Linear sweep voltabsorptometry with a finite diffusion space. J Electroanal Chem 332:63–71

Chapter 6
Models Involving One-Dimensional Convection-Diffusion

In this chapter we continue the discussion of the IEs arising for electroanalytical models dependent on one spatial coordinate, started in Chap. 5. However, in addition to the diffusion transport of dynamic distributed species, assumed in Chap. 5, we take into account convection. The effect of convection, on the derivation process of the IEs, is addressed in Sect. 6.1. The determination of the concentration–production rate relationships for the typical convection-diffusion PDEs is described in Sect. 6.2. Section 6.3 involves a literature overview of the published IEs corresponding to the above conditions.

6.1 Derivation of the IEs

The one-dimensional convection-diffusion PDEs are similar to the pure diffusion PDEs discussed in Chap. 5, in that a separate concentration–production rate relationship exists for every dynamic distributed species. Consequently, the derivation of the IEs proceeds in an analogous way. However, the general form of the concentration–production rate relationships may differ somewhat from Eq. (5.6), and is given by

$$c_j^{\dagger} = c_j^{\dagger}(t) = c_j^{\star} + \int_0^t \mathscr{K}_j(t,\tau) \left[p_j^{\mathrm{het}}(\tau) - p_j^{\mathrm{het},\star} \right] \mathrm{d}\tau \,. \tag{6.1}$$

In Eq. (6.1) $p_j^{\mathrm{het},\star}$ denotes the species production rate at the interface studied, in the initial state of the system. Formally, $p_j^{\mathrm{het},\star}$ may be nonzero, if the convection rate at the interface studied is not zero (see Sect. 6.2.1 below), although physically such a situation is unclear, and should probably be regarded as an approximation. One expects that for realistic convection fields $p_j^{\mathrm{het},\star} = 0$, as is the case with the convection-diffusion systems in Sects. 6.2.2–6.2.4.

L.K. Bieniasz, *Modelling Electroanalytical Experiments by the Integral Equation Method*, Monographs in Electrochemistry, DOI 10.1007/978-3-662-44882-3_6

6.2 Concentration–Production Rate Relationships

The concentration–production rate relationships for convection-diffusion PDEs strongly depend on the type of the convection field. Below we present the relationships currently known from the literature. Only a few of the conceivable convection-diffusion fields have been handled thus far by the IE method.

6.2.1 *Constant Convection Velocity*

The simplest possible one-dimensional convection-diffusion transport arises when planar diffusion in a semi-infinite spatial domain is coupled with convection having just a constant velocity along the Cartesian coordinate x. The concentration–flux relationship corresponding to this case was published by Oldham [23], with the suggestion that the case may be applicable to fluid flows through porous electrodes, although no references were provided to support the statement. But, no matter how realistic the assumption of the constant convection velocity may be, it is interesting to consider this simple case as an example showing the degree of complication that is expected when convection is present. Let the convection velocity be

$$v(x) = -v_0 \, , \tag{6.2}$$

where $v_0 \geq 0$, so that the general reaction-convection-diffusion PDE (2.8) becomes

$$\frac{\partial c_j(x,t)}{\partial t} = D_j \frac{\partial^2 c_j(x,t)}{\partial x^2} + v_0 \frac{\partial c_j(x,t)}{\partial x} \, . \tag{6.3}$$

The initial and boundary conditions are

$$c_j(x,0) = c_j^\star \, , \tag{6.4}$$

$$c_j(\infty,t) = c_j^\star \, . \tag{6.5}$$

We define an auxiliary variable

$$u_j = c_j - c_j^\star \, , \tag{6.6}$$

for which the incomplete IBVP (6.3)–(6.5) becomes

$$\frac{\partial u_j(x,t)}{\partial t} = D_j \frac{\partial^2 u_j(x,t)}{\partial x^2} + v_0 \frac{\partial u_j(x,t)}{\partial x} \, , \tag{6.7}$$

$$u_j(x,0) = 0 \, , \tag{6.8}$$

$$u_j(\infty,t) = 0 \, . \tag{6.9}$$

In view of Eq. (6.6), the flux $J_j(x,t)$ of the concentration is related to the flux $J_j^u(x,t)$ of the variable u_j by

$$J_j^u(x,t) = J_j(x,t) - J_j^\star(x,t) \tag{6.10}$$

where

$$J_j^u(x,t) = -D_j \frac{\partial u_j(x,t)}{\partial x} - v_0\, u_j(x,t) \,, \tag{6.11}$$

$$J_j(x,t) = -D_j \frac{\partial c_j(x,t)}{\partial x} - v_0\, c_j(x,t) \,, \tag{6.12}$$

and

$$J_j^\star(x,t) = -v_0\, c_j^\star = \text{const} \tag{6.13}$$

represents the initial flux of the concentration.

As the detailed derivation of the solution–flux relationship corresponding to Eqs. (6.7)–(6.9) was not published in Oldham [23], but only the final result was given, in Appendix B, Sect. B.1 we provide such a derivation. We show that the relationship between the solution u_j and its flux J_j^u at $x = 0$ takes the form [cf. Eqs. (B.17) and (B.18)]:

$$u_j(0,t) = \int_0^t \mathcal{K}_j^{\text{conv}}(t,\tau)\; J_j^u(0,\tau)\; \mathrm{d}\tau \,, \tag{6.14}$$

where

$$\mathcal{K}_j^{\text{conv}}(t,\tau) = D_j^{-1/2} \left\langle \frac{\exp\left[-\upsilon_j^2(t-\tau)\right]}{[\pi(t-\tau)]^{1/2}} + \upsilon_j \left\{1 + \text{erf}\left[\upsilon_j(t-\tau)^{1/2}\right]\right\} \right\rangle \tag{6.15}$$

with

$$\upsilon_j = \frac{v_0}{2D_j^{1/2}} \tag{6.16}$$

is the relevant integral transformation kernel. The kernel is of the convolution type. In view of Eqs. (2.46), (6.6), and (6.10), the concentration–production rate relationship is:

$$c_j^\dagger(t) = c_j(0,t) = c_j^\star + \int_0^t \mathcal{K}_j^{\text{conv}}(t,\tau)\; \left[p_j^{\text{het}}(\tau) - p_j^{\text{het},\star}\right]\; \mathrm{d}\tau \,. \tag{6.17}$$

As can be seen from Eq. (6.15), the kernel $\mathscr{K}_j^{\text{conv}}(t,\tau)$ consists of three additive terms. One term is the kernel for planar diffusion in a semi-infinite domain, multiplied by an exponential, the second term is simply constant, and the third term is proportional to the error function erf(·) defined by Eq. (5.78). When the convection velocity v_0 (and consequently v_j) is sufficiently large, the contribution from the first term can be neglected in Eq. (6.15), whereas the contributions from the second and third terms become practically identical, so that Eq. (6.17) becomes

$$c_j^{\dagger}(t) = c_j(0,t) = c_j^{\star} + 2D_j^{-1/2} v_j \int_0^t \left[p_j^{\text{het}}(\tau) - p_j^{\text{het},\star} \right] \mathrm{d}\tau \, . \tag{6.18}$$

Apart from the above transient solutions, a non-trivial steady state solution of Eq. (6.7), satisfying $\partial u_j(x,t)/\partial t \approx 0$ is also possible. As we show in Appendix B, Sect. B.1, at steady state $J_j^u(x,t) = 0$, which [in view of Eqs. (2.46) and (6.10)] implies that $p_j^{\text{het}}(t) - p_j^{\text{het},\star} = 0$. Hence, the steady state corresponds to the initial state.

6.2.2 *Dropping Mercury Electrode*

As we have indicated in Sect. 2.6.1, the transport to the DME is most commonly described by means of the expanding plane model. According to this model, any jth distributed species obeys the convection-diffusion PDE (2.12):

$$\frac{\partial c_j(x,t)}{\partial t} = D_j \frac{\partial^2 c_j(x,t)}{\partial x^2} + \frac{2x}{3t} \frac{\partial c_j(x,t)}{\partial x} \, , \tag{6.19}$$

for which the initial and boundary conditions are

$$c_j(x,0) = c_j^{\star} \, , \tag{6.20}$$

$$c_j(\infty,t) = c_j^{\star} \, . \tag{6.21}$$

The expanding planar interface (mercury drop surface) is located at $x = 0$. Below we consider the electrolyte phase, for which $x \in (0,\infty)$.

We define an auxiliary variable

$$u_j = c_j - c_j^{\star} \, , \tag{6.22}$$

for which the incomplete IBVP (6.19)–(6.21) becomes

$$\frac{\partial u_j(x,t)}{\partial t} = D_j \frac{\partial^2 u_j(x,t)}{\partial x^2} + \frac{2x}{3t} \frac{\partial u_j(x,t)}{\partial x} , \tag{6.23}$$

$$u_j(x,0) = 0 , \tag{6.24}$$

$$u_j(\infty,t) = 0 . \tag{6.25}$$

In view of Eq. (6.22), the flux $J_j(x,t)$ of the concentration is related to the flux $J_j^u(x,t)$ of the variable u_j by

$$J_j^u(x,t) = J_j(x,t) - J_j^\star(x,t) \tag{6.26}$$

where

$$J_j^u(x,t) = -D_j \frac{\partial u_j(x,t)}{\partial x} - \frac{2x}{3t} u_j(x,t) , \tag{6.27}$$

$$J_j(x,t) = -D_j \frac{\partial c_j(x,t)}{\partial x} - \frac{2x}{3t} c_j(x,t) , \tag{6.28}$$

and

$$J_j^\star(x,t) = -\frac{2x}{3t} c_j^\star . \tag{6.29}$$

The analytical solution–flux relationship at the interface studied, resulting from Eqs. (6.23)–(6.25), has been obtained in a number of papers (see, for example, [13, 14, 17, 18]). In Appendix B, Sect. B.2 we follow these works to show that the relationship between the solution u_j and its flux J_j^u at $x = 0$ takes the form [cf. Eqs. (B.35) and (B.36)]:

$$u_j(0,t) = \int_0^t \mathscr{K}_j^{\mathrm{ep}}(t,\tau)\; J_j^u(0,\tau)\, \mathrm{d}\tau , \tag{6.30}$$

where

$$\mathscr{K}_j^{\mathrm{ep}}(t,\tau) = D_j^{-1/2} \left(\frac{3}{7}\pi\right)^{-1/2} \tau^{2/3} \left(t^{7/3} - \tau^{7/3}\right)^{-1/2} \tag{6.31}$$

is the integral transformation kernel representing the expanding plane model. Note that in contrast to the pure diffusion kernels discussed in Chap. 5, the kernel (6.31) is not a convolution kernel (when considered as a function of variables t and τ).

Consequently, in view of Eqs. (2.46), (6.22), and (6.26), the concentration–production rate relationship at the expanding plane is:

$$c_j^{\dagger}(t) = c_j(0,t) = c_j^{\star} + \int_0^t \mathscr{K}_j^{\mathrm{ep}}(t,\tau)\; p_j^{\mathrm{het}}(\tau)\; \mathrm{d}\tau \,, \tag{6.32}$$

since $p_j^{het,\star} = J_j^{\star}(0,t) = 0$ in this case. A discussion useful for the generalisation of the above formulae onto hypothetical interfaces expanding according to power laws other than the dropping mercury $t^{2/3}$ rate law was provided by Oldham [22].

6.2.3 Rotating Disk Electrode

As we have indicated in Sect. 2.6.2, in order to obtain solution–flux relationships pertinent to convection-diffusion at the RDE, one possible approach is to assume a finite spatial domain with a permeable second boundary. Examples of this approach to the RDE modelling have been listed in Sect. 5.3.2.1. However, as in real experiments the spatial domain is large and can be formally regarded as semi-infinite, it seems conceptually more elegant to assume that the convection-diffusion PDE (2.15):

$$\frac{\partial c_j(x,t)}{\partial t} = D_j \frac{\partial^2 c_j(x,t)}{\partial x^2} + \beta\, x^2 \frac{\partial c_j(x,t)}{\partial x} \tag{6.33}$$

(with $\beta = 0.51023\,\omega^{3/2}\nu^{-1/2}$) is accompanied by the initial and boundary conditions

$$c_j(x,0) = c_j^{\star} \,, \tag{6.34}$$

$$c_j(\infty,t) = c_j^{\star} \,. \tag{6.35}$$

We define an auxiliary variable

$$u_j = c_j - c_j^{\star} \,, \tag{6.36}$$

for which the incomplete IBVP (6.33)–(6.35) becomes

$$\frac{\partial u_j(x,t)}{\partial t} = D_j \frac{\partial^2 u_j(x,t)}{\partial x^2} + \beta\, x^2 \frac{\partial u_j(x,t)}{\partial x} \,, \tag{6.37}$$

$$u_j(x,0) = 0 \,, \tag{6.38}$$

$$u_j(\infty,t) = 0 \,. \tag{6.39}$$

In view of Eq. (6.36), the flux $J_j(x,t)$ of the concentration is related to the flux $J_j^u(x,t)$ of the variable u_j by

$$J_j^u(x,t) = J_j(x,t) - J_j^\star(x,t) \tag{6.40}$$

where

$$J_j^u(x,t) = -D_j \frac{\partial u_j(x,t)}{\partial x} - \beta\, x^2\, u_j(x,t)\,, \tag{6.41}$$

$$J_j(x,t) = -D_j \frac{\partial c_j(x,t)}{\partial x} - \beta\, x^2\, c_j(x,t)\,, \tag{6.42}$$

and

$$J_j^\star(x,t) = -\beta\, x^2\, c_j^\star \tag{6.43}$$

An analytical but approximate solution–flux relationship at the interface studied, resulting from Eqs. (6.37)–(6.39), has been obtained by Filinovskii and Kiryanov [8]. In Appendix B, Sect. B.3 we follow those authors, and present detailed derivations of their approximation. The derivations based on [8] are approximate because of a simplifying assumption that significant variations of $u_j(x,t)$ are limited to a small neighbourhood of $x = 0$. Under this assumption some term arising in the process of deriving the solution–flux relationship is neglected, which simplifies the treatment, but the result obtained is not exact. A more realistic solution–flux relationship can only be obtained numerically, by the procedure indicated in Sect. B.3. Independently of the approach taken, the solution–flux relationship is [cf. Eq. (B.71)]:

$$u_j(0,t) = \int\limits_0^t \mathcal{K}_j^{\mathrm{rd}}(t,\tau)\; J_j(0,\tau)\; \mathrm{d}\tau\,, \tag{6.44}$$

where $\mathcal{K}_j^{\mathrm{rd}}(t,\tau)$ is the integral transformation kernel of the convolution type, characteristic of the RDE. Therefore, we can write [cf. Eqs. (B.43) and (B.72)]:

$$\mathcal{K}_j^{\mathrm{rd}}(t,\tau) = D_j^{-1/2} \upsilon_j\, \varphi_j^{\mathrm{rd}}\left[\upsilon_j^2 (t-\tau)\right]\,, \tag{6.45}$$

where $\varphi_j^{\mathrm{rd}}(\vartheta)$ is a suitable function of one variable, and

$$\upsilon_j = \left(D_j^{1/2} \beta\right)^{1/3}\,. \tag{6.46}$$

However, by analogy to kernels previously discussed in Chap. 5, it is more convenient to express $\mathscr{K}_j^{\mathrm{rd}}(t,\tau)$ in the form:

$$\mathscr{K}_j^{\mathrm{rd}}(t,\tau) = D_j^{-1/2}\left\{[\pi(t-\tau)]^{-1/2} - \upsilon_j\,\mathrm{krd}\left[\upsilon_j(t-\tau)^{1/2}\right]\right\}\,, \tag{6.47}$$

where krd(·) is a special function expressing the effect of rotating disk convection. The name "krd" is an acronym of "kernel for rotating disk".

According to Filinovskii and Kiryanov [8], the function $\varphi_j^{\mathrm{rd}}(\vartheta)$ is approximated by the formulae [cf. Eqs. (B.64) and (B.67)]:

$$\varphi_j^{\mathrm{rd}}(\vartheta) \approx \mathscr{L}^{-1}\left\{-\frac{\mathrm{Ai}\,(s)}{\mathrm{Ai}'\,(s)}\right\} = \mathscr{L}^{-1}\left\{\frac{\mathrm{K}_{1/3}(\frac{2}{3}s^{3/2})}{s^{1/2}\,\mathrm{K}_{2/3}(\frac{2}{3}s^{3/2})}\right\}\,. \tag{6.48}$$

In Eq. (6.48) the Laplace transformation is between the ϑ and s domains, Ai(·) is the Airy function [1], Ai′(·) is the derivative of Ai(·), and $\mathrm{K}_{1/3}(\cdot)$ and $\mathrm{K}_{2/3}(\cdot)$ are modified Bessel functions of the second kind and fractional orders [1]. Consequently, krd(·) is approximated by:

$$\mathrm{krd}(z) = \mathrm{krd}(\vartheta^{1/2}) \approx \mathscr{L}^{-1}\left\{\frac{\mathrm{K}_{2/3}(\frac{2}{3}s^{3/2}) - \mathrm{K}_{1/3}(\frac{2}{3}s^{3/2})}{s^{1/2}\,\mathrm{K}_{2/3}(\frac{2}{3}s^{3/2})}\right\}\,. \tag{6.49}$$

As will be seen in the next section, the approximant in Eq. (6.49) is identical to the function kct(·) occurring in the kernel function for channel and tubular electrodes. To show how large the errors are in the approximation of Filinovskii and Kiryanov, Fig. 6.1 presents a comparison of functions krd(·) and kct(·). The difference between these functions is sufficiently big to visibly affect simulation results. By considering

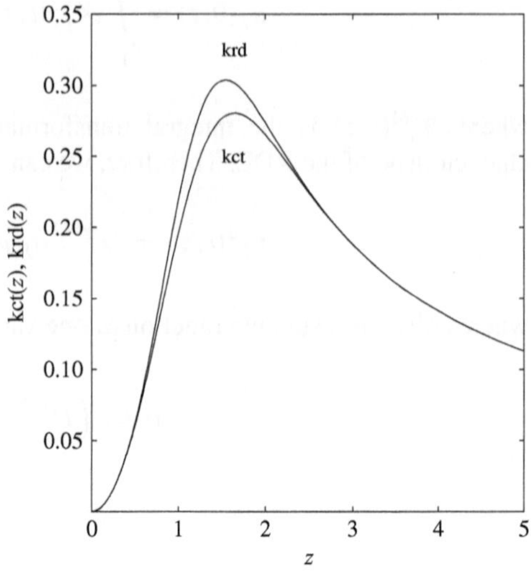

Fig. 6.1 Comparison of functions krd(·) and kct(·), representing the effect of convection in the case of RDE(s), and channel and tubular electrodes, respectively

models of current step chronopotentiometry and potential step chronoamperometry at the RDE, Krylov and Babak [11] have shown that calculations employing the approximation (6.49) lead to unsatisfactory predictions of the concentrations. The inaccuracies tend to increase with time t. Similar conclusions were obtained by Viswanathan and Cheh [29].

In view of Eqs. (2.46), (6.36), (6.40), and (6.44), the concentration–production rate relationship at the RDE is:

$$c_j^{\dagger}(t) = c_j(0,t) = c_j^{\star} + \int_0^t \mathscr{K}_j^{\mathrm{rd}}(t,\tau)\; p_j^{\mathrm{het}}(\tau)\; \mathrm{d}\tau \; . \tag{6.50}$$

since $p_j^{\mathrm{het},\star} = J_j^{\star}(0,t) = 0$ in this case.

A steady state satisfying $\partial u_j(x,t)/\partial t \approx 0$ is possible for Eq. (6.37), and the relevant equation was obtained without any simplifying assumptions (see, for example, Nişancioğlu and Newman [21]). A detailed derivation is provided in Appendix B, Sect. B.3. It is shown [cf. Eq. (B.82)] that at steady state

$$u_j(0,t) = D_j^{-1/2} \upsilon_j^{-1}\, 3^{-2/3} \Gamma(1/3)\, J_j(0,t) \; , \tag{6.51}$$

where $\Gamma(\cdot)$ is the Euler gamma function [1]. In view of Eqs. (2.46), (6.36), and (6.40), this gives

$$c_j^{\dagger}(t) = c_j(0,t) = c_j^{\star} + D_j^{-1/2} \upsilon_j^{-1}\, 3^{-2/3} \Gamma(1/3)\, p_j^{\mathrm{het}}(t) \; . \tag{6.52}$$

Equation (6.52) is the concentration–production rate relationship for the steady state.

6.2.4 Channel and Tubular Electrodes

Although the convection-diffusion PDEs (2.17) and (2.19) for channel and tubular electrodes depend on two spatial coordinates, it is possible to devise approximate IE-based models resulting from these PDEs, similar to those obtainable for spatially one-dimensional models. This was accomplished by Singh, Dutt, and co-workers, in a series of papers [6, 7, 24–26] devoted to the modelling of linear potential sweep and cyclic voltammetry at tubular electrodes. The Lévêque approximation (2.21) and (2.22) was adopted. According to the proposal from [6, 7, 24–26], currently known as the "Singh and Dutt approximation", the axial concentration gradient in Eq. (2.19) can be approximated by the "average gradient":

$$\frac{\partial c_j(r,z,t)}{\partial z} \approx \frac{c_j(r,z,t) - c_j^{\star}}{l} \; . \tag{6.53}$$

As a result, the spatial variable z is eliminated from Eq. (2.19), so that c_j depends only on r and t. By replacing r with a new spatial variable x equal to the distance from the electrode surface:

$$x = r_0 - r \,, \tag{6.54}$$

and neglecting the cylindrical curvature effect, we then obtain from Eq. (2.19)

$$\frac{\partial c_j(x,t)}{\partial t} = D_j \frac{\partial^2 c_j(x,t)}{\partial x^2} - \gamma x \left[c_j(x,t) - c_j^\star \right] \,, \tag{6.55}$$

where

$$\gamma = \frac{2v_0}{r_0 l} \,. \tag{6.56}$$

An analogous modification can be applied to the PDE (2.17) for channel electrodes. If we define

$$x = h - y \tag{6.57}$$

and

$$\gamma = \frac{2v_0}{hl} \,, \tag{6.58}$$

we obtain an approximate PDE identical to Eq. (6.55). As in this approach r_0 or h are assumed to be (infinitely) large compared with the diffusion layer thickness, Eq. (6.55) is accompanied by the initial and boundary conditions analogous to those for semi-infinite spatial domains:

$$c_j(x,0) = c_j^\star \,, \tag{6.59}$$

$$c_j(\infty,t) = c_j^\star \,. \tag{6.60}$$

The validity of the Singh and Dutt approximation has been extensively examined by Compton and co-workers [3, 4]. It has been concluded that although the approximation is rather crude, it leads to surprisingly accurate theoretical predictions, possibly due to some error cancellation. Consequently, the use of Singh and Dutt approximation has been recommended in [3, 4] (at least for the modelling of linear potential sweep and cyclic voltammetry, and/or as a preliminary tool before more exact but computationally expensive simulation techniques are applied).

We define an auxiliary variable

$$u_j = c_j - c_j^\star \,, \tag{6.61}$$

for which the incomplete IBVP (6.55), (6.59), and (6.60) becomes

$$\frac{\partial u_j(x,t)}{\partial t} = D_j \frac{\partial^2 u_j(x,t)}{\partial x^2} - \gamma \, x \, u_j(x,t) \,, \tag{6.62}$$

$$u_j(x,0) = 0 \,, \tag{6.63}$$

$$u_j(\infty,t) = 0 \,. \tag{6.64}$$

The Singh and Dutt approximation thus formally replaces the convection-diffusion PDEs (2.17) or (2.19) by the quasi-reaction-diffusion PDE (6.62), in which the reaction rate constant is proportional to the distance x from the electrode. As the convection is eliminated, the fluxes $J_j(x,t)$ of the concentration and $J_j^u(x,t)$ of the variable u_j are identical:

$$J_j(x,t) = -D_j \frac{\partial c_j(x,t)}{\partial x} = -D_j \frac{\partial u_j(x,t)}{\partial x} = J_j^u(x,t) \,. \tag{6.65}$$

An analytical solution–flux relationship resulting from Eqs. (6.62)–(6.64) was obtained in [6, 7, 24–26]. In Appendix B, Sect. B.4, we follow these works and present detailed derivations of the relationship between the solution u_j and its flux J_j^u at $x = 0$. The relationship is [cf. Eq. (B.98)]:

$$u_j(0,t) = \int_0^t \mathcal{K}_j^{\mathrm{ct}}(t,\tau) \; J_j^u(0,\tau) \; \mathrm{d}\tau \,, \tag{6.66}$$

where $\mathcal{K}_j^{\mathrm{ct}}(t,\tau)$ is the relevant integral transformation kernel of the convolution type. The kernel function turns out to be formally identical to the approximate kernel for the RDE, discussed in Sect. 6.2.3, but with the parameter γ in place of β, that is

$$\mathcal{K}_j^{\mathrm{ct}}(t,\tau) = D_j^{-1/2} \upsilon_j \, \varphi_j^{\mathrm{ct}} \left[\upsilon_j^2 (t-\tau) \right] \,, \tag{6.67}$$

where [cf. Eqs. (B.89) and (B.97)]

$$\varphi_j^{\mathrm{ct}}(\vartheta) = \mathcal{L}^{-1} \left\{ -\frac{\mathrm{Ai}\,(s)}{\mathrm{Ai}'\,(s)} \right\} = \mathcal{L}^{-1} \left\{ \frac{\mathrm{K}_{1/3}(\frac{2}{3}s^{3/2})}{s^{1/2}\,\mathrm{K}_{2/3}(\frac{2}{3}s^{3/2})} \right\} \,, \tag{6.68}$$

and

$$\upsilon_j = \left(D_j^{1/2} \gamma \right)^{1/3} \,. \tag{6.69}$$

In Eq. (6.68) the Laplace transformation is between the ϑ and s domains, Ai(·) is the Airy function [1], Ai′(·) is the derivative of Ai(·), and $\mathrm{K}_{1/3}(\cdot)$ and $\mathrm{K}_{2/3}(\cdot)$ are modified Bessel functions of the second kind and fractional orders [1]. The form (6.67) of the kernel was originally obtained [6,7,24–26], but a more convenient equivalent form, similar to the kernels previously discussed in Chap. 5, is:

$$\mathscr{K}_j^{\mathrm{ct}}(t,\tau) = D_j^{-1/2}\left\{[\pi(t-\tau)]^{-1/2} - \upsilon_j\,\mathrm{kct}\left[\upsilon_j(t-\tau)^{1/2}\right]\right\}\,, \tag{6.70}$$

where kct(·) is a special function defined by

$$\mathrm{kct}(z) = \mathrm{kct}(\vartheta^{1/2}) = \mathscr{L}^{-1}\left\{\frac{\mathrm{K}_{2/3}(\frac{2}{3}s^{3/2}) - \mathrm{K}_{1/3}(\frac{2}{3}s^{3/2})}{s^{1/2}\,\mathrm{K}_{2/3}(\frac{2}{3}s^{3/2})}\right\}\,. \tag{6.71}$$

The name "kct" is an acronym of "kernel for channel and tubular electrodes".

In view of Eqs. (2.46), (6.61), (6.65), and (6.66), the concentration–production rate relationship at the electrode is:

$$c_j^{\dagger}(t) = c_j(0,t) = c_j^{\star} + \int_0^t \mathscr{K}_j^{\mathrm{ct}}(t,\tau)\; p_j^{\mathrm{het}}(\tau)\;\mathrm{d}\tau\,, \tag{6.72}$$

since $p_j^{\mathrm{het},\star} = J_j^{\star}(0,t) = 0$ in this case.

A steady state satisfying $\partial u_j(x,t)/\partial t \approx 0$ is possible for Eq. (6.62). The steady state has been observed experimentally [6,24], but a relevant theoretical description was not provided. In Appendix B, Sect. B.4 the analytical solution for such a steady state is derived, and it is shown that at steady state Eq. (6.66) takes the following limiting form [cf. Eq. (B.107)]:

$$u_j(0,t) = D_j^{-1/2}\upsilon_j^{-1}\,3^{-1}\frac{\Gamma(1/3)}{\Gamma(2/3)}\,J_j^{u}(0,t)\,, \tag{6.73}$$

where $\Gamma(\cdot)$ is the Euler gamma function [1]. In view of Eqs. (2.46), (6.61), and (6.65), this gives

$$c_j^{\dagger}(t) = c_j(0,t) = c_j^{\star} + D_j^{-1/2}\upsilon_j^{-1}\,3^{-1}\frac{\Gamma(1/3)}{\Gamma(2/3)}\,p_j^{\mathrm{het}}(t)\,. \tag{6.74}$$

Equation (6.74) is the concentration–production rate relationship for the steady state.

6.3 Literature Examples

The following examples of IE-based models are limited to the instances of one-dimensional convection-diffusion transport unaffected by homogeneous reactions, in the absence of dynamic localised species and other complications. Examples involving such additional phenomena will be listed in further Chaps. 8, 9, and 10.

6.3.1 Expanding Plane

Smutek [27] obtained expressions for the electrolytic current for the non-equilibrium reversible reaction

$$X_1 + n\,e^- \rightleftarrows X_2 \tag{6.75}$$

involving dynamic species X_1 and X_2, distributed in the electrolyte. Matsuda [13], and Matsuda and Ayabe [17, 18] presented the theory of polarographic waves for the non-equilibrium reaction (6.75). Matsuda [14] developed the theory of polarography under conditions of small amplitude periodic potential perturbations. The effect of alternating potential amplitude in AC polarography was studied by Delmastro and Smith [5]. Matsuda [15] considered the effect of the partially blocked electrode surface (covered by a surface-active substance), on polarographic currents. In a later article [16] he presented a theory of current–potential curves in pulse polarography, for reaction (6.75) involving dynamic species X_1 and X_2, distributed in the electrolyte or in the expanding DME. He also considered differential pulse polarography. Lovrić and Branica [12] simulated drop life-time dependence of the current in differential pulse polarography. Aoki and Osteryoung [2] presented a theory of differential pulse polarography, employing a non-standard IE approach, in which the IE was formulated and solved in the Laplace space, rather than in the time domain.

Komorsky and Lovrić [10] presented a theory of DC and AC polarography for an EEE reaction scheme involving three consecutive, non-equilibrium reversible reactions (6.75), with all species distributed in an electrolyte. Niki et al. [20] simulated pulse polarography for a series of consecutive equilibrium one-electron transfers, with all species distributed in an electrolyte.

An interesting example was presented by Matsuda et al. [19], who analysed current–potential curves for an ion transfer reaction across a liquid | liquid interface. In this example, one liquid phase was in the form of a growing droplet, and the liquid | liquid interface was moving according to the expanding plane model.

6.3.2 Rotating Disk Electrode

Filinovskii and Kiryanov [8] modelled potential step chronoamperometry at the RDE under limiting current conditions, assuming an electron transfer reaction

$$X_1 + n\,e^- \rightleftarrows \text{products} \tag{6.76}$$

with species X_1 distributed in the electrolyte. They employed the kernel approximation (6.48). Girina et al. [9] obtained analogous solutions for linear potential sweep voltammetry, assuming the equilibrium reaction

$$X_1 + n\,e^- \rightleftarrows X_2 \tag{6.77}$$

with dynamic species X_1 and X_2, distributed in the electrolyte, and having equal diffusion coefficients. The approximate kernel (6.48) was also used by Valdes and Cheh [28] for the convolution analysis of potential sweep voltammetric data obtained at the RDE, assuming equilibrium, non-equilibrium reversible, or irreversible reaction (6.77).

6.3.3 Tubular Electrodes

The Singh and Dutt approximate IE approach was employed to obtain predictions for linear potential sweep and cyclic voltammetry at tubular electrodes, for the reaction

$$X_1 + n\,e^- \rightleftarrows X_2 \tag{6.78}$$

involving dynamic species X_1 and X_2, distributed in the electrolyte. In a series of papers [6, 7, 24–26] equilibrium, non-equilibrium reversible, and irreversible reaction (6.78) was considered.

References

1. Abramowitz M, Stegun IA (1972) Handbook of mathematical functions. Dover Publications, New York
2. Aoki K, Osteryoung J (1980) Theory of differential pulse polarography at expanding or stationary planar electrodes for quasi-reversible or totally irreversible reactions. J Electroanal Chem 110:19–36
3. Compton RG, Unwin PR (1989) A critical assessment of the theories for linear sweep voltammetry of reversible electrode processes at a channel electrode. J Electroanal Chem 264:27–36
4. Compton RG, Pilkington MBG, Stearn GM, Unwin PR (1987) Mass transport to channel and tubular electrodes: the "Singh and Dutt approximation". J Electroanal Chem 238:43–66
5. Delmastro JR, Smith DE (1967) Some considerations on the effect of alternating potential amplitude in AC polarography. J Electroanal Chem 14:261–268
6. Dutt J, Singh T (1985) Linear sweep voltammetry at the tubular electrode. J Electroanal Chem 182:259–266
7. Dutt J, Singh T (1986) Linear sweep voltammetry at the tubular graphite electrode. Part III. Quasi-reversible process. J Electroanal Chem 207:41–48
8. Filinovskii VYu, Kiryanov VA (1964) On the theory of non-stationary convective diffusion at the rotating disk electrode. Dokl Akad Nauk SSSR 156:1412–1415 (in Russian)
9. Girina GP, Filinovskii VYu, Feoktistov LG (1967) Dependence of the current on potential sweep rate and disk electrode rotation rate in the case of linearly changing potential. Elektrohim 3:941–945 (in Russian)
10. Komorsky Š, Lovrić M (1980) Simple EEE mechanism at DME. J Electroanal Chem 112:169–174
11. Krylov VS, Babak VN (1971) Unsteady diffusion towards the rotating disk surface. Elektrohim 7:649–654 (in Russian)

12. Lovrić M, Branica M (1985) Drop life-time dependence of current density in differential pulse polarography. J Electroanal Chem 183:107–122
13. Matsuda H (1957) Beiträge zur Theorie der polarographischen Stromstärke. Allgemeine Formel der diffusionsbedingten Stromstärke und ihre Anwendung. Z Elektrochem 61:489–506
14. Matsuda H (1958) Zur Theorie der Wechselspannungs-Polarographie. Z Elektrochem 62:977–989
15. Matsuda H (1967) A contribution to the theory of polarographic diffusion currents influenced by the adsorption of surface-active substances. Rev Polarogr 14:87–92
16. Matsuda H (1980) A contribution to the theory of pulse polarography. I. Theory of current–potential curves. Bull Chem Soc Jpn 53:3439–3446
17. Matsuda H, Ayabe Y (1955) Theoretical analysis of polarographic waves. I. Reduction of simple metal ions. Bull Chem Soc Jpn 28:422–429
18. Matsuda H, Ayabe Y (1956) Theoretical analysis of polarographic waves. II. Reduction of complex metal ions. Bull Chem Soc Jpn 29:134–140
19. Matsuda H, Yamada Y, Kanamori K, Kudo Y, Takeda Y (1991) On the facilitation effect of neutral macrocyclic ligands on the ion transfer across the interface between aqueous and organic solutions. I. Theoretical equation of ion-transfer-polarographic current–potential curves and its experimental verification. Bull Chem Soc Jpn 64:1497–1508
20. Niki K, Kobayashi Y, Matsuda H (1984) Determination of macroscopic standard potentials of a molecule with a reversible n-consecutive one-electron transfer process. Application to a tetra-heme protein: cytochrome c_3. J Electroanal Chem 178:333–341
21. Nişancioğlu K, Newman J (1974) Transient convective diffusion to a disk electrode. J Electroanal Chem 50:23–29
22. Oldham KB (1969) A unified treatment of electrolysis at an expanding mercury electrode. Anal Chem 41:936–945
23. Oldham KB (1991) Interrelation of current and concentration at electrodes. J Appl Electrochem 21:1068–1072
24. Singh T, Dutt J (1985) Cyclic voltammetry at the tubular graphite electrode. Reversible process (theory). J Electroanal Chem 190:65–73
25. Singh T, Dutt J (1985) Linear sweep voltammetry at the tubular graphite electrode. Part II. Totally irreversible process. J Electroanal Chem 196:35–42
26. Singh T, Dutt J, Kumari B, Dewan RK (1987) Cyclic voltammetry at the tubular electrode. Part II. Quasi-reversible and totally irreversible processes. J Electroanal Chem 238:33–41
27. Smutek M (1955) Kinetics of electrode processes. XIII. A contribution to the theory of slow electrode processes. Collect Cecoslov Chem Commun 20:247–251
28. Valdes JL, Cheh HY (1984) The application of convolution potential sweep voltammetry at a rotating disc electrode to simple electron transfer kinetics. J Electroanal Chem 181:65–81
29. Viswanathan K, Cheh HY (1979) The application of pulsed potential and pulsed current to a rotating disc electrode system. J Appl Electrochem 9:537–543

Chapter 7
Models Involving Two- and Three-Dimensional Diffusion

In this chapter we consider IE formulations of electroanalytical models described by systems of diffusion PDEs defined over two- and three-dimensional spatial domains. We assume that one of the spatial directions, in which diffusion can proceed, is perpendicular to the interface studied, and that the models do not involve homogeneous reactions, dynamic localised species, nor other complications. Up to date, there have been only few attempts to derive and solve IEs corresponding to such two- and three-dimensional models. Three main approaches have been proposed for this purpose in the literature. Two of them apply in the case of spatial domains that are regular in the sense that they can be represented as Cartesian products of single intervals along each coordinate axis. The intervals can be infinite, semi-infinite, or finite. The third approach is applicable to arbitrary spatial domains. We briefly present these approaches in Sect. 7.1. Then, in Sects. 7.2 and 7.3 we elaborate more on two of these approaches. Literature examples are provided in Sect. 7.4. In general, the treatment of spatially two- and three-dimensional models is more difficult and less developed, compared to one-dimensional models.

7.1 Derivation of the IEs

Consider a controlled potential experiment, in which a set of $N_{\mathrm{r}}^{\mathrm{het}}$ heterogeneous reactions takes place at an interface studied, among $N_{\mathrm{s}}^{\mathrm{distr}}$ distributed species $\mathrm{X}_1, \ldots, \mathrm{X}_{N_{\mathrm{s}}^{\mathrm{distr}}}$ present in spatially extended phase(s) adjacent to this interface. Let $N_{\mathrm{s}}^{\mathrm{dd}}$ of the species be dynamic distributed species subject to two- or three-dimensional diffusion PDEs, accompanied by appropriate initial and boundary conditions.

Let us first consider the case when a coordinate system exists, in which the spatial domain is a Cartesian product of single intervals along each coordinate axis. To fix attention, we focus on two frequently encountered types of interfaces, for which the

L.K. Bieniasz, *Modelling Electroanalytical Experiments by the Integral Equation Method*, Monographs in Electrochemistry, DOI 10.1007/978-3-662-44882-3_7

relevant coordinate systems are depicted in Figs. 2.3 and 2.4: the interface consisting of infinite electrode band(s) embedded in an insulator, and the interface consisting of concentric electrode disk and/or ring(s) embedded in an insulator. In both cases the spatial domains are two-dimensional and semi-infinite or infinite. Diffusion to band(s) is governed by the PDE resulting from Eq. (2.8) written in the Cartesian coordinate system (cf. Table 2.1):

$$\frac{\partial c_j(x,z,t)}{\partial t} = D_j \left[\frac{\partial^2 c_j(x,z,t)}{\partial x^2} + \frac{\partial^2 c_j(x,z,t)}{\partial z^2} \right], \tag{7.1}$$

and diffusion to disk/ring(s) is governed by the PDE resulting from Eq. (2.8) written in the cylindrical coordinate system:

$$\frac{\partial c_j(r,z,t)}{\partial t} = D_j \left[\frac{\partial^2 c_j(r,z,t)}{\partial r^2} + \frac{1}{r}\frac{\partial c_j(r,z,t)}{\partial r} + \frac{\partial^2 c_j(r,z,t)}{\partial z^2} \right]. \tag{7.2}$$

Equations (7.1) or (7.2) must be accompanied by initial conditions, which we assume to correspond to the initial equilibrium state (see Sect. 2.10):

$$c_j(x,z,0) = c_j^\star \tag{7.3}$$

for diffusion to band(s), or

$$c_j(r,z,0) = c_j^\star \tag{7.4}$$

for diffusion to disk/ring(s). The initial concentration $c_j^\star$ can be positive or zero. Boundary conditions at the boundaries other than the interface studied are:

$$c_j(\pm\infty,z,t) = c_j^\star, \tag{7.5}$$

$$c_j(x,\infty,t) = c_j^\star \tag{7.6}$$

for diffusion to band(s), or

$$c_j(\infty,z,t) = c_j^\star, \tag{7.7}$$

$$\left.\frac{\partial c_j(r,z,t)}{\partial r}\right|_{r=0} = 0, \tag{7.8}$$

$$c_j(r,\infty,t) = c_j^\star \tag{7.9}$$

for diffusion to disk/ring(s). Boundary conditions must also be provided at the interface studied.

Similarly to models dependent on one spatial coordinate, discussed in Chap. 5, the boundary conditions at the interface studied are generally given by Eq. (2.54), with matrices $\overline{V}$ and $\overline{Z}$ resulting from the partial inversion of Eq. (2.53). However, there is an important difference: the heterogeneous reaction rates and interfacial species production rates now depend not only on time but also on one or two spatial coordinates along the interface. Similarly, the stoichiometric matrix $\overline{N}^{\text{dyn,het}}$ and matrices $\overline{V}$ and $\overline{Z}$ depend on the location along the interface, because there are usually different boundary conditions at different parts of the interface (electrodes, insulators).

For the conversion of this sort of IBVPs into IEs two approaches have been proposed, which are called Procedures 1 and 2 below.

Procedure 1 The first approach, due to Cope, Tallman, and co-workers [9–16, 18, 27, 28, 41] applies to selected models of controlled potential experiments, and consists of the following steps: (a) Laplace transformation of the IBVP; (b) Expression of the Laplace transforms $\hat{c}_j^{\dagger}$ of the interfacial concentrations in terms of the surface integrals of the Laplace transforms $\hat{p}_j^{\text{het}}$ of the production rates (equal to the fluxes normal to the interface studied); (c) Combination of the integrals with the expressions for $\hat{c}_j^{\dagger}$ obtained from Laplace transformed boundary conditions at the interface studied, resulting in IEs for $\hat{p}_j^{\text{het}}$. Step (b) can be accomplished, for example, by using Fourier transforms and the convolution theorem (3.23). The procedure must be followed by numerical solution of the IEs, and a numerical inversion of the Laplace transformed $\hat{p}_j^{\text{het}}$ to obtain production rates in the time domain. Finally, the production rates must be integrated over the interface to obtain Faradaic current(s). The Cope–Tallman approach has provided important theoretical predictions for a number of (micro)electrode configurations. However, the approach is mathematically advanced, may require fluency in complex calculus, and is probably not readily accessible to average electrochemists. In addition, it seems to be limited to rather simple electroanalytical experiments and reaction schemes. The need to operate on Laplace transforms of the interfacial concentrations restricts the approach to situations when such transforms are obtainable from the boundary conditions. In practice this means that only potential step chronoamperometric experiments can be modelled. This limitation was relaxed in [12, 13, 27, 41] by showing that the solutions obtained for potential step chronoamperometry can be re-used to calculate any controlled potential transients for a simple equilibrium electron transfer, if diffusion coefficients are equal. This applies to any electrode geometry. More complicated reaction schemes may not be tractable, though. The IEs arising from this approach have integrals over interfaces, in contrast to the IEs from Chap. 5, in which the integration is over time. The IEs are also singular, which requires special solution methods. A further difficulty is presented by the numerical inversion of the Laplace transforms, which is known to be a sensitive task, usually requiring multi-precision environments for reliable calculations [1].

Procedure 2 As a remedy against the difficulties inherent in the Cope–Tallman approach, Mirkin and Bard [29, 30] suggested an alternative procedure, which is analogous to the procedure used in the case of spatially one-dimensional models, previously described in Chap. 5. A similar approach was earlier used by Aoki et al. [5, 6], in the context of convection-diffusion transport at channel electrodes, although the IE formulation was not their primary goal. Also Daschbach [17] employed a similar approach. In the Mirkin and Bard procedure one strives to obtain integral concentration–production rate relationships in the time (and space) domains rather than in the Laplace (and space) domains. In the absence of homogeneous reactions, assumed here, a separate concentration–production rate relationship is envisaged for every dynamic distributed species. As is shown in Sect. 7.2 below, in the case of two-dimensional spatial domains these relationships take the general form:

$$c_j^{\dagger} = c_j^{\dagger}(\xi, t) = c_j^{\star} + \int_{\mathbb{D}} \mathscr{K}_j\left(\begin{bmatrix} \xi \\ t \end{bmatrix}, \begin{bmatrix} \zeta \\ \tau \end{bmatrix}\right) p_j^{\text{het}}(\zeta, \tau)\,\mathrm{d}\zeta\,\mathrm{d}\tau\,, \tag{7.10}$$

where $\mathscr{K}_j\left(\begin{bmatrix} \xi \\ t \end{bmatrix}, \begin{bmatrix} \zeta \\ \tau \end{bmatrix}\right)$ is a suitable integral transformation kernel, consistent with the definition (3.3) of linear integral transformations. The kernel depends on the diffusion coefficient of the jth species, the geometry of the spatial domain, the system of coordinates used, and boundary conditions at other interfaces. The kernel arguments are: the vector variable $[\xi, t]^{\mathrm{T}}$ and the dummy integration vector variable $[\zeta, \tau]^{\mathrm{T}}$. By ξ we denote here a coordinate along the interface studied. Thus, in particular, $\xi = x$ for the diffusion to band(s), and $\xi = r$ for diffusion to disk/ring(s). Consequently, we have $\vec{r}^{\text{ het}} = \vec{r}^{\text{ het}}(\xi, t)$, $\vec{p}^{\text{ het}} = \vec{p}^{\text{ het}}(\xi, t)$, $\overline{N}^{\text{ dyn,het}} = \overline{N}^{\text{ dyn,het}}(\xi)$, $\overline{V} = \overline{V}(\xi)$, and $\overline{Z} = \overline{Z}(\xi)$. The integration domain $\mathbb{D}$ is a Cartesian product of the spatial interval of the ξ coordinate, and time interval $[0, t]$. In a similar way, in the case of three-dimensional spatial domains, $c_j^{\dagger}$ and other variables listed above depend on two spatial coordinates along the interface studied, and $\mathbb{D}$ is a Cartesian product of two spatial intervals, and the above time interval. Hence, the conversion of an IBVP into IEs eliminates one spatial coordinate, the one perpendicular to the interface studied.

As an example of Procedure 2, we consider a controlled potential experiment for an electron transfer reaction

$$\mathrm{X}_1 + n\,\mathrm{e}^- \rightleftarrows \mathrm{X}_2 \tag{7.11}$$

taking place between dynamic distributed species X_1 and X_2 at an interface corresponding to an array of electrodes (such as the band(s) or a disk and ring(s) seen in Figs. 2.3 and 2.4), assuming a two-dimensional spatial domain. The relevant IEs can be derived by a straightforward reformulation of Eqs. (5.19), (5.21), and (5.23) for the spatially one-dimensional case. One only has to replace one-dimensional integrals by double integrals, remembering about the additional dependence of the various variables on ξ. Thus, Eq. (5.19) for the non-equilibrium reversible electron transfer (7.11) would be transformed into:

$$\Psi(\xi,t) - k^0 \Bigg\langle f_f(\xi,t) \left\{ c_1^\star - \int_{\mathbb{D}} \mathscr{K}_1\left(\begin{bmatrix} \xi \\ t \end{bmatrix}, \begin{bmatrix} \zeta \\ \tau \end{bmatrix} \right) \Psi(\zeta,\tau)\,\mathrm{d}\zeta\,\mathrm{d}\tau \right\}$$

$$- f_b(\xi,t) \left\{ c_2^\star + \int_{\mathbb{D}} \mathscr{K}_2\left(\begin{bmatrix} \xi \\ t \end{bmatrix}, \begin{bmatrix} \zeta \\ \tau \end{bmatrix} \right) \Psi(\zeta,\tau)\,\mathrm{d}\zeta\,\mathrm{d}\tau \right\} \Bigg\rangle = 0\,. \tag{7.12}$$

Equation (5.21) for the irreversible electron transfer (7.11) would be transformed into:

$$\Psi(\xi,t) - k^0\, f_f(\xi,t) \left\{ c_1^\star - \int_{\mathbb{D}} \mathscr{K}_1\left(\begin{bmatrix} \xi \\ t \end{bmatrix}, \begin{bmatrix} \zeta \\ \tau \end{bmatrix} \right) \Psi(\zeta,\tau)\,\mathrm{d}\zeta\,\mathrm{d}\tau \right\} = 0\,, \tag{7.13}$$

and Eq. (5.23) for the equilibrium electron transfer (7.11) would be transformed into:

$$\left\{ c_1^\star - \int_{\mathbb{D}} \mathscr{K}_1\left(\begin{bmatrix} \xi \\ t \end{bmatrix}, \begin{bmatrix} \zeta \\ \tau \end{bmatrix} \right) \Psi(\zeta,\tau)\,\mathrm{d}\zeta\,\mathrm{d}\tau \right\}$$

$$- \left\{ c_2^\star + \int_{\mathbb{D}} \mathscr{K}_2\left(\begin{bmatrix} \xi \\ t \end{bmatrix}, \begin{bmatrix} \zeta \\ \tau \end{bmatrix} \right) \Psi(\zeta,\tau)\,\mathrm{d}\zeta\,\mathrm{d}\tau \right\} \exp\left\{ \frac{nF}{RT} \left[E(\xi,t) - E^0 \right] \right\} = 0\,. \tag{7.14}$$

In Eqs. (7.12)–(7.14)

$$\Psi(\xi,t) = -p_1^{\mathrm{het}}(\xi,t) = p_2^{\mathrm{het}}(\xi,t)\,, \tag{7.15}$$

and the variables $E(\xi,t)$, $f_f(\xi,t)$, and $f_b(\xi,t)$ are considered dependent on ξ, because in the case of electrode arrays different electrodes in the array can be polarised differently. Equation (5.17) for the current has to be replaced by equations (one for every electrode):

$$\frac{I(t)}{F} = n \int_{\mathbb{D}_\xi} p_1^{\mathrm{het}}(\xi,t)\,\mathrm{d}\xi = -n \int_{\mathbb{D}_\xi} \Psi(\xi,t)\,\mathrm{d}\xi\,, \tag{7.16}$$

where the domain $\mathbb{D}_\xi$ corresponds to the particular electrode. Each of the Eqs. (7.12)–(7.14) can be viewed as a system of infinitely many IEs, one for every ξ value. The IEs in these systems are coupled, owing to the spatial integration.

Procedure 3 Apart from Procedures 1 and 2, a more generally applicable procedure is known under the name of the boundary integral method (BIM), or boundary integral equation method. This procedure is often used in conjunction with finite-element type discretisations, in which case it is also called the boundary element method (BEM). The BIM appears particularly useful in cases of complicated shapes of the spatial domain, when Procedures 1 and 2 cannot be applied. There is a review of the BIM applications to electroanalysis, written by Fan et al. [19], so that we do not go much into details here. The BIM was used mostly to solve models of steady state experiments (see, for example, [8, 20–23, 25, 32, 35–40, 43]) which are of secondary interest in the present book. Transient experiments were modelled by Qiu et al. [24,31,33,34] and Träuble et al. [42]. There are several possible variants of transient BIM simulations. One of them was utilised by Qiu et al. [34], by following an earlier non-electrochemical paper by Wrobel and Brebbia [44]. In this variant one makes use of a specific integral solution–flux relationship to replace an IBVP by a boundary IE or IEs, avoiding any need of volume discretisations (see Sect. 7.3 below for more information on this variant). Another variant, used in [24, 31, 42], makes use of the so-called dual reciprocity method, where the volume discretisation is not entirely avoided.

Procedure 4 In addition to the above three major procedures, for some multidimensional models it may be possible to obtain approximate representations in the form of one-dimensional IEs. One such approximation, referring to controlled potential experiments at stationary disk microelectrodes, was reported by Jin et al. [26,45], by using a former work of Aoki and Osteryoung [3] (and forgetting to mention another related work by Aoki et al. [4]). Unfortunately, papers [26, 45] seem to contain errors and undefined parameters (parameter p is undefined in [26], and although the Reader is directed to [45] for explanations of undefined symbols, p does not appear at all in [45]). It is therefore problematic to reproduce or reuse these results.

7.2 The Mirkin and Bard Approach

Owing to its analogy to the IE formulations of spatially one-dimensional models discussed in Chaps. 5 and 6, the Mirkin and Bard approach [29, 30] may seem particularly appealing and relatively uncomplicated, so that it is pursued in more detail below.

The crucial element of the approach is the determination (analytically, if possible) of the integral concentration–production rate relationships, characteristic of a particular geometry of the spatial domain. Mirkin and Bard [29, 30] obtained such relationships for a few particular electrode configurations, including those from Figs. 2.3 and 2.4. For the purpose of the derivations it is convenient to introduce the usual auxiliary variable:

$$u_j = c_j - c_j^\star \,, \tag{7.17}$$

for which the incomplete IBVPs to be solved, resulting from Eqs. (7.1)–(7.9), take the following forms. For diffusion to band(s):

$$\frac{\partial u_j(x,z,t)}{\partial t} = D_j \left[\frac{\partial^2 u_j(x,z,t)}{\partial x^2} + \frac{\partial^2 u_j(x,z,t)}{\partial z^2} \right] , \tag{7.18}$$

$$u_j(x,z,0) = 0 , \tag{7.19}$$

$$u_j(\pm\infty,z,t) = 0 , \tag{7.20}$$

$$u_j(x,\infty,t) = 0 , \tag{7.21}$$

and for diffusion to disk/ring(s):

$$\frac{\partial u_j(r,z,t)}{\partial t} = D_j \left[\frac{\partial^2 u_j(r,z,t)}{\partial r^2} + \frac{1}{r}\frac{\partial u_j(r,z,t)}{\partial r} + \frac{\partial^2 u_j(r,z,t)}{\partial z^2} \right] , \tag{7.22}$$

$$u_j(r,z,0) = 0 , \tag{7.23}$$

$$u_j(\infty,z,t) = 0 , \tag{7.24}$$

$$\left. \frac{\partial u_j(r,z,t)}{\partial r} \right|_{r=0} = 0 , \tag{7.25}$$

$$u_j(r,\infty,t) = 0 . \tag{7.26}$$

In Appendix C we follow Mirkin and Bard [29, 30] and present detailed analytical solutions of the IBVPs (7.18)–(7.21) and (7.22)–(7.26), by employing the Laplace, combined with Fourier, or Hankel transforms. As a result, we obtain the following relationships between the solution u_j and its flux component J_j along the coordinate z, at the interface studied [cf. Eqs. (C.22), (C.23), (C.44), and (C.45)]. For diffusion to band(s):

$$u_j(x,0,t) = \int_{\mathbb{D}} \mathcal{K}_j^{\mathrm{b}}\left(\begin{bmatrix} x \\ t \end{bmatrix}, \begin{bmatrix} \zeta \\ \tau \end{bmatrix} \right) J_j(\zeta,0,\tau)\, \mathrm{d}\zeta\, \mathrm{d}\tau , \tag{7.27}$$

with

$$\mathcal{K}_j^{\mathrm{b}}\left(\begin{bmatrix} x \\ t \end{bmatrix}, \begin{bmatrix} \zeta \\ \tau \end{bmatrix} \right) = D_j^{-1} \left[2\pi(t-\tau)\right]^{-1} \exp\left[-\frac{(x-\zeta)^2}{4D_j(t-\tau)} \right] \tag{7.28}$$

and integration domain $\mathbb{D} = (-\infty,\infty) \times [0,t)$. For diffusion to disk/ring(s):

$$u_j(r,0,t) = \int_{\mathbb{D}} \mathcal{K}_j^{\mathrm{dr}}\left(\begin{bmatrix} r \\ t \end{bmatrix}, \begin{bmatrix} \zeta \\ \tau \end{bmatrix} \right) J_j(\zeta,0,\tau)\, \mathrm{d}\zeta\, \mathrm{d}\tau , \tag{7.29}$$

with

$$\mathscr{K}_j^{\mathrm{dr}}\left(\begin{bmatrix} r \\ t \end{bmatrix}, \begin{bmatrix} \zeta \\ \tau \end{bmatrix}\right) = D_j^{-3/2} \frac{\zeta \exp\left[-\frac{\zeta^2+r^2}{4D_j(t-\tau)}\right] \mathrm{I}_0\left[\frac{r\zeta}{2D_j(t-\tau)}\right]}{2\pi^{1/2}(t-\tau)^{3/2}} \tag{7.30}$$

and integration domain $\mathbb{D} = (0, \infty) \times [0, t)$. In Eq. (7.30) $\mathrm{I}_0(\cdot)$ denotes the modified Bessel function of the first kind and order zero [2]. For some other geometries and transport conditions (such as in particular, convection–diffusion at channel electrodes), instead of using two different integral transformations, one might also employ a double Laplace transform [5, 6].

The spatial intervals of integration in Eqs. (7.27) and (7.29) are formally infinite, but they reduce to one or more finite intervals corresponding to the electrode locations, because the reactant flux J_j is zero at insulators, so that the insulators do not contribute to the integrals. A possibility for handling electrodes of arbitrary shapes, provided they are co-planar with an insulator plane in which they are embedded, was indicated [29].

At first sight the temporal singularities present in the kernels (7.28) and (7.30) appear stronger than the weak singularity of the kernels typical for spatially one-dimensional models. However, a closer examination reveals that for sufficiently small $\varepsilon > 0$ the double integrals exhibit the following limiting behaviour:

$$\begin{aligned} &\int_{-\infty}^{\infty} \int_{t-\varepsilon}^{t} \mathscr{K}_j^{\mathrm{b}}\left(\begin{bmatrix} x \\ t \end{bmatrix}, \begin{bmatrix} \zeta \\ \tau \end{bmatrix}\right) J_j(\zeta, 0, \tau)\, \mathrm{d}\zeta\, \mathrm{d}\tau \\ &\approx \int_{t-\varepsilon}^{t} D_j^{-1/2} [\pi(t-\tau)]^{-1/2} \left[\int_{-\infty}^{\infty} \delta(x-\zeta)\, J_j(\zeta, 0, \tau)\, \mathrm{d}\zeta\right] \mathrm{d}\tau \\ &= \int_{t-\varepsilon}^{t} D_j^{-1/2} [\pi(t-\tau)]^{-1/2} J_j(x, 0, \tau)\, \mathrm{d}\tau\,, \end{aligned} \tag{7.31}$$

$$\begin{aligned} &\int_{0}^{\infty} \int_{t-\varepsilon}^{t} \mathscr{K}_j^{\mathrm{dr}}\left(\begin{bmatrix} r \\ t \end{bmatrix}, \begin{bmatrix} \zeta \\ \tau \end{bmatrix}\right) J_j(\zeta, 0, \tau)\, \mathrm{d}\zeta\, \mathrm{d}\tau \\ &\approx \int_{t-\varepsilon}^{t} D_j^{-1/2} [\pi(t-\tau)]^{-1/2} \left[\int_{0}^{\infty} (\zeta/r)^{1/2}\, \delta(r-\zeta)\, J_j(\zeta, 0, \tau)\, \mathrm{d}\zeta\right] \mathrm{d}\tau \\ &= \int_{t-\varepsilon}^{t} D_j^{-1/2} [\pi(t-\tau)]^{-1/2} J_j(r, 0, \tau)\, \mathrm{d}\tau \end{aligned} \tag{7.32}$$

(for any $r > 0$). In Eqs. (7.31) and (7.32) $\delta(\cdot)$ denotes the Dirac delta. Hence, in both cases we have a weak temporal singularity, analogous to that for planar diffusion in a one-dimensional semi-infinite spatial domain.

7.3 The Boundary Integral Method

The BIM variant from Qiu et al. [34] relies on a general integral solution–flux relationship which we derive in Appendix D. The relationship refers to the three-dimensional diffusion PDE:

$$\frac{\partial c_j(\boldsymbol{x},t)}{\partial t} = D_j \ \triangle \ c_j(\boldsymbol{x},t) \tag{7.33}$$

with initial condition

$$c_j(\boldsymbol{x},0) = c_j^\star \,, \tag{7.34}$$

defined over a spatial domain Ω of the vector $\boldsymbol{x} = [x,\ y,\ z]^{\mathrm{T}}$ of Cartesian coordinates, surrounded by a surface S. It takes the form [cf. Eqs. (D.15) and (D.18)]:

$$\lambda c_j(\boldsymbol{x},t) = c^\star \int\limits_{\Omega} \mathscr{G}_j\left(\begin{bmatrix}\boldsymbol{x}\\ t\end{bmatrix}, \begin{bmatrix}\boldsymbol{\chi}\\ \tau\end{bmatrix}\right) \mathrm{d}\Omega \Bigg|_{\tau=0}$$
$$+ D_j \int\limits_0^t \oint\limits_S \left[\mathscr{G}_j\left(\begin{bmatrix}\boldsymbol{x}\\ t\end{bmatrix}, \begin{bmatrix}\boldsymbol{\chi}\\ \tau\end{bmatrix}\right) \frac{\partial c_j(\boldsymbol{\chi},\tau)}{\partial n} - c_j(\boldsymbol{\chi},\tau) \frac{\partial \mathscr{G}_j\left(\begin{bmatrix}\boldsymbol{x}\\ t\end{bmatrix}, \begin{bmatrix}\boldsymbol{\chi}\\ \tau\end{bmatrix}\right)}{\partial n} \right] \mathrm{d}s\, \mathrm{d}\tau \tag{7.35}$$

with

$$\mathscr{G}_j\left(\begin{bmatrix}\boldsymbol{x}\\ t\end{bmatrix}, \begin{bmatrix}\boldsymbol{\chi}\\ \tau\end{bmatrix}\right) = \begin{cases} \left[4\pi D_j (t-\tau)\right]^{-3/2} \exp\left[-\frac{(x-\xi)^2+(y-\eta)^2+(z-\zeta)^2}{4D_j(t-\tau)}\right] & \text{for } t > \tau \\ 0 & \text{for } t < \tau \end{cases} \tag{7.36}$$

and $\lambda = 1/2$ for $\boldsymbol{x} \in S$. In Eqs. (7.35) and (7.36) $\boldsymbol{\chi} = [\xi,\ \eta,\ \zeta]^{\mathrm{T}}$ and τ denote integration variables, and $\mathrm{d}\Omega = \mathrm{d}\xi\, \mathrm{d}\eta\, \mathrm{d}\zeta$. Equation (7.35) relates concentrations $c_j(\boldsymbol{x},t)$ at the surface S and flux components $-D_j\ \partial c_j(\boldsymbol{x},t)/\partial n$ normal to the surface. It has to be combined with boundary conditions at the surface S, resulting in IEs to be solved for the boundary values of $c_j(\boldsymbol{x},t)$ and $-D_j\ \partial c_j(\boldsymbol{x},t)/\partial n$. Although a three-dimensional domain Ω was assumed in Appendix D, Eq. (7.35) holds also in the case of a two-dimensional domain Ω surrounded by a closed curve

S. In such a case

$$\mathscr{G}_j\left(\begin{bmatrix} \boldsymbol{x} \\ t \end{bmatrix}, \begin{bmatrix} \boldsymbol{\chi} \\ \tau \end{bmatrix}\right) = \begin{cases} [4\pi D_j(t-\tau)]^{-1} \exp\left[-\frac{(x-\xi)^2+(y-\eta)^2}{4D_j(t-\tau)}\right] & \text{for } t > \tau \\ 0 & \text{for } t < \tau \end{cases} \tag{7.37}$$

must be taken, with $\boldsymbol{x} = [x, y]^{\mathrm{T}}$ and $\boldsymbol{\chi} = [\xi, \eta]^{\mathrm{T}}$.

Depending on the shape of the spatial domain, a non-trivial steady state solution may be possible. In Appendix D we show that at steady state the solution–flux relationship takes the form [cf. Eqs. (D.23) and (D.24)]:

$$\lambda c_j(\boldsymbol{x},t) = \oint\limits_S \left[\mathscr{G}(\boldsymbol{x},\boldsymbol{\chi}) \frac{\partial c_j(\boldsymbol{\chi},t)}{\partial n} - c_j(\boldsymbol{\chi},t) \frac{\partial \mathscr{G}(\boldsymbol{x},\boldsymbol{\chi})}{\partial n}\right] \mathrm{d}s \,, \tag{7.38}$$

where

$$\mathscr{G}(\boldsymbol{x},\boldsymbol{\chi}) = \left\{4\pi\left[(x-\xi)^2 + (y-\eta)^2 + (z-\zeta)^2\right]^{1/2}\right\}^{-1} \tag{7.39}$$

for a three-dimensional spatial domain. In the case of a two-dimensional domain Ω surrounded by a closed curve S,

$$\mathscr{G}(\boldsymbol{x},\boldsymbol{\chi}) = -(4\pi)^{-1} \ln\left[(x-\xi)^2 + (y-\eta)^2\right] \tag{7.40}$$

must be taken. The dependence of c_j on time t is retained in Eq. (7.38), because boundary conditions may depend on time, even at steady state. We note that, in the case of the two- and three-dimensional spatial domains, the steady state concentration–flux relationships are no longer purely algebraic (as was the case with one-dimensional models), but they involve spatial integrals such as in Eq. (7.38). Equation (7.38) combined with boundary conditions along the surface (or curve) S gives IE(s) from which the unknown steady state boundary concentrations and fluxes can be determined.

7.4 Literature Examples

The Cope and Tallman approach was used to simulate potential step chronoamperometry for an electron transfer reaction

$$\mathrm{X}_1 + n\,\mathrm{e}^- \rightleftarrows \mathrm{X}_2 \tag{7.41}$$

involving dynamic distributed species X_1 and X_2 under limiting current conditions at band (micro)electrodes [9, 11, 13–15], stationary disk (micro)electrodes [10, 11, 13, 28], ring (micro)electrodes [11, 13, 16], and tubular band electrodes

[12, 13, 18]. Linear potential sweep voltammetry and square wave voltammetry for an equilibrium electron transfer (7.41) was considered by Kalapathy et al. [27] and Tallman [41], respectively, assuming ring (micro)electrodes.

The Mirkin and Bard approach of Sect. 7.2 was used to formulate and solve IEs describing: potential step chronoamperometry and cyclic voltammetry at a single (micro)disk and a single (micro)band electrode, generator and collector transients at double band electrodes, steady state and voltammogram-like curves in a scanning electrochemical microscope (SECM) experiment (for details, see [29, 30]). Further simulations concerning SECM experiments were performed by Bard et al. [7]. A non-equilibrium reversible electron transfer (7.41) was assumed.

The BIM was used for transient simulations of the potential step chronoamperometry assuming limiting current conditions of reaction (7.41) at (micro)band and (micro)cylinder electrodes [31] and also at (micro)band, (micro)cylinder, (micro)hemisphere, and single and double (micro)disk electrodes [33]. Other examples include: cyclic voltammetry at immiscible liquid | liquid interface of different topography [34], voltammetry of electroactive oil droplets [24], and transient measurements by the SECM [42].

References

1. Abate J, Valkó PP (2004) Multi-precision Laplace transform inversion. Int J Numer Methods Eng 60:979–993
2. Abramowitz M, Stegun IA (1972) Handbook of mathematical functions. Dover Publications, New York
3. Aoki K, Osteryoung J (1984) Formulation of the diffusion-controlled current at very small stationary disk electrodes. J Electroanal Chem 160:335–339
4. Aoki K, Koji A, Tokuda K, Matsuda H, Ostryoung J (1984) Linear sweep voltammetry at very small stationary disk electrodes. J Electroanal Chem 171:219–230
5. Aoki K, Tokuda K, Matsuda H (1985) Hydrodynamic voltammetry at channel electrodes. Part VII. Current transients at double channel electrodes. J Electroanal Chem 195:229–249
6. Aoki K, Tokuda K, Matsuda H (1986) Hydrodynamic voltammetry at channel electrodes. Part VIII. Theory of reversible voltammograms for chronoamperometry and linear sweep voltammetry. J Electroanal Chem 209:247–258
7. Bard AJ, Mirkin MV, Unwin PR, Wipf DO (1992) Scanning electrochemical microscopy. 12. Theory and experiment of the feedback mode with finite heterogeneous electron-transfer kinetics and arbitrary substrate size. J Phys Chem 96:1861–1868
8. Burchardt M, Träuble M, Wittstock G (2009) Digital simulation of scanning electrochemical microscopy approach curves to enzyme films with Michaelis–Menten kinetics. Anal Chem 81:4857–4863
9. Coen S, Cope DK, Tallman DE (1986) Diffusion current at a band electrode by an integral equation method. J Electroanal Chem 215:29–48
10. Cope DK, Tallman DE (1990) Transient behavior at planar microelectrodes. Diffusion current at the disk electrode by the integral equation method. J Electroanal Chem 285:79–84
11. Cope DK, Tallman DE (1990) Transient behavior at planar microelectrodes. A comparison of diffusion current at ring, band and disk electrodes. J Electroanal Chem 285:85–92
12. Cope DK, Tallman DE (1991) Representation of reversible current for arbitrary electrode geometries and arbitrary potential modulation. J Electroanal Chem 303:1–15

13. Cope DK, Tallman DE (1995) An integral equation method for computing the transient current at microelectrodes. J Electroanal Chem 396:265–275
14. Cope DK, Tallman DE (1999) Transient behavior at planar microelectrodes. High efficiency algorithm for an integral equation method at a band electrode. Electrochem Soc Proc 99–5:82–89
15. Cope DK, Scott CH, Kalapathy U, Tallman DE (1990) Transient behavior at planar microelectrodes. Diffusion current at a band electrode by an integral equation method. Part II. J Electroanal Chem 280:27–35
16. Cope DK, Scott CH, Tallman DE (1990) Transient behavior at planar microelectrodes. Diffusion current at ring electrodes by the integral equation method. J Electroanal Chem 285:49–69
17. Daschbach JL (1991) Time dependent responses at disk electrodes. In: Montenegro MI, Queirós MA, Daschbach JL (eds) Microelectrodes: theory and applications. Kluwer, Dordrecht, pp 67–82
18. Engblom SO, Cope DK, Tallman DE (1996) Diffusion current at the tubular band electrode by the integral equation method. J Electroanal Chem 406:23–31
19. Fan TH, Mayle EJ, Kottke PA, Fedorov AG (2006) Simulation of electroanalysis using the boundary integral method. Trends Anal Chem 25:52–65
20. Fulian Q, Fisher AC (1998) Computational electrochemistry: the boundary element method. J Phys Chem B 102:9647–9652
21. Fulian Q, Fisher AC, Denuault G (1999) Applications of the boundary element method in electrochemistry: scanning electrochemical microscopy. J Phys Chem B 103:4387–4392
22. Fulian Q, Fisher AC, Denuault G (1999) Applications of the boundary element method in electrochemistry: scanning electrochemical microscopy. Part 2. J Phys Chem B 103:4393–4398
23. Fulian Q, Williams NA, Fisher AC (1999) Computational electrochemistry: three-dimensional boundary element simulations of double electrode geometries. Electrochem Commun 1:124–127
24. Fulian Q, Ball JC, Marken F, Compton RG, Fisher AC (2000) Voltammetry of electroactive oil droplets. Part I: numerical modelling for three mechanistic models using the dual reciprocity finite element method. Electroanalysis 12:1012–1016
25. Hume EC III, Deen WM, Brown RA (1984) Mass transfer analysis of electrodeposition through polymeric masks. J Electrochem Soc 131:1251–1258
26. Jin BK, Zhang JR, Zhang ZX (1996) Theory and application of cyclic voltammetry for measurement of fast electrode kinetics at microdisk electrode. Chin J Chem 14:338–347
27. Kalapathy U, Tallman DE, Hagen S (1992) Linear sweep voltammetry at ring microelectrodes. The reversible case. J Electroanal Chem 325:65–81
28. Mahon PJ, Oldham KB (2004) The transient current at the disk electrode under diffusion control: a new determination by the Cope–Tallman method. Electrochim Acta 49:5041–5048
29. Mirkin MV, Bard AJ (1992) Multidimensional integral equations. Part 1. A new approach to solving microelectrode diffusion problems. J Electroanal Chem 323:1–27
30. Mirkin MV, Bard AJ (1992) Multidimensional integral equations: a new approach to solving microelectrode diffusion problems. Part 2. Applications to microband electrodes and the scanning electrochemical microscope. J Electroanal Chem 323:29–51
31. Qiu FL, Fisher AC (2000) The dual reciprocity method: simulation of potential step voltammetry at microelectrodes. Electrochem Commun 2:738–742
32. Qiu FL, Fisher AC (2001) The boundary element method: applications to steady-state voltammetric simulations within domains extending to infinity. Electrochem Commun 3:117–121
33. Qiu FL, Fisher AC (2003) The boundary element method: chronoamperometric simulations at microelectrodes. Electrochem Commun 5:87–93
34. Qiu FL, Fisher AC, Henley IE, Dryfe RAW (2003) The boundary element method: the simulation of voltammetry at immiscible liquid/liquid interfaces. Electrochem Commun 5:169–174

35. Qiu FL, Fisher AC, Walker AB, Peter LM (2003) The distribution of photoinjected electrons a dye-sensitized nanocrystalline TiO_2 solar cell modelled by a boundary element method. Electrochem Commun 5:711–716
36. Sklyar O, Wittstock G (2002) Numerical simulations of complex nonsymmetrical 3D systems for scanning electrochemical microscopy using the boundary element method. J Phys Chem B 106:7499–7508
37. Sklyar O, Ufheil J, Heinze J, Wittstock G (2003) Application of the boundary element method numerical simulations for characterization of heptode ultramicroelectrodes in SECM experiments. Electrochim Acta 49:117–128
38. Sklyar O, Kueng A, Kranz C, Mizaikoff B, Lugstein A, Bertagnolli E, Wittstock G (2005) Numerical simulation of scanning electrochemical microscopy experiments with frame-shaped integrated atomic force microscopy-SECM probes using the boundary element method. Anal Chem 77:764–771
39. Sklyar O, Treutler TH, Vlachopoulos N, Wittstock G (2005) The geometry of nanometer-sized electrodes and its influence on electrolytic currents and metal deposition processes in scanning tunneling and scanning electrochemical microscopy. Surf Sci 597:181–195
40. Sklyar O, Träuble M, Zhao C, Wittstock G (2006) Modeling steady-state experiments with a scanning electrochemical microscope involving several independent diffusing species using the boundary element method. J Phys Chem B 110:15869–15877
41. Tallman DE (1994) Square wave voltammetry of reversible systems at ring microelectrodes. 1. Theoretical study. Anal Chem 66:557–565
42. Traeuble M, Sklyar O, Wittstock G (2005) Boundary element simulation for transient measurements in SECM. In: Selvadurai AP, Tan CL, Aliabadi MH (eds) Advances in boundary element techniques VI, EC Ltd., Eastleigh, pp 161–165
43. Träuble M, Kirchner CN, Wittstock G (2007) Nonlinear boundary conditions in simulations of electrochemical experiments using the boundary element method. In: Simos TE, Maroulis G (eds) Proceedings of the international conference on computational methods in science and engineering 2007. American Institute of Physics, pp 500–503
44. Wrobel LC, Brebbia CA (1981) A formulation of the boundary element method for axisymmetric transient heat conduction. Int J Heat Mass Transf 24:843–850
45. Zhang JR, Jin BK, Zhang ZX (1994) A kinetic equation for various electrochemical techniques at a finite disk electrode. J Anal Sci 10:11–15 (in Chinese)

Chapter 8
Models Involving Transport Coupled with Homogeneous Reactions

In Chaps. 5, 6, and 7 we have assumed that diffusion and possibly convection are the sole modes of transport, and that no homogeneous reactions between the distributed species accompany the transport. Although there are many electrochemical systems that obey these assumptions, there are also numerous systems in which homogeneous reactions play a crucial role and cannot be neglected. Therefore, in the present chapter we address the issue how to handle the homogeneous reactions in addition to the diffusion or convection–diffusion transport, by the IE method.

8.1 Derivation of the IEs

The treatment of homogeneous reactions by the IE method is generally a difficult matter. One reason for this is associated with the homogeneous reactions subject to nonlinear kinetic laws, such as the power rate law with reaction orders higher than one, or with Michaelis–Menten kinetics. The application of the Laplace transform is questionable in such cases, because the relevant reaction–diffusion PDEs are nonlinear. Consequently, the majority of the publications, describing the IEs for the reaction–transport problems, deals with first- and pseudo first-order homogeneous reactions only. The treatment of such reactions is outlined in Sects. 8.2–8.4. Nonlinear homogeneous kinetics have been handled thus far only in a number of special cases, by employing certain additional assumptions. These special treatments are outlined in Sect. 8.5.

The second difficulty is that homogeneous reactions acting between distributed species cause couplings between the concentrations of the different species in every point in space, because some species disappear and some other are formed. Therefore, when homogeneous reactions are present, the concentration of a particular species, at the interface studied, may depend on the production rates of other species. As a consequence, the concentration–production rate relationships

L.K. Bieniasz, *Modelling Electroanalytical Experiments by the Integral Equation Method*, Monographs in Electrochemistry, DOI 10.1007/978-3-662-44882-3_8

must generally be considered jointly for all dynamic distributed species. This is a complication compared to the transport in the absence of homogeneous reactions, where a separate concentration–production rate relationship exists for every species at the interfaces studied (cf. Chaps. 5, 6 and 7).

In the case of models dependent on one spatial coordinate, such joint relationships can usually be expressed as the vector–matrix analogue of Eq. (6.1):

$$\vec{c}^{\,\dagger} = \vec{c}^{\,\dagger}(t) = \vec{c}^{\,\star} + \int_0^t \overline{\mathcal{K}}(t,\tau)\,\left[\vec{p}^{\,\text{het}}(\tau) - \vec{p}^{\,\text{het},\star}\right]\,\mathrm{d}\tau\,, \tag{8.1}$$

where $\vec{c}^{\,\dagger} = \left[c_1^\dagger, \ldots, c_{N_s^{\text{dd}}}^\dagger\right]^{\text{T}}$ is the vector of the boundary concentrations of all N_s^{dd} dynamic distributed species at the interface studied, $\vec{p}^{\,\text{het}}(t)$ is the vector of the production rates of these species at the interface studied, $\vec{p}^{\,\text{het},\star}$ is the analogous vector in the initial state of the system (if non-vanishing), and $\overline{\mathcal{K}}(t,\tau)$ is the matrix of integral transformation kernels. Equation (8.1) allows one to express interfacial concentrations $\vec{c}^{\,\dagger}$ as integrals of the species production rates $\vec{p}^{\,\text{het}}(t)$. Substitution of these concentrations into boundary conditions (2.54) at the interface studied yields a system of IEs for the unknown production rates, in the way analogous to that in Chaps. 5 and 6. The determination of the matrix $\overline{\mathcal{K}}(t,\tau)$ presents the essential theoretical challenge associated with the use of the IE method for models involving homogeneous reactions, and we shall see below that this can be a difficult task.

Fortunately, in the theoretical description of quite many reaction–transport models it happens to be possible to replace the concentrations by an equivalent set of auxiliary variables u_j selected in such a way that the reaction–transport PDE system obtained for u_j consists of uncoupled PDEs, which substantially simplifies the treatment. In order to gain a basic understanding of the conditions under which the reaction–transport PDEs can be cast into the uncoupled form, and to prepare the ground for further discussion in Sects. 8.2 and 8.4, let us consider the case of (pseudo) first-order homogeneous reactions acting between N_s^{dd} dynamic distributed species and N_s^{sd} static distributed species. The general system (2.10) of the reaction–transport PDEs for the dynamic distributed species can be combined with Eq. (2.52), giving

$$\frac{\partial \vec{c}}{\partial t} = \overline{D}\,\Delta\,\vec{c} - \boldsymbol{v}\cdot\mathbf{grad}\,\vec{c} - \overline{K}\,\vec{c} + \vec{\sigma}\,. \tag{8.2}$$

We further assume that all initial concentrations are uniform in space, and that all homogeneous reactions are initially at equilibrium. Hence, the vector $\vec{c}^{\,\star} = \left[c_1^\star, \ldots, c_{N_s^{\text{dd}}}^\star\right]^{\text{T}}$ of initial concentrations cannot be entirely arbitrary, but its elements must obey the condition

$$\vec{\sigma} - \overline{K}\,\vec{c}^{\,\star} = \vec{0}\,. \tag{8.3}$$

Let us introduce a vector of auxiliary variables $\overrightarrow{u} = \left[u_1, \ldots, u_{N_s^{dd}}\right]^T$ linked to the concentrations by the linear transformation

$$\overrightarrow{u} = \overline{S}\,(\overrightarrow{c} - \overrightarrow{c}^{\,\star}) \tag{8.4}$$

represented by a certain invertible matrix $\overline{S} = \left\{S_{i,j}\right\}_{N_s^{dd} \times N_s^{dd}}$, so that

$$\overrightarrow{c} = \overrightarrow{c}^{\,\star} + \overline{S}^{\,-1}\,\overrightarrow{u}\,, \tag{8.5}$$

where $\overline{S}^{\,-1} = \left\{S_{i,j}^{\text{inv}}\right\}_{N_s^{dd} \times N_s^{dd}}$ is the matrix inverse of $\overline{S}$. After inserting $\overrightarrow{c}$ given by Eq. (8.5) into Eq. (8.2), and applying matrix $\overline{S}$ to both sides of the resulting equation, we obtain after some manipulation

$$\frac{\partial \overrightarrow{u}}{\partial t} = \tilde{\overline{D}}\,\Delta\,\overrightarrow{u} - \boldsymbol{v}\cdot\mathbf{grad}\,\overrightarrow{u} - \tilde{\overline{K}}\,\overrightarrow{u} \tag{8.6}$$

with the initial condition

$$t = 0 \Rightarrow \overrightarrow{u} = \overrightarrow{0}\,, \tag{8.7}$$

where

$$\tilde{\overline{D}} = \overline{S}\;\overline{D}\;\overline{S}^{\,-1}\,, \tag{8.8}$$

$$\tilde{\overline{K}} = \overline{S}\;\overline{K}\;\overline{S}^{\,-1} \tag{8.9}$$

are the matrices of diffusion coefficients and rate constants for the transformed reaction–transport equation (8.6).

In a similar way, we can obtain the expression for the vector of transformed fluxes. The vector of concentration fluxes is defined by the Eq. (2.9):

$$\overrightarrow{J} = -\overline{D}\,\mathbf{grad}\,\overrightarrow{c} + \boldsymbol{v}\,\overrightarrow{c}\,. \tag{8.10}$$

As was discussed in Chap. 6, in the presence of convection this vector generally does not vanish when $\overrightarrow{c} = \overrightarrow{c}^{\,\star}$, so that we have to distinguish the initial vector of fluxes:

$$\overrightarrow{J}^{\,\star} = -\overline{D}\,\mathbf{grad}\,\overrightarrow{c}^{\,\star} + \boldsymbol{v}\,\overrightarrow{c}^{\,\star} = \boldsymbol{v}\,\overrightarrow{c}^{\,\star}\,. \tag{8.11}$$

By subtracting Eqs. (8.10) and (8.11) we obtain

$$\overrightarrow{J} - \overrightarrow{J}^{\,\star} = -\overline{D}\,\mathbf{grad}\,(\overrightarrow{c} - \overrightarrow{c}^{\,\star}) + \boldsymbol{v}\,(\overrightarrow{c} - \overrightarrow{c}^{\,\star})\,. \tag{8.12}$$

Multiplication of both sides of Eq. (8.12) by $\overline{S}$ gives, in view of Eqs. (8.4) and (8.8):

$$\overline{S}\,(\vec{J} - \vec{J}^{\,\star}) = -\tilde{\overline{D}}\,\mathbf{grad}\,\vec{u} + v\,\vec{u}\,. \tag{8.13}$$

Let us denote

$$\vec{J}^{\,u} = -\tilde{\overline{D}}\,\mathbf{grad}\,\vec{u} + v\,\vec{u}\,. \tag{8.14}$$

Vector $\vec{J}^{\,u}$ represents the transformed vector of flux vectors, that is the fluxes of the variables $\vec{u}$. The fact is indicated by adding the superscript "u". Equation (8.13) can therefore be rewritten as

$$\vec{J}^{\,u} = \overline{S}\,(\vec{J} - \vec{J}^{\,\star})\,. \tag{8.15}$$

This is a flux analogue of Eq. (8.4) for concentrations. From Eq. (8.15) we get the flux analogue of Eq. (8.5):

$$\vec{J} = \vec{J}^{\,\star} + \overline{S}^{\,-1}\,\vec{J}^{\,u}\,. \tag{8.16}$$

If Eq. (8.6) is to represent uncoupled PDEs, matrix $\overline{S}$ must be such that $\tilde{\overline{D}}$ and $\tilde{\overline{K}}$ are simultaneously diagonal. One obvious situation (albeit possibly not the only one) when such a transformation exists is when the following two conditions (a) and (b) are simultaneously satisfied: (a) all diffusion coefficients are equal; and (b) matrix $\overline{K}$ is similar to a diagonal matrix. Condition (a) implies that

$$\overline{D} = D\,\overline{1}\,, \tag{8.17}$$

where $\overline{1}$ is the unit matrix and D is the common diffusion coefficient, so that $\tilde{\overline{D}}$ is then diagonal:

$$\tilde{\overline{D}} = D\,\overline{S}\,\overline{1}\,\overline{S}^{\,-1} = D\,\overline{1} = \overline{D}\,. \tag{8.18}$$

Condition (b) implies that $\tilde{\overline{K}}$ is diagonal, if for $\overline{S}$ we take the relevant matrix of the similarity transformation. In such a case the diagonal elements $\tilde{K}_{j,j}$ ($j = 1, \ldots, N_s^{dd}$) of $\tilde{\overline{K}}$ are equal to the eigenvalues of $\overline{K}$. In practice it also happens that the PDE system (8.2) can be decomposed into subsystems such that the PDEs are coupled only within the subsystems. Conditions (a) and (b) need not then apply to the whole system, but individually to every subsystem. Hence, the diffusion coefficients may be identical within the subsystems, but different in different subsystems.

Examples of systems, for which the above de-coupling of the reaction–transport PDEs is possible, are provided by the following standard reaction schemes involving (pseudo) first-order homogeneous reactions; the EC scheme:

$$\left.\begin{array}{c} X_1 + e^- \rightleftarrows X_2 \\ X_2 \xrightarrow{k_f} X_3 \end{array}\right\} , \tag{8.19}$$

(assuming equal diffusion coefficients of species X_2 and X_3); the catalytic EC scheme:

$$\left.\begin{array}{c} X_1 + e^- \rightleftarrows X_2 \\ X_2 \xrightarrow{k_f} X_1 \end{array}\right\} , \tag{8.20}$$

(assuming equal diffusion coefficients of X_1 and X_2); and the CE scheme:

$$\left.\begin{array}{c} X_1 \underset{k_b}{\overset{k_f}{\rightleftarrows}} X_2 \\ X_2 + e^- \rightleftarrows X_3 \end{array}\right\} , \tag{8.21}$$

(assuming equal diffusion coefficients of X_1 and X_2). Only species X_1 is initially present in the schemes (8.19) and (8.20), and only species X_1 and X_2 are initially present at positive equilibrium in the scheme (8.21). Table 8.1 provides matrices $\overline{D}$, $\overline{K}$, $\overline{S}$, $\overline{S}^{-1}$, $\tilde{\overline{D}}$ and $\tilde{\overline{K}}$, and vectors $\vec{c}^{\,\star}$ for schemes (8.19)–(8.21) under assumptions adopted.

Once we have a system (8.6) of uncoupled PDEs, we can obtain solution–flux relationships for every u_j separately, from every individual PDE. As the transformation (8.4) is linear, the mutually interrelated concentration–production rate relationships for the various species are then easy to obtain from those for the u_j variables. The problem thus reduces to obtaining solution–flux relationships for a single reaction–transport PDE. This task is discussed in Sect. 8.2. The largely unresolved problem of handling the general case of coupled (pseudo) first-order homogeneous reaction–transport PDEs (8.2) is addressed in Sect. 8.4.

8.2 A Single Reaction–Transport PDE, First-Order Homogeneous Reactions

In view of the preceding discussion, let us assume that all homogeneous reactions are of (pseudo) first order, and that a suitable set of auxiliary variables u_j exists, such that we have an uncoupled system of reaction–transport PDEs for u_j. We thus have to consider a single reaction–transport PDE that in accordance with Eq. (8.6)

Table 8.1 Matrices $\overline{D}, \overline{K}, \overline{S}, \overline{S}^{-1}, \tilde{\overline{D}}$ and $\tilde{\overline{K}}$, and vectors $\overrightarrow{c}^{\star}$ for schemes (8.19)–(8.21)

Matrix/vector	Scheme (8.19)	Scheme (8.20)	Scheme (8.21)
$\overline{D}$	$\begin{bmatrix} D_1 & 0 & 0 \\ 0 & D & 0 \\ 0 & 0 & D \end{bmatrix}$	$\begin{bmatrix} D & 0 \\ 0 & D \end{bmatrix}$	$\begin{bmatrix} D & 0 & 0 \\ 0 & D & 0 \\ 0 & 0 & D_3 \end{bmatrix}$
$\overline{K}$	$\begin{bmatrix} 0 & 0 & 0 \\ 0 & k_f & 0 \\ 0 & -k_f & 0 \end{bmatrix}$	$\begin{bmatrix} 0 & -k_f \\ 0 & k_f \end{bmatrix}$	$\begin{bmatrix} k_f & -k_b & 0 \\ -k_f & k_b & 0 \\ 0 & 0 & 0 \end{bmatrix}$
$\overline{S}$	$\begin{bmatrix} 1 & 0 & 0 \\ 0 & 1 & 0 \\ 0 & 1 & 1 \end{bmatrix}$	$\begin{bmatrix} 1 & 1 \\ 0 & 1 \end{bmatrix}$	$\begin{bmatrix} 1 & 1 & 0 \\ 1 & -k_b/k_f & 0 \\ 0 & 0 & 1 \end{bmatrix}$
$\overline{S}^{-1}$	$\begin{bmatrix} 1 & 0 & 0 \\ 0 & 1 & 0 \\ 0 & -1 & 1 \end{bmatrix}$	$\begin{bmatrix} 1 & -1 \\ 0 & 1 \end{bmatrix}$	$\begin{bmatrix} \frac{k_b}{k_f+k_b} & \frac{k_f}{k_f+k_b} & 0 \\ \frac{k_f}{k_f+k_b} & \frac{-k_f}{k_f+k_b} & 0 \\ 0 & 0 & 1 \end{bmatrix}$
$\tilde{\overline{D}}$	$\begin{bmatrix} D_1 & 0 & 0 \\ 0 & D & 0 \\ 0 & 0 & D \end{bmatrix}$	$\begin{bmatrix} D & 0 \\ 0 & D \end{bmatrix}$	$\begin{bmatrix} D & 0 & 0 \\ 0 & D & 0 \\ 0 & 0 & D_3 \end{bmatrix}$
$\tilde{\overline{K}}$	$\begin{bmatrix} 0 & 0 & 0 \\ 0 & k_f & 0 \\ 0 & 0 & 0 \end{bmatrix}$	$\begin{bmatrix} 0 & 0 \\ 0 & k_f \end{bmatrix}$	$\begin{bmatrix} 0 & 0 & 0 \\ 0 & k_f + k_b & 0 \\ 0 & 0 & 0 \end{bmatrix}$
$\overrightarrow{c}^{\star}$	$\begin{bmatrix} c_1^{\star} \\ 0 \\ 0 \end{bmatrix}$	$\begin{bmatrix} c_1^{\star} \\ 0 \end{bmatrix}$	$\begin{bmatrix} c_1^{\star} \\ (k_f/k_b)\,c_1^{\star} \\ 0 \end{bmatrix}$

can be written in the general form

$$\frac{\partial u_j}{\partial t} = \tilde{D}_j \, \Delta \, u_j - \boldsymbol{v} \cdot \textbf{grad}\, u_j - \tilde{K}_{j,j}\, u_j \, . \tag{8.22}$$

The corresponding flux of the variable u_j is [cf. Eq. (8.14)]:

$$\boldsymbol{J}_j^u = -\tilde{D}_j \, \textbf{grad}\, u_j + \boldsymbol{v}\, u_j \, . \tag{8.23}$$

In agreement with Eq. (8.4) we further assume the initial condition

$$t = 0 \Rightarrow u_j = 0 \, , \tag{8.24}$$

and boundary conditions at infinity (in the case of the semi-infinite spatial domains) or at the second boundary (in the case of finite spatial domains) analogous to those considered in Sects. 5.2 and 5.3, i.e. either

$$u_j = 0 \tag{8.25}$$

or

$$\mathbf{grad}\, u_j = 0\,. \tag{8.26}$$

In order to obtain concentration–flux relationships for the incomplete IBVP (8.22), (8.24), and (8.25) or (8.26), one can follow at least two approaches. One approach (exemplified by [40, 43, 55, 71, 73, 116]) relies on the variable substitution:

$$u_j = w_j \, \exp(-\tilde{K}_{j,j}\, t)\,. \tag{8.27}$$

Temporal differentiation of Eq. (8.27) gives

$$\frac{\partial u_j}{\partial t} = -w_j\, \tilde{K}_{j,j}\, \exp(-\tilde{K}_{j,j}\, t) + \exp(-\tilde{K}_{j,j}\, t)\, \frac{\partial w_j}{\partial t}\,, \tag{8.28}$$

and, since the Laplace and gradient operators do not involve temporal derivatives (cf. Table 2.1),

$$\tilde{D}_j\, \triangle\, u_j - \boldsymbol{v} \cdot \mathbf{grad}\, u_j = \exp(-\tilde{K}_{j,j}\, t)\, \left[\tilde{D}_j\, \triangle\, w_j - \boldsymbol{v} \cdot \mathbf{grad}\, w_j\right]\,. \tag{8.29}$$

After substituting Eqs. (8.28) and (8.29) into Eq. (8.22), and simplifying the exponential factor, we find that the substitution (8.27) reduces the reaction–transport equation (8.22) to a pure transport PDE for the variable w_j:

$$\frac{\partial w_j}{\partial t} = \tilde{D}_j\, \triangle\, w_j - \boldsymbol{v} \cdot \mathbf{grad}\, w_j\,. \tag{8.30}$$

At the same time, substitution (8.27) turns Eq. (8.23) into

$$\boldsymbol{J}_j^u = \exp(-\tilde{K}_{j,j}\, t)\, \boldsymbol{J}_j^w\,, \tag{8.31}$$

where

$$\boldsymbol{J}_j^w = -\tilde{D}_j\, \mathbf{grad}\, w_j + \boldsymbol{v}\, w_j \tag{8.32}$$

is the flux of w_j.

Combination of Eqs. (8.24) and (8.27) gives the initial condition for w_j:

$$t = 0 \Rightarrow w_j = 0\,, \tag{8.33}$$

which is analogous to the initial condition for u_j. Similarly, combination of the boundary conditions (8.25) or (8.26) for u_j with Eq. (8.27) gives analogous boundary conditions for w_j:

$$w_j = 0\,, \tag{8.34}$$

or

$$\mathbf{grad}\, w_j = 0\,. \tag{8.35}$$

This does not mean that the variable w_j obeys the same boundary conditions at the interface studied, as the variable u_j does. For this, a more complicated substitution of variables would be necessary (for reaction–diffusion PDEs this issue was studied by Danckwerts [24]). However, Eqs. (8.30), (8.33), and (8.34) or (8.35) are sufficient to expect that the solution–flux relationship for w_j at the interface studied is identical to what one would obtain for u_j in the absence of the homogeneous reaction term $-\tilde{K}_{j,j}\, u_j$.

Let us see what this means in the case of models dependent on one spatial coordinate. In view of the results from Chaps. 5 and 6, the solution–flux relationships for the variable w_j at the interface studied can then be written in the general form

$$w_j(0,t) = \int_0^t \mathcal{K}_j(t,\tau)\; J_j^w(0,\tau)\; d\tau\,, \tag{8.36}$$

valid for planar diffusion or for convection–diffusion models, or in the form

$$w_j(r_0,t) = \int_0^t \mathcal{K}_j(t,\tau)\; J_j^w(r_0,\tau)\; d\tau\,, \tag{8.37}$$

valid for spherical and cylindrical diffusion, where $\mathcal{K}_j(t,\tau)$ is one of the kernels discussed there, that is $\mathcal{K}_j^{\mathrm{p}}(t,\tau)$, $\mathcal{K}_j^{\mathrm{s}}(t,\tau)$, $\mathcal{K}_j^{\mathrm{c}}(t,\tau)$, $\mathcal{K}_j^{\mathrm{plp}}(t,\tau)$, $\mathcal{K}_j^{\mathrm{ct}}(t,\tau)$, etc., or any other suitable kernel, calculated assuming the diffusion coefficient $\tilde{D}_j$. By combining Eqs. (8.36) and (8.37) with Eqs. (8.27) and (8.31), we obtain the following solution–flux relationship for the variable u_j. For planar diffusion and for convection–diffusion models considered:

$$u_j(0,t) = \int_0^t \tilde{\mathcal{K}}_j(t,\tau)\; J_j^u(0,\tau)\; d\tau\,, \tag{8.38}$$

and for spherical and cylindrical diffusion:

$$u_j(r_0,t) = \int_0^t \tilde{\mathcal{K}}_j(t,\tau)\; J_j^u(r_0,\tau)\; d\tau\,, \tag{8.39}$$

where

$$\tilde{\mathcal{K}}_j(t,\tau) = \exp\left[-\tilde{K}_{j,j}\,(t-\tau)\right]\, \mathcal{K}_j(t,\tau)\,. \tag{8.40}$$

Thus, we conclude that in the presence of the single homogeneous reaction term $-\tilde{K}_{j,j}\, u_j$ in the transport PDE (8.22), the integral transformation kernels valid for variables u_j without homogeneous reactions must be multiplied by $\exp\left[-\tilde{K}_{j,j}\,(t-\tau)\right]$. The resulting kernels are denoted below by the same symbols as the kernels obtained in the absence of homogeneous reactions, but distinguished with a tilde over the symbols. For example, for planar diffusion with a homogeneous reaction in a semi-infinite spatial domain:

$$\tilde{\mathscr{K}}_j^{\mathrm{p}}(t,\tau) = \tilde{D}_j^{-1/2} \exp\left[-\tilde{K}_{j,j}\,(t-\tau)\right] [\pi(t-\tau)]^{-1/2} \,. \tag{8.41}$$

For planar diffusion in finite spatial domain:

$$\begin{aligned}\tilde{\mathscr{K}}_j^{\mathrm{plp}}(t,\tau) = \tilde{D}_j^{-1/2} & \exp\left[-\tilde{K}_{j,j}\,(t-\tau)\right] \\ & \times \left\{[\pi(t-\tau)]^{-1/2} - \tilde{\mu}_j\,\mathrm{kplp}\left[\tilde{\mu}_j\,(t-\tau)^{1/2}\right]\right\} ,\end{aligned} \tag{8.42}$$

or

$$\begin{aligned}\tilde{\mathscr{K}}_j^{\mathrm{pli}}(t,\tau) = \tilde{D}_j^{-1/2} & \exp\left[-\tilde{K}_{j,j}\,(t-\tau)\right] \\ & \times \left\{[\pi(t-\tau)]^{-1/2} + \tilde{\mu}_j\,\mathrm{kpli}\left[\tilde{\mu}_j\,(t-\tau)^{1/2}\right]\right\} ,\end{aligned} \tag{8.43}$$

with $\tilde{\mu}_j$ defined by Eq. (5.119), assuming the diffusion coefficient $\tilde{D}_j$. For convection–diffusion to an expanding plane, with a homogeneous reaction:

$$\tilde{\mathscr{K}}_j^{\mathrm{ep}}(t,\tau) = \tilde{D}_j^{-1/2} \left(\frac{3}{7}\pi\right)^{-1/2} \exp\left[-\tilde{K}_{j,j}\,(t-\tau)\right] \tau^{2/3} \left(t^{7/3} - \tau^{7/3}\right)^{-1/2} , \tag{8.44}$$

etc.

Entirely analogous are kernels for spatially two-dimensional and three-dimensional models in the Mirkin and Bard approach (see Sect. 7.2). For example, for diffusion to band(s), with a homogeneous reaction:

$$\begin{aligned}\tilde{\mathscr{K}}_j^{\mathrm{b}}\left(\begin{bmatrix} x \\ t \end{bmatrix}, \begin{bmatrix} \zeta \\ \tau \end{bmatrix}\right) &= \tilde{D}_j^{-1}\, [2\pi(t-\tau)]^{-1} \\ & \times \exp\left[-\tilde{K}_{j,j}\,(t-\tau)\right] \exp\left[-\frac{(x-\zeta)^2}{4\tilde{D}_j(t-\tau)}\right] ,\end{aligned} \tag{8.45}$$

and for diffusion to disk/ring(s), with a homogeneous reaction:

$$\tilde{\mathscr{K}}_j^{\mathrm{dr}}\left(\begin{bmatrix} r \\ t \end{bmatrix}, \begin{bmatrix} \zeta \\ \tau \end{bmatrix}\right) = \tilde{D}_j^{-3/2} \exp\left[-\tilde{K}_{j,j}\,(t-\tau)\right] \frac{\zeta \exp\left[-\frac{\zeta^2+r^2}{4\tilde{D}_j(t-\tau)}\right] \mathrm{I}_0\left[\frac{r\zeta}{2\tilde{D}_j(t-\tau)}\right]}{2\pi^{1/2}(t-\tau)^{3/2}} . \tag{8.46}$$

A second possible approach to obtaining the solution–flux relationships for Eq. (8.22) relies on the direct application of the Laplace transform, in the way analogous to the derivations shown in Appendices A and B. This approach has been rather predominant in the literature, and it seems unavoidable in the case of coupled reaction–transport PDEs (see Sect. 8.4 below). However, in the case of the PDEs that can be reduced to uncoupled reaction–transport PDEs, discussed in the present section, this approach presents an unnecessary complication, and should be avoided, the more so because it does not always work (in particular, it cannot be used in the case of convection–diffusion to an expanding plane). Nevertheless, for the convenience of the Reader, in Appendix E, Sect. E.1 we present example derivations of the relationships (8.38) and (8.39) by the Laplace transform method. The derivations reveal that in the presence of the single homogeneous reaction term $-\tilde{K}_{j,j}\, u_j$ in Eq. (8.22) any explicit occurrence of the Laplace variable s is always in the form of the sum $s + \tilde{K}_{j,j}$. As the addition of a constant to the variable s in the Laplace transform of an original function is equivalent to the multiplication of the original function by the exponential function [cf. Eq. (3.16) in Sect. 3.2], any kernel corresponding to $\tilde{K}_{j,j} \neq 0$ is equal to the kernel corresponding to $\tilde{K}_{j,j} = 0$, times $\exp\left[-\tilde{K}_{j,j}(t-\tau)\right]$.

After deriving the solution–flux relationships for all u_j, $j = 1, \ldots, N_{\mathrm{s}}^{\mathrm{dd}}$, one can finally calculate the corresponding concentration–production rate relationships. In the case of models dependent on one spatial coordinate, by introducing a diagonal matrix $\overline{\tilde{\mathscr{K}}}(t,\tau) = \mathrm{diag}\left[\tilde{\mathscr{K}}_1(t,\tau), \ldots, \tilde{\mathscr{K}}_{N_{\mathrm{s}}^{\mathrm{dd}}}(t,\tau)\right]$ of the kernels $\tilde{\mathscr{K}}_j(t,\tau)$, the relationship between the vectors $\vec{u}(0,t)$ and $\vec{J}^{\,u}(0,t)$ can be written in the following way. For planar diffusion and for convection–diffusion models considered:

$$\vec{u}(0,t) = \int_0^t \overline{\tilde{\mathscr{K}}}(t,\tau)\, \vec{J}^{\,u}(0,\tau)\, \mathrm{d}\tau\,, \tag{8.47}$$

and for spherical and cylindrical diffusion:

$$\vec{u}(r_0,t) = \int_0^t \overline{\tilde{\mathscr{K}}}(t,\tau)\, \vec{J}^{\,u}(r_0,\tau)\, \mathrm{d}\tau\,. \tag{8.48}$$

By combining Eqs. (8.47) or (8.48) with Eqs. (8.4) and (8.15) we then obtain, respectively,

$$\vec{c}(0,t) = \vec{c}^{\,\star} + \int_0^t \overline{\mathscr{K}}(t,\tau)\left[\vec{J}(0,\tau) - \vec{J}^{\,\star}\right] \mathrm{d}\tau \tag{8.49}$$

and

$$\vec{c}(r_0,t) = \vec{c}^{\,\star} + \int_0^t \overline{\mathscr{K}}(t,\tau)\left[\vec{J}(r_0,\tau) - \vec{J}^{\,\star}\right] \mathrm{d}\tau \,, \tag{8.50}$$

where

$$\overline{\mathscr{K}}(t,\tau) = \overline{S}^{-1}\,\overline{\tilde{\mathscr{K}}}(t,\tau)\,\overline{S} \tag{8.51}$$

is the matrix of kernel functions representing the relationships between the concentrations and their fluxes at the interface studied. Equations (8.49) and (8.50) are seen to be special cases of the general equation (8.1), as the fluxes at the interface represent species production rates [cf. Eq. (2.46)].

Obviously, by comparison of Eqs. (8.9) and (8.51) we see that matrix $\overline{\mathscr{K}}(t,\tau)$ has the same sparsity structure as the matrix $\overline{K}$, so that in general it is not diagonal. This proves that the concentration of any particular species may be related to the fluxes of all remaining species, when homogeneous reactions are present.

Apart from the effect of $\tilde{K}_{j,j} \neq 0$ on the integral transformation kernels characteristic of transient conditions, an important consequence of the presence of the homogeneous reaction terms, in the uncoupled reaction–transport equations for variables u_j, is the possibility of the occurrence of steady states in which the transport term is compensated by the reaction term, so that $\partial u_j/\partial t \approx 0$. These steady states are somewhat similar to another steady state characteristic of spherical diffusion without homogeneous reactions in semi-infinite domain, previously mentioned in Sect. 5.2.1, but now they may occur for all types of geometry and transport conditions. In Appendix E, Sect. E.2 we derive some of the solution–flux relationships for such steady states. The derivations are restricted to one-dimensional semi-infinite spatial domains, and they require that $\tilde{K}_{j,j} > 0$. In particular, we obtain the following relationships [cf. Eqs. (E.37), (E.44), and (E.52)]. For planar diffusion with reaction:

$$u_j(0,t) = \tilde{D}_j^{-1/2}\,\tilde{K}_{j,j}^{-1/2}\,J_j^u(0,t)\,. \tag{8.52}$$

For spherical diffusion with reaction:

$$u_j(r_0,t) = \tilde{D}_j^{-1/2}\left(\tilde{\varrho}_j + \tilde{K}_{j,j}^{1/2}\right)^{-1} J_j^u(r_0,t)\,. \tag{8.53}$$

When $\tilde{K}_{j,j}^{1/2} \gg \tilde{\varrho}_j$, Eq. (8.53) converges to Eq. (8.52) for planar diffusion with reaction. When $\tilde{K}_{j,j}^{1/2} \ll \tilde{\varrho}_j$, Eq. (8.53) becomes analogous to Eq. (5.89) specific for the steady state spherical diffusion without reaction. For cylindrical diffusion with reaction:

$$u_j(r_0,t) = \tilde{D}_j^{-1/2}\,\tilde{K}_{j,j}^{-1/2}\,\frac{\mathrm{K}_0(\tilde{K}_{j,j}^{1/2}/\tilde{\varrho}_j)}{\mathrm{K}_1(\tilde{K}_{j,j}^{1/2}/\tilde{\varrho}_j)}\,J_j^u(r_0,t)\,. \tag{8.54}$$

According to the asymptotic expansions for the Bessel functions $\mathrm{K}_0(\cdot)$ and $\mathrm{K}_1(\cdot)$ (cf. Eq. (9.7.2) in [1]), for large real z the ratio $\mathrm{K}_0(z)/\mathrm{K}_1(z) \sim 1-(1/2)\,z^{-1}+(3/8)\,z^{-2}-(3/8)\,z^{-3}+\ldots \approx 1$. Therefore, for $\tilde{K}_{j,j}^{1/2} \gg \tilde{\varrho}_j$ Eq. (8.54) becomes analogous to the Eq. (8.52) for planar diffusion with reaction. For small real z we have in turn $\mathrm{K}_0(z) \sim -\ln(z)$ and $\mathrm{K}_1(z) \sim z^{-1}$. Therefore, for $\tilde{K}_{j,j}^{1/2} \ll \tilde{\varrho}_j$ Eq. (8.54) becomes

$$u_j(r_0,t) = \tilde{D}_j^{-1/2}\,\tilde{\varrho}_j^{-1}\,\ln(\tilde{\varrho}_j\,\tilde{K}_{j,j}^{-1/2})\,J_j^u(r_0,t)\,. \tag{8.55}$$

The determination of the steady state solution–flux relationships becomes more sophisticated in the case of diffusion with homogeneous reaction(s) in two- and three-dimensional spatial domains. Similarly to the case of purely diffusional steady states mentioned in Sect. 7.1, it may be necessary to use, e.g. the BIM to derive such relationships. Examples of such calculations were reported by Bender and Stone [19] and Lucas et al. [51].

8.3 Literature Examples, Transport Kernels Multiplied by $\exp[-k_j\,(t-\tau)]$

Based on the results of the previous section, we expect that kernel terms resulting from multiplying a kernel for uncomplicated one-dimensional transport, by exponential factors $\exp\left[-k_j\,(t-\tau)\right]$, where k_j is a certain coefficient related to the rate constants of homogeneous reactions, should be particularly frequent in the IEs, when the presence of homogeneous reactions is taken into account. In this section we list a number of IE-based models published in the literature, and containing such kernels, for previously discussed transport conditions. In the majority of these examples equal diffusion coefficients of the dynamic distributed species were assumed. Sections 8.3.1–8.3.6 contain examples of models defined on one-dimensional spatial domain. Section 8.3.7 contains examples of models defined on spatial domains having a higher dimensionality.

8.3.1 Planar Diffusion, Semi-Infinite Spatial Domain

IE-based models utilising kernels $\exp\left[-k_j\,(t-\tau)\right]\,\mathscr{K}_j^{\mathrm{p}}(t,\tau)$, referring to planar diffusion in semi-infinite spatial domains, are encountered relatively frequently.

For the EC reaction scheme:

$$X_1 + n\,e^- \rightleftarrows X_2\,, \tag{8.56}$$

$$X_2 \rightleftarrows X_3 \tag{8.57}$$

involving electron transfer reaction (8.56) and homogeneous reaction (8.57), and dynamic distributed species X_1, X_2, and X_3, there are the following examples. Kumar and Birke [47] considered potential step chronoamperometry, assuming non-equilibrium reversible reactions (8.56) and (8.57). Ohsaka et al. [79] considered double potential step chronoamperometry, assuming non-equilibrium reversible homogeneous reaction (8.57). Nicholson and Shain [69] presented and solved IEs describing linear potential sweep/cyclic voltammetry for equilibrium electron transfer (8.56) and non-equilibrium reversible or irreversible reaction (8.57). The case with irreversible reaction (8.57) was re-solved by Rudolph [90]. Schwarz and Shain [105], and Lundquist and Nicholson [52] discussed a theory of potential step chronoamperometry followed by linear potential sweep voltammetry, assuming equilibrium electron transfer (8.56), and irreversible homogeneous reaction (8.57). Savéant [94] presented a theory of cyclic voltammetry with asymmetrical potential sweeps. O'Dea et al. [77] and Garay and Lovrić [30] obtained a theory of square wave voltammetry, assuming equilibrium or non-equilibrium reversible electron transfer (8.56), and non-equilibrium reversible homogeneous reaction (8.57). Smith [114] discussed a theory of AC polarography, assuming non-equilibrium reversible reactions (8.56) and (8.57). Theory of cyclic staircase voltammetry was presented by Murphy et al. [60], assuming equilibrium reaction (8.56) and non-equilibrium reversible reaction (8.57).

For the catalytic EC reaction scheme:

$$X_1 + n\,e^- \rightleftarrows X_2\,, \tag{8.58}$$

$$X_2 \rightleftarrows X_1 \tag{8.59}$$

involving electron transfer (8.58) and homogeneous reaction (8.59), and dynamic distributed species X_1 and X_2 having identical diffusion coefficients [possible static species participating in reaction (8.59) are omitted], there are the following examples. Savéant and Vianello [97, 101], Nicholson and Shain [69], and also Davčeva et al. [25], presented and solved IEs describing linear potential sweep/cyclic voltammetry. Equilibrium or irreversible electron transfer (8.58) and irreversible homogeneous reaction (8.59) were assumed. Rudolph [90] completed these results for non-equilibrium reversible electron transfer (8.58). Savéant [94] presented a theory of cyclic voltammetry with asymmetrical potential sweeps. O'Dea et al. [77], Mirčeski et al. [58], and Davčeva et al. [25] obtained a theory of square wave voltammetry, covering the cases of equilibrium or irreversible electron transfer (8.58), and irreversible or non-equilibrium reversible homogeneous reaction (8.59). Davčeva et al. [25] considered also potential step chronoamperometry. Smith [114] discussed a theory of AC polarography, assuming non-equilibrium reversible electron transfer (8.58) and irreversible homogeneous reaction (8.59). He also discussed [115] the theory of the Faradaic impedance, as studied by four small amplitude techniques. Theory of cyclic staircase voltammetry was presented by Murphy et al. [60], assuming equilibrium reaction (8.58) and non-equilibrium reversible reaction (8.59).

For the CE reaction scheme:

$$X_1 \rightleftarrows X_2 , \tag{8.60}$$

$$X_2 + n\,e^- \rightleftarrows X_3 \tag{8.61}$$

involving homogeneous reaction (8.60) and electron transfer (8.61), and dynamic distributed species X_1, X_2 and X_3, there are the following examples. Savéant and Vianello [99] and Nicholson and Shain [69] presented and solved IEs describing linear potential sweep/cyclic voltammetry. Non-equilibrium reversible homogeneous reaction (8.60), and equilibrium or irreversible electron transfer (8.61) were assumed. Rudolph [90] re-solved some of the IEs. O'Dea et al. [77] and Garay and Lovrić [30] obtained a theory of square wave voltammetry, covering the cases of non-equilibrium reversible homogeneous reaction (8.60) and equilibrium or non-equilibrium reversible electron transfer (8.61). Smith [114] discussed a theory of AC polarography, assuming non-equilibrium reversible reactions (8.60) and (8.61). He also discussed [115] the theory of the Faradaic impedance, as studied by four small amplitude techniques. Theory of cyclic staircase voltammetry was presented by Murphy et al. [60], assuming equilibrium reaction (8.61) and non-equilibrium reversible reaction (8.60).

For the ECE reaction scheme:

$$X_1 + n_1\,e^- \rightleftarrows X_2 , \tag{8.62}$$

$$X_2 \rightarrow X_3 , \tag{8.63}$$

$$X_3 + n_2\,e^- \rightleftarrows X_4 \tag{8.64}$$

involving electron transfers (8.62) and (8.64), and irreversible homogeneous reaction (8.63) there are the following examples. Savéant [93] and Nicholson and Shain [70] presented IEs describing linear potential sweep/cyclic voltammetry, under various assumptions regarding equilibrium or non-equilibrium of the electron transfers. Their theory was later extended by Stuart and Foulkes [117] onto the reaction scheme in which reactions (8.63) and (8.64) were replaced by several analogous parallel reactions. Savéant [94] presented a theory of cyclic voltammetry with asymmetrical potential sweeps. O'Dea et al. [78] obtained a theory of square wave voltammetry, assuming equilibrium electron transfers (8.62) and (8.64).

There are also examples corresponding to other reaction schemes. Polcyn and Shain [85] reported a theory of linear potential sweep/cyclic voltammetry for the catalytic ECE reaction scheme:

$$X_1 + n_1\,e^- \rightleftarrows X_2\,, \tag{8.65}$$

$$X_2 \rightarrow X_1\,, \tag{8.66}$$

$$X_2 + n_2\,e^- \rightleftarrows X_3 \tag{8.67}$$

under various assumptions regarding equilibrium or non-equilibrium of the electron transfers. Shuman and Shain [107] considered IEs describing cyclic voltammetry for the CEE scheme:

$$X_1 \rightleftarrows X_2\,, \tag{8.68}$$

$$X_2 + n\,e^- \rightleftarrows X_3\,, \tag{8.69}$$

$$X_1 + n\,e^- \rightarrow X_3\,. \tag{8.70}$$

Non-equilibrium reversible homogeneous reaction (8.68), equilibrium electron transfer (8.69) and irreversible electron transfer (8.70) were assumed. Andrieux et al. [15] considered cyclic voltammetry for the ECC reaction scheme:

$$X_1 + e^- \rightleftarrows X_2\,, \tag{8.71}$$

$$X_2 + X_3 \rightarrow X_4 + \text{products}\,, \tag{8.72}$$

$$X_2 + X_4 \rightarrow X_1 + \text{products}\,, \tag{8.73}$$

obtaining IEs for some limiting cases. Maran et al. [53] considered linear potential sweep voltammetry for the EEC reaction scheme:

$$2X_1 + 2e^- \rightarrow X_2\,, \tag{8.74}$$

$$X_2 + 2e^- \rightarrow X_3\,, \tag{8.75}$$

$$X_2 \rightarrow X_4\,. \tag{8.76}$$

Balducci and Costa [18] discussed cyclic voltammetry for a four-member square scheme involving equilibrium electron transfers. Andrieux and Savéant [8] used IEs in their theory of cyclic voltammetry for several complex reaction schemes that can be classified as extended ECE schemes. Hung et al. [42] developed a theory of AC polarography for several reaction schemes involving electron transfers coupled with multistep homogeneous reactions. Smith [115] referred to a theory of the Faradaic impedance, for a complicated reaction scheme, but details were not provided.

Gonzalez et al. [34] considered the theory of radical polarograms obtained in a photo-electrochemical experiment. A complicated reaction scheme was assumed.

All the above examples refer to experiments in which processes occurring at an electrode | electrolyte interface were the subject of the studies. A unique example of IEs describing homogeneous reactions accompanying charge transfer at a liquid | liquid interface was provided by Quentel et al. [87]. Reaction schemes resembling EC and CE schemes were considered.

Finally, we mention again (see Chap. 5, Sect. 5.2.2.1) that IEs involving kernels $\exp\left[-k_j\,(t-\tau)\right]\,\mathscr{K}^{\mathrm{p}}_j(t,\tau)$ are also encountered in the theory of the effect of partially covered electrodes on linear potential sweep and cyclic voltammetry [35].

8.3.2 Spherical Diffusion, Semi-Infinite Spatial Domain

Compared to planar diffusion, kernels $\exp\left[-k_j\,(t-\tau)\right]\,\mathscr{K}^{\mathrm{s}}_j(t,\tau)$ referring to spherical diffusion in semi-infinite spatial domains are encountered less frequently.

For the EC reaction scheme:

$$X_1 + n\,e^- \rightleftarrows X_2\,, \tag{8.77}$$

$$X_2 \rightleftarrows X_3 \tag{8.78}$$

involving dynamic distributed species X_1, X_2, and X_3, Olmstead and Nicholson [81] presented a theory of linear potential sweep voltammetry, assuming equilibrium electron transfer (8.77) and irreversible homogeneous reaction (8.78). Zakharov and Bakanov [121] reported IEs describing general current–potential curves. Calvente et al. [23] derived and solved an IE describing potential step chronoamperometry, assuming non-equilibrium reversible electron transfer (8.77), and irreversible homogeneous reaction (8.78). Garay and Lovrić [32] provided a theory of square wave voltammetry.

For the catalytic EC reaction scheme:

$$X_1 + n\,e^- \rightleftarrows X_2\,, \tag{8.79}$$

$$X_2 \rightarrow \frac{a}{b}\,X_1\,, \tag{8.80}$$

where a/b is a fraction of the regenerated depolariser, Delmastro and Booman [27] considered potential step chronoamperometry.

For the CE reaction scheme:

$$X_1 \rightleftarrows X_2\,, \tag{8.81}$$

$$X_2 + n\,e^- \rightleftarrows X_3 \tag{8.82}$$

involving homogeneous reaction (8.81) and electron transfer (8.82), and dynamic distributed species X_1, X_2, and X_3, there are the following examples. Delmastro and Booman [27] and Lovrić and Tur'yan [50] considered potential step chronoamperometry. Zakharov and Pnev [122] reported IEs describing general current–potential curves. Garay and Lovrić [31] considered square wave voltammetry.

8.3.3 Cylindrical Diffusion, Semi-Infinite Spatial Domain

There are even fewer examples utilising kernels $\exp\left[-k_j\,(t-\tau)\right]\,\mathcal{K}_j^{\mathrm{c}}(t,\tau)$ referring to cylindrical diffusion in semi-infinite spatial domains.

For the EC reaction scheme:

$$X_1 + n\,e^- \rightleftarrows X_2\,, \tag{8.83}$$

$$X_2 \rightleftarrows X_3 \tag{8.84}$$

involving dynamic distributed species X_1, X_2, and X_3, Zakharov and Bakanov [121] reported IEs describing general current–potential curves.

Similar results for the CE reaction scheme:

$$X_1 \rightleftarrows X_2\,, \tag{8.85}$$

$$X_2 + n\,e^- \rightleftarrows X_3 \tag{8.86}$$

involving dynamic distributed species X_1, X_2, and X_3 were obtained by Zakharov and Pnev [122]. In Zakharov et al. [121, 122] the kernels for cylindrical diffusion were approximated by truncated series expansions.

8.3.4 Planar Diffusion, Finite Spatial Domain

Kernel functions corresponding to planar diffusion in a finite spatial domain in the presence of homogeneous reactions were derived by Keller and Reinmuth [45], assuming either a permeable or impermeable second boundary. Homogeneous reactions preceding, following, or parallel to (catalytic) electron transfers were considered, and the effect of the reactions on linear potential sweep voltammograms was estimated.

Ng and Cheh [65–67] presented a theory of linear potential sweep voltammetry at an RDE, by using the approximation of planar diffusion in a thin electrolyte layer with a permeable second boundary, and taking into account homogeneous reactions. The following three reaction schemes were discussed.

The EC scheme:

$$X_1 + n\,e^- \rightleftarrows X_2\,, \tag{8.87}$$

$$X_2 \rightarrow X_3 \tag{8.88}$$

involving a non-equilibrium reversible electron transfer (8.87) and irreversible homogeneous reaction (8.88) among dynamic distributed species X_1, X_2, and X_3, was considered by Ng and Cheh [66].

The catalytic EC scheme:

$$X_1 + n\,e^- \rightleftarrows X_2\,, \tag{8.89}$$

$$X_2 + X_{3,s} \rightarrow X_1 \tag{8.90}$$

involving a non-equilibrium reversible electron transfer (8.89) and irreversible homogeneous reaction (8.90) among dynamic distributed species X_1 and X_2, and static species $X_{3,s}$, was considered by Ng and Cheh [65].

The CE scheme:

$$X_1 \rightleftarrows X_2\,, \tag{8.91}$$

$$X_2 + n\,e^- \rightleftarrows X_3 \tag{8.92}$$

involving a reversible homogeneous reaction (8.91), and non-equilibrium reversible electron transfer (8.92) among dynamic distributed species X_1, X_2, and X_3, was considered by Ng and Cheh [67].

The thin layer model of the RDE was also used by Tokuda and Matsuda [120] to obtain a theory of AC voltammetry for ten reaction schemes (CE, EC, EC catalytic, and several CEC, CCE scheme, and ECC schemes) involving electron transfers and homogeneous reactions.

The theory of linear potential sweep voltammetry for the CE reaction scheme (8.91) and (8.92) taking place in a thin planar layer with an impermeable second boundary was discussed by Roizenblat et al. [89]. Reversible homogeneous reaction (8.91) and irreversible electron transfer (8.92) were assumed. In addition, diffusion of species X_1 was neglected.

8.3.5 Expanding Plane

For kernels $\exp\left[-k_j\,(t-\tau)\right]\,\mathscr{K}_j^{\mathrm{ep}}(t,\tau)$ there are several examples referring to the theory of DC polarography.

Nishihara [72] analysed polarographic currents for the catalytic EC reaction scheme:

$$X_1 + n\,e^- \rightleftarrows X_2\,, \tag{8.93}$$

$$X_2 \rightleftarrows X_1 \tag{8.94}$$

involving electron transfer (8.93) and irreversible homogeneous reaction (8.94).

Matsuda [55] discussed DC polarographic current–potential curves assuming equilibrium electron transfer (8.93) and non-equilibrium reversible reaction (8.94).

Nishihara and Matsuda [73, 75] reported the theory of polarography for the CE reaction scheme:

$$X_1 \rightleftarrows X_2 , \tag{8.95}$$

$$X_2 + n\,e^- \rightleftarrows X_3 \tag{8.96}$$

involving homogeneous reaction (8.95) and electron transfer (8.96).

Matsuda [55] discussed DC polarographic current–potential curves assuming equilibrium electron transfer (8.96) and non-equilibrium reversible reaction (8.95).

Sánchez Maestre et al. [92] presented a model of pulse polarography assuming non-equilibrium reversible reactions (8.95) and (8.96), and steady state condition [Eq. (8.52)] for one of the auxiliary variables.

Nicholson et al. [71], Kastening [44], Sobel and Smith [116], and Matsuda [55] considered the DC polarographic theory for the ECE reaction scheme:

$$X_1 + n\,e^- \to X_2 , \tag{8.97}$$

$$X_2 \rightleftarrows X_3 , \tag{8.98}$$

$$X_3 + n\,e^- \to X_4 \tag{8.99}$$

involving electron transfers (8.97) and (8.99), separated by the homogeneous reaction (8.98).

McCord and Smith [56, 57] presented a theory of AC polarography for the CCE reaction scheme:

$$X_1 \rightleftarrows X_2 , \tag{8.100}$$

$$X_2 \rightleftarrows X_3 , \tag{8.101}$$

$$X_3 + n\,e^- \rightleftarrows X_4 . \tag{8.102}$$

Nishihara et al. [76] obtained equations for AC polarography for the EC reaction scheme:

$$X_1 + n\,e^- \rightleftarrows X_2 , \tag{8.103}$$

$$X_2 \to X_3 \tag{8.104}$$

involving equilibrium electron transfer (8.103), and irreversible follow-up homogeneous reaction (8.104).

Nishihara and Matsuda [74] elaborated a simplified theory of DC polarography, for a complex reaction scheme involving three parallel electron transfer reactions coupled by (pseudo) first-order homogeneous reactions. They assumed steady state conditions for some of the auxiliary variables, which eliminated the kernels $\exp\left[-k_j\,(t-\tau)\right]\,\mathcal{K}_j^{\mathrm{ep}}(t,\tau)$ from the IEs.

8.3.6 Tubular Electrodes

For kernels $\exp\left[-k_j\,(t-\tau)\right]\,\mathscr{K}_j^{\mathrm{ct}}(t,\tau)$ there are examples of IEs for linear potential sweep voltammetry only, developed mostly by Singh, Dutt, and co-workers, and consistent with the Singh–Dutt approximation discussed in Sect. 6.2.4.

In particular, the EC reaction scheme:

$$X_1 \rightleftarrows X_2 + n\,e^- \,, \tag{8.105}$$

$$X_2 \rightarrow \text{products} \tag{8.106}$$

with equilibrium electron transfer (8.105) and irreversible follow-up homogeneous reaction (8.106) was considered by Singh et al. [110]. Another EC scheme:

$$X_1 + n\,e^- \rightleftarrows X_2 \,, \tag{8.107}$$

$$X_2 \rightleftarrows X_3 \tag{8.108}$$

involving equilibrium electron transfer (8.107) and non-equilibrium reversible homogeneous reaction (8.108) was studied by Singh et al. [111] and Teja [119].

Catalytic EC reaction scheme:

$$X_1 + n\,e^- \rightleftarrows X_2 \,, \tag{8.109}$$

$$X_2 + X_{3,\mathrm{s}} \rightarrow X_1 \tag{8.110}$$

involving equilibrium electron transfer (8.109) and irreversible homogeneous reaction (8.110) with a static species $X_{3,\mathrm{s}}$ was discussed by Dutt et al. [28]. Analogous theory for irreversible reaction (8.109) was obtained by Singh et al. [113].

The CE reaction scheme:

$$X_1 \rightleftarrows X_2 \,, \tag{8.111}$$

$$X_2 \rightleftarrows X_3 + n\,e^- \tag{8.112}$$

involving reversible homogeneous reaction (8.111) and equilibrium or irreversible electron transfer (8.112) was considered by Singh et al. [112].

The ECE scheme:

$$X_1 + n_1\,e^- \rightleftarrows X_2 \,, \tag{8.113}$$

$$X_2 \rightarrow X_3 \,, \tag{8.114}$$

$$X_3 + n_2\,e^- \rightleftarrows X_4 \tag{8.115}$$

involving two equilibrium electron transfers (8.113) and (8.115), and irreversible homogeneous reaction (8.114) was studied by Singh et al. [108, 109].

8.3.7 Two- and Three-Dimensional Diffusion

IEs for a controlled potential SECM experiment and the EC reaction scheme

$$X_1 + n\,e^- \rightleftarrows X_2\,, \tag{8.116}$$

$$X_2 \rightarrow X_3 \tag{8.117}$$

involving a non-equilibrium reversible electron transfer (8.116) and an irreversible homogeneous reaction (8.117) were formulated by Mirkin and Bard [59].

8.4 Coupled Reaction–Planar Diffusion PDEs, First-Order Homogeneous Reactions

The task of obtaining concentration–flux relationships in the general case of Eq. (8.2), that is without assuming that the PDEs are equivalent or transformable to a system of uncoupled PDEs, does not seem to have been analysed so far in the literature. The only available discussion is that of the present author [22], restricted to the planar diffusion transport without convection, in a semi-infinite one-dimensional spatial domain. There is also one published example of the concentration–production relationships that require some additional kernel functions, apart from those discussed in the previous sections, as a result of the PDE coupling by homogeneous reactions. It is the example of the catalytic EC reaction scheme (8.20) under the assumption of planar diffusion with unequal diffusion coefficients of species X_1 and X_2, and semi-infinite spatial domain. The example was considered by Bieniasz [20, 21], in connection with the modelling of linear potential sweep voltammetry. Incomplete formulae for the concentration–flux relationships for this example were also reported by Oldham [80]. Below we outline the results of Bieniasz [22].

For planar diffusion in a one-dimensional spatial domain, Eq. (8.2) becomes

$$\frac{\partial \vec{c}(x,t)}{\partial t} = \overline{D}\,\frac{\partial^2 \vec{c}(x,t)}{\partial x^2} - \overline{K}\,\vec{c}(x,t) + \vec{\sigma}\,. \tag{8.118}$$

The initial condition is

$$\vec{c}(x,0) = \vec{c}^{\,\star}\,, \tag{8.119}$$

[where $\vec{c}^{\,\star}$ obeys Eq. (8.3)]. Assuming a semi-infinite spatial domain, the boundary condition at infinity is

$$\vec{c}(\infty,t) = \vec{c}^{\,\star}\,. \tag{8.120}$$

We introduce the following vector of auxiliary variables [note that the definition is now different from Eq. (8.4)]:

$$\vec{u}(x,t) = \vec{c}(x,t) - \vec{c}^{\,\star}, \tag{8.121}$$

so that the incomplete IBVP (8.118)–(8.120) is transformed into

$$\frac{\partial \vec{u}(x,t)}{\partial t} = \overline{D}\,\frac{\partial^2 \vec{u}(x,t)}{\partial x^2} - \overline{K}\,\vec{u}(x,t), \tag{8.122}$$

$$\vec{u}(x,0) = \vec{0}, \tag{8.123}$$

$$\vec{u}(\infty,t) = \vec{0}. \tag{8.124}$$

In Appendix E, Sect. E.3 we follow Bieniasz [22] and demonstrate that the general expression for the solution–flux relationships for $\vec{u}(x,t)$, resulting from Eqs. (8.122)–(8.124), takes the form

$$\vec{u}(0,t) = \int_0^t \overline{\mathscr{K}}(t,\tau)\,\vec{J}^{\,u}(0,\tau)\,\mathrm{d}\tau, \tag{8.125}$$

where $\overline{\mathscr{K}}(t,\tau)$ is an $N_\mathrm{s}^\mathrm{dd} \times N_\mathrm{s}^\mathrm{dd}$ matrix of kernel functions, and the vector of (one-dimensional) flux elements $J_j^u(0,t) = -D_j\,\partial u_j(x,t)/\partial x\big|_{x=0}$ of the variables $\vec{u}(x,t)$ at the interface studied is denoted by $\vec{J}^{\,u}(0,t) = \left[J_1^u(0,t),\,\ldots,\,J_{N_\mathrm{s}^\mathrm{dd}}^u(0,t)\right]^\mathrm{T}$. In view of Eq. (8.121) the fluxes $\vec{J}^{\,u}(x,t)$ of $\vec{u}(x,t)$ and $\vec{J}(x,t)$ of $\vec{c}(x,t)$ are identical, so that from Eqs. (8.121) and (8.125) we obtain

$$\vec{c}(0,t) = \vec{c}^{\,\star} + \int_0^t \overline{\mathscr{K}}(t,\tau)\,\vec{J}(0,\tau)\,\mathrm{d}\tau. \tag{8.126}$$

The matrix $\overline{\mathscr{K}}(t,\tau)$ is shown to depend on the spectral properties of the complex square matrix $\overline{B}(s)$ defined by

$$\overline{B}(s) = \overline{D}^{\,-1}\,(s\,\overline{1} + \overline{K}) \tag{8.127}$$

with s denoting the Laplace variable and $\overline{1}$ denoting the unit matrix. Specifically,

$$\overline{\mathscr{K}}(t,\tau) = \overline{\varphi}(t-\tau), \tag{8.128}$$

where the matrix $\overline{\varphi}(\vartheta)$ of functions is given by

$$\overline{\varphi}(\vartheta) = \mathscr{L}^{-1}\left\{\overline{U}(s)\,\overline{W}^{-1}(s)\right\} . \tag{8.129}$$

In Eq. (8.129) the Laplace transformation is between the ϑ and s domains, and the matrices $\overline{U}(s)$ and $\overline{W}(s)$ are related to $\overline{B}(s)$ in the following way. The columns of $\overline{U}(s)$ are formed by linearly independent generalised eigenvectors of $\overline{B}(s)$, and the columns of $\overline{W}(s)$ are formed by the same eigenvectors of $\overline{B}(s)$, multiplied by certain additional matrices that depend on s and square roots of the eigenvalues of $\overline{B}(s)$. Detailed procedures and formulae for $\overline{U}(s)$ and $\overline{W}(s)$ are in Appendix E, Sect. E.3 (see also [22]). Matrices $\overline{U}(s)$ and $\overline{W}(s)$ strongly depend on the numbers of distinct eigenvalues and ordinary eigenvectors of $\overline{B}(s)$ (which in turn depend on s). The variable s can often be selected sufficiently large so that the most favourable situation of all eigenvalues distinct can be approached, but in general there is no guarantee that all eigenvalues and/or ordinary eigenvectors are distinct.

The picture emerging from this analysis is generally very complicated, and implies a number of difficulties. It is well known that the elements of the generalised eigenvectors may depend on the eigenvalues, so that one expects that the eigenvalues of $\overline{B}(s)$ may appear in the formulae for $\overline{U}(s)$. The square roots of the eigenvalues enter also $\overline{W}(s)$, as was stated. The elements of $\overline{U}(s)\,\overline{W}^{-1}(s)$ are therefore rational functions of the square roots of the eigenvalues of $\overline{B}(s)$, and of the variable s. The eigenvalues of $\overline{B}(s)$, in turn, may be complicated functions of s, and there is even no guarantee that closed-form expressions for the eigenvalues exist. As the eigenvalues are roots of characteristic equations, which are polynomial equations, the existence of the closed-form expressions for the roots is guaranteed only up to $N_{\mathrm{s}}^{\mathrm{dd}} = 4$. Furthermore, even for $N_{\mathrm{s}}^{\mathrm{dd}} \leq 4$, the expressions for the eigenvalues may be so complicated that obtaining the inverse Laplace transform (3.14) in Eq. (8.129) is difficult. If no closed-form expressions for the eigenvalues exist, only numerical inversion of the Laplace transform (8.129) may be feasible. Even if closed-form expressions for the elements of $\overline{\mathscr{K}}(t,\tau)$ are obtainable, it seems impossible to predict a complete and re-usable set of such expressions; clearly, a variety of new kernel functions may arise, different from the kernels $\tilde{D}_j^{-1/2} \exp\left[-\tilde{K}_{j,j}\,(t-\tau)\right] [\pi(t-\tau)]^{-1/2}$ that occur in the case when the PDEs can be de-coupled (as we have shown in Sect. 8.2). Particular reaction schemes have to be analysed individually.

In order to illustrate some of the above predictions and difficulties, below we analyse two examples of the simplest conceivable coupled systems of reaction–diffusion PDEs, involving only two dynamic distributed species. For $N_{\mathrm{s}}^{\mathrm{dd}} = 2$ matrix $\overline{K}$ is generally full:

$$\overline{K} = \begin{bmatrix} K_{1,1} & K_{1,2} \\ K_{2,1} & K_{2,2} \end{bmatrix} . \tag{8.130}$$

In general, there are also distinct diffusion coefficients $D_1 \neq D_2$. The corresponding matrix $\overline{B}(s)$ is

$$\overline{B}(s) = \begin{bmatrix} \frac{K_{1,1}+s}{D_1} & \frac{K_{1,2}}{D_1} \\ \frac{K_{2,1}}{D_2} & \frac{K_{2,2}+s}{D_2} \end{bmatrix} . \tag{8.131}$$

The characteristic equation for the eigenvalues $\lambda(s)$ of $\overline{B}(s)$ is

$$D_1 D_2 \, \lambda(s)^2 - [D_1(s + K_{2,2}) + D_2(s + K_{1,1})] \, \lambda(s) \\ +(s + K_{1,1})(s + K_{2,2}) - K_{1,2}K_{2,1} = 0 , \tag{8.132}$$

so that

$$\lambda(s) = \frac{1}{2}\left(\frac{s + K_{1,1}}{D_1} + \frac{s + K_{2,2}}{D_2} \pm \frac{\Delta^{1/2}}{D_1 D_2}\right) , \tag{8.133}$$

where

$$\Delta = (D_1 - D_2)\, s^2 + 2(D_1 - D_2)(D_1 K_{2,2} - D_2 K_{1,1})\, s \\ + (D_1 K_{2,2} - D_2 K_{1,1})^2 + 4 D_1 D_2 \, K_{1,2} K_{2,1} \, . \tag{8.134}$$

The number of distinct eigenvalues depends on Δ. Analysis of Eq. (8.134) reveals that when $D_1 = D_2 = D$ the dependence of Δ on s vanishes, so that

$$\Delta = D^2[(K_{2,2} - K_{1,1})^2 + 4\, K_{1,2}K_{2,1}] \, . \tag{8.135}$$

Consequently, the eigenvalues (8.133) are then linear functions of s, and their multiplicity depends only on the structure of matrix $\overline{K}$. This is the situation previously discussed in Sect. 8.2. In the opposite case of $D_1 \neq D_2$ there are two distinct and isolated points s_1 and s_2 in the Laplace plane, given by

$$s_{1/2} = \frac{(D_2 - D_1)(D_1 K_{2,2} - D_2 K_{1,1}) \pm 2\left[-(D_2 - D_1)^2 D_1 D_2 \, K_{1,2} K_{2,1}\right]^{1/2}}{(D_2 - D_1)^2} \tag{8.136}$$

(which can be either real or complex numbers, depending on the signs of $K_{1,2}$ and $K_{2,1}$), at which $\Delta = 0$, and consequently there is only a single eigenvalue, and a single ordinary eigenvector $\vec{u}^{\,(1)}$of $\overline{B}(s)$:

$$\vec{u}^{\,(1)} = \left[1, -\frac{D_1}{2K_{1,2}}\left(\frac{s_{1/2} + K_{1,1}}{D_1} - \frac{s_{1/2} + K_{2,2}}{D_2}\right)\right]^{\mathrm{T}} . \tag{8.137}$$

In order to avoid this degeneracy, we can take $\Re\{s\} > \max(\Re\{s_1\}, \Re\{s_2\})$, for which $\Delta \neq 0$ so that (in agreement with the general predictions of Sect. E.3 in Appendix E) we get distinct eigenvalues and distinct ordinary eigenvectors. However, the eigenvalues $\lambda(s)$ are then generally complicated, nonlinear functions of s and it is unlikely that closed-form formulae for $\overline{\mathscr{K}}(t,\tau)$ can be obtained in this case.

As a first example, let us take the hypothetical catalytic CEC reaction scheme:

$$\left.\begin{array}{c} X_0 \underset{k_b}{\overset{k_f}{\rightleftarrows}} X_1 \\ X_1 + e^- \rightleftarrows X_2 \\ X_2 \overset{k}{\rightarrow} X_1 \end{array}\right\}, \tag{8.138}$$

in which X_0 is a static distributed species, and X_1 and X_2 are dynamic distributed species. We have $K_{1,1} = k_b$, $K_{1,2} = -k$, $K_{2,1} = 0$, and $K_{2,2} = k$. If we assume an identical diffusion coefficient D of X_1 and X_2, then in agreement with Eq. (8.133), as long as $k \neq k_b$, we have a trivial case of two distinct eigenvalues $\lambda_1(s) = (s + k_b)/D$ and $\lambda_2(s) = (s+k)/D$, with corresponding distinct ordinary eigenvectors $\vec{u}_1^{(1)} = [1, 0]^T$ and $\vec{u}_2^{(1)} = \left[\frac{kD}{k_b - k}, 1\right]^T$. Consequently, the procedure of Sect. E.3 in Appendix E gives

$$\overline{U}(s)\,\overline{W}^{-1}(s) = \begin{bmatrix} [D(s+k_b)]^{-1/2} & \frac{kD}{k-k_b}\left\{[D(s+k_b)]^{-1/2} - [D(s+k)]^{-1/2}\right\} \\ 0 & [D(s+k)]^{-1/2} \end{bmatrix}. \tag{8.139}$$

In view of Eqs. (8.128) and (8.129), the inverse Laplace transformation (3.14) of Eq. (8.139) yields

$$\mathscr{K}_{1,1}(t,\tau) = D^{-1/2} \exp\left[-k_b\,(t-\tau)\right] \left[\pi(t-\tau)\right]^{-1/2}, \tag{8.140}$$

$$\mathscr{K}_{1,2}(t,\tau) = \frac{kD}{k-k_b}\left\{D^{-1/2} \exp\left[-k_b\,(t-\tau)\right] \left[\pi(t-\tau)\right]^{-1/2} \right.$$
$$\left. - D^{-1/2} \exp\left[-k\,(t-\tau)\right] \left[\pi(t-\tau)\right]^{-1/2}\right\}, \tag{8.141}$$

$$\mathscr{K}_{2,1}(t,\tau) = 0, \tag{8.142}$$

$$\mathscr{K}_{2,2}(t,\tau) = D^{-1/2} \exp\left[-k\,(t-\tau)\right] \left[\pi(t-\tau)\right]^{-1/2}. \tag{8.143}$$

Thus, there are no new kernel terms, apart from those known from Sect. 8.2. This is well expected, because when $k \neq k_b$ matrix $\overline{K}$ is similar to a diagonal matrix. The situation becomes diametrally different if it happens that $k = k_b$. In such a case there is only one eigenvalue $\lambda_1(s) = (s+k)/D$, and one ordinary eigenvector $\vec{u}_1^{(1)} = [1, 0]^T$. We therefore have to determine a generalised second-order eigenvector $\vec{u}_1^{(2)}$ of $\overline{B}(s)$, which is $\vec{u}_1^{(2)} = [0, 1]^T$. The procedure of

Sect. E.3 in Appendix E then gives

$$\overline{U}(s)\,\overline{W}^{-1}(s) = \begin{bmatrix} [D(s+k)]^{-1/2} & \frac{kD^{1/2}}{2(s+k)^{3/2}} \\ 0 & [D(s+k)]^{-1/2} \end{bmatrix} . \tag{8.144}$$

Inverse Laplace transformation (3.14) of Eq. (8.144) yields

$$\mathscr{K}_{1,1}(t,\tau) = D^{-1/2}\exp[-k\,(t-\tau)]\,[\pi(t-\tau)]^{-1/2} , \tag{8.145}$$

$$\mathscr{K}_{1,2}(t,\tau) = k(D/\pi)^{1/2}\exp[-k\,(t-\tau)]\,(t-\tau)^{1/2} , \tag{8.146}$$

$$\mathscr{K}_{2,1}(t,\tau) = 0 , \tag{8.147}$$

$$\mathscr{K}_{2,2}(t,\tau) = D^{-1/2}\exp[-k\,(t-\tau)]\,[\pi(t-\tau)]^{-1/2} . \tag{8.148}$$

As can be seen, $\mathscr{K}_{1,2}(t,\tau)$ is a new kernel function arising in this example, not predicted in Sect. 8.2. This kernel arises, because when $k = k_b$ matrix $\overline{K}$ is no longer similar to a diagonal matrix.

As a second example, we take the catalytic EC reaction scheme:

$$\left.\begin{aligned} X_1 + e^- &\rightleftarrows X_2 \\ X_2 &\underset{k_b}{\overset{k_f}{\rightleftarrows}} X_1 \end{aligned}\right\} , \tag{8.149}$$

for which $K_{1,1} = k_b$, $K_{1,2} = -k_f$, $K_{2,1} = -k_b$, and $K_{2,2} = k_f$. If we assume different diffusion coefficients $D_1 \neq D_2$, then Eq. (8.133) for the eigenvalues is too complicated to allow one to obtain closed-form formulae for $\overline{\mathscr{K}}(t,\tau)$. The situation becomes more tractable if $k_b = 0$, so that we effectively have an irreversible follow-up homogeneous reaction (still admitting the possibility of $D_1 \neq D_2$, so that the procedure of Sect. 8.2 cannot be applied). This case has been shown solvable in Bieniasz [20, 21]. Equation (8.133) then gives two distinct eigenvalues $\lambda_1(s) = s/D_1$ and $\lambda_2(s) = (s+k_f)/D_2$, and corresponding distinct ordinary eigenvectors $\vec{u}_1^{(1)} = [1,\,0]^T$ and $\vec{u}_2^{(1)} = [1,\,s/k_f - \gamma(s+k_f)/k_f]^T$, where $\gamma = D_1/D_2$. The procedure of Sect. E.3 in Appendix E then gives

$$\overline{U}(s)\,\overline{W}^{-1}(s) = \begin{bmatrix} (D_1\,s)^{-1/2} & k_f\frac{[D_2(s+k_f)]^{-1/2} - \gamma^{1/2}(D_2\,s)^{-1/2}}{s-\gamma(s+k_f)} \\ 0 & [D_2(s+k_f)]^{-1/2} \end{bmatrix} . \tag{8.150}$$

Inverse Laplace transformation (3.14) of Eq. (8.150) yields

$$\mathscr{K}_{1,1}(t,\tau) = D_1^{-1/2}\,[\pi(t-\tau)]^{-1/2} , \tag{8.151}$$

$$\mathscr{K}_{2,1}(t,\tau) = 0 , \tag{8.152}$$

$$\mathscr{K}_{2,2}(t,\tau) = D_2^{-1/2}\exp[-k_f\,(t-\tau)]\,[\pi(t-\tau)]^{-1/2} , \tag{8.153}$$

and

$$\mathcal{K}_{1,2}(t,\tau) = D_2^{-1/2} \left(\frac{k_\mathrm{f}}{1-\gamma}\right)^{1/2} \left\{ \mathrm{erex}\left[\left(\frac{k_\mathrm{f}\gamma}{1-\gamma}\right)^{1/2} (t-\tau)^{1/2}\right] - \exp\left[-k_\mathrm{f}\,(t-\tau)\right] \mathrm{erex}\left[\left(\frac{k_\mathrm{f}}{1-\gamma}\right)^{1/2} (t-\tau)^{1/2}\right]\right\} \tag{8.154}$$

for $\gamma < 1$,

$$\mathcal{K}_{1,2}(t,\tau) = D_1^{-1/2}\,\{1 - \exp\left[-k_\mathrm{f}\,(t-\tau)\right]\}\,\left[\pi(t-\tau)\right]^{-1/2} \tag{8.155}$$

for $\gamma = 1$, and

$$\mathcal{K}_{1,2}(t,\tau) = D_2^{-1/2} \left(\frac{k_\mathrm{f}}{\gamma-1}\right)^{1/2} \frac{2}{\pi^{1/2}} \left\{ \mathrm{daw}\left[\left(\frac{k_\mathrm{f}\gamma}{\gamma-1}\right)^{1/2} (t-\tau)^{1/2}\right] - \exp\left[-k_\mathrm{f}\,(t-\tau)\right] \mathrm{daw}\left[\left(\frac{k_\mathrm{f}}{\gamma-1}\right)^{1/2} (t-\tau)^{1/2}\right]\right\} \tag{8.156}$$

for $\gamma > 1$. In Eq. (8.154) erex(·) is given by Eq. (5.76), and in Eq. (8.156) daw(·) is the Dawson integral

$$\mathrm{daw}(z) = \exp(-z^2) \int_0^z \exp(\zeta^2)\,\mathrm{d}\zeta\,. \tag{8.157}$$

Kernels $\mathcal{K}_{1,1}(t,\tau)$, $\mathcal{K}_{2,2}(t,\tau)$, and $\mathcal{K}_{1,2}(t,\tau)$ for $\gamma = 1$ correspond to the predictions of Sect. 8.2, where equal diffusion coefficients were assumed. However, new kernels $\mathcal{K}_{1,2}(t,\tau)$ arise in this example, when $\gamma \neq 1$. They contain terms having general forms $\exp[-a(t-\tau)]\,\mathrm{erex}\{[b(t-\tau)]^{1/2}\}$ and $\exp[-a(t-\tau)]\,\mathrm{daw}\{[b(t-\tau)]^{1/2}\}$, where a and b are suitable coefficients.

8.5 Transport with Nonlinear Homogeneous Reaction Kinetics

There is a number of approaches to handling nonlinear homogeneous reaction kinetics in the IE method, documented in the literature. Most of these approaches involve some sort of simplifications or approximations, and therefore have a limited applicability. The approaches have been published using models defined over one-dimensional spatial domains, although they can be extended to multiple dimensions. We discuss them in the following Sects. 8.5.1–8.5.4.

8.5.1 Equilibrium State Assumption

One possibility is to make use of equilibrium assumptions to lump homogeneous reactions with heterogeneous reactions within the reaction scheme considered. In this way one actually eliminates the need to determine and use the concentration–production rate relationships for distributed species subject to nonlinear reaction–transport PDEs. This approach was used by Shuman [106] to derive IEs describing cyclic voltammetry and corresponding to the following EC reaction scheme at a planar electrode:

$$\left.\begin{array}{c} X_1 + n\,e^- \rightleftarrows X_2 \\ m\,X_2 \rightleftarrows q\,X_3 \end{array}\right\} , \tag{8.158}$$

in which m and q are small integers, not simultaneously equal unity, and species X_3 is not electroactive. Pure diffusion in a semi-infinite spatial domain was assumed, but the approach might well be applied to other kinds of transport. Reference [106] seems to contain errors, notably there is some lack of clarity around Eqs. (2) and (3) in that reference, with regard to the symbols and assumptions. However, from the final IEs it appears that Shuman used the following idea. A complete modelling of electroanalytical experiments involving scheme (8.158) would require a solution of three PDEs for the concentrations c_1, c_2, and c_3 of species X_1, X_2, and X_3. The PDE for c_1 would be a diffusion equation, whereas those for c_2 and c_3 would be nonlinear reaction–diffusion equations containing reaction terms with large (but finite) rate constants. If we assume that the homogeneous reaction is fast and reversible, the assumption of a virtual equilibrium state of this reaction is applied, implying the following approximate relationship between the concentrations of X_2 and X_3:

$$c_3^q(x,t)/c_2^m(x,t) = K , \tag{8.159}$$

where K is the equilibrium constant of the homogeneous reaction. Therefore, the PDE for c_2 need not to be considered at all, whereas the PDE for c_3 reduces to a diffusion PDE, since the reaction terms cancel. Effectively, this amounts to replacing scheme (8.158) by the single electrochemical reaction

$$m\,X_1 + mn\,e^- \rightleftarrows q\,X_3 . \tag{8.160}$$

Of course, the higher-order non-elementary heterogeneous reaction (8.160) does not present any difficulty for the IE method.

A similar assumption of homogeneous equilibrium state was used by other authors. Fatouros et al. [29] applied it in their study of non-equilibrium reversible electron transfer associated with an equilibrium follow-up dimerisation reaction, while discussing potential step chronoamperometry and linear potential sweep voltammetry. Puy et al. [86] used it in connection with the modelling of normal pulse polarography. Hapiot et al. [36] employed it for another reaction scheme involving electron transfer and homogeneous dimerisation, while considering cyclic voltammetry. Lovrić et al. [46, 49] applied it to the modelling of square wave voltammetry.

The above approach may seem simple and useful, but in the opinion of the present author it should be avoided. In the case of reaction–transport PDEs the validity of the assumption of homogeneous equilibria is debatable. Reaction terms involving large rate constants cause the PDEs to be singularly perturbed (see, for example, [41, 84] for the discussion of singularly perturbed problems). This usually implies the existence of local narrow boundary or interior layers (in electrochemistry known as reaction layers), inside which the homogeneous reaction IS NOT in equilibrium. Only outside the reaction layer do the concentrations obey Eq. (8.159). Thus, the crucial question is how big an error is committed by neglecting the reaction layer(s). This is not clear in Shuman [106]. Species X_3 is not electroactive, so that its concentration gradient at the electrode | electrolyte interface must be zero, whereas the gradient of the electroactive species X_2 is nonzero. Consequently, equilibrium condition (8.159) certainly cannot hold in the neighbourhood of the interface. Hence, a reaction layer must be present there.

8.5.2 Steady State Assumption

Another approach to dealing with nonlinear homogeneous kinetics consists in employing steady state assumptions. Some early formulations of this idea can be found in the works of Tachi and Senda [118] (who modelled polarographic current for an ECE scheme with a fast homogeneous reaction, assuming the expanding plane convection–diffusion), and Savéant and Vianello [98, 100] and Nicholson [68] (who modelled linear potential sweep for EC and CE schemes under planar diffusion conditions). This approach has found application in numerous further theoretical studies by the IE method [2, 4–7, 9–11, 13, 14, 17, 26, 37, 48, 61–63, 95, 96, 102–104], mostly devoted to linear potential sweep and cyclic voltammetry. The approach is applicable in the situations when model PDEs involve reaction–transport PDEs, containing nonlinear homogeneous kinetic terms, which are practically at steady state and can be regarded as second-order ODEs. The ODEs must be solvable analytically. It is then possible to combine the analytically obtained steady state solution–flux relationships for the ODEs with the boundary conditions at the interface studied. Together with other, integral solution–production rate relationships, obtained for remaining PDEs, this gives IEs describing transient experiments. Such IEs are nonlinear. Most of the literature examples of this approach refer to planar diffusion in semi-infinite spatial domains, but the approach is not necessarily restricted to such conditions, as long as appropriate analytical concentration–flux relationships (see below) can be obtained.

The approach can be most transparently explained by considering a system involving one or more irreversible homogeneous reactions

$$m_j \, \mathrm{X}_j \xrightarrow{k_j} \text{product(s)} \,, \tag{8.161}$$

with species X_j being initially absent transient intermediates that do not participate in any other homogeneous reactions. Let the stoichiometric coefficients m_j be integers greater than or equal 2, and let the reactions follow the power rate law with reaction orders equal to the stoichiometric coefficients (see Sect. 2.7). The case of $m_j = 2$ (dimerisation or disproportionation) is particularly important; reactions with $m_j = 3$ are rather rare, and those with $m_j > 3$ rather unlikely. Assume further that the PDEs for the concentrations of the product(s) of all reactions (8.161) need not be included in the mathematical description (which happens, for example, if the reaction scheme studied does not contain any further reactions involving these product(s)). Under these assumptions, the only nonlinear PDEs that have to be taken into account are the reaction–transport PDEs for the concentrations c_j of species X_j. The EC reaction scheme:

$$\left.\begin{array}{c} \mathrm{X}_1 + \mathrm{e}^- \rightleftarrows \mathrm{X}_2 \\ 2\,\mathrm{X}_2 \xrightarrow{k} \text{product(s)} \end{array}\right\} , \tag{8.162}$$

discussed in [63, 68, 98], is an example when the above assumptions are fulfilled. In the case of planar diffusion in a semi-infinite spatial domain, the nonlinear reaction–transport PDEs for species X_j take the form

$$\frac{\partial c_j(x,t)}{\partial t} = D_j \frac{\partial^2 c_j(x,t)}{\partial x^2} - m_j \, k_j \, c_j(x,t)^{m_j} \,, \tag{8.163}$$

with the initial condition

$$c_j(x,0) = 0 \,, \tag{8.164}$$

and the boundary condition

$$c_j(\infty,t) = 0 \,. \tag{8.165}$$

If the rate constant k_j is large, a steady state is approached, in which the rate of diffusion is nearly compensated by the rate of the homogeneous reaction in Eq. (8.163), so that the steady state condition $\partial c_j(x,t)/\partial t \approx 0$ can be assumed. The steady state is analogous to those indicated in Sect. 8.2 for the linear reaction–diffusion equations. In Appendix E, Sect. E.4, we present an analytical solution of the incomplete IBVP given by Eqs. (8.163)–(8.165), under the steady state condition. The following relationship between the concentration and production rate of species X_j at the interface is obtained [cf. Eq. (E.99)]:

$$c_j(0,t) = \left[\frac{m_j+1}{2}\left(m_j\,k_j\,D_j\right)^{-1}\right]^{1/(m_j+1)} p_j^{\mathrm{het}}(t)^{2/(m_j+1)} \,. \tag{8.166}$$

As a specific example, let us consider a controlled potential experiment for the reaction scheme (8.162) with an equilibrium electron transfer reaction. In this case $p_1^{\mathrm{het}}(t) = J_1(0,t)$, $p_2^{\mathrm{het}}(t) = J_2(0,t)$, $p_1^{\mathrm{het}}(t) + p_2^{\mathrm{het}}(t) = 0$, and the Faradaic current is $I(t) = FA\,J_1(0,t)$. The integral solution–production rate relationship for species X_1 is given by Eq. (5.6) with kernel (5.72):

$$c_1(0,t) = c_1^{\star} + D_1^{-1/2}\int_0^t [\pi(t-\tau)]^{-1/2}\, p_1^{\mathrm{het}}(\tau)\,\mathrm{d}\tau \,, \tag{8.167}$$

whereas for species X_2 Eq. (8.166) gives

$$c_2(0,t) = \left[\frac{3}{2}\,(2k\,D_2)^{-1}\right]^{1/3} p_2^{\mathrm{het}}(t)^{2/3} \,. \tag{8.168}$$

Let us define the unknown function as $\Psi(t) = -I(t)(FA)^{-1}$. By combining Eqs. (8.167) and (8.168) with the Nernst equation $c_1(0,t) = c_2(0,t) \exp\left\{F(RT)^{-1}\left[E(t)-E^0\right]\right\}$ we obtain the following nonlinear second kind Volterra IE for $\Psi(t)$:

$$c_1^{\star} - D_1^{-1/2}\int_0^t [\pi(t-\tau)]^{-1/2}\,\Psi(\tau)\,\mathrm{d}\tau$$
$$-\left[\frac{3}{2}\,(2k\,D_2)^{-1}\right]^{1/3} \exp\left\{\frac{F}{RT}\left[E(t)-E^0\right]\right\}\Psi(t)^{2/3} = 0 \,. \tag{8.169}$$

Relationships analogous to Eq. (8.166), but valid for spherical and cylindrical interfaces cannot be obtained analytically. However, if the homogeneous reaction is very fast, the reaction layer is very thin compared to the radii of such interfaces, so that the concentration–production rate relationship (8.166) for planar interfaces should be a good approximation. Olmstead and Nicholson [82] used such an approximation in the case of the theory of disproportionation reaction at spherical electrodes. Aoki et al. [16] used a similar approximation while considering voltammetry for a second-order catalytic reaction scheme at microcylinder electrodes. A different attempt was made by Delmastro [26], who suggested an approximate relationship for $m_j = 2$ and one-dimensional spherical diffusion in semi-infinite spatial domain, while considering the IE arising in the theory of polarographic currents for the second-order regeneration (catalytic) EC reaction scheme:

$$\left.\begin{aligned} X_1 + e^- &\rightleftarrows X_2 \\ 2\,X_2 &\overset{k}{\rightarrow} X_1 + X_3 \end{aligned}\right\} . \tag{8.170}$$

This reaction scheme involves three distributed species X_1, X_2, and X_3, but the concentration of X_3 has no effect on the current and need not be considered. Delmastro defined auxiliary variables $u_1 = 2\,c_1 + c_2$ and $u_2 = c_2$, and assumed identical diffusion coefficient D of X_1 and X_2, which resulted in $k_1 = 0$, $k_2 = k$, and $\varrho_1 = \varrho_2 = D^{1/2}/r_0$ [cf. Eq. (5.75)]. The steady state condition $\partial u_2(r,t)/\partial t \approx 0$ was assumed. Delmastro suggested that the solution–flux relationship for the variable u_2 at a spherical interface can be approximated by Eq. (8.166) (with $m_2 = 2$), valid for a planar interface, provided that the following replacement is made in this equation:

$$\left(2\,k_2\,c_1^\star\right)^{1/2} \leftarrow \left(2\,k_2\,c_1^\star\right)^{1/2} + \varrho_2\,. \tag{8.171}$$

The replacement was inspired by the relationship (8.53) characteristic of the first-order homogeneous reaction–spherical diffusion equation, where the terms $\tilde{K}_{j,j}^{1/2}$ and $\tilde{\varrho}_j$ occur together as a sum $\tilde{K}_{j,j}^{1/2} + \tilde{\varrho}_j$. Although such modified Eq. (8.166) was found useful for the purposes of the Delmastro study [26], its general validity is rather problematic. As $u_2 = c_2$, a correct relationship between the solution and flux of u_2 at the interface should depend exclusively on the parameters characteristic of the incomplete IBVP (8.163)–(8.165) for c_2. Expression (8.171) does not satisfy this expectation, since it involves the initial concentration $c_1^\star$ of species X_1, which does not occur in the PDE for u_2, nor in the initial and boundary conditions at infinity, for u_2.

8.5.3 *The Gerischer Approximation*

There exists a special linearising approximation that can be used to obtain IEs in the presence of nonlinear homogeneous kinetics, for a class of electroanalytical experiments characterised by small perturbations. In such experiments a certain reference state of an electrochemical system is slightly perturbed by a rapidly varying signal (for example a sinusoidal perturbation of the electrode potential), and the response to this perturbation is studied. The reference state can be, for example, an equilibrium state, a steady state, or a state corresponding to another perturbation, which must be large and slowly varying. The linearising approximation is attributed to Gerischer [33], and it was used by Hayes et al. [38, 39] and Ružić et al. [91], to derive IEs pertinent to the theory of AC polarography.

In order to present the idea of the Gerischer approximation, let us focus on a jth dynamic distributed species X_j subject to convection–diffusion transport with homogeneous reactions. To fix attention, assume that the homogeneous reaction is

$$m_j\, X_j \xrightarrow{k_j} \text{product(s)}\,, \tag{8.172}$$

and it obeys the power rate law with the reaction order equal to the stoichiometric coefficient $m_j > 1$ (see Sect. 2.7). Consequently, the perturbed state is described by the following PDE:

$$\frac{\partial c_j}{\partial t} = D_j\, \Delta\, c_j - \boldsymbol{v} \cdot \mathbf{grad}\, c_j - m_j\, k_j\, c_j^{m_j}\,. \tag{8.173}$$

Although in Eq. (8.173) we assume only one nonlinear term for simplicity, several kinetic terms and/or kinetic term(s) dependent on concentrations of several different species present no bigger difficulty and can be handled in a similar way. The reference state is generally described by an analogous PDE:

$$\frac{\partial \tilde{c}_j}{\partial t} = D_j\, \Delta\, \tilde{c}_j - \boldsymbol{v} \cdot \mathbf{grad}\, \tilde{c}_j - m_j\, k_j\, \tilde{c}_j^{m_j}\,, \tag{8.174}$$

where $\tilde{c}_j$ is the concentration in the reference state. Equation (8.174) may simplify if the reference state is a steady state or an equilibrium state. Assuming a small departure from the reference state, the homogeneous kinetic term can be linearised:

$$c_j^{m_j} \approx \tilde{c}_j^{m_j} + \left.\frac{\mathrm{d}\left(c_j^{m_j}\right)}{\mathrm{d}\, c_j}\right|_{c_j=\tilde{c}_j} \left(c_j - \tilde{c}_j\right) = \tilde{c}_j^{m_j} + m_j\, \tilde{c}_j^{m_j-1}\left(c_j - \tilde{c}_j\right)\,. \tag{8.175}$$

Multivariate linearisation has to be used in cases when the homogeneous kinetic terms depend on concentrations of several different species. By subtracting Eqs. (8.173) and (8.174), taking into account Eq. (8.175), and introducing an auxiliary variable

$$u_j = c_j - \tilde{c}_j\,, \tag{8.176}$$

one obtains:

$$\frac{\partial u_j}{\partial t} = D_j\, \Delta\, u_j - \boldsymbol{v} \cdot \mathbf{grad}\, u_j - m_j^2\, k_j\, \tilde{c}_j^{m_j-1}\, u_j\,. \tag{8.177}$$

Equation (8.177) is linear with respect to the variable u_j, and can be handled by the IE method in the same manner as described in Sects. 8.1–8.4, provided that $\tilde{c}_j$ occurring in Eq. (8.177) can be regarded as constant. This happens, for example, in the case of one-dimensional diffusion transport in semi-infinite spatial domains, if the time scale of the small perturbation is much smaller than the time scale of the evolution of the reference state. In such a case $\tilde{c}_j$ can be taken equal to the momentary boundary value of the concentration (in the reference state), at the interface studied.

8.5.4 Conversion to IDEs

Yet another way of handling nonlinear homogeneous reaction kinetics relies on converting the IBVPs to IDEs, rather than to IEs. This approach was suggested by Mastragostino et al. [54], and further elaborated by Olmstead et al. [83] and Nadjo et al. [64]. A similar procedure was also employed for the convolutive analysis of experimental data by Amatore et al. [3]. Further examples were reported by Andrieux et al. [12] and Renault et al. [88]. Planar diffusion in semi-infinite spatial domains was assumed in [3, 12, 54, 64, 88], and spherical in [83], but the approach is not restricted to such assumptions.

The idea of this approach can be formulated as follows. Consider an IBVP describing an electroanalytical experiment. Assume that the problem involves PDEs for the concentrations c_j of distributed species, or equivalent PDEs for auxiliary variables u_j that depend on the concentrations. In cases when some of these PDEs are linear, the solution–flux relationships for such PDEs may be easy to obtain, in the same manner as previously described in Sects. 5.2, 8.2, and 8.4. By combining these relationships with the boundary conditions at the interface studied, we obtain one or more IEs which are incomplete in the sense that they still involve unknowns (c_j, u_j, or their fluxes at the interface) subject to the remaining, nonlinear PDEs. In other words, we obtain a coupled system of IEs and PDEs, i.e. an IDE system, which has to be solved simultaneously. The IDE system has identical solutions as the original IBVP (but not vice versa, since not all spatial concentration profiles are determined from the IDE system). Therefore, the nonlinear homogeneous kinetics is handled rigorously in this approach, without using any simplifying assumption or approximation. However, it is most likely that the solution of the IDE system obtained can only be performed numerically.

As an example, let us consider a controlled potential experiment for the EC reaction scheme discussed in [83]:

$$\left.\begin{array}{c} X_1 + e^- \rightleftarrows X_2 \\ 2\,X_2 \xrightarrow{k} \text{products} \end{array}\right\} \tag{8.178}$$

with an equilibrium electron transfer, assuming planar diffusion in a semi-infinite spatial domain. The IBVP is:

$$\frac{\partial c_1(x,t)}{\partial t} = D_1 \frac{\partial^2 c_1(x,t)}{\partial x^2} \,, \tag{8.179}$$

$$\frac{\partial c_2(x,t)}{\partial t} = D_2 \frac{\partial^2 c_2(x,t)}{\partial x^2} - 2\,k\,c_2(x,t)^2 \,, \tag{8.180}$$

$$c_1(x,0) = c_1^\star \, , \tag{8.181}$$

$$c_2(x,0) = 0 \, , \tag{8.182}$$

$$c_1(0,t) = c_2(0,t) \exp\left\{\frac{F}{RT}\left[E(t) - E^0\right]\right\} \, , \tag{8.183}$$

$$D_1 \frac{\partial c_1(x,t)}{\partial x}\bigg|_{x=0} + D_2 \frac{\partial c_2(x,t)}{\partial x}\bigg|_{x=0} = 0 \, , \tag{8.184}$$

$$c_1(\infty,t) = c_1^\star \, , \tag{8.185}$$

$$c_2(\infty,t) = 0 \, . \tag{8.186}$$

The Faradaic current response $I(t)$ to the electrode potential perturbation $E(t)$ is to be determined:

$$I(t) = -F\,A\,D_1 \,\frac{\partial c_1(x,t)}{\partial x}\bigg|_{x=0} \, . \tag{8.187}$$

The PDE (8.179) is linear, so that the concentration–production rate relationship resulting from Eqs. (8.179), (8.181), and (8.185) can be written in the form consistent with Eqs. (5.86) and (8.187):

$$c_1(0,t) = c_1^\star + D_1^{-1/2} \int_0^t [\pi(t-\tau)]^{-1/2}\, \frac{I(\tau)}{FA}\, \mathrm{d}\tau \, . \tag{8.188}$$

The IDE system obtained from the IBVP (8.179)–(8.187) therefore is:

$$\frac{\partial c_2(x,t)}{\partial t} = D_2 \frac{\partial^2 c_2(x,t)}{\partial x^2} - 2\,k\,c_2(x,t)^2 \, , \tag{8.189}$$

$$c_2(x,0) = 0 \, , \tag{8.190}$$

$$c_1^\star + D_1^{-1/2} \int_0^t [\pi(t-\tau)]^{-1/2}\, \frac{I(\tau)}{FA}\, \mathrm{d}\tau = c_2(0,t) \exp\left\{\frac{F}{RT}\left[E(t) - E^0\right]\right\} \, , \tag{8.191}$$

$$\frac{I(t)}{FA} = D_2 \,\frac{\partial c_2(x,t)}{\partial x}\bigg|_{x=0} \, , \tag{8.192}$$

$$c_2(\infty,t) = 0 \, , \tag{8.193}$$

System (8.189)–(8.193) has to be solved for $I(t)$ and $c_2(x,t)$.

References

1. Abramowitz M, Stegun IA (1972) Handbook of mathematical functions. Dover Publications, New York
2. Amatore C, Savéant JM (1977) ECE and disproportionation. Part V. Stationary state general solution application to linear sweep voltammetry. J Electroanal Chem 85:27–46
3. Amatore C, Nadjo L, Savéant JM (1978) Convolution and finite difference approach. Application to cyclic voltammetry and spectroelectrochemistry. J Electroanal Chem 90:321–331
4. Amatore C, Savéant JM, Thiebault A (1979) Electrochemically induced chemical reactions. Kinetics of competition with electron transfer. J Electroanal Chem 103:303–320
5. Amatore C, Gareil M, Savéant JM (1983) Homogeneous vs. heterogeneous electron transfer in electrochemical reactions. Application to the electrohydrogenation of anthracene and related reactions. J Electroanal Chem 147:1–38
6. Amatore C, Garreau D, Hammi M, Pinson J, Savéant JM (1985) Kinetic analysis of reversible electrodimerization reactions by the combined use of double potential step chronoamperometry and linear sweep voltammetry. Application to the reduction of 9-cyanoanthracene. J Electroanal Chem 184:1–24
7. Amatore C, Savéant JM, Combellas C, Robveille S, Thiébault A (1985) Electrochemically induced reactions: kinetics of the competition with homogeneous electron transfer in non-catalytic systems. Application to the substitution of 4-bromobenzophenone by cyanide ions in liquid ammonia. J Electroanal Chem 184:25–40
8. Andrieux CP, Savéant JM (1989) Detection of electroactive transient intermediates of electrochemical reactions by steady state and time-dependent techniques. Competition between heterogeneous electron transfer and homogeneous chemical reactions. J Electroanal Chem 267:15–32
9. Andrieux CP, Nadjo L, Savéant JM (1970) Electrodimerization 1. One-electron irreversible dimerization. Diagnostic criteria and rate determination procedures for voltammetric studies. J Electroanal Chem 26:147–186
10. Andrieux CP, Nadjo L, Savéant JM (1973) Electrodimerization. VII. Electrode and solution electron transfers in the radical–substrate coupling mechanism. Discriminative criteria from the other mechanisms in voltammetric studies (linear sweep, rotating disc, polarography). J Electroanal Chem 42:223–242
11. Andrieux CP, Blocman C, Dumas-Bouchiat JM, M'Halla F, Savéant JM (1980) Homogeneous redox catalysis of electrochemical reactions Part V. Cyclic voltammetry. J Electroanal Chem 113:19–40
12. Andrieux CP, Hapiot P, Savéant JM (1987) Mechanism of superoxide ion disproportionation in aprotic solvents. J Am Chem Soc 109:3768–3775
13. Andrieux CP, Gallardo I, Savéant JM (1989) Outer-sphere electron-transfer reduction of alkyl halides. A source of alkyl radicals or of carbanions? Reduction of alkyl radicals. J Am Chem Soc 111:1620–1626
14. Andrieux CP, Grzeszczuk M, Savéant JM (1991) Electrochemical generation and reduction of organic free radicals. α-hydroxybenzyl radicals from the reduction of benzaldehyde. J Am Chem Soc 113:8811–8817
15. Andrieux CP, Delgado G, Savéant JM, Su KB (1993) Improvement and estimation of precision in the cyclic voltammetric determination of rate constants and activation parameters of coupled homogeneous second order reactions. Reaction of n-butyl bromide with anthracene anion radicals. J Electroanal Chem 348:141–154
16. Aoki K, Ishida M, Tokuda K (1988) Voltammetry at microcylinder electrodes. Part VI. Second-order catalytic reaction of Fe(edta) with H_2O_2. J Electroanal Chem 245:39–50
17. Arévalo M, Farnia G, Severin G, Vianello E (1987) Kinetic analysis of a slow electron transfer coupled with a father–son reaction. J Electroanal Chem 220:201–211

18. Balducci G, Costa G (1993) The four-member square scheme in cyclic voltammetry: general solution for Nernstian electron transfers. J Electroanal Chem 348:355–365
19. Bender MA, Stone HA (1993) An integral equation approach to the study of the steady state current at surface microelectrodes. J Electroanal Chem 351:29–55
20. Bieniasz L (1984) Influence of diffusion coefficient ratio D_O/D_R on potential-step chronoamperometric and linear voltammetric current at stationary planar electrodes in the case of a pseudo-first-order EC catalytic reaction scheme. J Electroanal Chem 170:77–87
21. Bieniasz LK (1985) Linear voltammetric current functions for a pseudo-first-order EC catalytic reaction scheme with $D_O \neq D_R$: series expansion algorithm. J Electroanal Chem 188:13–20
22. Bieniasz LK (2011) Analysis of the applicability of the integral equation method in the theory of transient electroanalytical experiments for homogeneous reaction–diffusion systems: the case of planar electrodes. J Electroanal Chem 657:91–97
23. Calvente JJ, Andreu R, Roldán E, Domínguez M (1994) On the simultaneous evaluation of charge transfer kinetics and adsorption: reduction of parabanic acid in low acidity xM HCl + $(2 - x)$M LiCl mixtures. J Electroanal Chem 366:105–125
24. Danckwerts PV (1951) Absorption by simultaneous diffusion and chemical reaction into particles of various shapes and into falling drops. Trans Faraday Soc 47:1014–1023
25. Davčeva M, Mirčeski V, Komorsky-Lovrić Š (2011) Evaluation of the antioxidative activity by measuring the rate of the homogeneous oxidation reaction with ferroceniumdimethanol cation. Comparative analysis of glutathione and ascorbic acid. Int J Electrochem Sci 6:2718–2729
26. Delmastro JR (1969) Theory of polarographic kinetic currents for second-order regeneration reactions at spherical electrodes. II. Numerical solution of the integral equations for steady-state behavior. Anal Chem 41:747–753
27. Delmastro JR, Booman GL (1969) Polarographic kinetic currents for first-order preceding and regeneration reactions at spherical electrodes. Anal Chem 41:1409–1420
28. Dutt J, Chhabra J, Singh T (1989) Linear sweep voltammetry at tubular electrodes. Part IV. Catalytic reactions. J Electroanal Chem 273:69–78
29. Fatouros N, Chemla M, Amatore C, Savéant JM (1984) Slow charge transfer associated with a fast equilibriated follow-up dimerization reaction. J Electroanal Chem 172:67–81
30. Garay F, Lovrić M (2002) Square-wave voltammetry of quasi-reversible electrode processes with coupled homogeneous chemical reactions. J Electroanal Chem 518:91–102
31. Garay F, Lovrić M (2002) Square-wave voltammetry of quasi-reversible CE reactions at spherical microelectrodes. Electroanalysis 14:1635–1643
32. Garay F, Lovrić M (2002) Quasi-reversible EC reactions at spherical microelectrodes analysed by square-wave voltammetry. J Electroanal Chem 527:85–92
33. Gerischer H (1951) Wechselstrompolarisation von Elektroden mit einem potentialbestimmenden Schritt beim Gleichgewichtspotential I. Z Phys Chem 198:286–313
34. Gonzalez J, Hapiot P, Konovalov V, Savéant JM (1999) Investigating the reduction characteristics of transient free radicals by laser-pulse electron photo-injection—mechanism diagnostic criteria and determination of reactivity parameters from time-resolved experiments. J Electroanal Chem 463:157–189
35. Gueshi T, Tokuda K, Matsuda H (1979) Voltammetry at partially covered electrodes. Part II. Linear potential sweep and cyclic voltammetry. J Electroanal Chem 101:29–38
36. Hapiot P, Audebert P, Monnier K, Pernaut JM, Garcia P (1994) Electrochemical evidence of π-dimerization with short thiophene oligomers. Chem Mater 6:1549–1555
37. Harima Y, Kurihara H, Nishiki Y, Aoyagui S, Tokuda K, Matsuda H (1982) Dissociation kinetics of the solvated electron species in methylamine and methylamine–ammonia mixtures. Can J Chem 60:445–455
38. Hayes JW, Ružić I, Smith DE, Booman GL, Delmastro JR (1974) Fundamental harmonic A.C. polarography with disproportionation following the charge transfer step. Theory and experimental results with the U(VI)/U(V) couple. J Electroanal Chem 51:245–267

39. Hayes JW, Ružić I, Smith DE, Booman GL, Delmastro JR (1974) Fundamental harmonic A.C. polarography with irreversible dimerization following the charge transfer step. Theory and experimental results with the benzaldehyde system. J Electroanal Chem 51:269–285
40. Henke KH, Hans W (1955) Reaktionskinetisch bedingte polarographische Stromstärke. 3. Mitteilung: Dem Elektrodenprozeß nachgelagerte chemische Reaktionen unter Rückbildung des Depolarisators. Z Elektrochem 59:676–680
41. Holmes MH (1995) Introduction to perturbation methods. Springer, New York
42. Hung HL, Delmastro JR, Smith DE (1964) Alternating current polarography of electrode processes with coupled homogeneous chemical reactions. III. Theory for systems with multistep first-order chemical reactions. J Electroanal Chem 7:1–25
43. Kant R, Rangarajan SK (1989) Chronopotentiometry with power-law perturbation functions at an expanding plane electrode with and without a preceding blank period for systems with a coupled first-order homogeneous chemical reaction. J Electroanal Chem 265:39–65
44. Kastening B (1969) Note on the polarographic theory for an ECE mechanism. Anal Chem 41:1142–1144
45. Keller HE, Reinmuth WH (1972) Theory of potential scan voltammetry with finite diffusion. Kinetics and other complications. Anal Chem 44:1167–1178
46. Komorsky-Lovrić Š, Lovrić M (2012) Theory of square-wave voltammetry of two electron reduction with the intermediate that is stabilized by complexation. Electrochim Acta 69:60–64
47. Kumar VT, Birke RL (1993) Evaluation of electrochemical parameters for an EC mechanism from a global analysis of current–potential–time data: application to reductive cleavage of methylcobalamin. Anal Chem 65:2428–2436
48. Lexa D, Rentien P, Savéant JM, Xu F (1985) Methods for investigating the mechanistic and kinetic role of ligand exchange reactions in coordination electrochemistry. Cyclic voltammetry of chloroiron(III)tetraphenylporphyrin in dimethylformamide. J Electroanal Chem 191:253–279
49. Lovrić M, Jadreško D, Komorsky-Lovrić Š (2013) Theory of square-wave voltammetry of electrode reaction followed by the dimerization of product. Electrochim Acta 90:226–231
50. Lovrić M, Tur'yan YI (2003) A model of CE mechanism on spherical electrodes. Croat Chem Acta 76:189–197
51. Lucas SK, Sipcic R, Stone HA (1997) An integral equation solution for the steady-state current at a periodic array of surface microelectrodes. SIAM J Appl Math 57:1615–1638
52. Lundquist JT Jr, Nicholson RS (1968) Theory of the potential step–linear scan electrolysis method with a comparison of rate constants determined electrochemically and by classical methods. J Electroanal Chem 16:445–456
53. Maran F, Severin MG, Vianello E, D'Angeli F (1993) Self-protonation of electrogenerated carbanions. Competition between electrode reduction and chemical decay of the conjugate base of the substrate. J Electroanal Chem 352:43–50
54. Mastragostino M, Nadjo L, Savéant JM (1968) Disproportionation and ECE mechanisms—I. Theoretical analysis. Relationships for linear sweep voltammetry. Electrochim Acta 13:721–749
55. Matsuda H (1974) Contributions to the theory of D.C. polarographic current–potential curves for electrode processes involving a chemical reaction. J Electroanal Chem 56:165–175
56. McCord TG, Smith DE (1968) Alternating current polarography: an extension of the general theory for systems with coupled first-order homogeneous chemical reactions. Anal Chem 40:1959–1966
57. McCord TG, Smith DE (1968) Second harmonic alternating current polarography: a general theory for systems with coupled first-order homogeneous chemical reactions. Anal Chem 40:1967–1970
58. Mirčeski V, Bobrowski A, Zarebski J, Spasovski F (2010) Electrocatalysis of the first and second kind: theoretical and experimental study in conditions of square-wave voltammetry. Electrochim Acta 55:8696–8703

59. Mirkin MV, Bard AJ (1992) Multidimensional integral equations: a new approach to solving microelectrode diffusion problems. Part 2. Applications to microband electrodes and the scanning electrochemical microscope. J Electroanal Chem 323:29–51
60. Murphy MM, O'Dea JJ, Arn D, Osteryoung JG (1990) Theory of cyclic staircase voltammetry for first-order coupled reactions. Anal Chem 62:903–909, with correction in Anal Chem 62:1904
61. Nadjo L, Savéant JM (1971) Dimerization, disproportionation and E.C.E. mechanisms in the reduction of aromatic carbonyl compounds in alkaline media. J Electroanal Chem 33:419–451
62. Nadjo L, Savéant JM (1973) Electrodimerization. VIII. Role of proton transfer reactions in the mechanism of electrohydrodimerization. Formal kinetics for voltammetric studies (linear sweep, rotating disc, polarography). J Electroanal Chem 44:327–366
63. Nadjo L, Savéant JM (1973) Linear sweep voltammetry: Kinetic control by charge transfer and/or secondary chemical reactions. I. Formal kinetics. J Electroanal Chem 48:113–145
64. Nadjo L, Savéant JM, Su KB (1985) Homogeneous redox catalysis of multielectron electrochemical reactions. Part II. Competition between homogeneous electron transfer and addition on the catalyst. J Electroanal Chem 196:23–34
65. Ng SLL, Cheh HY (1985) The application of linear sweep voltammetry to a rotating disk electrode with a catalytic process. J Electrochem Soc 132:93–98
66. Ng SLL, Cheh HY (1985) The application of linear sweep voltammetry to a rotating disk electrode with a kinetic process. J Electrochem Soc 132:98–102
67. Ng SLL, Cheh HY (1986) The application of linear sweep voltammetry to a rotating disk electrode for electrode reactions preceded by a chemical reaction. J Electrochem Soc 133:1385–1388
68. Nicholson RS (1965) Some examples of the numerical solution of nonlinear integral equations. Anal Chem 37:667–671
69. Nicholson RS, Shain I (1964) Theory of stationary electrode polarography. Single scan and cyclic methods applied to reversible, irreversible, and kinetic systems. Anal Chem 36:706–723.
70. Nicholson RS, Shain I (1965) Theory of stationary electrode polarography for a chemical reaction coupled between two charge transfers. Anal Chem 37:178–190
71. Nicholson RS, Wilson JM, Olmstead ML (1966) Polarographic theory for an ECE mechanism. Application to reduction of p-nitrosophenol. Anal Chem 38:542–545
72. Nishihara C (1982) Note on approximate equations of DC polarographic current–potential curves for a catalytic wave. J Electroanal Chem 134:171–176
73. Nishihara C, Matsuda H (1974) Contributions to the theory of kinetic and catalytic currents in polarography. J Electroanal Chem 51:287–293
74. Nishihara C, Matsuda H (1977) A contribution to the theory of D.C. polarographic current–voltage curves. Electrode reactions with three parallel charge transfer processes. J Electroanal Chem 85:17–26
75. Nishihara C, Matsuda H (1979) Note on approximate equations of DC polarographic current–potential curves for electrode reactions coupled with a preceding chemical reaction. J Electroanal Chem 103:261–265
76. Nishihara C, Satô GP, Matsuda H (1975) A.C. polarography of the Nernstian charge transfer followed by an irreversible chemical reaction. J Electroanal Chem 63:131–138
77. O'Dea JJ, Osteryoung J, Osteryoung RA (1981) Theory of square wave voltammetry for kinetic systems. Anal Chem 53:695–701
78. O'Dea JJ, Wikiel K, Osteryoung J (1990) Square-wave voltammetry for ECE mechanism. J Phys Chem 94:3628–3636
79. Ohsaka T, Sotomura T, Matsuda H, Oyama N (1983) Double potential step chronoamperometry for reversible follow-up chemical reactions. Application to the aquation kinetics of bis(ethylenediaminemonoacetato)cobalt(II). Bull Chem Soc Jpn 56:3065–3072
80. Oldham KB (1986) Convolution: a general electrochemical procedure implemented by a universal algorithm. Anal Chem 58:2296–2300

81. Olmstead ML, Nicholson RS (1967) Theoretical evaluation of effects of electrode sphericity on stationary electrode polarography. Case of a chemical reaction following reversible electron transfer. J Electroanal Chem 14:133–141
82. Olmstead ML, Nicholson RS (1969) Cyclic voltammetry theory for the disproportionation reaction and spherical diffusion. Anal Chem 41:862–864
83. Olmstead ML, Hamilton RG, Nicholson RS (1969) Theory of cyclic voltammetry for a dimerization reaction initiated electrochemically. Anal Chem 41:260–267
84. O'Malley RE Jr (1974) Introduction to singular perturbations. Academic Press, New York
85. Polcyn DS, Shain I (1966) Theory of stationary electrode polarography for a multistep charge transfer with catalytic (cyclic) regeneration of the reactant. Anal Chem 38:376–382
86. Puy J, Salvador J, Galceran J, Esteban M, Díaz-Cruz JM, Mas F (1993) Voltammetry of labile metal-complex systems with induced reactant adsorption. Theoretical analysis for any ligand-to-metal ratio. J Electroanal Chem 360:1–25
87. Quentel F, Mirčeski V, L'Her M, Stankoska K (2012) Assisted ion transfer at organic film-modified electrodes. J Phys Chem C 116:22885–22892
88. Renault C, Andrieux CP, Tucker RT, Brett MJ, Balland V, Limoges B (2012) Unraveling the mechanism of catalytic reduction of O_2 by microperoxidase-11 adsorbed within a transparent 3D-nanoporous ITO film. J Am Chem Soc 134:6834–6844
89. Roizenblat EM, Kosogov AA, Kolmanovich VYu (1989) Voltamperometry under conditions of limited diffusion. Irreversible process with a preceding chemical reaction. Elektrohim 25:201–207 (in Russian)
90. Rudolph M (1990) An algorithm of general application for the digital simulation of electrochemical processes. J Electroanal Chem 292:1–7
91. Ružić I, Smith DE, Feldberg SW (1974) On the influence of coupled homogeneous redox reactions on electrode processes in D.C. and A.C. polarography. I. Theory for two independent electrode reactions coupled with a homogeneous redox reaction. J Electroanal Chem 52:157–192
92. Sánchez Maestre M, Muñoz E, Avila JL, Camacho L (1992) Application of Matsuda's pulse polarography theory to electrode processes coupled to very fast chemical reactions: study of the CE mechanism by differential pulse polarography. Electrochim Acta 37:1129–1134
93. Saveant JM (1967) ECE mechanisms as studied by polarography and linear sweep voltammetry. Electrochim Acta 12:753–766
94. Saveant JM (1967) Cyclic voltammetry with asymmetrical potential scan: a simple approach to mechanisms involving moderately fast chemical reactions. Electrochim Acta 12:999–1030
95. Savéant JM, Su KB (1984) Homogeneous redox catalysis of electrochemical reaction Part VI. Zone diagram representation of the kinetic regimes. J Electroanal Chem 171:341–349
96. Savéant JM, Su KB (1985) Homogeneous redox catalysis of multielectron electrochemical reactions. Part I. Competition between heterogeneous and homogeneous electron transfer. J Electroanal Chem 196:1–22
97. Savéant JM, Vianello E (1960) Recherches sur les courants catalytiques en polarographie–oscillographique á balayage linéaire de tension. Etude théorique. In: Advances in polarography. Pergamon Press, London, pp 367–374
98. Savéant JM, Vianello E (1963) Étude de la polarisation chimique en régime de variation linéaire du potentiel. Cas d'une désactivation spontanée, rapide et irréversible du produit de la réduction. CR Hebd Séances Acad Sci Paris 256:2597–2600
99. Savéant JM, Vianello E (1963) Potential-sweep chronoamperometry theory of kinetic currents in the case of a first order chemical reaction preceding the electron-transfer process. Electrochim Acta 8:905–923
100. Savéant JM, Vianello E (1964) Chronoampérométrie á variation linéaire de potentiel. Étude de la polarisation chimique dans le cas d'une désactivation rapide et spontanée du dépolarisant. CR Hebd Séances Acad Sci Paris 259:4017–4020
101. Savéant JM, Vianello E (1965) Potential-sweep chronoamperometry: kinetic currents for first-order chemical reaction parallel to electron-transfer process (catalytic currents). Electrochim Acta 10:905–920

102. Savéant JM, Vianello E (1967) Potential-sweep voltammetry: general theory of chemical polarization. Electrochim Acta 12:629–646
103. Savéant JM, Vianello E (1967) Potential-sweep voltammetry: theoretical analysis of monomerization and dimerization mechanisms. Electrochim Acta 12:1545–1561
104. Savéant JM, Xu F (1986) First- and second-order chemical–electrochemical mechanisms. Extraction of standard potential, equilibrium and rate constants from linear sweep voltammetric curves. J Electroanal Chem 208:197–217
105. Schwarz WM, Shain I (1966) A potential step–linear scan method for investigating chemical reactions initiated by a charge transfer. J Phys Chem 70:845–852
106. Shuman MS (1970) Calculation of stationary electrode polarograms for cases of reversible chemical reactions other than first-order reactions following a reversible charge transfer. Anal Chem 42:521–523
107. Shuman MS, Shain I (1969) Study of the chemical reaction preceding reduction of cadmium nitrilotriacetic acid complexes using stationary electrode polarography. Anal Chem 41:1818–1825
108. Singh RP, Singh T (2011) Effect of electrons involved in charge transfer reactions in ECE processes under linear sweep voltammetry: a theoretical model. Model Assist Stat Appl 6:261–266
109. Singh RP, Singh T, Dutt J, Atamjyot (2000) Mass transfer to tubular electrodes: ECE process. J Math Chem 27:183–190
110. Singh T, Dutt J, Kaur S (1991) Cyclic and linear sweep voltammetry at tubular electrodes. Part V. Kinetic processes. J Electroanal Chem 304:17–30
111. Singh T, Singh RP, Dutt J (1995) Mass transport to tubular electrodes. J Math Chem 17:335–346
112. Singh T, Singh RP, Dutt J (1998) Mass transport to tubular electrodes. Part 2: CE process. J Math Chem 23:297–308
113. Singh T, Singh RP, Dutt J (2000) Linear sweep voltammetry of irreversible charge transfer coupled with irreversible catalytic reaction under diffusion–convection control: An integral equation approach. Indian J Pure Appl Math 31:363–374
114. Smith DE (1963) Alternating current polarography of electrode processes with coupled homogeneous chemical reactions. I. Theory for systems with first-order preceding, following, and catalytic chemical reactions. Anal Chem 35:602–609
115. Smith DE (1964) Theory of the Faradaic impedance. Relationship between Faradaic impedances for various small amplitude alternating current techniques. Anal Chem 36:962–970
116. Sobel HR, Smith DE (1970) D.C. polarography: on the theory for the current–potential profile with an ECE mechanism. J Electroanal Chem 26:271–284
117. Stuart A, Foulkes F (1988) Study of ECE reactions with multiple parallel chemical steps using linear scan voltammetry. Electrochim Acta 33:1411–1424
118. Tachi I, Senda M (1960) Polarographic current of stepwise electrode process involving chemical reaction. In: Longmuir IS (ed) Advances in polarography, vol 2. Pergamon Press, Oxford, pp 454–464
119. Teja HS (2012) Mass transport of visco-elastic electrodes. Res Anal Eval 4:40–42
120. Tokuda K, Matsuda H (1979) Theory of A.C. voltammetry at a rotating disk electrode. Part III. Redox-electrode reactions coupled with first-order chemical reactions. J Electroanal Chem 95:147–157
121. Zakharov MS, Bakanov VI (1975) Voltammetry under conditions of semi-infinite spherical and cylindrical diffusion. Electrode process followed by a first order chemical reaction. Izv Tomsk Politekh Inst 197:30–33 (in Russian)
122. Zakharov MS, Pnev VV (1975) Voltammetry with an arbitrary current (electrode potential) form, under conditions of semi-infinite spherical and cylindrical diffusion. I. Electrode processes preceded by a first order chemical reaction. Tr Tyumen Ind Inst 32:3–11 (in Russian)

Chapter 9
Models Involving Distributed and Localised Species

In this chapter we discuss electroanalytical models in which dynamic distributed species have to be taken into account simultaneously with dynamic localised species. Models of this kind are frequently encountered in organic electrochemistry, or in studies of electrocatalytic reactions, deposition/precipitation of solid phases (or layers), or modified electrodes.

We have seen in Chap. 4 that in the case of models involving dynamic localised species alone one can obtain a surprising diversity of the IEs, depending on the choice of unknown variables and/or on the procedure of deriving the IEs. One can expect that the combination of these diverse approaches to handling dynamic localised species, with the various possible concentration–production rate relationships for dynamic distributed species, discussed in Chaps. 5, 6, 7, and 8, can lead to an even bigger variety of equations. Indeed, the literature reveals that a particular electrochemical system involving dynamic distributed and dynamic localised species can often be described in a number of alternative ways. As a result one obtains different, but equivalent equations. The equations can be IEs involving integrals with kernels already known from Chaps. 4, 5, 6, 7, and 8, but also IDEs of various forms. The appearance of new kernels, not predicted in previous chapters, is also possible, as well as the occurrence of (generally rare in electrochemistry) nonlinear IEs involving integrals given by Eq. (3.1), in which integrands are nonlinear expressions of the unknown function $\Psi(t)$. Section 9.1 discusses all these details of the IE or IDE derivation for the present category of models.

There are many published IE- or IDE-based spatially one-dimensional models of electroanalytical experiments for systems involving distributed and localised species. Most of these IEs or IDEs describe controlled potential experiments. However, there have also been a few IEs or IDEs describing controlled current experiments. A brief summary of all these literature models is given in Sect. 9.2.

L.K. Bieniasz, *Modelling Electroanalytical Experiments by the Integral Equation Method*, Monographs in Electrochemistry, DOI 10.1007/978-3-662-44882-3_9

9.1 Derivation of the IEs or IDEs

Consider an experiment, in which a set of N_r^{het} heterogeneous reactions takes place at an interface studied, among N_s^{dd} dynamic distributed species and N_s^{dl} dynamic localised species. We focus here on spatially one-dimensional models, because there are currently no literature examples of higher dimensional models. However, the procedures described below can be extended to multiple dimensions. The dynamic distributed species may belong to any one of the two spatially extended phases adjacent to the interface, and they are subject to one-dimensional PDEs of diffusion, convection–diffusion, (homogeneous) reaction–diffusion, or reaction–convection–diffusion type. The dynamic localised species are present at the interface only. Apart from these species, the system may also involve static (distributed and/or localised) species, but this possibility does not influence the present discussion. Similarly to the experiments discussed in Chaps. 4, 5, 6, and 8, the derivation of the IEs (or IDEs) for the present kind of models is based on Eq. (2.54) with matrices $\overline{V}$ and $\overline{Z}$ resulting from the partial inversion of Eq. (2.53), possibly followed by the application of additional assumptions. These can be equilibrium assumption(s) referring to some of the heterogeneous reactions, or steady state assumption(s) referring to some of the dynamic species.

In view of the discussion in Sects. 2.7 and 2.9, the elements of the vector $\vec{p}^{\,het}(t)$ are now given by Eqs. (2.46), (2.47), and (2.49), depending on the type of the dynamic species (distributed, localised, and electrons exchanged). It is convenient to distinguish subvectors $\vec{p}^{\,dd,het}(t)$ and $\vec{p}^{\,dl,het}(t)$ of $\vec{p}^{\,het}(t)$, that group together interfacial production rates of dynamic distributed and dynamic localised species, respectively (it is not required that all species of one type precede all species of the second type in vector $\vec{p}^{\,het}(t)$; the species can be ordered in any way). Hence, from Eq. (2.47) we have:

$$\vec{p}^{\,dl,het}(t) = \frac{d\vec{\Gamma}(t)}{dt} , \tag{9.1}$$

or, by analogy with Eq. (4.12):

$$\vec{\Gamma}(t) = \vec{\Gamma}^{\star} + \int_0^t \vec{p}^{\,dl,het}(\tau)\, d\tau , \tag{9.2}$$

where $\vec{\Gamma}(t)$ is the vector of the concentrations of the dynamic localised species. Simultaneously, for the dynamic distributed species Eq. (8.1) now takes the form:

$$\vec{c}^{\,\dagger}(t) = \vec{c}^{\,\star} + \int_0^t \overline{\mathcal{K}}(t,\tau) \left[\vec{p}^{\,dd,het}(\tau) - \vec{p}^{\,dd,het,\star}\right] d\tau , \tag{9.3}$$

where $\vec{c}^{\,\dagger}(t)$ is the vector of the boundary concentrations of the dynamic distributed species at the interface studied, and $\overline{\mathscr{K}}(t,\tau)$ is the matrix of suitable kernel functions representing transport and homogeneous reactions. In agreement with the previous notation, initial concentrations are denoted by $\vec{\Gamma}^{\,\star}$ and $\vec{c}^{\,\star}$, and $\vec{p}^{\,\mathrm{dd,het},\star}$ is the vector of initial values of $\vec{p}^{\,\mathrm{dd,het}}(t)$. Equations (9.1)–(9.3) are necessary for the procedures of deriving the IEs or IDEs. Below we attempt to describe, in a systematic way, the various derivation procedures, most frequently encountered in the literature. A former review by Lovrić [52] can be suggested as an introduction to this subject.

Procedure 1 In this procedure one assumes that $\vec{p}^{\,\mathrm{dd,het}}(t)$ and $\vec{\Gamma}(t)$ are to play the role of the unknown functions of time. The mere substitution of Eqs. (9.1) and (9.3) into Eq. (2.54) then gives IDE(s) from which these unknowns can be determined. Equation system (2.54) may contain one additional equation, from which model responses (current or potential) are to be calculated. We obtain IDEs, because $\vec{p}^{\,\mathrm{dd,het}}(t)$ occurs under integrals in Eq. (9.3), whereas $\vec{\Gamma}(t)$ occurs both directly (in the reaction rate expressions or equilibrium equations) and in the derivative form $\mathrm{d}\vec{\Gamma}(t)/\mathrm{d}t$. This procedure was used by Reinmuth [105]. Its further applications can be found in [4, 42, 107, 108, 114, 115]. Although the procedure is generally applicable, we do not recommend it, because the solution of the IDEs is usually more involved than the solution of the IEs. For example, discretisations of the derivatives must be provided, in the IDE case, in addition to the discretisations of the integrals, sufficient for IEs. Also initial conditions for the unknowns present in the derivative form must be supplied, in contrast to the initial values of other unknowns. The equally general Procedure 4, described below, entirely avoids all these problems.

Procedure 2 A somewhat simpler variant of Procedure 1 is obtained if, owing to appropriate model assumptions, elements of matrix $\overline{\mathbf{Z}}$ acting on $\vec{p}^{\,\mathrm{dl,het}}(t)$ are zero [with the allowed exception of one equation in Eq. (2.54), which serves for determining the model response, if such equation is present], and which can still involve elements of $\mathrm{d}\vec{\Gamma}(t)/\mathrm{d}t$. Such model assumptions can be equilibrium assumptions for selected reactions, steady state assumptions for selected intermediates, or replacements of some of the structural conservation relationships by equivalent AEs. For example, the structural conservation relationship involving exclusively production rates of localised species:

$$\sum_l a_l \frac{\mathrm{d}\Gamma_l(t)}{\mathrm{d}t} = 0\,, \tag{9.4}$$

with numerical coefficients a_l, which is differential in nature, can be replaced by the equivalent AE:

$$\sum_l a_l \Gamma_l(t) = \sum_l a_l \Gamma_l^{\star}\,. \tag{9.5}$$

In such cases, the AEs present in Eq. (2.54) allow one to calculate $\vec{\Gamma}(t)$ as a function of $\vec{c}^{\,\dagger}(t)$. Substitution of $\vec{\Gamma}(t)$ into the remaining equations then gives IEs for $\vec{p}^{\,\mathrm{dd,het}}(t)$, instead of the IDEs obtainable in the general case of Procedure 1. The treatment of the electroanalytical models by this procedure is therefore almost identical to the models discussed in Chaps. 5, 6, and 8, for systems involving exclusively dynamic distributed species. This procedure was used by White and Lawson [119]. Its further applications are in [3,6,12,13,35,42,43,123].

Procedure 3 Another variant of Procedure 1, resulting in somewhat different IDEs, can be applied when (one or more) integral(s) having the form:

$$Y(t) = \int_0^t \mathscr{K}(t,\tau)\, \frac{\mathrm{d}\Gamma_i(\tau)}{\mathrm{d}\tau}\, \mathrm{d}\tau \tag{9.6}$$

occurs in the IDEs obtained by Procedure 1, where $\mathscr{K}(t,\tau)$ is a weakly singular kernel of the convolution type. Such conditions are satisfied by the majority of kernels considered in Chaps. 5, 6, and 8, with the exception of the kernel for the expanding plane model of the DME. Since for the convolution kernel $\mathscr{K}(t,\tau) = \varphi(t-\tau)$, where $\varphi(t)$ is a certain function of time, the application of the Laplace transformation (3.12), and of the convolution theorem (3.17), to Eq. (9.6) gives:

$$\begin{aligned} Y(t) &= \mathscr{L}^{-1}\left\{\hat{\varphi}(s)\left[s\hat{\Gamma}_i(s) - \Gamma_i^\star\right]\right\} \\ &= \mathscr{L}^{-1}\left\{s\hat{\varphi}(s)\left[\hat{\Gamma}_i(s) - \frac{\Gamma_i^\star}{s}\right] - 0\right\} = \frac{\mathrm{d}}{\mathrm{d}t}\int_0^t \mathscr{K}(t,\tau)\,\left[\Gamma_i(\tau) - \Gamma_i^\star\right]\,\mathrm{d}\tau\,. \end{aligned} \tag{9.7}$$

In proving the equivalence (9.7), one takes into account formula (3.18) with $n = 1$, the weak singularity of $\mathscr{K}(t,\tau)$, and the boundedness of $\Gamma_i(t) - \Gamma_i^\star$ at $t \to 0^+$, which together imply that $\lim_{t\to 0^+} \int_0^t \mathscr{K}(t,\tau)\,\left[\Gamma_i(\tau) - \Gamma_i^\star\right]\,\mathrm{d}\tau = 0$. By using Eq. (9.7), (some of) the temporal derivatives $\mathrm{d}\Gamma_i(t)/\mathrm{d}t$ in the IDEs obtained by Procedure 1 are replaced by the temporal derivatives of the integrals $\int_0^t \mathscr{K}(t,\tau)\,\left[\Gamma_i(\tau) - \Gamma_i^\star\right]\,\mathrm{d}\tau$. Such a modification of Procedure 1 was used by Lovrić [46], and it remained in use mostly by Lovrić and co-workers [47,50,53–55,59,60,64,84,125–127].

A different attempt to get rid of the temporal derivative $\mathrm{d}\Gamma_i(t)/\mathrm{d}t$ under the integral (9.6) was reported by Nicholson [82], who employed integration by parts:

$$Y(t) = \varphi(t-\tau)\,\Gamma_i(\tau)\big|_{\tau=0}^{\tau=t} - \int_0^t \frac{\mathrm{d}\varphi(t-\tau)}{\mathrm{d}\tau}\Gamma_i(\tau)\,\mathrm{d}\tau\,. \tag{9.8}$$

For the kernel $\mathscr{K}(t,\tau) = [\pi(t-\tau)]^{-1/2}$, related to planar diffusion in a semi-infinite spatial domain, Eq. (9.8) becomes

$$Y(t) = [\pi(t-\tau)]^{-1/2}\, \Gamma_i(\tau)\Big|_{\tau=0}^{\tau=t} - \frac{1}{2}\pi^{-1/2} \int\limits_0^t (t-\tau)^{-3/2}\, \Gamma_i(\tau)\, \mathrm{d}\tau\,. \tag{9.9}$$

The problem with Eq. (9.9) is that the kernel $(t-\tau)^{-3/2}$ in the integral in Eq. (9.9) is strongly singular (cf. Sect. 3.1), and also the first term in the right-hand side of Eq. (9.9) is divergent when $\tau \to t$. It is not clear how this problem was handled by Nicholson [82], but it seems that the calculations reported there are erroneous. Wrong predictions of Nicholson [82] were identified by White and Lawson [119]. We therefore do not recommend the route based on Eq. (9.8).

Procedure 4 Compared to Procedures 1, 2, and 3, a better idea is to consider $\vec{p}^{\,\mathrm{dd,het}}(t)$ and $\vec{p}^{\,\mathrm{dl,het}}(t)$ as unknown functions of time. This is a generally applicable approach. It can be viewed as a combination of the procedure from Chaps. 5, 6, or 8 (for dynamic distributed species), with Procedure 3 of Chap. 4 (for dynamic localised species). After substituting expressions (9.2) and (9.3) for $\vec{\Gamma}(t)$ and $\vec{c}^{\,\dagger}(t)$ into Eq. (2.54), one obtains IEs, from which $\vec{p}^{\,\mathrm{dd,het}}(t)$ and $\vec{p}^{\,\mathrm{dl,het}}(t)$ can be determined. Since numerical solution algorithms (see Chap. 12) usually yield approximations both to the unknown functions and to their integrals present in the IEs, by solving the IEs one also obtains $\vec{c}^{\,\dagger}(t)$ and $\vec{\Gamma}(t)$. This procedure was used by Wopschall and Shain [120], and it has become quite popular among various authors [5, 8, 24, 34, 62, 63, 72, 74, 86, 87, 121, 122]. The procedure can be recommended as the most transparent and unsophisticated one. It elegantly combines integrals having kernels characteristic of transport with homogeneous reactions, with integrals having kernel $\mathscr{K}(t,\tau) = 1$, characteristic of localised species. Therefore, numerical algorithms developed for the above kernels can be easily re-used, without a need to adapt them to any new kernels.

Procedure 5 Apart from the generally applicable Procedures 1 and 4, there have been a few alternative procedures valid in special situations. One of them makes use of the relationship (9.3) inverted in a particular way. The inversion may be not easy for an arbitrary kernel function, but it becomes rather straightforward in the case of planar diffusion in a semi-infinite spatial domain. In accordance with the discussion in Sect. 5.2 [cf. Eqs. (5.71) and (5.86)], we then have for any jth dynamic distributed species:

$$c_j^{\dagger}(t) = c_j^{\star} + D_j^{-1/2} \int\limits_0^t [\pi(t-\tau)]^{-1/2}\, p_j^{\mathrm{dd,het}}(\tau)\, \mathrm{d}\tau\,. \tag{9.10}$$

We divide the Laplace transform of Eq. (9.10) by $s^{1/2}$, and re-transform, which in view of the convolution theorem (3.17) and formula (3.19) gives:

$$\int_0^t [\pi(t-\tau)]^{-1/2} \left[c_j^\dagger(\tau) - c_j^\star\right] \mathrm{d}\tau = D_j^{-1/2} \int_0^t p_j^{\mathrm{dd,het}}(\tau)\, \mathrm{d}\tau \,, \tag{9.11}$$

or, after analytical integration of $c_j^\star$ with the kernel $[\pi(t-\tau)]^{-1/2}$:

$$\int_0^t [\pi(t-\tau)]^{-1/2}\, c_j^\dagger(\tau)\, \mathrm{d}\tau = 2c_j^\star \pi^{-1/2} t^{1/2} + D_j^{-1/2} \int_0^t p_j^{\mathrm{dd,het}}(\tau)\, \mathrm{d}\tau \,. \tag{9.12}$$

Equations (9.11) or (9.12) are useful if the reaction scheme at the interface studied involves equilibrium heterogeneous non-electrochemical reactions, so that $c_j^\dagger(t)$ is related to (some elements of) $\overrightarrow{\Gamma}(t)$ by AEs. For example, the equation of an adsorption isotherm may hold, between the jth dynamic distributed species and the ith dynamic localised species:

$$K\, c_j^\dagger(t) = f\left(\Gamma_i(t)\right) , \tag{9.13}$$

where K is the adsorption equilibrium constant, and function $f(\cdot)$ defines the particular isotherm (see Sect. 2.7). Usually another equation also exists in equation system (2.54), linking $p_j^{\mathrm{dd,het}}(t)$ and $p_i^{\mathrm{dl,het}}(t) = \mathrm{d}\Gamma_i(t)/\mathrm{d}t$. Such equation can take the form:

$$p_j^{\mathrm{dd,het}}(t) + a\, \frac{\mathrm{d}\Gamma_i(t)}{\mathrm{d}t} = b(t) \,, \tag{9.14}$$

where a is a numerical coefficient, and $b(t)$ denotes jointly terms such as production rates of other species or (possibly) some reaction rates. Consequently, one obtains from Eqs. (9.12), (9.13), and (9.14):

$$\begin{aligned} K^{-1} \int_0^t [\pi(t-\tau)]^{-1/2}\, f\left(\Gamma_i(\tau)\right) \mathrm{d}\tau - 2c_j^\star \pi^{-1/2} t^{1/2} \\ = -a\, D_j^{-1/2} \int_0^t \frac{\mathrm{d}\Gamma_i(\tau)}{\mathrm{d}t}\, \mathrm{d}\tau + D_j^{-1/2} \int_0^t b(\tau)\, \mathrm{d}\tau \\ = -a\, D_j^{-1/2} \left[\Gamma_i(t) - \Gamma_i^\star\right] + D_j^{-1/2} \int_0^t b(\tau)\, \mathrm{d}\tau \,. \end{aligned} \tag{9.15}$$

Equation (9.15) is an IE, in which one unknown is $\Gamma_i(t)$. Other unknowns may be present in $b(t)$. As the function $f(\cdot)$ is usually nonlinear, the IE(s) obtained by this procedure may be nonlinear in the sense of Eq. (3.1). This procedure was used by Levich et al. [41], who considered planar diffusion in a semi-infinite spatial domain, and convection–diffusion to an expanding plane, but the description of the derivations was very brief. Holub and Němec [30, 31] used this procedure with some modifications, to consider approximately spherical diffusion in a semi-infinite spatial domain, and DMEs. Rampazzo [104] presented a detailed outline of this procedure for planar diffusion in a semi-infinite spatial domain, with a focus on the adsorption subject to Frumkin isotherm. The procedure was also used by Guidelli and Pezzatini [22], the group of researchers from the University of Barcelona [16, 20, 65–68, 98, 99, 101–103, 112], and by Fujioka [15]. Van Leeuwen and co-workers [32, 33, 116] used a time-differentiated Eq. (9.11) for their IE derivation, and extended this equation to spherical diffusion, and expanding plane convection–diffusion.

Procedure 6 In cases when the reaction scheme involves equilibrium adsorption reactions subject to the linear (Henry) isotherms, a special procedure can be used, leading to the appearance of integrals with new kernel functions. In this procedure concentrations $\Gamma_i(t)$ of localised species, that are in adsorption equilibria, form one set of unknown functions. The remaining unknowns are production rates of other species, or perhaps some reaction rates. The procedure may not be easy to perform for arbitrary transport conditions, but it becomes rather straightforward in the case of planar diffusion in a semi-infinite spatial domain. In agreement with these conditions, let us assume Eqs. (9.2), (9.10), and (9.14) to hold, whereas Eq. (9.13) is replaced by the Henry isotherm:

$$K\, c_j^{\dagger}(t) = \Gamma_i(t)\,. \tag{9.16}$$

We apply the Laplace transformation (3.12) to Eqs. (9.2), (9.10), (9.14), and (9.16), and combine the transforms obtained, which gives

$$\begin{aligned} \hat{c}_j^{\dagger}(s) &= \frac{c_j^{\star}}{s} + D_j^{-1/2}\frac{\hat{p}_j^{\,\mathrm{dd,het}}(s)}{s^{1/2}} \\ &= \frac{c_j^{\star}}{s} + D_j^{-1/2}\frac{\hat{b}(s) - a\left[s\hat{\Gamma}_i(s) - \Gamma_i^{\star}\right]}{s^{1/2}} \\ &= \frac{c_j^{\star}}{s} + D_j^{-1/2}\frac{\hat{b}(s) - aKs\left[\hat{c}_j^{\dagger}(s) - \frac{c_j^{\star}}{s}\right] - a\left(Kc_j^{\star} - \Gamma_i^{\star}\right)}{s^{1/2}}\,. \end{aligned} \tag{9.17}$$

Equation (9.17) is rearranged to give

$$\hat{c}_j^{\dagger}(s) = \frac{c_j^{\star}}{s} + \frac{1}{aK}\,\frac{\hat{b}(s)}{s^{1/2}\left(s^{1/2} + \frac{D_j^{1/2}}{aK}\right)} - \frac{c_j^{\star} - \frac{\Gamma_i^{\star}}{K}}{s^{1/2}\left(s^{1/2} + \frac{D_j^{1/2}}{aK}\right)}\,, \tag{9.18}$$

or, after re-transforming:

$$c_j^\dagger(t) = c_j^\star + \int_0^t \mathscr{K}(t,\tau)\, b(\tau)\, \mathrm{d}\tau - \left(c_j^\star - \frac{\Gamma_i^\star}{K}\right) \mathrm{erex}\left(\frac{D_j^{1/2}}{aK}\, t^{1/2}\right), \tag{9.19}$$

where

$$\mathscr{K}(t,\tau) = D_j^{-1/2} \left(\frac{D_j^{1/2}}{aK}\right) \mathrm{erex}\left[\frac{D_j^{1/2}}{aK}(t-\tau)^{1/2}\right], \tag{9.20}$$

with erex(·) defined by Eq. (5.76). Expression (9.19) consists of two components. The first component is a convolution integral with the kernel (9.20), which links the interfacial concentration $c_j^\dagger(t)$ with the function $b(t)$. The second component is a specific function of time. Note that despite Eq. (9.16), the factor $c_j^\star - \Gamma_i^\star/K$ in the second component is not necessarily zero, as the initial concentration $\Gamma_i^\star$ may not be in equilibrium with $c_j^\star$. For example, in the case of DMEs it is often assumed that $c_j^\star > 0$ and $\Gamma_i^\star = 0$, if $t = 0$ corresponds to the drop formation moment (cf. Sect. 2.10). However, if $\Gamma_i^\star$ is an equilibrium concentration, then the second component vanishes. Substitution of Eq. (9.19) into Eq. (2.54) gives IE(s) containing the kernel (9.20), from which the unknown functions present in expression $b(t)$ can be determined. The unknowns $\Gamma_i(t)$ are obtained from Eq. (9.16). The kernel term involving function erex(z) is already known from our discussion of spherical diffusion in Sect. 5.2 [cf. Eq. (5.74)], but expression (9.20) involves a different parameter ($D_j^{1/2} a^{-1} K^{-1}$). The advantage of Procedure 6, over Procedure 4, is debatable, although in favourable circumstances Procedure 6 may reduce the number of IEs to be solved, or simplify the solution process (see examples below). Procedure 6 was used in a number of studies by Lovrić and co-authors, including (among others) Gulaboski, Mirčeski, Murray, and Solis [18, 19, 23, 38–40, 48, 49, 51, 56, 57, 61, 71, 73, 75, 76, 79, 81], and also by Osteryoung and co-workers [83, 113].

Procedure 7 It sometimes happens, especially if the reaction scheme at the interface studied involves non-equilibrium heterogeneous reactions, that selected ODEs or DAEs contained in Eq. (2.54) can be solved analytically, in whole or in part. By proceeding in the way somewhat similar to Procedure 2 in Chap. 4, one can substitute these analytical solutions into the remaining equations. As a result one obtains IEs or IDEs with integrands and kernels that are hard to predict or describe in a general manner, as the procedure is strongly problem-dependent. We therefore do not recommend this procedure, unless one can clearly prove its advantage over Procedure 4, for any particular purpose. Examples of IEs obtained by such procedure(s) are found in [7, 9, 29, 58, 77, 110].

We illustrate Procedures 1–6 by three examples. As a first example, let us consider an experiment with the electrode at which an equilibrium adsorption reaction takes place (see, for example, Reinmuth [105] or Rampazzo [104]):

$$X_1 \rightleftarrows X_{2,\mathrm{ad}} \, . \tag{9.21}$$

Reaction (9.21) involves a dynamic distributed species X_1, and dynamic localised species $X_{2,\mathrm{ad}}$. We assume that neither X_1 nor its adsorbed form $X_{2,\mathrm{ad}}$ undergoes further reactions. This is neither a controlled potential nor a controlled current experiment, since no electrochemistry is involved. However, the experiment can be followed by an electrochemical stripping analysis of the deposited adsorbate at a controlled potential or current, by measurements of the differential capacitance of the double layer, or by other measurements, for example spectro(electro)chemical. The stoichiometric matrix $\overline{N}^{\,\mathrm{dyn,het}}$ in Eq. (2.53) is

$$\overline{N}^{\,\mathrm{dyn,het}} = \begin{bmatrix} -1 \\ 1 \end{bmatrix} . \tag{9.22}$$

The vector $\overrightarrow{r}^{\,\mathrm{het}}(t) = \left[r_1^{\mathrm{het}}(t)\right]^{\mathrm{T}}$ of reaction rates comprises a single element: the rate of reaction (9.21), and the vector of production rates is $\overrightarrow{p}^{\,\mathrm{het}}(t) = \left[p_1^{\mathrm{het}}(t),\, p_2^{\mathrm{het}}(t)\right]^{\mathrm{T}} = \left[J_1^{\perp}(t),\, \mathrm{d}\Gamma_2(t)/\mathrm{d}t\right]^{\mathrm{T}}$. As there is no current flowing, electrons exchanged are not taken into account. Partial inversion of Eq. (2.53) with matrix (9.22) gives matrices $\overline{V}$ and $\overline{Z}$:

$$\overline{V} = \begin{bmatrix} 1 \\ 0 \end{bmatrix} , \tag{9.23}$$

$$\overline{Z} = \begin{bmatrix} -1 & 0 \\ 1 & 1 \end{bmatrix} . \tag{9.24}$$

The resulting first equation in equation system (2.54) is replaced by the adsorption equilibrium equation:

$$K\, c_1^{\dagger}(t) = f\left(\Gamma_2(t)\right) , \tag{9.25}$$

and the second equation in equation system (2.54) is the structural conservation relationship:

$$J_1^{\perp}(t) + \frac{\mathrm{d}\Gamma_2(t)}{\mathrm{d}t} = 0 \tag{9.26}$$

of the form of Eq. (9.14) with $a = 1$ and $b(t) = 0$. By applying Procedure 1, we take $\Psi(t) = \Gamma_2(t)$ as the unknown function, so that from Eqs. (9.3), (9.25), and (9.26) we obtain a single (generally nonlinear) IDE:

$$K \left[c_1^{\star} - \int_0^t \mathscr{K}_1(t,\tau)\, \frac{\mathrm{d}\Psi(\tau)}{\mathrm{d}\tau}\, \mathrm{d}\tau \right] - f\left(\Psi(t)\right) = 0 \, , \tag{9.27}$$

with initial condition $\Psi(0) = \Gamma_2^\star$, where $\mathscr{K}_1(t,\tau)$ is the kernel representing the transport of X_1. Procedure 2 is not applicable to this model. Procedure 3 gives an alternative single (generally nonlinear) IDE:

$$K\left\{c_1^\star - \frac{\mathrm{d}}{\mathrm{d}t}\int\limits_0^t \mathscr{K}_1(t,\tau)\,\left[\Psi(\tau) - \Gamma_2^\star\right]\,\mathrm{d}\tau\right\} - f\left(\Psi(t)\right) = 0\,, \tag{9.28}$$

where again $\Psi(t) = \Gamma_2(t)$ is assumed. In Procedure 4 we take $\Psi(t) = \mathrm{d}\Gamma_2(t)/\mathrm{d}t = -J_1^{\perp}(t)$ as the unknown function, so that Eq. (2.54) turns into the single (generally nonlinear) IE:

$$K\left[c_1^\star - \int\limits_0^t \mathscr{K}_1(t,\tau)\,\Psi(\tau)\,\mathrm{d}\tau\right] - f\left(\Gamma_2^\star + \int\limits_0^t \Psi(\tau)\,\mathrm{d}\tau\right) = 0\,. \tag{9.29}$$

Assuming that $\mathscr{K}_1(t,\tau) = D_1^{-1/2}\,[\pi(t-\tau)]^{-1/2}$, and by taking $\Psi(t) = \Gamma_2(t)$ as an unknown function, Procedure 5 allows one to present Eq. (9.29) in the form of the single IE:

$$K^{-1}\int\limits_0^t [\pi(t-\tau)]^{-1/2}\;f\left(\Psi(\tau)\right)\,\mathrm{d}\tau + D_j^{-1/2}\,\left[\Psi(t) - \Gamma_2^\star\right] - 2c_1^\star\pi^{-1/2}t^{1/2} = 0\,. \tag{9.30}$$

Finally, by applying Procedure 6, we take $\Psi(t) = \Gamma_2(t)$ as the only unknown. As $b(t) = 0$, the integral term in Eq. (9.19) vanishes entirely, so that the combination of Eqs. (9.19) and (9.25) simply gives an analytical expression for $\Psi(t)$, valid for the Henry isotherm:

$$\Psi(t) = K\,c_1^\star - \left(K\,c_1^\star - \Gamma_2^\star\right)\,\mathrm{erex}\left(\frac{D_1^{1/2}}{K}\,t^{1/2}\right)\,. \tag{9.31}$$

Equation (9.31) means that for this example Procedure 6 actually yields the solution without requiring IE solving.

As a second example, let us consider a controlled potential experiment for an equilibrium adsorption reaction, followed by an irreversible electron transfer between the electrode and the adsorbed intermediate (see, for example, [61]):

$$X_1 \rightleftarrows X_{2,\mathrm{ad}}\,, \tag{9.32}$$

$$X_{2,\mathrm{ad}} + n\,\mathrm{e}^- \rightarrow \text{products.} \tag{9.33}$$

The stoichiometric matrix $\overline{N}^{\text{dyn,het}}$ in Eq. (2.53) is

$$\overline{N}^{\text{dyn,het}} = \begin{bmatrix} 1 & 0 \\ 1 & -1 \\ 0 & -n \end{bmatrix} . \tag{9.34}$$

Electrons exchanged are taken into account as dynamic species. Therefore, the vector of production rates is $\vec{p}^{\,\text{het}}(t) = \left[p_1^{\text{het}}(t), p_2^{\text{het}}(t), p_3^{\text{het}}(t)\right]^{\text{T}} = \left[J_1^{\perp}(t), \mathrm{d}\Gamma_2(t)/\mathrm{d}t, I(t)(FA)^{-1}\right]^{\text{T}}$. The vector of reaction rates is $\vec{r}^{\,\text{het}}(t) = \left[r_1^{\text{het}}(t), r_2^{\text{het}}(t)\right]^{\text{T}}$. Partial inversion of the corresponding Eq. (2.53) gives Eq. (2.54) with matrices $\overline{V}$ and $\overline{\mathbf{Z}}$:

$$\overline{V} = \begin{bmatrix} 1 & 0 \\ 0 & 1 \\ 0 & 0 \end{bmatrix} , \tag{9.35}$$

$$\overline{\mathbf{Z}} = \begin{bmatrix} -1 & 0 & 0 \\ -1 & -1 & 0 \\ -n & -n & 1 \end{bmatrix} . \tag{9.36}$$

The resulting first equation in equation system (2.54) is replaced by the adsorption equilibrium equation:

$$K\, c_1^{\dagger}(t) = f\left(\Gamma_2(t)\right) . \tag{9.37}$$

The second equation in equation system (2.54) is the reaction rate equation for reaction (9.33):

$$k_2^0\, \Gamma_2(t) \exp\left\{-\alpha_2^{\text{f}} \frac{nF}{RT}\left[E(t) - E_2^0\right]\right\} + J_1^{\perp}(t) + \frac{\mathrm{d}\Gamma_2(t)}{\mathrm{d}t} = 0 , \tag{9.38}$$

and the last equation in equation system (2.54) is the structural conservation relationship

$$nJ_1^{\perp}(t) + n\frac{\mathrm{d}\Gamma_2(t)}{\mathrm{d}t} - \frac{I(t)}{FA} = 0 , \tag{9.39}$$

that can be cast into the form of Eq. (9.14) with $a = 1$ and $b(t) = I(t)(nFA)^{-1}$.

To apply Procedure 1, we can take $\vec{\Psi}(t) = [\Psi_1(t), \Psi_2(t)]^{\text{T}} = \left[J_1^{\perp}(t), \Gamma_2(t)\right]^{\text{T}}$ as unknown functions. With this choice, Eqs. (9.37) and (9.38) yield a (generally nonlinear) IDE system:

$$K\left[c_1^\star + \int_0^t \mathscr{K}_1(t,\tau)\,\Psi_1(\tau)\,\mathrm{d}\tau\right] - f\left(\Psi_2(t)\right) = 0\,, \tag{9.40}$$

$$k_2^0\,\Psi_2(t)\,\exp\left\{-\alpha_2^{\mathrm{f}}\frac{nF}{RT}\left[E(t)-E_2^0\right]\right\} + \Psi_1(t) + \frac{\mathrm{d}\Psi_2(t)}{\mathrm{d}t} = 0\,, \tag{9.41}$$

with initial condition for $\Psi_2(t)$ only: $\Psi_2(0) = \Gamma_2^\star$. After determining $\vec{\Psi}(t)$, the Faradaic current is calculated from Eq. (9.39):

$$\frac{I(t)}{FA} = n\Psi_1(t) + n\frac{\mathrm{d}\Psi_2(t)}{\mathrm{d}t}\,. \tag{9.42}$$

Procedure 2 is applicable, if a steady state condition $\mathrm{d}\Gamma_2(t)/\mathrm{d}t \approx 0$ is assumed for the intermediate $X_{2,\mathrm{ad}}$. Neglect of $\mathrm{d}\Psi_2(t)/\mathrm{d}t$ in Eq. (9.41) then gives an IE system, instead of an IDE system, but retaining of $\mathrm{d}\Psi_2(t)/\mathrm{d}t$ in Eq. (9.42) may still be necessary. Procedure 3 is not applicable for the above choice of unknowns. However, if we choose $\vec{\Psi}(t) = [\Psi_1(t), \Psi_2(t)]^{\mathrm{T}} = \left[\Gamma_2(t), I(t)(FA)^{-1}\right]^{\mathrm{T}}$ as unknowns, we obtain equations alternative to Eqs. (9.40) and (9.41), which can be transformed by Procedure 3 into the following system of one (generally nonlinear) IDE and one AE:

$$K\left\{c_1^\star + \frac{1}{n}\int_0^t \mathscr{K}_1(t,\tau)\,\Psi_2(\tau)\,\mathrm{d}\tau - \frac{\mathrm{d}}{\mathrm{d}t}\int_0^t \mathscr{K}_1(t,\tau)\,\left[\Psi_1(\tau)-\Gamma_2^\star\right]\,\mathrm{d}\tau\right\}$$

$$-f\left(\Psi_1(t)\right) = 0\,, \tag{9.43}$$

$$k_2^0\,\Psi_1(t)\,\exp\left\{-\alpha_2^{\mathrm{f}}\frac{nF}{RT}\left[E(t)-E_2^0\right]\right\} + \frac{\Psi_2(t)}{n} = 0\,, \tag{9.44}$$

with initial condition for $\Psi_1(t)$ only: $\Psi_1(0) = \Gamma_2^\star$. The dimensionless Faradaic current is obtainable immediately from Eqs. (9.43) and (9.44) as $\Psi_2(t)$. In the case of a potential step chronoamperometric experiment, $\Psi_2(t)$ can be calculated from Eq. (9.44), and substituted into Eq. (9.43), so that we get a single IDE for $\Psi_1(t)$. If $E(t)$ is not constant, such a substitution is also possible, but one then obtains a kernel dependent on the $E(t)$ function, which is not convenient, because most of the numerical algorithms for IEs are built for specific, pre-defined kernels (see Chap. 12). To apply Procedure 4, we choose $\vec{\Psi}(t) = [\Psi_1(t), \Psi_2(t)]^{\mathrm{T}} = \left[J_1^{\perp}(t), \mathrm{d}\Gamma_2(t)/\mathrm{d}t\right]^{\mathrm{T}}$ as unknown functions, so that Eqs. (9.37) and (9.38) become an IE system:

$$K\left[c_1^\star + \int_0^t \mathscr{K}_1(t,\tau)\,\Psi_1(\tau)\,\mathrm{d}\tau\right] - f\left(\Gamma_2^\star + \int_0^t \Psi_2(\tau)\,\mathrm{d}\tau\right) = 0\,, \tag{9.45}$$

$$k_2^0\left[\Gamma_2^\star + \int_0^t \Psi_2(\tau)\,\mathrm{d}\tau\right]\exp\left\{-\alpha_2^{\mathrm{f}}\frac{nF}{RT}\left[E(t)-E_2^0\right]\right\} + \Psi_1(t) + \Psi_2(t) = 0\,, \tag{9.46}$$

(which is generally nonlinear). After determining $\vec{\Psi}(t)$, the Faradaic current is calculated from Eq. (9.39):

$$\frac{I(t)}{FA} = n\Psi_1(t) + n\Psi_2(t)\,. \tag{9.47}$$

Procedure 5 can be applied assuming $\vec{\Psi}(t) = [\Psi_1(t), \Psi_2(t)]^{\mathrm{T}} = [\Gamma_2(t), I(t)\,(FA)^{-1}]^{\mathrm{T}}$ as unknown functions. For planar diffusion in a semi-infinite spatial domain, combination of Eq. (9.12) for species X_1 with Eqs. (9.37), (9.38), and (9.39), gives a system of one (generally nonlinear) IE and one AE:

$$K^{-1}\int_0^t [\pi(t-\tau)]^{-1/2}\, f\left(\Psi_1(\tau)\right)\,\mathrm{d}\tau + D_1^{-1/2}\left[\Psi_1(t) - \Gamma_2^\star\right]$$
$$-\frac{D_1^{-1/2}}{n}\int_0^t \Psi_2(\tau)\,\mathrm{d}\tau - 2c_1^\star\pi^{-1/2}t^{1/2} = 0\,, \tag{9.48}$$

$$k_2^0\,\Psi_1(t)\exp\left\{-\alpha_2^{\mathrm{f}}\frac{nF}{RT}\left[E(t)-E_2^0\right]\right\} + \frac{\Psi_2(t)}{n} = 0\,. \tag{9.49}$$

The dimensionless Faradaic current is obtained immediately from Eqs. (9.48) and (9.49), as the solution $\Psi_2(t)$. Similarly to Procedure 3, the unknown $\Psi_2(t)$ can be eliminated in the case of the constant potential. Procedure 6 can be applied in the case of the Henry isotherm. By taking $\Psi(t) = I(t)(FA)^{-1}$ as an unknown function, and considering Eq. (9.39) as the relationship (9.14), combination of Eqs. (9.19) and Eq. (9.39) gives a single, linear IE:

$$k_2^0\,K\left\{c_1^\star + \int_0^t \frac{1}{nK}\,\mathrm{erex}\left[\frac{D_1^{1/2}}{K}(t-\tau)^{1/2}\right]\Psi(\tau)\,\mathrm{d}\tau\right.$$
$$\left.-\left(c_1^\star - \frac{\Gamma_2^\star}{K}\right)\mathrm{erex}\left(\frac{D_1^{1/2}}{K}t^{1/2}\right)\right\}\exp\left\{-\alpha_2^{\mathrm{f}}\frac{nF}{RT}\left[E(t)-E_2^0\right]\right\} + \frac{\Psi(t)}{n} = 0\,. \tag{9.50}$$

As the third example, let us consider a controlled current experiment for the reaction scheme (9.32) and (9.33) (see, in particular, [106]). In this example, Eqs. (9.37) and (9.39), with a known perturbation $I(t)$ serve as a basis for formulating IEs, whereas Eq. (9.38) is combined with Eq. (9.39) to give

$$k_2^0\, \Gamma_2(t)\, \exp\left\{-\alpha_2^{\mathrm{f}} \frac{nF}{RT}\left[E(t) - E_2^0\right]\right\} = -\frac{I(t)}{nFA}\,, \tag{9.51}$$

from which the potential–time response $E(t)$ can be derived. In Procedure 1 we take as unknowns $\vec{\Psi}(t) = [\Psi_1(t), \Psi_2(t)]^{\mathrm{T}} = \left[J_1^{\perp}(t), \Gamma_2(t)\right]^{\mathrm{T}}$, so that Eqs. (9.37) and (9.39) form a (generally nonlinear) IDE system:

$$K\left[c_1^\star + \int_0^t \mathscr{K}_1(t,\tau)\, \Psi_1(\tau)\, \mathrm{d}\tau\right] - f\left(\Psi_2(t)\right) = 0\,, \tag{9.52}$$

$$\Psi_1(t) + \frac{\mathrm{d}\Psi_2(t)}{\mathrm{d}t} - \frac{I(t)}{nFA} = 0\,, \tag{9.53}$$

with an initial condition for $\Psi_2(t)$ only: $\Psi_2(0) = \Gamma_2^\star$. If the product $\mathscr{K}_1(t,\tau)\, I(\tau)$ is analytically integrable, it is convenient to replace Eqs. (9.52) and (9.53) by a single (generally nonlinear) IDE for $\Psi_2(t)$:

$$K\left[c_1^\star + \int_0^t \mathscr{K}_1(t,\tau)\, \frac{I(\tau)}{nFA}\, \mathrm{d}\tau - \int_0^t \mathscr{K}_1(t,\tau)\, \frac{\mathrm{d}\Psi_2(\tau)}{\mathrm{d}\tau}\, \mathrm{d}\tau\right] - f\left(\Psi_2(t)\right) = 0\,, \tag{9.54}$$

because the first integral in Eq. (9.54) is then a known analytical function of t. By solving Eqs. (9.52) and (9.53), or Eq. (9.54), we obtain $\Psi_2(t)$ and set it into Eq. (9.51) to get the potential response:

$$E(t) = E_2^0 + \frac{RT}{\alpha_2^{\mathrm{f}} nF} \ln\left[-\frac{nFA}{I(t)}\, k_2^0\, \Psi_2(t)\right]. \tag{9.55}$$

Procedure 2 can be applied if one assumes a steady state condition $\mathrm{d}\Gamma_2(t)/\mathrm{d}t \approx 0$. The IDE (9.54) then reduces to the AE for $\Psi_2(t)$:

$$K\left[c_1^\star + \int_0^t \mathscr{K}_1(t,\tau)\, \frac{I(\tau)}{nFA}\, \mathrm{d}\tau\right] - f\left(\Psi_2(t)\right) = 0\,. \tag{9.56}$$

The solution $\Psi_2(t)$ is again substituted into Eq. (9.55). By applying Procedure 3, Eq. (9.54) is replaced by the (generally nonlinear) IDE:

$$K\left\{c_1^\star + \int_0^t \mathcal{K}_1(t,\tau)\,\frac{I(\tau)}{nFA}\,\mathrm{d}\tau - \frac{\mathrm{d}}{\mathrm{d}t}\int_0^t \mathcal{K}_1(t,\tau)\,\left[\Psi_2(\tau) - \Gamma_2^\star\right]\,\mathrm{d}\tau\right\}$$
$$-f\left(\Psi_2(t)\right) = 0\,, \qquad (9.57)$$

with an initial condition $\Psi_2(0) = \Gamma_2^\star$. Equation (9.57) is solved for $\Psi_2(t)$, and the result is set into Eq. (9.55). Procedure 4 assumes $\vec{\Psi}(t) = [\Psi_1(t), \Psi_2(t)]^{\mathrm{T}} = \left[J_1^{\perp}(t), \mathrm{d}\Gamma_2(t)/\mathrm{d}t\right]^{\mathrm{T}}$, so that Eqs. (9.37) and (9.39) turn into a (generally nonlinear) IE system:

$$K\left[c_1^\star + \int_0^t \mathcal{K}_1(t,\tau)\,\Psi_1(\tau)\,\mathrm{d}\tau\right] - f\left(\Gamma_2^\star + \int_0^t \Psi_2(\tau)\,\mathrm{d}\tau\right) = 0\,, \qquad (9.58)$$

$$\Psi_1(t) + \Psi_2(t) - \frac{I(t)}{nFA} = 0\,. \qquad (9.59)$$

System (9.58) and (9.59) can be reduced to a single IE, by two different routes. If the product $\mathcal{K}_1(t,\tau)\,I(\tau)$ is analytically integrable, we can calculate $\Psi_1(t)$ from Eq. (9.59), and set it into Eq. (9.58), which gives

$$K\left[c_1^\star + \int_0^t \mathcal{K}_1(t,\tau)\,\frac{I(\tau)}{nFA}\,\mathrm{d}\tau - \int_0^t \mathcal{K}_1(t,\tau)\,\Psi_2(\tau)\,\mathrm{d}\tau\right]$$
$$-f\left(\Gamma_2^\star + \int_0^t \Psi_2(\tau)\,\mathrm{d}\tau\right) = 0\,. \qquad (9.60)$$

For planar diffusion in a semi-infinite spatial domain, Eq. (9.60) is equivalent to Eq. (A23) in Reinmuth [106]. After solving Eq. (9.60) for $\Psi_2(t)$, the electrode potential is obtained from

$$E(t) = E_2^0 + \frac{RT}{\alpha_2^{\mathrm{f}} nF}\ln\left\{-\frac{nFA}{I(t)}\,k_2^0\left[\Gamma_2^\star + \int_0^t \Psi_2(\tau)\,\mathrm{d}\tau\right]\right\}\,. \qquad (9.61)$$

Alternatively, if $I(\tau)$ is analytically integrable, we can calculate $\Psi_2(t)$ from Eq. (9.59), and set it into Eq. (9.58), which gives

$$K\left[c_1^\star + \int_0^t \mathscr{K}_1(t,\tau)\,\Psi_1(\tau)\,\mathrm{d}\tau\right] - f\left(\Gamma_2^\star + \int_0^t \frac{I(\tau)}{nFA}\,\mathrm{d}\tau - \int_0^t \Psi_1(\tau)\,\mathrm{d}\tau\right) = 0\,. \tag{9.62}$$

After solving Eq. (9.62) for $\Psi_1(t)$, the electrode potential is obtained from

$$E(t) = E_2^0 + \frac{RT}{\alpha_2^{\mathrm{f}} nF}\ln\left\{-\frac{nFA}{I(t)}\,k_2^0\left[\Gamma_2^\star + \int_0^t \frac{I(\tau)}{nFA}\,\mathrm{d}\tau - \int_0^t \Psi_1(\tau)\,\mathrm{d}\tau\right]\right\}\,. \tag{9.63}$$

Procedure 5 is applicable assuming a single unknown $\Psi(t) = \Gamma_2(t)$. For planar diffusion in a semi-infinite spatial domain, combination of Eqs. (9.15), (9.37) and (9.39) gives a single (generally nonlinear) IE:

$$K^{-1}\int_0^t [\pi(t-\tau)]^{-1/2}\, f(\Psi(\tau))\,\mathrm{d}\tau + D_1^{-1/2}\left[\Psi(t) - \Gamma_2^\star\right]$$
$$-2c_1^\star \pi^{-1/2} t^{1/2} - D_1^{-1/2}\int_0^t \frac{I(\tau)}{nFA}\,\mathrm{d}\tau = 0\,. \tag{9.64}$$

After solving it for $\Psi(t)$, the electrode potential is calculated from

$$E(t) = E_2^0 + \frac{RT}{\alpha_2^{\mathrm{f}} nF}\ln\left\{-\frac{nFA}{I(t)}\,k_2^0\left[\Gamma_2^\star + \int_0^t \Psi(\tau)\,\mathrm{d}\tau\right]\right\}\,, \tag{9.65}$$

where the integral of $\Psi(t)$ must be additionally evaluated, as it does not occur in the IE (9.64). Equation (9.64) is equivalent to Eq. (A.24) in [106]. Finally, Procedure 6 gives an explicit expression for $c_1^\dagger(t)$:

$$c_1^\dagger(t) = c_1^\star + \int_0^t \frac{1}{K}\,\mathrm{erex}\left[\frac{D_1^{1/2}}{K}(t-\tau)^{1/2}\right]\frac{I(\tau)}{nFA}\,\mathrm{d}\tau$$
$$-\left(c_1^\star - \frac{\Gamma_2^\star}{K}\right)\mathrm{erex}\left(\frac{D_1^{1/2}}{K}\,t^{1/2}\right)\,. \tag{9.66}$$

Equation (9.66) can be useful if the integral contained in this equation can be evaluated analytically. Numerical integration would not present any advantage over other procedures, especially Procedure 4. By combining Eq. (9.66) with Eq. (9.37) one obtains the AE for the unknown $\Psi(t) = \Gamma_2(t)$:

$$K\left\{c_1^\star + \int_0^t \frac{1}{K}\,\mathrm{erex}\left[\frac{D_1^{1/2}}{K}(t-\tau)^{1/2}\right]\frac{I(\tau)}{nFA}\,\mathrm{d}\tau - \left(c_1^\star - \frac{\Gamma_2^\star}{K}\right)\mathrm{erex}\left(\frac{D_1^{1/2}}{K}t^{1/2}\right)\right\} - f\left(\Psi(t)\right) = 0 \,. \tag{9.67}$$

The electrode potential is calculated from

$$E(t) = E_2^0 + \frac{RT}{\alpha_2^{\mathrm{f}} nF}\ln\left[-\frac{nFA}{I(t)}k_2^0\,\Psi(t)\right] . \tag{9.68}$$

9.2 Literature Examples

We summarise here the IE- or IDE-based, spatially one-dimensional, models of electrochemical systems containing localised and distributed species, that are available in the literature. We begin with models involving very simple reaction schemes, and gradually pass to models involving more and more complicated schemes. A few of the models listed below were previously reviewed by Lovrić [52].

9.2.1 One-Reaction Schemes

The simplest models deal with a single heterogeneous non-electrochemical reaction

$$\mathrm{X_1} \rightleftarrows \mathrm{X_{2,ad}} \tag{9.69}$$

involving a dynamic distributed species X_1, and a dynamic localised species $X_{2,\mathrm{ad}}$ (adsorbed at the electrode). It is assumed that the electrode studied is initially free of $X_{2,\mathrm{ad}}$, but it subsequently becomes covered by this adsorbate. Assuming equilibrium reaction (9.69), the relevant theory describing the temporal evolution of such a system was given by Reinmuth [105], Levich et al. [41], Rampazzo [104], Vogel [117, 118], Sluyters-Rehbach and Sluyters [115], Miller [69], and Risović [109]. Miller [69], Bhugun and Anson [7], and Kankare and Vinokurov [36] provided a theory for a non-equilibrium reversible reaction (9.69). Langmuir isotherm or reaction rate expressions consistent at equilibrium with the Langmuir isotherm were assumed in [7, 36, 69, 105, 115]. Temkin isotherm was assumed in [117, 118]. Frumkin isotherm was assumed in [104], whereas a modified Flory–Huggins isotherm (including lateral interactions between adsorbed molecules) was assumed in [109]. A general isotherm was approximately handled in [41]. Transport was by planar diffusion in a semi-infinite spatial domain [7, 36, 41, 104, 105, 109, 115]; spherical diffusion [36, 117, 118]; cylindrical diffusion [36]; and by convection–diffusion to an expanding plane [69, 105, 115]. Podgaetskii [88] considered a somewhat opposite problem of determining the amount of $X_{2,ad}$, initially present on the electrode, from the measurements of the differential capacitance of the double layer. He assumed planar diffusion in a semi-infinite spatial domain, Frumkin isotherm, and linear potential sweep voltammetry. He later extended his theory to spherical diffusion [91] and galvanostatic experiments [92]. Most of the above works were motivated by electrochemistry, but some of them are also applicable to non-electrochemical systems. It should also be mentioned that there exist analogous IE-based models for reaction (9.69) taking place at non-electrochemical interfaces. Examples of such models are found in studies of liquid | air interfaces [25, 70] or diffusion of impurities in metals [93, 94]. Further similar IEs arise in some heat conduction problems [90].

Holub [29] considered a single reaction of a substance X_1 transported by diffusion to the electrode, with an adsorbed substance $X_{2,ad}$, initially covering the electrode:

$$X_1 + X_{2,ad} \rightarrow \text{products}\,. \tag{9.70}$$

Planar diffusion in a semi-infinite spatial domain, and kinetic equations consistent with the Langmuir isotherm were assumed. The resulting equations were shown to be equivalent to the equations describing Langmuir adsorption without desorption.

Another class of very simple models deals with a single electrochemical reaction

$$X_1 \pm n\,e^- \rightleftarrows X_{2,ad} \tag{9.71}$$

involving a dynamic distributed species X_1, and a dynamic localised species $X_{2,ad}$ (e.g. adsorbed or otherwise deposited or precipitated at the electrode). Nicholson [82] presented a theory (containing errors, see Sect. 9.1) of linear potential sweep voltammetry for equilibrium reaction (9.71), assuming planar diffusion in a semi-infinite spatial domain. The theory was later reconsidered by Gao and Roy [17]. White and Lawson [119] discussed linear potential sweep voltammetry for a non-equilibrium reversible reaction (9.71) subject to Butler–Volmer kinetics, assuming planar or spherical diffusion in a semi-infinite spatial domain. Pnev and Parubotchaya [85] modelled inversion voltammetry under analogous assumptions (but for planar diffusion only). A linear dependence of the adsorption energy, on the electrode coverage by the adsorbed reactant, was assumed. Klimatchev et al. [37] discussed a method for determining the activity of the deposited species, from experimental voltammograms. Pnev et al. [86, 87] analysed inverse polarography for equilibrium and irreversible reaction (9.71), assuming Langmuir-type reaction rate and equilibrium expressions. Planar diffusion in a semi-infinite spatial domain was assumed. Roizenblat and Reznik [110] considered an electrodissolution of metals concentrated in a very thin, planar mercury film electrode, by DC and AC voltammetry. Hence, the species present in the mercury film was formally considered as a quasi-localised species (see Sect. 2.9). Later they modelled also inversion voltammetry under similar assumptions [111]. Lovrić et al. [62] analysed potential step chronoamperometry and square wave voltammetry for equilibrium and irreversible reaction (9.71), assuming planar diffusion in a semi-infinite spatial domain.

Xiang [122] considered the theory of current step chronopotentiometry, potential step chronoamperometry, and linear potential sweep voltammetry for a single, but somewhat more complicated irreversible electrochemical reaction:

$$X_1 + X_{2,s} \rightarrow X_{3,ad} + n\,e^- \tag{9.72}$$

involving one dynamic distributed species X_1, one static species $X_{2,s}$ (species belonging to the solid electrode phase), and one dynamic localised species $X_{3,ad}$ (adsorbed). Planar diffusion in a semi-infinite spatial domain was assumed. The discussion was motivated by the etching metal phase analysis. Lovrić et al. [63] and Mirčeski and Lovrić [74] discussed square wave voltammetry and potential step chronoamperometry for another similar reaction:

$$X_{1,dep} + 2e^- \rightleftarrows X_{2,s} + X_3 \tag{9.73}$$

involving one dynamic localised species $X_{1,dep}$ (deposited solid phase of a mercury salt) and $X_{2,s}$ (static species belonging to the liquid mercury electrode phase), and one distributed species X_3. Planar diffusion in a semi-infinite spatial domain was assumed.

9.2.2 Two-Reaction Schemes

Controlled potential experiments for the two-reaction scheme:

$$X_1 \rightleftarrows X_{2,ad}\,, \tag{9.74}$$

$$X_1 + n\,e^- \rightleftarrows X_3\,, \tag{9.75}$$

comprising an electron transfer between dynamic distributed species X_1 and X_3, and an adsorption of the reactant X_1, were a subject of several theoretical studies [15, 32, 33, 53–55, 58, 64, 84, 113, 116, 126]. In some studies reaction (9.74) was assumed to be an equilibrium reaction, subject to the Henry isotherm [32, 33, 113, 116], the Langmuir isotherm [15], the Frumkin isotherm [53–55, 64, 84, 126], or to the conditions of full coverage [32, 33, 116]. Lovrić and Komorsky-Lovrić [58] assumed a non-equilibrium reversible reaction (9.74) consistent with the Henry isotherm at equilibrium. Reaction (9.75) was either an equilibrium reaction [53–55,58,64,84,126], or irreversible reaction [15,113]. Planar diffusion was considered in [58, 84, 113, 116], and spherical in [32, 53–55, 64, 126]. Expanding plane convection–diffusion was considered in [33]. Convection–diffusion to an RDE, modelled by a thin layer diffusion was assumed by Fujioka [15], A semi-infinite spatial domain was assumed in all remaining cases. The theory was obtained for DC polarography [53], normal pulse voltammetry [113], or polarography [32, 33, 116], pulse polarography [64, 126], differential pulse polarography [64, 84, 126], linear potential sweep voltammetry [15], cyclic staircase voltammetry [54], and square wave voltammetry [55,58].

Reaction scheme (9.74) and (9.75) was also considered in the studies of the effect of adsorption on chronopotentiometric transition times, by Reinmuth [106], who assumed a general adsorption isotherm, and Podgaetskii and Filinovskii [14,95,96], who assumed adsorption isotherm with two plateau [95], Frumkin isotherm [96], and Henry isotherm [14].

Another two-reaction scheme with the adsorption of the reactant of the electrochemical reaction:

$$X_1 \rightleftarrows X_{2,ad} , \tag{9.76}$$

$$X_{2,ad} + n\,e^- \rightleftarrows X_3 , \tag{9.77}$$

is an alternative to the scheme (9.74) and (9.75), in which the electron transfer reaction proceeds between dynamic localised species $X_{2,ad}$ (which is an adsorbed form of the dynamic distributed reactant X_1), and distributed species X_3. When reactions (9.74) and (9.76) are equilibrium reactions, the two schemes (9.74) and (9.75), and (9.76) and (9.77) are experimentally indistinguishable. Controlled potential experiments for the scheme (9.76) and (9.77) were a subject of several theoretical studies [18, 41, 51, 56, 57, 61, 83, 113]. In [18, 51, 56, 57, 61, 83] a theory of square wave voltammetry was discussed. In [41] polarography was considered, and in [113] normal pulse voltammetry was discussed. Equilibrium reaction (9.76) subject to the Henry isotherm was assumed. Planar diffusion in a semi-infinite spatial domain was assumed in [18,51,56,57,61,83,113], and convection–diffusion to an expanding plane in [41]. Reaction (9.77) was either non-equilibrium reversible [18,51,56,57], or irreversible [41,61,83,113]. The number n of electrons transferred was equal 1 in [51, 56, 57, 61, 83].

Although this was not clearly stated, Holub [26–28] probably assumed a generalisation of the scheme (9.76) and (9.77) in his theory of a potentiostatic experiment at a mercury electrode. Equilibrium reaction (9.76) subject to the Henry isotherm and a first-order non-equilibrium follow-up reaction that might be an electron transfer (9.77) were considered. External spherical diffusion was assumed in Holub [27]. Convection–diffusion to an expanding plane was assumed in Holub [26, 28]. The transport PDEs were converted to pure planar diffusion PDEs prior to formulating IEs.

Aoki and Kato [5] presented a theory of cyclic voltammetry for the reaction scheme:

$$X_1 \rightleftarrows X_{2,ad} , \tag{9.78}$$

$$X_{2,ad} + n\,e^- \rightleftarrows X_{3,ad} , \tag{9.79}$$

which differs from the scheme (9.76) and (9.77) by the electroreduction product $X_{3,ad}$, which is now adsorbed, instead of being distributed in the electrolyte. Reaction (9.78) was non-equilibrium reversible, with reaction rate equations consistent with the Henry isotherm at equilibrium, whereas reaction (9.80) was an equilibrium reaction. Planar diffusion in a semi-infinite spatial domain was assumed.

Mirčeski et al. [78] modelled square wave voltammetry for the reaction scheme:

$$X_1 \rightleftarrows X_{2,ad} , \tag{9.80}$$

$$X_{2,ad} + X_{3,s} \rightleftarrows X_{4,ad} + 2\,e^- , \tag{9.81}$$

involving dynamic distributed species X_1, static localised species $X_{3,s}$ (mercury), and dynamic localised species $X_{2,ad}$ and $X_{4,ad}$. Planar diffusion in a semi-infinite spatial domain was assumed. Equilibrium reaction (9.80) subject to the Henry isotherm, and non-equilibrium reversible reaction (9.81) were considered.

The theory of cyclic voltammetry for the two-reaction scheme with the adsorption of the electrochemical reaction product:

$$X_1 + n\,e^- \rightleftarrows X_2 , \tag{9.82}$$

$$X_2 \rightleftarrows X_{3,ad} \tag{9.83}$$

involving dynamic distributed species X_1 and X_2, and a dynamic localised species $X_{3,ad}$, was studied by Hulbert and Shain [34]. Equilibrium reaction (9.82) and non-equilibrium reversible reaction (9.83), with reaction rate expressions consistent at equilibrium with the Langmuir isotherm, were considered. Planar diffusion in a semi-infinite spatial domain was assumed.

A two-reaction scheme comprising equilibrium electron transfers:

$$X_{1,s} \rightleftarrows X_2 + e^- , \tag{9.84}$$

$$X_{1,s} + X_3 \rightleftarrows X_{4,ad} + e^- , \tag{9.85}$$

where $X_{1,s}$ is a static localised species (solid material), X_2 and X_3 are dynamic distributed species, and $X_{4,ad}$ is a dynamic localised species, was considered by Casadio [12], as a model of an anodic metal dissolution inhibited by a redox adsorption. Casadio developed a theory of cyclic voltammetry, assuming planar diffusion in a semi-infinite spatial domain. In a subsequent paper [13] the theory was extended to a non-equilibrium reversible reaction (9.84), and an equilibrium reaction (9.85) subject to a Temkin-type isotherm.

A two-reaction heterogeneous catalytic EC scheme:

$$X_{1,ad} + e^- \rightleftarrows X_{2,ad} , \tag{9.86}$$

$$X_{2,ad} + X_3 \rightarrow X_{1,ad} + X_4 , \tag{9.87}$$

with equilibrium reaction (9.86) and irreversible reaction (9.87) was discussed by Andrieux and Savéant [3], assuming planar diffusion of X_3 and X_4 in a semi-infinite spatial domain. Predictions for cyclic voltammetry were obtained. Aoki et al. [6] considered a mirror image of the scheme (9.86) and (9.87), for oxidation:

$$X_{1,ad} \rightleftarrows X_{2,ad} + e^- , \tag{9.88}$$

$$X_{2,ad} + X_3 \rightleftarrows X_{1,ad} + X_4 , \tag{9.89}$$

assuming equilibrium reaction (9.88) and non-equilibrium reversible reaction (9.89), other conditions identical. The theory of linear potential sweep and cyclic voltammetry was obtained. Bieniasz et al. [11] extended the theories from [3, 6] onto spherical diffusion in a semi-infinite spatial domain.

Mirčeski et al. [80] considered the theory of square wave voltammetry for the reaction scheme:

$$X_1 \rightleftarrows X_2 , \tag{9.90}$$

$$X_2 \rightleftarrows X_{3,ad} + n\,e^- , \tag{9.91}$$

involving non-equilibrium reversible homogeneous reaction (9.90) and non-equilibrium reversible heterogeneous deposition reaction (9.91) between dynamic distributed species X_1 and X_2, and dynamic localised species $X_{3,ad}$ (insoluble mercury salt). Planar diffusion in a semi-infinite spatial domain was assumed.

Yet another two-reaction scheme, of EE type, was studied by Lovrić [45], assuming a potentiostatic method and transport by planar diffusion in a semi-infinite spatial domain, or expanding plane convection–diffusion. The scheme was:

$$X_1 + e^- \rightleftarrows X_{2,ad} , \tag{9.92}$$

$$X_{2,ad} + e^- \rightleftarrows X_3 , \tag{9.93}$$

where X_1 and X_3 are dynamic distributed species, and $X_{2,ad}$ is a dynamic localised intermediate.

9.2.3 *Three-Reaction Schemes*

Controlled potential transient experiments for the scheme comprising reactions (9.74) and (9.75), plus one additional reaction of product adsorption:

$$X_1 \rightleftarrows X_{2,\mathrm{ad}} , \tag{9.94}$$

$$X_1 + n\,\mathrm{e}^- \rightleftarrows X_3 , \tag{9.95}$$

$$X_3 \rightleftarrows X_{4,\mathrm{ad}} , \tag{9.96}$$

were discussed theoretically by many authors [20, 22, 46, 65–67, 98–100, 107, 108, 112, 114, 120], and some of their results were compared by Sluyters-Rehbach and Sluyters [114]. All reactions (9.94), (9.95), and (9.96) were assumed to be equilibrium reactions. Henry isotherms of reactions (9.94) and (9.96) were assumed in [22, 46], Langmuir isotherms in [20, 66, 67, 98, 107, 108, 112, 120], Frumkin isotherms in [22, 98, 108], and general isotherms in [65, 99, 100]. Planar diffusion in a semi-infinite spatial domain was assumed in [20, 22, 46, 100, 107, 108, 120], spherical diffusion in a semi-infinite spatial domain was assumed in [20, 22, 108], convection–diffusion to an expanding plane was assumed in [22, 65–67, 98, 99, 108, 112]. The electroanalytical techniques considered include: potentiostatic methods such as chronocoulometry [20, 22, 65, 98, 107, 108] and double potential step chronocoulometry [100], cyclic voltammetry [120], pulse polarography [46, 66, 99], differential pulse polarography [112], normal pulse polarography [67], and flux-jump technique [22]. The IEs obtained in [20] were also used by Guaus et al. [21]. Podgaetskii [89] considered the effect of adsorption on polarographic curves for the scheme (9.94), (9.95), and (9.96). He assumed irreversible reaction (9.95), coupled Frumkin isotherms for equilibrium reactions (9.94) and (9.96), and expanding plane convection–diffusion.

Podgaetskii and Filinovskii [97] calculated chronopotentiometric transition times assuming planar diffusion in a semi-infinite spatial domain, and coupled Frumkin isotherms for equilibrium reactions (9.94) and (9.96).

An alternative scheme comprising reactions (9.78) and (9.80), plus one additional reaction of product desorption:

$$X_1 \rightleftarrows X_{2,\mathrm{ad}} , \tag{9.97}$$

$$X_{2,\mathrm{ad}} + \mathrm{e}^- \rightleftarrows X_{3,\mathrm{ad}} , \tag{9.98}$$

$$X_{3,\mathrm{ad}} \rightleftarrows X_4 , \tag{9.99}$$

was studied by Lovrić and co-workers [40, 56, 57], assuming planar diffusion in a semi-infinite spatial domain, equilibrium reactions (9.97) and (9.99) subject to the Henry isotherms, and non-equilibrium reversible reaction (9.98). A theory of square wave voltammetry was obtained.

There have also been other three-reaction schemes studied by the IE method. Lovrić and Komorsky-Lovrić [59] considered the following EE reaction scheme with adsorption of the intermediate:

$$X_1 + e^- \rightleftarrows X_2 \,, \tag{9.100}$$

$$X_2 \rightleftarrows X_{3,ad} \,, \tag{9.101}$$

$$X_2 + e^- \rightleftarrows X_4 \,. \tag{9.102}$$

All reactions (9.100), (9.101), and (9.102) were equilibrium reactions. Langmuir isotherm was assumed for reaction (9.101). Theoretical predictions for square wave voltammetry have been obtained, assuming planar diffusion in a semi-infinite spatial domain. A reaction scheme with parallel electron transfers, consuming both the distributed reactant and its adsorbed form:

$$X_1 \rightleftarrows X_{2,ad} \,, \tag{9.103}$$

$$X_1 + n\,e^- \rightleftarrows X_3 \,, \tag{9.104}$$

$$X_{2,ad} + n\,e^- \rightleftarrows X_3 \,, \tag{9.105}$$

was considered by Lovrić and Komorsky-Lovrić [38, 47]. Irreversible or non-equilibrium reversible electron transfers (9.104) and (9.105) were assumed, and the adsorption reaction (9.103) was subject to the Henry isotherm. Theoretical predictions for pulse polarography were obtained, assuming planar diffusion in a semi-infinite spatial domain. Mirčeski et al. [81] discussed a somewhat similar scheme, but with product adsorption:

$$X_{1,s} \rightleftarrows X_2 + n\,e^- \,, \tag{9.106}$$

$$X_2 \rightleftarrows X_{3,ad} \,, \tag{9.107}$$

$$X_{1,s} \rightleftarrows X_{3,ad} + n\,e^- \,, \tag{9.108}$$

where $X_{1,s}$ is a static species (solid material), X_2 is a dynamic distributed species, and $X_{3,ad}$ is a dynamic localised species. Equilibrium reaction (9.107) subject to the Henry isotherm, and non-equilibrium reversible reactions (9.106) and (9.108) were assumed. A possibility of interactions between adsorbate molecules was also taken into account. Theoretical predictions for square wave voltammetry were obtained, assuming planar diffusion in a semi-infinite spatial domain.

There have also been models with three-reaction schemes including homogeneous reactions, in addition to heterogeneous reactions, such as the EC scheme with reactant adsorption:

$$X_1 \rightleftarrows X_{2,ad} \,, \tag{9.109}$$

$$X_1 + e^- \rightleftarrows X_3 \,, \tag{9.110}$$

$$X_3 \rightarrow X_4 \,, \tag{9.111}$$

considered by Wopschall and Shain [121]. In [121] equilibrium reaction (9.109) subject to the Langmuir isotherm, equilibrium reaction (9.110), and irreversible homogeneous reaction (9.111) were assumed. Predictions for cyclic voltammetry were obtained, assuming planar diffusion in a semi-infinite spatial domain. The scheme (9.109)–(9.111) was also considered by Mirčeski and Lovrić [75]. In [75] equilibrium reaction (9.109) subject to the Henry isotherm, non-equilibrium reversible reaction (9.110), and irreversible reaction (9.111) were assumed. Predictions for square wave voltammetry were obtained, assuming planar diffusion in a semi-infinite spatial domain. Another three-reaction scheme:

$$X_1 + e^- \rightarrow X_2 \,, \tag{9.112}$$

$$X_2 \rightarrow X_{3,\mathrm{ad}} \,, \tag{9.113}$$

$$X_2 \rightarrow X_4 \,, \tag{9.114}$$

involving an electron transfer (9.112) with product adsorption (9.113), and homogeneous reaction (9.114), was considered by Bhugun and Savéant [8], as a model of self-inhibition of an electrochemical process. Assuming all reactions irreversible, and planar diffusion in a semi-infinite spatial domain, a relevant theory was obtained, of cyclic voltammetry and preparative scale electrolysis. The IEs obtained were later used by Allongue et al. [2], and Yang and Wang [124]. Bhugun and Savéant [9] (and Bhugun et al. [10]) considered also a model of self-inhibition in the catalytic ECC scheme:

$$X_1 + e^- \rightleftarrows X_2 \,, \tag{9.115}$$

$$X_2 \rightarrow X_1 + m\, X_3 \,, \tag{9.116}$$

$$X_3 \rightleftarrows X_{4,\mathrm{ad}} \,. \tag{9.117}$$

Predictions for cyclic voltammetry were obtained, assuming equilibrium reaction (9.115), irreversible homogeneous reaction (9.116), and non-equilibrium reversible reaction (9.117), together with planar diffusion in a semi-infinite spatial domain. A catalytic ECC reaction scheme involving immobilised redox enzyme $X_{3,\mathrm{ad}}$:

$$X_1 + e^- \rightleftarrows X_2 \,, \tag{9.118}$$

$$X_2 + X_{3,\mathrm{ad}} \rightleftarrows X_{4,\mathrm{ad}} \,, \tag{9.119}$$

$$X_{4,\mathrm{ad}} \rightarrow X_1 + X_{3,\mathrm{ad}} \,, \tag{9.120}$$

was considered by Limoges and Savéant [42], assuming planar diffusion in a semi-infinite spatial domain. Predictions for cyclic voltammetry were obtained.

9.2.4 More Complicated Reaction Schemes

An extension of the reaction scheme (9.94)–(9.96), consisting of four reactions, and including separate electron transfers between distributed and localised reactants and products:

$$X_1 \rightleftarrows X_{2,\mathrm{ad}} , \tag{9.121}$$

$$X_1 + e^- \rightleftarrows X_3 , \tag{9.122}$$

$$X_3 \rightleftarrows X_{4,\mathrm{ad}} , \tag{9.123}$$

$$X_{2,\mathrm{ad}} + e^- \rightleftarrows X_{4,\mathrm{ad}} , \tag{9.124}$$

was studied by Lovrić [49], assuming non-equilibrium reversible electron transfers, and equilibrium adsorption reactions subject to the Henry isotherms. Planar diffusion in a semi-infinite spatial domain was assumed. Predictions for differential pulse voltammetry were obtained. The same author [48] considered scheme (9.121)–(9.124) extended with two further electrochemical cross-reactions between distributed and localised reactants:

$$X_1 \rightleftarrows X_{2,\mathrm{ad}} , \tag{9.125}$$

$$X_1 + n\,e^- \rightleftarrows X_3 , \tag{9.126}$$

$$X_3 \rightleftarrows X_{4,\mathrm{ad}} , \tag{9.127}$$

$$X_{2,\mathrm{ad}} + n\,e^- \rightleftarrows X_{4,\mathrm{ad}} , \tag{9.128}$$

$$X_1 + n\,e^- \rightleftarrows X_{4,\mathrm{ad}} , \tag{9.129}$$

$$X_{2,\mathrm{ad}} + n\,e^- \rightleftarrows X_3 , \tag{9.130}$$

at the same assumptions. Predictions for pulse polarography were obtained.

Mirčeski [71] considered an EC scheme with the adsorption of both reactant and product:

$$X_1 \rightleftarrows X_{2,\mathrm{ad}} , \tag{9.131}$$

$$X_1 + n\,e^- \rightleftarrows X_3 , \tag{9.132}$$

$$X_3 \rightleftarrows X_{4,\mathrm{ad}} , \tag{9.133}$$

$$X_3 \rightarrow X_5 , \tag{9.134}$$

assuming equilibrium reactions (9.131) and (9.133) subject to Henry isotherms, non-equilibrium reversible reaction (9.132), and irreversible homogeneous reaction (9.134). Planar diffusion in a semi-infinite spatial domain was assumed. Predictions for square wave voltammetry were obtained. Subsequently, Mirčeski and Lovrić [76] considered an extension of the reaction scheme (9.131)–(9.134):

$$X_1 \rightleftarrows X_{2,ad} , \tag{9.135}$$

$$X_1 + n\,e^- \rightleftarrows X_3 , \tag{9.136}$$

$$X_3 \rightleftarrows X_{4,ad} , \tag{9.137}$$

$$X_3 \rightarrow X_5 , \tag{9.138}$$

$$X_{2,ad} + n\,e^- \rightleftarrows X_{4,ad} , \tag{9.139}$$

$$X_{4,ad} \rightarrow X_{6,ad} , \tag{9.140}$$

under analogous assumptions. Mirčeski and Quentel [77] discussed also the catalytic reaction scheme with adsorption, and a surface regeneration reaction competing with a homogeneous one:

$$X_1 + n\,e^- \rightleftarrows X_2 , \tag{9.141}$$

$$X_2 \rightarrow X_1 , \tag{9.142}$$

$$X_1 \rightleftarrows X_{3,ad} , \tag{9.143}$$

$$X_2 \rightleftarrows X_{4,ad} , \tag{9.144}$$

$$X_{3,ad} + n\,e^- \rightleftarrows X_{4,ad} , \tag{9.145}$$

$$X_{4,ad} \rightarrow X_{3,ad} . \tag{9.146}$$

Assumptions analogous to those in [71, 76] were adopted. A different catalytic reaction scheme:

$$X_{1,ad} + n\,e^- \rightleftarrows X_{2,ad} , \tag{9.147}$$

$$X_{1,ad} \rightleftarrows X_{3,ad} , \tag{9.148}$$

$$X_{2,ad} \rightleftarrows X_{4,ad} , \tag{9.149}$$

$$m\,X_{2,ad} + X_5 \rightarrow m\,X_{1,ad} + X_6 , \tag{9.150}$$

with equilibrium surface reactions (9.147)–(9.149), and an irreversible reaction (9.150), was considered by Jaworski and Cox [35], in their modelling of linear potential sweep voltammetry. Planar diffusion in a semi-infinite spatial domain was assumed. A number of similar, but more complicated schemes were considered at analogous assumptions by Xie and Anson [123], and Limoges et al. [43].

A number of similarly or even more complicated schemes have been studied in the publications that we list below, but we refer the Reader to these publications for details of the schemes. In particular, a scheme representing induced reactant adsorption in a metal–polyelectrolyte system was studied by Puy et al. [16, 68, 101–103], assuming planar diffusion in a semi-infinite spatial domain. Predictions were obtained for pulse polarography and normal pulse polarography. Potential-independent and potential-dependent adsorption parameters were assumed. Some further complicated schemes associated with the studies of anion induced adsorption of metal ions, complexation reactions, or electrochemistry of mercury salts were considered in [19, 23, 24, 39, 50, 60, 72, 73, 79, 125, 127]. Planar diffusion in a semi-infinite spatial domain, and adsorption subject to the Henry isotherms were assumed, except for [125], where Frumkin isotherm was considered. Predictions for DC polarography [50, 125], normal pulse polarography [60], differential pulse polarography [127], and square wave voltammetry [19, 23, 24, 39, 72, 73, 79] were obtained. Complicated enzymatic reaction schemes, for redox enzymes immobilised on electrodes, were modelled by Limoges et al. [44] and Andrieux et al. [4], assuming planar diffusion in a semi-infinite spatial domain, and potential step chronoamperometry or cyclic voltammetry. A possibility of Michaelis–Menten kinetics was adopted for heterogeneous reactions, which resulted in nonlinear IEs or IDEs. Further IE-based modelling of electron transfer and transport in enzyme layers was reported by Agnès et al. [1].

References

1. Agnès A, Demaille C, Moiroux J, Savéant JM, Bourdillon C (2007) Electron transfer and transport in ordered enzyme layers. In: Bard AJ, Stratmann M (eds) Encyclopedia of electrochemistry, vol 10. Wiley-VCH, Weinheim, pp 763–812
2. Allongue P, Delamar M, Desbat B, Fagebaume O, Hitmi R, Pinson J, Savéant JM (1997) Covalent modification of carbon surfaces by aryl radicals generated from the electrochemical reduction of diazonium salts. J Am Chem Soc 119:201–207
3. Andrieux CP, Saveant JM (1978) Heterogeneous (chemically modified electrodes, polymer electrodes) vs. homogeneous catalysis of electrochemical reactions. J Electroanal Chem 93:163–168
4. Andrieux CP, Limoges B, Marchal D, Savéant JM (2006) Redox enzymes immobilized on electrodes with solution cosubstrates. General procedure for simulation of time-resolved catalytic responses. Anal Chem 78:3138–3143
5. Aoki K, Kato N (1988) Analysis of the cyclic voltammograms associated with deposition or precipitation of the electrochemical product. J Electroanal Chem 245:51–60
6. Aoki K, Tokuda K, Matsuda H (1986) Linear sweep and cyclic voltammetry for electrocatalysis at modified electrodes with very thin films. J Electroanal Chem 199:69–79
7. Bhugun I, Anson FC (1997) A generalised treatment of the dynamics of the adsorption of Langmuirian systems at stationary or rotating disk electrodes. J Electroanal Chem 439:1–6
8. Bhugun I, Savéant JM (1995) Derivatization of surfaces and self-inhibition in irreversible electrochemical reactions: cyclic voltammetry and preparative-scale electrolysis. J Electroanal Chem 395:127–131
9. Bhugun I, Savéant JM (1996) Self-inhibition in catalytic processes: cyclic voltammetry. J Electroanal Chem 408:5–14
10. Bhugun I, Lexa D, Savéant JM (1996) Catalysis of the electrochemical reduction of carbon dioxide by iron(0) porphyrins: Synergystic effect of weak Brönsted acids. J Am Chem Soc 118:1769–1776
11. Bieniasz LK, González J, Molina Á, Laborda E (2010) Theory of linear sweep/cyclic voltammetry for the electrochemical reaction mechanism involving a redox catalyst couple attached to a spherical electrode. Electrochim Acta 56:543–552

12. Casadio S (1976) Cyclic voltammetry for reversible anodic dissolution of the electrode material inhibited by a redox adsorption process. J Electroanal Chem 67:123–128
13. Casadio S (1976) Cyclic voltammetry for mixed diffusion-charge transfer control of simple anodic dissolution and passivation of the electrode material. J Electroanal Chem 72:243–250
14. Filinovskii VYu, Podgaetskii EM (1972) Calculation of the "transition" time at a non-equilibrium initial surface coverage. Elektrohim 8:1492–1496 (in Russian)
15. Fujioka J (1987) Integral equations in linear sweep voltammetry of irreversible reactions with adsorption at stationary and rotating disk electrodes. J Electrochem Soc 134:1446–1450
16. Galceran J, Salvador J, Puy J, Cecília J, Esteban M, Mas F (1995) Influence of adsorption on calibration curves in normal pulse polarography. Anal Chim Acta 305:273–284
17. Gao R, Roy D (1994) Effects of diffusion-limited mass transfer on metal underpotential deposition voltammograms. J Appl Electrochem 24:1276–1278
18. Garay F, Solis V (1999) Square wave stripping voltammetry of Cd-oxine complexes; surface redox reactions. J Electroanal Chem 476:165–170
19. Garay F, Solis V, Lovrić M (1999) Cathodic stripping square wave voltammetry of Cu(II)-oxine complexes. A mechanistic study. J Electroanal Chem 478:17–24
20. Guaus E, Mas F, Puy J, Sanz F (1987) A formalism for performing chronocoulometry at a stationary planar or spherical electrode. Part I. Theoretical treatment with Langmuirian adsorption. J Electroanal Chem 224:1–26
21. Guaus E, Sanz F, Sluyters-Rehbach M, Sluyters JH (1994) Study of reactant adsorption in the system Cd(II)KBr(1 M)/Hg by chronocoulometry. J Electroanal Chem 368:307–314
22. Guidelli R, Pezzatini G (1977) Diffusion with reactant adsorption at potentiostated stationary and dropping electrodes. J Electroanal Chem 84:211–234
23. Gulaboski R, Mirčeski V, Komorsky-Lovrić Š (2002) Square-wave voltammetry of a second order cathodic stripping process coupled by adsorption of the reacting ligand. Electroanalysis 14:345–355
24. Gulaboski R, Mirčeski V, Komorsky-Lovrić Š, Lovrić M (2004) Square-wave voltammetry of cathodic stripping reactions. Diagnostic criteria, redox kinetic measurements, and analytical applications. Electroanalysis 16:832–842
25. Hansen RS (1960) The theory of diffusion controlled absorption kinetics with accompanying evaporation. J Phys Chem 64:637–641
26. Holub K (1966) Surface reaction of an adsorbed substance transported by diffusion towards the dropping electrode. Collect Cecoslov Chem Commun 31:1461–1474
27. Holub K (1966) Surface reaction of an adsorbed substance transported by diffusion to a spherical surface. Collect Cecoslov Chem Commun 31:1655–1665
28. Holub K (1968) Current–time dependence for charge transfer processes with reactant adsorption at a dropping electrode. J Electroanal Chem 16:433–436
29. Holub K (1973) The reaction of an adsorbed substance with a substance transported by diffusion to the electrode. Kinetics of Langmuir adsorption without desorption coupled with diffusion. J Electroanal Chem 48:253–264
30. Holub K, Němec L (1966) The analog method for solution of problems involving diffusion to the electrode. I. Diffusion to a sphere with Langmuirian adsorption solved by analog method. J Electroanal Chem 11:1–11
31. Holub K, Němec L (1968) The analog method for solution of problems involving diffusion to the electrode. II. Kinetics of adsorption controlled by diffusion to the plane, spherical, and dropping electrodes for an arbitrary adsorption isotherm. J Electroanal Chem 18:209–214
32. Holub K, Van Leeuwen HP (1984) Influence of reactant adsorption on limiting currents in normal pulse polarography. Part II. Theory for the stationary, spherical electrode. J Electroanal Chem 162:55–65
33. Holub K, Van Leeuwen HP (1985) Influence of reactant adsorption on limiting currents in normal pulse polarography. Part IV. Theory for the expanding drop electrode. J Electroanal Chem 191:281–292
34. Hulbert MH, Shain I (1970) Rate-controlled adsorption of product in stationary electrode polarography. Anal Chem 42:162–171

35. Jaworski RK, Cox JA (1991) Effect of coupled chemical reactions involving the mediator species on voltammetry at a modified electrode. J Electroanal Chem 297:93–101
36. Kankare J, Vinokurov IA (1999) Kinetics of Langmuirian adsorption onto planar, spherical, and cylindrical surfaces. Langmuir 15:5591–5599
37. Klimatchev GV, Polyakova VN, Kaplin AA (1986) A method for determining the metal activity in an amalgam and solid deposit, from a volt-amperometric dissolution curve. Elektrohim 22:1234–1237 (in Russian)
38. Komorsky-Lovrić Š, Lovrić M (1985) Reactant adsorption in pulse polarography. Part III. Quasi-reversible charge transfers. J Electroanal Chem 190:1–20
39. Komorsky-Lovrić Š, Lovrić M (1989) Theory of square-wave stripping voltammetry with adsorptive accumulation. Fresenius Z Anal Chem 335:289–294
40. Komorsky-Lovrić Š, Lovrić M (1995) Square-wave voltammetry of quasi-reversible surface redox reactions. J Electroanal Chem 384:115–122
41. Levich VG, Khaikin BI, Belokolos ED (1965) Establishment of the adsorption equilibrium on flat and dropping electrodes and the irreversible electrochemical conversion of adsorbed substances. Elektrohim 1:1273–1279 (in Russian)
42. Limoges B, Savéant JM (2003) Cyclic voltammetry of immobilized redox enzymes. Interference of steady-state and non-steady state Michaelis–Menten kinetics of the enzyme-redox cosubstrate system. J Electroanal Chem 549:61–70
43. Limoges B, Moiroux J, Savéant JM (2002) Kinetic control by the substrate and the cosubstrate in electrochemically monitored redox enzymatic immobilized systems. Catalytic responses in cyclic voltammetry and steady state techniques. J Electroanal Chem 521:8–15
44. Limoges B, Savéant JM, Yazidi D (2003) Quantitative analysis of catalysis and inhibition at horseradish peroxidase monolayers immobilized on an electrode surface. J Am Chem Soc 125:9192–9203
45. Lovrić M (1983) EE mechanism with immobilisation of the intermediate. J Electroanal Chem 144:45–57
46. Lovrić M (1984) Reactant adsorption in pulse polarography. J Electroanal Chem 170:143–173
47. Lovrić M (1984) Reactant adsorption in pulse polarography. Part II. Totally irreversible redox reactions. J Electroanal Chem 181:35–49
48. Lovrić M (1986) Reactant adsorption in pulse polarography. Part IV. Influences of product adsorption. J Electroanal Chem 197:49–61
49. Lovrić M (1987) Irreversibility and reactant adsorption in differential pulse polarography. J Electroanal Chem 218:77–91
50. Lovrić M (1989) Influence of anion-induced adsorption on D.C. polarography of metal ions. Anal Chim Acta 218:7–23
51. Lovrić M (1991) A quasi-reversible maximum in square-wave voltammetry. Sov Electrochem 27:168–173
52. Lovrić M (1996) Simulation of electrochemical problems by numerical integration. Russ J Electrochem 32:988–995
53. Lovrić M (1999) A minimum separating diffusion and adsorption waves in polarography using a static mercury drop electrode. J Electroanal Chem 465:30–36
54. Lovrić M (2002) Reactant adsorption in cyclic staircase voltammetry on spherical microelectrodes. Reversible redox reactions. Anal Bioanal Chem 373:781–786
55. Lovrić M (2002) Theory of square-wave voltammetry of a reversible redox reaction complicated by the reactant adsorption. Electroanalysis 14:405–414
56. Lovrić M, Branica M (1987) Square-wave voltammetric peak current enhancements by adsorption and reversibility of the redox reaction. J Electroanal Chem 226:239–251
57. Lovrić M, Komorsky-Lovrić Š (1988) Square-wave voltammetry of an adsorbed reactant. J Electroanal Chem 248:239–253
58. Lovrić M, Komorsky-Lovrić Š (2010) Theory of reverse scan square-wave voltammetry influenced by the kinetics of reactant adsorption. Cent Eur J Chem 8:513–518
59. Lovrić M, Komorsky-Lovrić Š (2012) Theory of square-wave voltammetry of two-electron reduction with the adsorption of intermediate. Int J Electrochem ID596268

60. Lovrić M, Zelić M (1991) Anion induced adsorption in pulse polarography. J Electroanal Chem 316:315–328
61. Lovrić M, Komorsky-Lovrić Š, Murray RW (1988) Adsorption effects in square-wave voltammetry of totally irreversible redox reactions. Electrochim Acta 33:739–744
62. Lovrić M, Komorsky-Lovrić Š, Bond AM (1991) Theory of square-wave stripping voltammetry and chronoamperometry of immobilized reactants. J Electroanal Chem 319:1–18
63. Lovrić M, Pižeta I, Komorsky-Lovrić Š (1992) A square-wave voltammetry in a cathodic stripping mode. Electroanalysis 4:327–337
64. Lovrić M, Zelić M, Komorsky-Lovrić Š (2001) Theoretical analysis of pulse and differential pulse polarography of reversible redox reaction complicated by reactant adsorption. Collect Cecoslov Chem Commun 66:423–433
65. Mas F, Puy J, Sanz F, Virgili J (1983) Potentiostatic reversible reaction when both reactant and product are adsorbed at the dropping mercury electrode. Part I. Theoretical treatment. J Electroanal Chem 158:217–230
66. Mas F, Puy J, Sanz F, Virgili J (1985) Study of a simple redox system with adsorption of both reactant and product at the DME when a time dependent potential is applied. Pulse polarography. Part II. Langmuirian adsorption. J Electroanal Chem 183:41–56
67. Mas F, Puy J, Sanz F, Virgili J (1985) Study of a simple redox system with adsorption of both reactant and product at the DME when a time dependent potential is applied. Pulse polarography. Part IV. Langmuirian adsorption in normal pulse polarography. J Electroanal Chem 183:73–89
68. Mas F, Puy J, Díaz-Cruz JM, Esteban M, Casassas E (1992) Induced reactant adsorption in normal pulse polarography of labile metal polyelectrolyte systems. Part 1. Study of current–potential relationship assuming potential-independent adsorption parameters. J Electroanal Chem 326:299–316
69. Miller R (1980) Zur Adsorptionskinetik an der Oberfläche wachsenden Tropfen. Colloid Polym Sci 258:179–185
70. Miller R, Lunkenheimer K (1978) Zur Adsorptionskinetik an fluiden Phasengrenzen. Z Phys Chem (Leipzig) 259:863–868
71. Mirčeski V (2001) Square-wave voltammetry of an EC reaction of a partly adsorbed redox couple. J Electroanal Chem 508:138–149
72. Mirčeski V, Lovrić M (1998) A cathodic stripping square-wave voltammetry of a second-order redox reaction and its application to the mercury–cysteine system. Electroanalysis 10:976–984
73. Mirčeski V, Lovrić M (1999) Square-wave voltammetry of a cathodic stripping reaction complicated by adsorption of the reacting ligand. Anal Chim Acta 386:47–62
74. Mirčeski V, Lovrić M (1999) Quasireversible maximum in cathodic stripping square-wave voltammetry. Electroanalysis 11:984–989
75. Mirčeski V, Lovrić M (2000) Adsorption effects in square-wave voltammetry of an EC mechanism. Croat Chem Acta 73:305–329
76. Mirčeski V, Lovrić M (2004) EC mechanism of an adsorbed redox couple. Volume vs surface chemical reaction. J Electroanal Chem 565:191–202
77. Mirčeski V, Quentel F (2005) The role of adsorption in the catalytic electrode mechanism studied by means of square-wave voltammetry. J Electroanal Chem 578:25–35
78. Mirčeski V, Gulaboski R, Jordanoski B, Komorsky-Lovrić Š (2000) Square-wave voltammetry of 5-fluorouracil. J Electroanal Chem 490:37–47
79. Mirčeski V, Skrzypek S, Lovrić M (2009) Cathodic stripping voltammetry of uracil. Experimental and theoretical study under conditions of square-wave voltammetry. Electroanalysis 21:87–95

80. Mirčeski V, Guziejewski D, Ciesielski W (2011) Theoretical treatment of a cathodic stripping mechanism of an insoluble salt coupled with a chemical reaction in conditions of square wave voltammetry. Application to 6-mercaptopurine-9-D-riboside in the presence of Ni(II). Electroanalysis 23:1365–1375
81. Mirčeski V, Hocevar SB, Ogorevc B, Gulaboski R, Drangov I (2012) Diagnostics of anodic stripping mechanisms under square-wave voltammetry conditions using bismuth film substrates. Anal Chem 84:4429–4436
82. Nicholson MM (1957) Polarography of metallic monolayers. J Am Chem Soc 79:7–12
83. O'Dea JJ, Ribes A, Osteryoung JG (1993) Square-wave voltammetry applied to the totally irreversible reduction of adsorbate. J Electroanal Chem 345:287–301
84. Pižeta I, Lovrić M, Zelić M, Branica M (1991) Application of a Fourier transform method to the resolution enhancement of adsorption peaks in differential pulse polarography. J Electroanal Chem 318:25–38
85. Pnev VV, Parubotchaya KS (1979) Theory of inversion voltamperometry with depolariser adsorption. Diffusion controlled processes at linearly varying potential. Elektrohim 15:1804–1807 (in Russian)
86. Pnev VV, Popov GN (1975) Theory of inverse polarography on solid electrodes with adsorption of depolarizer in accordance with Langmuir equation. 2. Irreversible electrode processes. Sov Electrochem 11:961–962
87. Pnev VV, Popov GN, Zakharov MS (1975) Theory of inverse polarography on solid electrodes with adsorption of depolarizer in accordance with Langmuir equation. 1. Reversible electrode processes. Sov Electrochem 11:957–960
88. Podgaetskii EM (1973) Evaluation of the desorption from the capacitive part of a charging curve. Elektrohim 9:857–860 (in Russian)
89. Podgaetskii EM (1974) Effect of the simultaneous adsorption of a depolariser and product of an irreversible electrochemical reaction on the polarographic curve. Elektrohim 10:209–215 (in Russian)
90. Podgaetskii EM (1978) Construction of approximate solutions to a nonlinear integral equation with error estimate in heat and mass transfer problems. Ž Vyčisl Mat Fiz 18:226–232 (in Russian)
91. Podgaetskii EM (1979) Constructing a fragment of an adsorption isotherm using the θ, t−dependence. Elektrohim 15:1536–1539 (in Russian)
92. Podgaetskii EM (1979) Towards the theory of a desorption process under galvanostatic conditions. Elektrohim 15:1751–1756 (in Russian)
93. Podgaetskii EM (2011) Two-sided estimates for the solution of a nonlinear integral equation in a diffusion problem. Comput Math Math Phys 51:616–623
94. Podgaetskii EM (2011) Diffusion in solid plate taking into account internal adsorption of diffusing component. Prot Met Phys Chem Surf 47:286–290
95. Podgaetskii EM, Filinovskii VYu (1971) Chronopotentiometric measurement of adsorption in the case of an isotherm with two plateau. Elektrohim 7:1042–1047 (in Russian)
96. Podgaetskii EM, Filinovskii VYu (1971) Effect of the diffusion of an adsorbing depolarizer on "transition" time at large currents. Elektrohim 7:1856–1859 (in Russian)
97. Podgaetskii EM, Filinovskii VYu (1972) Calculation of the "transition" time accounting for the adsorption of the depolariser and product of an electrochemical reaction. Elektrohim 8:596–599 (in Russian)
98. Puy J, Mas F, Sanz F, Virgili J (1983) Potentiostatic reversible reaction when both reactant and product are adsorbed at the dropping mercury electrode. Part II. Langmuirian and Frumkinian adsorption. J Electroanal Chem 158:231–252
99. Puy J, Mas F, Sanz F, Virgili J (1985) Study of a simple redox system with adsorption of both reactant and product at the DME when a time dependent potential is applied. Pulse polarography. Part I. Theoretical treatment. J Electroanal Chem 183:27–39
100. Puy J, Pla M, Mas F, Sanz F (1988) Adsorption in double potential step chronocoulometry. J Electroanal Chem 241:89–104

101. Puy J, Mas F, Díaz-Cruz JM, Esteban M, Casassas E (1992) Induced reactant adsorption in metal–polyelectrolyte systems: pulse polarographic study. Anal Chim Acta 268:261–274
102. Puy J, Mas F, Díaz-Cruz JM, Esteban M, Casassas E (1992) Induced reactant adsorption in normal pulse polarography of labile metal+polyelectrolyte systems. Part 2. Study of the current–potential relationship assuming potential-dependent adsorption parameters. J Electroanal Chem 328:271–285
103. Puy J, Salvador J, Galceran J, Esteban M, Díaz-Cruz JM, Mas F (1993) Voltammetry of labile metal-complex systems with induced reactant adsorption. Theoretical analysis for any ligand-to-metal ratio. J Electroanal Chem 360:1–25
104. Rampazzo L (1969) Diffusion to a plane with adsorption according to Frumkin's isotherm. Electrochim Acta 14:733–739
105. Reinmuth WH (1961) Diffusion to a plane with Langmuirian adsorption. J Phys Chem 65:473–476
106. Reinmuth WH (1961) Chronopotentiometric transition times and their interpretation. Anal Chem 33:322–325
107. Reinmuth WH, Balasubramanian K (1972) Effects of Langmuirian adsorption on potentiostatic relaxation of diffusion-limited reversible electrode reactions. J Electroanal Chem 38:79–94
108. Reinmuth WH, Balasubramanian K (1972) Unified theory of reversible diffusion-limited electrode reactions with adsorption under potentiostatic conditions. J Electroanal Chem 38:271–281
109. Risović D (2006) Interfacial adsorption with molecular reorientation. J Electroanal Chem 588:122–128
110. Roizenblat EM, Reznik IM (1978) Electrodissolution of metals from a mercury film electrode under conditions of direct current and alternating current voltammetry. Elektrohim 15:233–239
111. Roizenblat EM, Reznik IM (1981) On accurate values of main peak parameters in inversion voltammetry on film electrodes. Ž Anal Him 36:48–53 (in Russian)
112. Sanz F, Puy J, Mas F, Virgili J (1985) Study of a simple redox system with adsorption of both reactant and product at the DME when a time dependent potential is applied. Pulse polarography. Part III. Langmuirian adsorption in differential pulse polarography. J Electroanal Chem 183:57–72
113. Seralathan M, Ribes A, O'Dea J, Osteryoung J (1991) Kinetics of reduction of adsorbed reactants in normal pulse voltammetry. J Electroanal Chem 306:195–211
114. Sluyters-Rehbach M, Sluyters JH (1977) A comparative study of approaches to the theory of reactant and/or product adsorption in D.C. polarography. J Electroanal Chem 75:371–385
115. Sluyters-Rehbach M, Sluyters JH (1977) A long-time series expansion as a solution of the diffusion problem in the case of Langmuirian adsorption at a dropping electrode. J Electroanal Chem 81:211–214
116. Van Leeuwen HP, Sluyters-Rehbach M, Holub K (1982) The influence of reactant adsorption on limiting currents in normal pulse polarography. The ranges of low and high concentrations. J Electroanal Chem 135:13–24
117. Vogel JJ (1973) Kinetik der durch eine Isotherme vom Tëmkin-Typus beschriebenen Adsorption, die durch die Diffusion zu einer ebenen respektive kugelförmigen Oberfläche begrenzt wird. Collect Cecoslov Chem Commun 38:976–978
118. Vogel JJ (1973) Vergleich einer gemessenen Abhängigkeit mit der Lösung der entsprechenden Integralgleichung durch Abschätzung der maximalen Abweichung ohne numerische Berechnung der Lösung. Collect Cecoslov Chem Commun 38:979–984
119. White N, Lawson F (1970) Potential sweep voltammetry of metal deposition and dissolution. Part I. Theoretical analysis. J Electroanal Chem 25:409–419
120. Wopschall RH, Shain I (1967) Effects of adsorption of electroactive species in stationary electrode polarography. Anal Chem 39:1514–1527
121. Wopschall RH, Shain I (1967) Adsorption effects in stationary electrode polarography with a chemical reaction following charge transfer. Anal Chem 39:1535–1542

122. Xiang M (1988) Application of an electroanalytical method to metal phase analysis. J Electroanal Chem 242:63–76
123. Xie Y, Anson FC (1996) Calculation of cyclic voltammetric responses for the catalytic reduction of substrates via intramolecular electron transfer within a catalyst–substrate complex formed on electrode surfaces. J Electroanal Chem 404:209–213
124. Yang N, Wang X (2007) Adsorption efficiency of poly(malachite green) films towards oxidation of ascorbic acid. Electrochim Acta 52:6962–6968
125. Zelić M, Lovrić M (1990) Bromide induced adsorption of lead ions on mercury electrodes. Electrochim Acta 35:1701–1706
126. Zelić M, Lovrić M (2003) The influence of electrolyte concentration on the parameters of the Frumkin isotherm in the $Cd^{2+} - I^{-}$ system. J Electroanal Chem 541:67–76
127. Zelić M, Lovrić M (2011) Components of the net current in differential pulse polarography. Part 2. Kinetics and adsorption. Electroanalysis 23:642–650

Chapter 10
Models Involving Additional Complications

In this chapter we consider the treatment, by the IE method, of additional complicating phenomena that may occur in electroanalytical models, apart from the phenomena already discussed in Chaps. 5–9. These additional complications are: the uncompensated Ohmic potential drop and double layer charging, and migration transport. The first two phenomena are considered in Sect. 10.1. Migration is briefly addressed in Sect. 10.2. We discuss almost exclusively models defined on one-dimensional spatial domains, because literature examples exist mostly for such models.

10.1 Uncompensated Ohmic Drop and Double Layer Charging

In the following analysis we focus on Eqs. (2.67), (2.69), and (2.70) for electrode | electrolyte interfaces only. Discussion of liquid | liquid interfaces, described by Eqs. (2.68), (2.69), and (2.71), would be analogous, provided that one replaces $E(t)$ by $\Delta\Phi(t)$, $E'(t)$ by $\Delta\Phi'(t)$, and R_U by $R_{U1} + R_{U2}$. We first consider the limiting cases when either the double layer charging or the Ohmic drop, is negligible, and later describe the general case. In all cases we describe how to modify theoretical models, formulated without considering the uncompensated Ohmic drop and the double layer charging, in order to take these effects into account.

L.K. Bieniasz, *Modelling Electroanalytical Experiments by the Integral Equation Method*, Monographs in Electrochemistry, DOI 10.1007/978-3-662-44882-3_10

10.1.1 Ohmic Drop Only

When the capacitive current $I_{\mathrm{C}}(t)$ is zero but R_{U} is significant, Eqs. (2.67) and (2.69) reduce to

$$E(t) = E'(t) + I_{\mathrm{F}}(t)\, R_{\mathrm{U}}\,, \tag{10.1}$$

$$I(t) = I_{\mathrm{F}}(t)\,. \tag{10.2}$$

This means that any IE-based model of a controlled potential experiment, formulated under the assumption of $R_{\mathrm{U}} = 0$ and $I_{\mathrm{C}}(t) = 0$, has to be modified by replacing all occurrences of $E(t)$ by $E'(t) = E(t) - I(t)\, R_{\mathrm{U}}$, and no modification of the occurrences of $I(t)$ is necessary. For example, the IE (5.21) obtained for an irreversible electron transfer (5.7) under conditions of planar diffusion in a one-dimensional semi-infinite spatial domain would become [46, 47]:

$$\Psi(t)\, \exp\left\{\alpha^{\mathrm{f}} \frac{nF}{RT}\left[E(t) + nFAR_{\mathrm{U}}\Psi(t) - E^0\right]\right\}$$
$$-k^0 \left\{c_1^{\star} - \int_0^t \left[D_1 \pi (t-\tau)\right]^{-1/2} \Psi(\tau)\, \mathrm{d}\tau\right\} = 0\,, \tag{10.3}$$

where we have defined one unknown $\Psi(t) = -p_1^{\mathrm{het}}(t) = p_2^{\mathrm{het}}(t) = -I(t)(nFA)^{-1}$. In a similar way, the IE (5.23) or (5.179), corresponding to an equilibrium electron transfer (5.7) under the same conditions, would become [2, 17, 41]:

$$\int_0^t \left[\pi(t-\tau)\right]^{-1/2} \Psi(\tau)\, \mathrm{d}\tau = \frac{c_1^{\star} D_1^{1/2}}{1 + \exp\left\{\frac{nF}{RT}\left[E(t) + nFAR_{\mathrm{U}}\Psi(t) - E_{1/2}\right]\right\}}\,, \tag{10.4}$$

where $\Psi(t)$ is defined identically. Inspection of Eqs. (10.3) and (10.4) leads to the following observations. Firstly, we note after Nicholson [41] that IEs (10.3) and (10.4) are nonlinear, even though they result from linear IEs (5.21) and (5.179). Secondly, we envisage mathematical inconsistencies in the IEs corresponding to equilibrium electron transfers, such as Eq. (10.4). The inconsistencies occur when $t \to 0^+$. If we assume that $\lim_{t\to 0+} \Psi(t)$ is finite, then the right-hand side of Eq. (10.4), and consequently the integral in Eq. (10.4) tend to nonzero finite values, which implies that $\Psi(t) \sim t^{-1/2}$ under the integral. This means a singularity, in contradiction with the finite limit assumed. Conversely, if we assume that $\Psi(t)$ is singular, that is $\lim_{t\to 0+} \Psi(t) = \infty$, then the right-hand side of Eq. (10.4) tends to zero, the integral in Eq. (10.4) tends to zero too, so that $\Psi(t)$ must be finite under the integral, in contradiction with the singularity assumed. This problem seems to have been overlooked in the literature, except for a brief comment in Bieniasz [8], and some analysis in Wein [51,52], which raises questions about the correctness of published solutions for equilibrium charge transfers. As the IE (10.3) is free from this inconsistency, it may be that the assumption of equilibrium charge transfers is non-physical when the uncompensated Ohmic drop needs to be taken into account.

Any IE-based model of a controlled current experiment, formulated under the assumption of $R_U = 0$ and $I_C(t) = 0$, is even easier to modify, since in view of Eq. (10.2) the IEs themselves remain unchanged. One only has to change the formula for the electrode potential response, by adding $I(t)R_U$ to the formula (where $I(t)$ is known in the case of controlled current experiments). For example, the IE (5.53) derived for the EE reaction scheme (5.24) and (5.25) remains unchanged, but formula (5.49) becomes:

$$E(t) = E_1^0 + \frac{RT}{n_1 F} \ln \frac{c_1^\dagger(t)}{c_2^\dagger(t)} + I(t)R_U \, . \tag{10.5}$$

Equation (10.5) describes the measurable potential response $E(t)$.

10.1.2 Double Layer Charging Only

When the resistance R_U is zero, but $I_C(t)$ is significant, Eqs. (2.67) and (2.70) reduce to

$$E'(t) = E(t) \tag{10.6}$$

and

$$I_C(t) = C_{DL} \frac{\mathrm{d}E(t)}{\mathrm{d}t} \, . \tag{10.7}$$

Any IE-based model of a controlled potential experiment, formulated under the assumption of $R_U = 0$ and $I_C(t) = 0$, has to be modified only by adding $I_C(t)$ given by Eq. (10.7) to the Faradaic current resulting from the model, in order to obtain the total current $I(t)$. No other changes in the model are needed. Assuming that the dependence of C_{DL} on $E(t)$ is known, the capacitive current (10.7) is easily calculated, since $E(t)$ is pre-defined in a controlled potential experiment. However, discontinuities or singularities of $I_C(t)$ are expected at time moments where $E(t)$ is not differentiable [1]. This refers, in particular, to the moments when potential steps are applied to the electrode, or when the direction of potential sweeps is changed in cyclic voltammetry.

More complicated is the modelling of controlled current experiments. Any model of such experiments, formulated under the assumption of $R_U = 0$ and $I_C(t) = 0$, has to be modified by replacing $I(t)$ by $I_F(t) = I(t) - I_C(t)$, and by including

Eq. (10.7) into the model. The inclusion can be performed in a variety of ways, out of which we describe the following two procedures (for yet different procedures see [6,7,14–16,33]).

Procedure 1 consists [36,41,42] in choosing $E(t)$ as an additional unknown. Given a model expression for $E(t)$, obtained when $R_U = 0$ and $I_C(t) = 0$, one replaces $I(t)$ by $I_F(t) = I(t) - C_{DL}\, dE(t)/dt$ in this expression, thereby obtaining an additional IDE for $E(t)$. To obtain $I_C(t)$ one differentiates the solution of this IDE (most likely numerically).

Procedure 2 is applicable when C_{DL} is assumed potential-independent. It consists [42] in choosing $I_C(t)$ as an additional unknown, and integrating Eq. (10.7):

$$E(t) = E(0) + C_{DL}^{-1} \int_0^t I_C(\tau)\, d\tau \, . \tag{10.8}$$

Comparison of Eq. (10.8) with the model expression for $E(t)$, modified by replacing $I(t)$ by $I_F(t) = I(t) - I_C(t)$, yields an additional IE for $I_C(t)$. As no IDE arises in this procedure, the procedure can be recommended.

As an example, let us consider a controlled current experiment for the electron transfer (5.7). When $R_U = 0$ and $I_C(t) = 0$, the potential response for irreversible reaction (5.7) is described by Eq. (5.47). Procedure 1 applied to Eq. (5.47) gives the IDE for $\Psi(t) = E(t)$:

$$\Psi(t) = E^0 - \frac{RT}{nF\alpha^f} \ln \frac{-\frac{I(t)}{nFA} + \frac{C_{DL}}{nFA}\frac{d\Psi(t)}{dt}}{k^0 \left[c_1^\star + \int_0^t \mathcal{K}_1(t,\tau)\, \frac{I(\tau)}{nFA}\, d\tau - \frac{C_{DL}}{nFA} \int_0^t \mathcal{K}_1(t,\tau)\, \frac{d\Psi(\tau)}{d\tau}\, d\tau \right]} \, . \tag{10.9}$$

Procedure 2 applied to Eq. (5.47) gives the IE for $\Psi(t) = I_C(t)(nFA)^{-1}$:

$$E(0) + C_{DL}^{-1} nFA \int_0^t \Psi(\tau)\, d\tau$$

$$= E^0 - \frac{RT}{nF\alpha^f} \ln \frac{-\frac{I(\tau)}{nFA} + \Psi(t)}{k^0 \left[c_1^\star + \int_0^t \mathcal{K}_1(t,\tau)\, \frac{I(\tau)}{nFA}\, d\tau - \int_0^t \mathcal{K}_1(t,\tau)\, \Psi(\tau)\, d\tau \right]} \, . \tag{10.10}$$

The potential response for equilibrium reaction (5.7) is described by Eq. (5.48). Procedure 1 applied to Eq. (5.48) gives the IDE for $\Psi(t) = E(t)$:

$$\Psi(t) = E^0 + \frac{RT}{nF} \ln \frac{c_1^\star + \int_0^t \mathcal{K}_1(t,\tau) \frac{I(\tau)}{nFA} \, d\tau - \frac{C_{DL}}{nFA} \int_0^t \mathcal{K}_1(t,\tau) \frac{d\Psi(\tau)}{d\tau} \, d\tau}{c_2^\star - \int_0^t \mathcal{K}_2(t,\tau) \frac{I(\tau)}{nFA} \, d\tau + \frac{C_{DL}}{nFA} \int_0^t \mathcal{K}_2(t,\tau) \frac{d\Psi(\tau)}{d\tau} \, d\tau} . \tag{10.11}$$

Procedure 2 applied to Eq. (5.48) gives the IE for $\Psi(t) = I_C(t)(nFA)^{-1}$:

$$E(0) + C_{DL}^{-1} nFA \int_0^t \Psi(\tau) \, d\tau$$

$$= E^0 + \frac{RT}{nF} \ln \frac{c_1^\star + \int_0^t \mathcal{K}_1(t,\tau) \frac{I(\tau)}{nFA} \, d\tau - \int_0^t \mathcal{K}_1(t,\tau) \Psi(\tau) \, d\tau}{c_2^\star - \int_0^t \mathcal{K}_2(t,\tau) \frac{I(\tau)}{nFA} \, d\tau + \int_0^t \mathcal{K}_2(t,\tau) \Psi(\tau) \, d\tau} . \tag{10.12}$$

We note that knowledge of $E(0)$ is required for formulating and solving Eqs. (10.9)–(10.12). Determining $E(0)$ may be difficult, especially in the case of Eqs. (10.11) and (10.12) for equilibrium electron transfer, which exhibit inconsistencies when $t \to 0^+$. The inconsistencies are similar to those mentioned in Sect. 10.1.1. They disappear when $E(0)$ is the equilibrium potential:

$$E(0) = E^0 + \frac{RT}{nF} \ln \frac{c_1^\star}{c_2^\star} . \tag{10.13}$$

In such a case $I_C(0) = I(0)$ [42]. The literature does not say how to select initial values in other cases, in particular when $c_2^\star = 0$. It seems that the assumption of equilibrium charge transfers is problematic also when double layer charging needs to be taken into account.

10.1.3 General Case

Let us pass now to the general case, when Ohmic drop and double layer charging occur simultaneously. Assume that IEs formulated in the absence of these phenomena have the production rates $p_j^{het}(t)$ as unknowns, and that the Faradaic current $I_F(t)$ can be expressed as a linear combination of these production rates. It is then natural to consider the capacitive current $I_C(t)$ as an additional unknown to be determined.

Any IE-based model of a controlled potential experiment, formulated under the assumption of $R_U = 0$ and $I_C(t) = 0$, can be modified by replacing $E(t)$ by $E'(t)$ resulting from Eq. (2.67), and by extending it with the following ODE for $I_C(t)$, obtained by combining Eqs. (2.67), (2.69), and (2.70):

$$I_C(t) = C_{DL}\frac{dE(t)}{dt} - R_U C_{DL}\left[\frac{dI_F(t)}{dt} + \frac{dI_C(t)}{dt}\right] . \tag{10.14}$$

In Eq. (10.14) $dE(t)/dt$ is a known expression (for controlled potential experiments), $dI_F(t)/dt$ can be expressed through the unknown production rates, and $I_C(t)$ is an unknown, too. We thus obtain an IDE system. For example, the IE (5.21) formulated for an irreversible electron transfer (5.7) under conditions of planar diffusion in a one-dimensional semi-infinite spatial domain, turns into the IDE system:

$$\Psi_1(t)\,\exp\left\langle \alpha^f \frac{nF}{RT}\left\{E(t) + nFAR_U\left[\Psi_1(t) + \Psi_2(t)\right] - E^0\right\}\right\rangle$$

$$-k^0\left\{c_1^\star - \int_0^t \left[D_1\pi(t-\tau)\right]^{-1/2}\Psi_1(\tau)\,d\tau\right\} = 0\,, \tag{10.15}$$

$$\frac{d\Psi_1(t)}{dt} + \frac{d\Psi_2(t)}{dt} + (C_{DL}R_U)^{-1}\,\Psi_2(t) + (nFAR_U)^{-1}\frac{dE(t)}{dt} = 0\,, \tag{10.16}$$

where we have defined two unknowns $\Psi_1(t) = -p_1^{het}(t) = p_2^{het}(t) = -I_F(t)(nFA)^{-1}$ and $\Psi_2(t) = -I_C(t)(nFA)^{-1}$. Similarly, from the IE (5.23) or (5.179), corresponding to an equilibrium electron transfer (5.7) under the same conditions, one obtains the following IDE system with the same unknowns:

$$\int_0^t \left[\pi(t-\tau)\right]^{-1/2}\,\Psi_1(\tau)\,d\tau$$

$$= \frac{c_1^\star D_1^{1/2}}{1 + \exp\left\langle \frac{nF}{RT}\left\{E(t) + nFAR_U\left[\Psi_1(t) + \Psi_2(t)\right] - E_{1/2}\right\}\right\rangle}\,, \tag{10.17}$$

$$\frac{d\Psi_1(t)}{dt} + \frac{d\Psi_2(t)}{dt} + (C_{DL}R_U)^{-1}\,\Psi_2(t) + (nFAR_U)^{-1}\frac{dE(t)}{dt} = 0\,. \tag{10.18}$$

In the case of equilibrium electron transfer it is actually possible to replace the IDE system (10.17) and (10.18) by a single IDE, as was shown by Imbeaux and Savéant [25, 26]. By introducing a new unknown $\Psi(t)$, related to the total current; $\Psi(t) =$

$\Psi_1(t) + \Psi_2(t) + C_{\mathrm{DL}}(nFA)^{-1}\mathrm{d}E(t)/\mathrm{d}t$, one obtains from Eqs. (10.17) and (10.18):

$$\int_0^t \frac{\Psi(\tau) + C_{\mathrm{DL}} R_{\mathrm{U}} \frac{\mathrm{d}\Psi(\tau)}{\mathrm{d}\tau} - C_{\mathrm{DL}}^2 R_{\mathrm{U}} \frac{\mathrm{d}^2 E(\tau)}{\mathrm{d}\tau^2}}{[\pi(t-\tau)]^{1/2}}\, \mathrm{d}\tau$$

$$= \frac{c_1^{\star} D_1^{1/2}}{1 + \exp\left\{\frac{nF}{RT}\left[E(t) + nFAR_{\mathrm{U}}\Psi(t) - C_{\mathrm{DL}} R_{\mathrm{U}} \frac{\mathrm{d}E(t)}{\mathrm{d}t} - E_{1/2}\right]\right\}} . \qquad (10.19)$$

The term involving $\mathrm{d}^2 E(\tau)/\mathrm{d}\tau^2$ vanishes when linear potential sweep/cyclic voltammetry is modelled, except for the time points where the potential–time function $E(t)$ is not differentiable.

Equations (10.17) and (10.18) suffer from the initial inconsistencies mentioned in Sects. 10.1.1 and 10.1.2. Imbeaux and Savéant [25, 26] tried to avoid this start-up problem in their study of linear potential sweep voltammetry, by assuming that initially $|I_{\mathrm{F}}(t)| \ll |I_{\mathrm{C}}(t)|$, and that the initial value of the dimensionless current is 10^{-7}. There are general physical arguments [1] suggesting that the initial potential step often accompanying controlled potential experiments leads in the first place to the capacitive current, but mathematical proofs of such assumptions for Eqs. (10.17)–(10.19) cannot be found in the literature.

Any IE-based model of a controlled current experiment, formulated under the assumption of $R_{\mathrm{U}} = 0$ and $I_{\mathrm{C}}(t) = 0$, can be modified in the following way. As an additional unknown one chooses $I_{\mathrm{C}}(t)$. One replaces all occurrences of $I(t)$ in the model by $I_{\mathrm{F}}(t) = I(t) - I_{\mathrm{C}}(t)$. One analytically differentiates the model expression for $E(t)$, and adds the resulting equation:

$$I_{\mathrm{C}}(t) = C_{\mathrm{DL}} \frac{\mathrm{d}E(t)}{\mathrm{d}t} \qquad (10.20)$$

to the model. Finally, one adds $I(t)\, R_{\mathrm{U}}$ to the model expression for $E(t)$. For example, Eqs. (5.53)–(5.57) for the EE reaction scheme (5.24) and (5.25) can be modified as follows. Equation (5.53) is left unchanged, but interfacial concentrations occurring in Eq. (5.53), and given by Eqs. (5.55)–(5.57) are replaced by

$$c_1^{\dagger}(t) = c_1^{\star} - \int_0^t \mathcal{K}_1(t,\tau)\, \Psi_1(\tau)\, \mathrm{d}\tau , \qquad (10.21)$$

$$c_2^{\dagger}(t) = c_2^{\star} + \int_0^t \mathcal{K}_2(t,\tau) \frac{I(\tau)}{n_2 FA}\, \mathrm{d}\tau - \int_0^t \mathcal{K}_2(t,\tau)\, \Psi_2(\tau)\, \mathrm{d}\tau$$

$$+ \frac{n_1 + n_2}{n_2} \int_0^t \mathcal{K}_2(t,\tau)\, \Psi_1(\tau)\, \mathrm{d}\tau , \qquad (10.22)$$

$$c_3^\dagger(t) = c_3^\star - \int_0^t \mathscr{K}_3(t,\tau)\,\frac{I(\tau)}{n_2FA}\,\mathrm{d}\tau + \int_0^t \mathscr{K}_3(t,\tau)\,\Psi_2(\tau)\,\mathrm{d}\tau$$

$$-\frac{n_1}{n_2}\int_0^t \mathscr{K}_3(t,\tau)\,\Psi_1(\tau)\,\mathrm{d}\tau, \tag{10.23}$$

where we have defined two unknowns $\Psi_1(t) = -p_1^{\mathrm{het}}(t)$ and $\Psi_2(t) = -I_{\mathrm{C}}(t)(n_2FA)^{-1}$. The additional IDE for $\Psi_2(t)$ is obtained by differentiating Eq. (5.49):

$$\Psi_2(t) = -\frac{C_{\mathrm{DL}}\frac{\mathrm{d}}{\mathrm{d}t}E(t)}{n_2FA} = \frac{C_{\mathrm{DL}}}{n_2FA}\,\frac{RT}{n_1F}\left[\frac{\frac{\mathrm{d}}{\mathrm{d}t}c_1^\dagger(t)}{c_1^\dagger(t)} - \frac{\frac{\mathrm{d}}{\mathrm{d}t}c_2^\dagger(t)}{c_2^\dagger(t)}\right], \tag{10.24}$$

where $c_1^\dagger(t)$ and $c_2^\dagger(t)$ are given by Eqs. (10.21) and (10.22). Finally, the equation for the measurable potential is obtained from Eq. (5.49) as

$$E(t) = E_1^0 + \frac{RT}{n_1F}\ln\frac{c_1^\dagger(t)}{c_2^\dagger(t)} + I(t)R_{\mathrm{U}}\,. \tag{10.25}$$

10.1.4 Literature Examples

For the effect of uncompensated Ohmic drop without double layer charging, the following examples of IE-based models are available. De Vries and Van Dalen [17] and Nicholson [41] considered the effect on linear potential sweep voltammograms, for an equilibrium electron transfer

$$X_1 + n\,e^- \rightleftarrows X_2\,, \tag{10.26}$$

assuming planar diffusion of dynamic distributed species X_1 and X_2 in a semi-infinite spatial domain. In [17] an initial equilibrium potential was assumed, whereas in [41] X_2 was initially absent. Roffia and Lavacchielli [46, 47] presented an analogous discussion for an irreversible reaction (10.26). Reinmuth [45] discussed the Ohmic loss in DC polarography, in which case the uncompensated resistance R_{U} depended on time, owing to the variable size of the DME. The expanding plane model of the DME was assumed. Another example of a variable R_{U} was considered by Fan et al. [19], in connection with the modelling of scanning electrochemical microscopic, potential step chronoamperometric and cyclic voltammetric experiments at polymer-covered electrodes. In this case R_{U} represented the resistance of the poly(ferrocene) polymer, assumed to be a linear function of the charge passed during the experiment. Planar diffusion in a finite spatial domain with an impermeable second boundary was assumed. Weidner and Fedkiw [50] analysed the effect of the Ohmic drop in linear potential sweep voltammetry, in the case of a cylindrical pore electrode. They assumed a spatially two-dimensional model, in which the potential drop was along the pore axis, whereas the diffusion was in the radial direction, inside the pore. Åberg and Sharp [2] discussed the Ohmic drop in potential step experiments, assuming a non-equilibrium reversible reaction (10.26), and planar diffusion in a semi-infinite spatial domain. Navarro-Laboulais et al. [40] considered a similar problem, in connection with a convolutive analysis of chronoamperograms. Wein [51, 52] presented a theory of potential step chronoamperometry, with the Ohmic potential drop taken into account, assuming equilibrium reaction (10.26), and planar diffusion in a semi-infinite spatial domain. Wein and Tovchigrechko [53, 54] discussed the effect of the Ohmic drop on voltage step transients in a two-electrode cell, assuming planar diffusion in a semi-infinite spatial domain, and non-equilibrium (10.26).

The Ohmic potential drop may well occur in an electrode material, instead of the electrolyte. Such a case was modelled by Bieniasz and Tomczyk [9], assuming reaction

$$\mathrm{X}_1 + n\,\mathrm{e}^- \rightleftarrows 2\mathrm{X}_2 \tag{10.27}$$

initially at a positive equilibrium, linear potential sweep voltammetry, and planar diffusion of species X_1 and X_2 in a semi-infinite electrolyte phase. Equations (5) and (8) in Bieniasz and Tomczyk [9] contain errors: there should be $\exp[-at + h\chi(at)]$ instead of $\exp[-at - h\chi(at)]$ in these equations.

The IE describing the effect of the Ohmic potential drop on linear potential sweep voltammetry for the equilibrium reaction of metal electrodeposition:

$$\mathrm{X}_1 + n\,\mathrm{e}^- \rightleftarrows \mathrm{X}_{2,\mathrm{s}} \tag{10.28}$$

was reported by White and Lawson [55], assuming planar diffusion in a semi-infinite spatial domain. In reaction (10.28) X_1 is a dynamic distributed species (metal ion), and $X_{2,s}$ is a static localised species (solid metal).

For the effect of double layer charging without Ohmic drop, the following examples of IE-based models are available. Let us begin with models independent of spatial coordinates. Galinker and Makovetskii [20] considered the influence of double layer charging in inversion voltammetry and inversion chronopotentiometry, for an irreversible electrodissolution reaction of a metal concentrated in a thin mercury film electrode. The reactant was formally considered as a quasi-localised species (see Sect. 2.9). Tezuka et al. [49] presented a simple model describing linear potential sweep voltammetry for a charge transfer involving exclusively localised species (oxidation of a polypyrrole film), in the presence of a capacitive-like current. Further models listed below depend on one spatial coordinate. Most of them refer to controlled current experiments. Matsuda [35] examined the influence of the capacitive current, on potential–time curves in controlled current oscillographic polarography, for a reaction scheme involving three independent electron transfer reactions. De Vries [15] considered the effect in current step chronopotentiometry for the equilibrium reaction (10.26), with dynamic species X_1 distributed in a semi-infinite electrolyte phase, and dynamic species X_2 distributed in a mercury film electrode. Planar diffusion was assumed. In other papers [14, 16] De Vries discussed also current step chronopotentiometry, and programmed current chronopotentiometry for the equilibrium reaction (10.26), with dynamic species X_1 and X_2 distributed in a semi-infinite electrolyte phase, and subject to planar diffusion. The latter problem was also studied by Olmstead and Nicholson [42] and Dračka [18]. Dračka [18] took into account an additional preceding homogeneous reaction as well. The theory of potential–time relaxation following a coulostatic impulse was discussed by Reinmuth [44], Nicholson [41], Astruc et al. [6, 7], and Kudirka and Enke [33]. Planar diffusion in a semi-infinite spatial domain was assumed. An electron transfer reaction (10.26) involving dynamic distributed species X_1 and X_2, as well as a few reaction schemes involving homogeneous reactions were treated. Matsuda and Aoyagui [36] analysed a modification of this coulostatic method. Kostin and Labyak [32] developed a model of the controlled current electrodeposition of some alloys, under conditions of periodic current pulses of rectangular form, when the capacitive current is significant. A rare example of a temperature perturbation experiment was presented by Harima and Aoyagui [23]. They studied an open circuit potential relaxation accompanying a linear variation of the temperature, for a non-equilibrium reversible reaction (10.26). Planar diffusion of X_1 and X_2 in a semi-infinite spatial domain was assumed. Safonov et al. [48] analysed variations of the electrode potential with time, after contacting a planar electrode with an electrolyte solution. It was assumed that the variations are caused by the simultaneous non-equilibrium reversible electrodissolution of the electrode metal, and double layer charging. Pototskaya and Yevtushenko [43] extended the model from [48] onto the case of a rough, but macroscopically planar electrode, assuming sinusoidal surface oscillations. The model was actually defined on a two-dimensional spatial domain, but the resulting IDEs were one-dimensional.

For the combined effects of the Ohmic drop and double layer charging the following examples of IE-based models are available. Imbeaux and Savéant [25] investigated the effects on linear potential sweep voltammograms for an equilibrium reaction (10.26) involving dynamic species X_1 and X_2 distributed in a semi-infinite electrolyte phase and subject to planar diffusion. In a later paper [26] they extended their treatment to several reaction schemes involving homogeneous reactions: the EC scheme and the ECE scheme with first-order homogeneous reactions, the EC scheme with a homogeneous dimerisation of the electron transfer product, and the EC scheme with a homogeneous disproportionation reaction. Garreau and Savéant [21, 22] extended the discussion from Imbeaux and Savéant [25], by considering additionally, in an equivalent circuit, an inductance that characterises the effect of the amplifier bandpass limitation. Andrieux et al. [4,5] presented a theory analogous to that of Imbeaux and Savéant [25], but for a non-equilibrium reversible electron transfer (10.26). Kostin [31] considered a model of a two-electrode electrolysis cell, seemingly (but tacitly) assuming that non-equilibrium reversible charge transfer reactions at the both electrodes proceed between one dynamic distributed species, and one static species. Either controlled current or controlled voltage experiments were discussed. Montella [37] studied the effect of replacing a differential double layer capacitance by a constant phase element, on linear potential sweep/cyclic voltammograms. He assumed a non-equilibrium reversible reaction (10.26), and planar diffusion of X_1 and X_2 in a semi-infinite spatial domain.

An interesting but unusual IE describing potential step chronoamperometry for the equilibrium electron transfer reaction

$$X_{1,\mathrm{ad}} \rightleftarrows X_{2,\mathrm{ad}} + n\,\mathrm{e}^- , \tag{10.29}$$

involving dynamic localised species $X_{1,\mathrm{ad}}$ and $X_{2,\mathrm{ad}}$, was presented by Yamada et al. [56]. Reaction (10.29) was taking place in a thin layer electrolytic cell, in the presence of the Ohmic drop and double layer charging. The model assumed spatial variations of the Ohmic potential drop along the electrode surface, so that the resulting IE involved integration over time and over the surface.

10.2 Electric Migration Effects

Electric migration transport cannot be handled rigorously by the IE method, at least not by the usual Laplace transform approach, owing to the nonlinearity of the Nernst–Planck equations (2.1). However, there have been a few studies where the role played by the electric potential distribution, in the ionic transport, was approximately accounted for in the IE-based models of electroanalytical experiments. We list these studies below.

Caselli and Maestro [10–13] analysed a controlled potential electrolysis at a membrane-coated liquid mercury electrode. Starting from the Nernst–Planck equations they assumed some linearising approximations to these PDEs, which allowed for IEs to be formulated.

Amatore and Lefrou [3] considered the theory of cyclic voltammetry at extremely high potential scan rates, at which the diffuse part of the double layer cannot be separated from the diffusion layer. Starting from the Nernst–Planck equations, they obtained an approximate IE for the Faradaic current, in which the effect of migration was accounted for by including the dependence of reaction rates on the electric potential at the outer Helmholtz plane.

Kakiuchi [29] presented an IE describing DC and AC responses of ion transfer across an oil | water interface. The interface was assumed to behave as a thin planar layer, with Nernst–Planck transport equations in the layer, and a linear potential profile. This assumption has led to a particular Goldman-type current–potential characteristic of the ion transfer through the interface [28], different from the standard Butler–Volmer characteristic. The current–potential characteristic was used as a boundary condition for diffusion at both sides of the interface, resulting in an IE for the Faradaic current.

Honeychurch and Rechnitz [24] analysed cyclic voltammetry for the equilibrium reduction of cytochrome c at monolayer covered electrodes. They assumed that the monolayer affects the electric potential profile in the vicinity of the electrode and alters the distance of closest approach of the reactant. The voltammograms were described by an IE similar to Eq. (10.4), in which an additional term was subtracted from the electrode potential $E(t)$. The additional term was approximately a quadratic function of time.

Models of transient experiments (mostly potential step chronoamperometry) at electrodes covered by polymer films were described by Malev et al. [34], Kondratiev et al. [30], and Ivanov et al. [27], assuming local electroneutrality. The migration transport PDEs were linearised, owing to the assumed small departures from equilibrium. This allowed the IE formalism to be applied.

Myland and Oldham [38, 39] investigated the effect of migration and changing (uncompensated) resistance on voltammetry, in the absence of supporting electrolyte. They considered a locally electroneutral solution of a binary electrolyte. The cation of the electrolyte was participating in an electron transfer reaction, leading to an uncharged reduction product. One-dimensional, semi-infinite, and finite spatial domains were considered. The particular situation studied by Myland and Oldham [38, 39] is rather exceptional, because it allows one to replace the nonlinear Nernst–Planck equations by an equivalent pure diffusion PDE. Consequently, concentration–flux relationships are obtainable analytically, and can be used to formulate IEs. The solution resistance, important for the model, can be calculated from the concentration-dependent local conductivity of the electrolyte solution, by an appropriate integration over space. Numerical integration was used.

References

1. Åberg S (1996) Measurement of uncompensated resistance and double layer capacitance during the course of a dynamic measurement: correction for IR drop and charging currents in arbitrary voltammetric techniques. J Electroanal Chem 419:99–103
2. Åberg S, Sharp M (1996) A theoretical expression for the effect of IR drop on the current response in a potential step experiment. J Electroanal Chem 403:31–38
3. Amatore C, Lefrou C (1990) Is cyclic voltammetry above a few hundred kilovolts per second still cyclic voltammetry? J Electroanal Chem 296:335–358
4. Andrieux CP, Garreau D, Hapiot P, Pinson J, Savéant JM (1988) Fast sweep cyclic voltammetry at ultra-microelectrodes. Evaluation of the method for fast electron-transfer kinetic measurements. J Electroanal Chem 243:321–335
5. Andrieux CP, Delgado G, Savéant JM, Su KB (1993) Improvement and estimation of precision in cyclic voltammetry: thermodynamic and kinetic determinations in chemically reversible systems. J Electroanal Chem 348:107–121
6. Astruc M, Bonastre J (1972) La polarographie a decharges. II. Étude théorique d'une réaction d'électrode totalement irréversible. J Electroanal Chem 36:435–446
7. Astruc M, Bonastre J, Royer R (1972) La polarographie a decharges. I. Étude théorique d'une réaction d'électrode réversible. J Electroanal Chem 34:211–225

8. Bieniasz LK (2010) Automatic simulation of cyclic voltammograms by the adaptive Huber method for weakly singular second kind Volterra integral equations. Electrochim Acta 55:721–728
9. Bieniasz LK, Tomczyk P (1993) Kinetics of the oxygen electrode reaction in molten (Li/Na) carbonate eutectic. Part VI: quantitative analysis of the linear scan voltammetric curves for the first reduction process at monocrystalline NiO electrodes. J Electroanal Chem 353:195–208
10. Caselli M, Maestro M (1985) Controlled-potential electrolysis at a membrane-coated electrode. Part I. Theoretical treatment. Bioelectrochem Bioenerg 14:275–291
11. Caselli M, Maestro M (1985) Controlled-potential electrolysis at a membrane-coated electrode. Part II. Numerical analysis and results. Electrochem Bioenerg 14:293–312
12. Caselli M, Maestro M (1985) A problem of charge flow across a permselective medium followed by discharge at a metallic electrode; a semianalytical approach. J Mol Struct THEOCHEM 120:337–342
13. Caselli M, Maestro M (1986) Controlled-potential electrolysis at a membrane-coated electrode. Part III. Effect of the electric field in the dielectric. Electrochem Bioenerg 15:353–370
14. De Vries WT (1968) Distortion of constant-current chronopotentiograms by double-layer charging. J Electroanal Chem 17:31–43
15. De Vries WT (1968) Double-layer charging in constant-current chronopotentiometry at a mercury-film electrode. J Electroanal Chem 19:41–53
16. De Vries WT (1968) Effect of double-layer charging in programmed-current chronopotentiometry. J Electroanal Chem 19:55–60
17. De Vries WT, Van Dalen E (1965) Distortion of linear-sweep polarograms by Ohmic drop. J Electroanal Chem 10:183–190
18. Dračka O (1969) Study of the kinetics of electrode processes by means of electrolysis with constant current. XIII. Correction for the influence of double layer charging in measurement of transition times. Collect Cecoslov Chem Commun 34:2627–2644
19. Fan FRF, Mirkin MV, Bard AJ (1994) Polymer films on electrodes. 25. Effect of polymer resistance on the electrochemistry of poly(vinylferrocene): scanning electrochemical microscopic, chronoamperometric, and cyclic voltammetric studies. J Phys Chem 98:1475–1481
20. Galinker EV, Makovetskii AL (1986) Effect of the electric double layer capacitance on the detection interval in inversion voltamperometry and chronopotentiometry. Elektrohim 22:1176–1180 (in Russian)
21. Garreau D, Savéant JM (1974) Linear sweep voltammetry—bandpass limitations and maximum usable sweep rate in the study of Faradaic processes. J Electroanal Chem 50:1–22
22. Garreau D, Savéant JM (1978) Resistance compensation and Faradaic instability in diffusion controlled processes. J Electroanal Chem 86:63–73
23. Harima Y, Aoyagui S (1977) Electrode potential relaxation following a rapid change of temperature. Part II. Theory. J Electroanal Chem 81:47–52
24. Honeychurch MJ, Rechnitz GA (1997) Cyclic voltammetry at monolayer covered electrodes: the effect of monolayers on the reduction of cytochrome c. J Phys Chem B 101:7472–7479
25. Imbeaux JC, Savéant JM (1970) Linear sweep voltammetry. Effect of uncompensated cell resistance and double layer charging on polarization curves. J Electroanal Chem 28:325–338
26. Imbeaux JC, Savéant JM (1971) Linear sweep voltammetry. Ohmic drop and chemical polarization. J Electroanal Chem 31:183–192
27. Ivanov VD, Kaplun MM, Kondratiev VV, Tikhomirova AV, Zigel VV, Yakovleva SV, Malev VV (2002) Charge transfer in iron and cobalt hexacyanoferrate films: effect of the film thickness and the counterion nature. Russ J Electrochem 38:173–181
28. Kakiuchi T (1992) Current–potential characteristic of ion transfer across the interface between two immiscible electrolyte solutions based on the Nernst–Planck equation. J Electroanal Chem 322:55–61
29. Kakiuchi T (1993) Dc and ac responses of ion transfer across an oil–water interface with a Goldman-type current–potential characteristic. J Electroanal Chem 344:1–12

30. Kondratiev VV, Tikhomirova AV, Malev VV (1999) Study of charge transport processes in Prussian-Blue film modified electrodes. Electrochim Acta 45:751–759
31. Kostin NA (1979) Mathematical models of processes and computer calculation of polarisation characteristics at non-stationary electrolysis. Elektrohim 15:1348–1351 (in Russian)
32. Kostin NA, Labyak OV (1995) Mathematical modelling of processes of pulse electrodeposition of alloys. Elektrohim 31:510–516 (in Russian)
33. Kudirka JM, Enke CG (1972) Use of higher overvoltages in coulostatic kinetic measurements. Anal Chem 44:614–615
34. Malev VV, Tikhomirova AV, Kondratiev VV, Rubashkin AA (1999) Chronoamperometric analysis of charge transfer processes in layers of Prussian Blue. Elektrohim 35:1184–1192 (in Russian)
35. Matsuda H (1956) Zur Theorie der Heyrovský-Forejtschen oszillographischen Polarographie. Z Elektrochem 60:617–626
36. Matsuda H, Aoyagui S (1978) A modified coulostatic pulse method for fast electrode processes. J Electroanal Chem 87:155–163
37. Montella C (2012) LSV/CV modelling of electrochemical reactions with interfacial CPE behaviour, using the generalised Mittag-Leffler function. J Electroanal Chem 667:38–47
38. Myland JC, Oldham KB (2002) Convolutive modelling in the absence of supporting electrolyte: coping with migration and changing resistance in predicting voltammetry. J Electroanal Chem 529:66–74
39. Myland JC, Oldham KB (2002) A model of cyclic voltammetry for a thin organic layer sandwiched between an electrode and an aqueous solution. Convolutive modelling in the absence of supporting electrolyte. J Electroanal Chem 530:1–9
40. Navarro-Laboulais J, Trijueque J, García-Jareño JJ, Benito D, Vicente V (1998) Determination of the electroactive area of graphite + polyethylene composite electrodes. Uncompensated resistance effects and convolution analysis of chronoamperograms. J Electroanal Chem 443:41–48
41. Nicholson RS (1965) Some examples of the numerical solution of nonlinear integral equations. Anal Chem 37:667–671
42. Olmstead ML, Nicholson RS (1968) Influence of double-layer charging in chronopotentiometry. J Phys Chem 72:1650–1656
43. Pototskaya VV, Yevtushenko NE (1995) Kinetics of the establishment of a stationary potential on a rough interfacial surface. Theoretical model. Elektrohim 31:174–180 (in Russian)
44. Reinmuth WH (1962) Theory of coulostatic impulse relaxation. Anal Chem 34:1272–1276
45. Reinmuth WH (1972) Theory of Ohmic loss in D.C. polarography. J Electroanal Chem 36:467–473
46. Roffia S (1968) Derivation of the kinetic parameters of irreversible electrode transfer processes from potential sweep voltammetry with uncompensated Ohmic drop. Ric Sci 38:1257–1259
47. Roffia S, Lavacchielli M (1969) Potential sweep voltammetry with uncompensated Ohmic potential drop. Irreversible charge transfer. J Electroanal Chem 22:117–125
48. Safonov VA, Damaskin BB, Choba MA (1989) Model description of the dependence of the electrode potential on the time of the contact with electrolyte solutions. Elektrohim 25:1432–1438 (in Russian)
49. Tezuka Y, Aoki K, Shinozaki K (1989) Kinetics of oxidation of polypyrrole-coated transparent electrodes by in situ linear sweep voltammetry and spectroscopy. Synth Met 30:369–379
50. Weidner JW, Fedkiw PS (1991) Effect of Ohmic, mass-transfer, and kinetic resistances on linear-sweep voltammetry in a cylindrical-pore electrode. J Electrochem Soc 138:2514–2526, with erratum in J Electrochem Soc 139:207
51. Wein O (1991) Effect of ohmic losses on potentiostatic transient response of a reversible redox system. J Appl Electrochem 21:1091–1094
52. Wein O (2010) Edge effects in voltage-step transient I. Ohmic losses in 1D approximation (revisited). Research Report ICPF No. 2010/3, Institute of Chemical Process Fundamentals ASCR, Prague

53. Wein O (2010) Voltage-step transient in redox systems II. 1D approximation (revisited). Research Report ICPF No. 2010/6, Institute of Chemical Process Fundamentals ASCR, Prague
54. Wein O, Tovchigrechko VV (2011) Voltage-step transient on circular electrodes. J Appl Electrochem 41:1065–1075
55. White N, Lawson F (1970) Potential sweep voltammetry of metal deposition and dissolution. Part I. Theoretical analysis. J Electroanal Chem 25:409–419
56. Yamada K, Kitamura F, Ohsaka T, Tokuda K (1996) Response time of thin-layer electrolytic cells to potential-step signals. J Electrochem Soc 143:4006–4012

Chapter 11
Analytical Solution Methods

In this chapter we review analytical methods that have been used to obtain solutions of the IEs or IDEs arising in electroanalytical chemistry. The classification of the methods considered here as "analytical" is to some extent arbitrary, because it is never possible to entirely avoid the use of numerical computing, while using the solutions obtained by these methods. In fact, it happens that the same methods are sporadically called "numerical" in the literature. However, owing to the involvement of various analytical transformations or procedures, in the process of deriving the solutions, it is reasonable to call them analytical, in order to distinguish them from purely numerical methods. In Sect. 11.1 a few analytical techniques, applicable to some one-dimensional IEs or IDEs, are presented. These techniques have a very limited application scope, but in cases when they are applicable they usually prove useful, and have received some attention.

11.1 Analytical Methods for One-Dimensional IEs or IDEs

Mathematics has developed a variety of analytical methods for solving one-dimensional IEs (see, for example, [15, 29, 30, 55, 56, 80] and the literature therein). Some of these methods have found application in electroanalytical modelling. It is important to note that these electrochemical applications are often quite original and non-standard. They should be of interest not only to electrochemists, but also to mathematicians, who are almost unaware of their existence.

L.K. Bieniasz, *Modelling Electroanalytical Experiments by the Integral Equation Method*, Monographs in Electrochemistry, DOI 10.1007/978-3-662-44882-3_11

11.1.1 Integral Solutions

The analytical technique discussed in this section is applicable to one-dimensional first kind Volterra IEs having the form:

$$\int_0^t \mathscr{K}(t,\tau)\,\Psi(\tau)\,\mathrm{d}\tau = f(t)\,, \tag{11.1}$$

where $\Psi(t)$ is an unknown solution, $f(t)$ is a known function of t, and $\mathscr{K}(t,\tau)$ is a convolution kernel. Therefore, we can write

$$\mathscr{K}(t,\tau) = \varphi(t-\tau)\,, \tag{11.2}$$

where $\varphi(\vartheta)$ with $\vartheta = t - \tau$ is a known function. The technique to be described was analysed and advocated relatively recently by Mirčeski and Tomovski [40], but examples of the integral solutions resulting from the technique were known earlier (see, for example, [14,22,34,36,58,59,64,68,81]), among other things dating back to the early works on the theory of linear potential sweep voltammetry. Looking from the mathematical perspective, the method can be viewed as a generalisation of the well-known analytical solution of the Abel IE [15,30]. The following description of the technique is loosely based on the discussion in Mirčeski and Tomovski [40], but we add some extensions.

Looking at the Eq. (11.1) one is tempted to consider the idea of trying to find the inverse kernel (cf. Chap. 3, Sect. 3.1) of the integral transformation in Eq. (11.1), with the aim of obtaining the solution in the form

$$\Psi(t) = \int_0^t \mathscr{K}^{\mathrm{inv}}(t,\tau)\,f(\tau)\,\mathrm{d}\tau\,, \tag{11.3}$$

where $\mathscr{K}^{\mathrm{inv}}(t,\tau)$ denotes the inverse kernel. However, as we shall see below, this cannot be done, although formulae somewhat similar to Eq. (11.3) can be obtained.

As the electrochemical IEs are usually obtained by using the Laplace transformation, it is most convenient to discuss the present solution technique in terms of this transformation. Hence, let us assume that the Laplace transforms of functions $\Psi(t)$, $f(t)$ and $\varphi(\vartheta)$ exist. By applying the Laplace transformation (3.12) to Eq. (11.1), the convolution theorem (3.17) yields:

$$\hat{\varphi}(s)\,\hat{\Psi}(s) - \hat{f}(s) = 0\,. \tag{11.4}$$

Consequently,

$$\hat{\Psi}(s) = [\hat{\varphi}(s)]^{-1}\,\hat{f}(s) \tag{11.5}$$

and

$$\Psi(t) = \mathscr{L}^{-1}\left\{[\hat{\varphi}(s)]^{-1}\,\hat{f}(s)\right\} . \tag{11.6}$$

Equation (11.6) may be considered an analytical solution of Eq. (11.1), in the sense that it is obtained by the methods of analysis, although a closed-form representation of the inverse Laplace transform (3.14) in Eq. (11.6) is rarely available, so that numerical inversion may be necessary. This is not a bad option, since a number of robust numerical inversion methods exist (see Chap. 3, Sect. 3.2, and the literature cited therein). However, one may prefer to look for alternatives requiring a deeper analytical insight.

Let us define $\hat{\varphi}_0(s) = [\hat{\varphi}(s)]^{-1}$. If $\mathscr{L}^{-1}\{\hat{\varphi}_0(s)\}$ existed, we might define the inverse kernel as $\mathscr{K}^{\text{inv}}(t,\tau) = \varphi_0(t-\tau)$, where $\varphi_0(\vartheta) = \mathscr{L}^{-1}\{\hat{\varphi}_0(s)\}$. By virtue of the convolution theorem (3.17), Eq. (11.6) might then be cast into the form (11.3). Unfortunately, $\mathscr{L}^{-1}\{\hat{\varphi}_0(s)\}$ is not likely to exist, because in view of the existence of $\mathscr{L}^{-1}\{\hat{\varphi}(s)\}$, function $\hat{\varphi}(s)$ must decay to zero for $s \to \infty$ (in fact we have seen in Chaps. 5, 6, and 8 that often $\hat{\varphi}(s) \sim s^{-1/2}$). Consequently, $\hat{\varphi}_0(s)$ diverges when $s \to \infty$, and cannot have the inverse Laplace transform (3.14). However, one can rewrite Eq. (11.5) in the equivalent alternative form:

$$\hat{\Psi}(s) = \hat{\varphi}_1(s)\,\left[s\,\hat{f}(s) - f(0^+)\right] + \hat{\varphi}_1(s)\,f(0^+) , \tag{11.7}$$

where

$$\hat{\varphi}_1(s) = [\hat{\varphi}(s)]^{-1}/s . \tag{11.8}$$

Owing to the division of $[\hat{\varphi}(s)]^{-1}$ by s in Eq. (11.8), function $\hat{\varphi}_1(s)$ is likely to possess the inverse Laplace transform $\varphi_1(\vartheta)$. Let us assume that a closed-form expression or computable approximation for this inverse exists. In addition, assume $f(t)$ is differentiable, so that in accordance with Eq. (3.18), $s\,\hat{f}(s) - f(0^+)$ represents the Laplace transform of the first derivative of $f(t)$. Consequently, by applying the convolution theorem (3.17) to Eq. (11.7) we obtain:

$$\Psi(t) = \int_0^t \varphi_1(t-\tau)\,\frac{\mathrm{d}f(\tau)}{\mathrm{d}\tau}\,\mathrm{d}\tau + f(0^+)\,\varphi_1(t) . \tag{11.9}$$

Equation (11.9) represents an analytical solution of Eq. (11.1) in the form of the integral of $\mathrm{d}f(t)/\mathrm{d}t$ rather than the integral of $f(t)$ seen in Eq. (11.3), with the integral transformation kernel $\varphi_1(t-\tau)$ which, similar to $\varphi(t-\tau)$, is of the convolution type. An extra term $f(0^+)\,\varphi_1(t)$ is added to the integral. Equation (11.9) avoids the problem of determining the inverse Laplace transform, present in Eq. (11.6), so that it may be regarded more attractive. However, for a majority of conceivable electroanalytical IEs, the integral present in Eq. (11.9) does not have a representation in terms of known functions, and needs to be evaluated numerically. This is not a trivial task, because the integral is usually improper.

Although such an option was not considered in Mirčeski and Tomovski [40], there exists a simple, but interesting generalisation of Eq. (11.9). Instead of dividing $[\hat{\varphi}(s)]^{-1}$ by s, as was done in Eq. (11.8), one might just as well divide by s^m, where m is any positive integer. Instead of Eq. (11.7) we would then write:

$$\hat{\Psi}(s) = \hat{\varphi}_m(s) \left[s^m \hat{f}(s) - s^{m-1} f(0^+) - s^{m-2} \left.\frac{\mathrm{d} f(t)}{\mathrm{d} t}\right|_{t=0^+} - \ldots - \left.\frac{\mathrm{d}^{m-1} f(t)}{\mathrm{d} t^{m-1}}\right|_{t=0^+} \right] + \hat{\varphi}_1(s) f(0^+) + \hat{\varphi}_2(s) \left.\frac{\mathrm{d} f(t)}{\mathrm{d} t}\right|_{t=0^+} + \ldots + \hat{\varphi}_m(s) \left.\frac{\mathrm{d}^{m-1} f(t)}{\mathrm{d} t^{m-1}}\right|_{t=0^+} , \tag{11.10}$$

where

$$\hat{\varphi}_k(s) = [\hat{\varphi}(s)]^{-1} / s^k \tag{11.11}$$

for $k = 1, \ldots, m$. Assuming that inverse Laplace transforms $\varphi_k(\vartheta)$ of all functions $\hat{\varphi}_k(s)$ exist, and that $f(t)$ is at least m times differentiable, the convolution theorem (3.17) applied to Eq. (11.10) gives:

$$\Psi(t) = \int_0^t \varphi_m(t-\tau) \frac{\mathrm{d}^m f(\tau)}{\mathrm{d}\tau^m} \mathrm{d}\tau + f(0^+) \varphi_1(t) + \left.\frac{\mathrm{d} f(t)}{\mathrm{d} t}\right|_{t=0^+} \varphi_2(t) + \ldots + \left.\frac{\mathrm{d}^{m-1} f(t)}{\mathrm{d} t^{m-1}}\right|_{t=0^+} \varphi_m(t) . \tag{11.12}$$

An integral solution of this kind, with $m = 2$, was proposed by Chryssoulakis et al. [7], for the IE describing cyclic voltammetry for an equilibrium electron transfer. Later, it was re-discovered by Suzuki et al. [76]. The advantage of Eq. (11.12), over Eq. (11.9), is that a singularity of the integrand is often present in Eq. (11.9), but it may not be present in Eq. (11.12) (see example below). Numerical integration is easier in the absence of the singularity. However, an increased sophistication of the analytical formulae is expected in Eq. (11.12) when m increases, especially for complicated kernel functions $\varphi(\vartheta)$.

The integral solution in yet another alternative form can be obtained in the case of IEs (11.1) with convolution kernels containing an exponential factor:

$$\mathscr{K}(t,\tau) = \exp[-k(t-\tau)]\, \varphi(t-\tau) . \tag{11.13}$$

A variety of such kernels arises when one-dimensional transport is accompanied by homogeneous reactions (see Chap. 8). In such cases, instead of expressing the solution by Eq. (11.9) with function φ_1 corresponding to the entire kernel (11.13), one can first make a substitution:

$$\tilde{\Psi}(t) = \Psi(t)\,\exp(kt)\,, \tag{11.14}$$

$$\tilde{f}(t) = f(t)\,\exp(kt)\,, \tag{11.15}$$

which reduces Eq. (11.1) to the IE:

$$\int_0^t \varphi(t-\tau)\,\tilde{\Psi}(\tau)\,\mathrm{d}\tau = \tilde{f}(t)\,. \tag{11.16}$$

The solution for $\tilde{\Psi}(t)$ is then obtained by using Eq. (11.9) with function φ_1 corresponding to function φ in Eq. (11.16):

$$\tilde{\Psi}(t) = \int_0^t \varphi_1(t-\tau)\,\frac{\mathrm{d}\tilde{f}(\tau)}{\mathrm{d}\tau}\,\mathrm{d}\tau + \tilde{f}(0^+)\,\varphi_1(t)\,. \tag{11.17}$$

By returning to the original functions $\Psi(t)$ and $f(t)$, Eq. (11.17) becomes:

$$\Psi(t) = \exp(-kt)\int_0^t \varphi_1(t-\tau)\,\frac{\mathrm{d}}{\mathrm{d}\tau}[\exp(k\tau)\,f(\tau)]\,\mathrm{d}\tau + \exp(-kt)\,f(0^+)\,\varphi_1(t)\,. \tag{11.18}$$

Savéant and Vianello [65, 66] used Eq. (11.18) to obtain integral solutions for linear potential sweep voltammetry in the case of the catalytic EC reaction scheme.

As a first example, let us consider any controlled potential experiment for the equilibrium electron transfer reaction

$$X_1 + n\,e^- \rightleftarrows X_2\,, \tag{11.19}$$

proceeding under conditions of planar diffusion with different diffusion coefficients $D_1 \neq D_2$, in a semi-infinite spatial domain. The relevant Abel IE (in dimensional form) is [cf. Eq. (5.179) in Chap. 5]:

$$\int_0^t [\pi(t-\tau)]^{-1/2}\,\Psi(\tau)\,\mathrm{d}\tau = f(t)\,, \tag{11.20}$$

with

$$f(t) = \frac{c_1^\star D_1^{1/2}}{1 + \exp\left\{\frac{nF}{RT}\left[E(t) - E_{1/2}\right]\right\}} . \tag{11.21}$$

In Eqs. (11.20) and (11.21) the unknown $\Psi(t)$ is the flux of species X_2 at the electrode, proportional to the absolute value of the Faradaic current, $E(t)$ is the applied electrode potential perturbation, and $E_{1/2}$ is the half-wave potential (5.180). In this example $\varphi(\vartheta) = (\pi\vartheta)^{-1/2}$ so that $\hat{\varphi}(s) = s^{-1/2}$. Consequently, $\hat{\varphi}_1(s) = s^{-1/2}$ and $\varphi_1(\vartheta) = (\pi\vartheta)^{-1/2}$. From Eq. (11.9), the solution is

$$\Psi(t) = \int_0^t [\pi(t-\tau)]^{-1/2} \frac{\mathrm{d}f(\tau)}{\mathrm{d}\tau} \,\mathrm{d}\tau + f(0^+)\,(\pi t)^{-1/2} . \tag{11.22}$$

If potential step chronoamperometry is considered, then $E(t)$ is constant, $\mathrm{d}f(t)/\mathrm{d}t$ and the integral vanish, and only the last term in Eq. (11.22) remains. Thus, we obtain the well-known Cottrell equation for the chronoamperometric current [9]:

$$\Psi(t) = f(0^+)\,(\pi t)^{-1/2} . \tag{11.23}$$

If linear potential sweep (or cyclic) voltammetry is considered, then $E(t)$ is a linear function of t (or a sawtooth function), and the integral term in Eq. (11.22) is essential. The last term then represents a residual singularity of the current which is formally present, but can be ignored in comparisons with experimental data. The integral solution (11.22) for the linear potential sweep voltammetry was obtained in [14, 36, 68]. The problem of numerically evaluating the integral in Eq. (11.22) was investigated by several authors, including De Vries and van Dalen [12], Ramamurthy and Rangarajan [57], Lether and Wenston [27], Moreno et al. [45], and Natarajan and Mohankumar [49]. Diverse methods to compute the integral in question were suggested, but since the studies were focused on a particular example of electroanalytical experiments, and their relevance for other examples is unknown, we do not present further details here. Interested Readers are referred to these references for such details.

For the same example, the generalised formula (11.12) with $m = 2$ predicts [7]:

$$\Psi(t) = 2\pi^{-1/2} \int_0^t (t-\tau)^{1/2} \frac{\mathrm{d}^2 f(\tau)}{\mathrm{d}\tau^2} \,\mathrm{d}\tau + f(0^+)\,(\pi t)^{-1/2} + 2\pi^{-1/2} \left.\frac{\mathrm{d}f(t)}{\mathrm{d}t}\right|_{t=0+} t^{1/2} . \tag{11.24}$$

As can be seen, the integrand singularity, previously present in Eq. (11.22), has been removed, so that a proper integral occurs in Eq. (11.24).

As a second example, let us consider, after Mirčeski and Tomovski [40] any controlled potential experiment for a catalytic EC reaction scheme involving an equilibrium n-electron transfer and a (pseudo) first-order irreversible homogeneous reaction with rate constant k. Assume planar diffusion with equal diffusion coefficients, in a semi-infinite spatial domain. The relevant IE (in dimensional form) is (see, for example, [50]):

$$\int_0^t \exp[-k(t-\tau)][\pi(t-\tau)]^{-1/2}\,\Psi(\tau)\,\mathrm{d}\tau = f(t)\,, \tag{11.25}$$

with

$$f(t) = \frac{c_1^{\star}\,D^{1/2}}{1+\exp\left\{\frac{nF}{RT}\,[E(t)-E^0]\right\}}\,, \tag{11.26}$$

where D denotes the common diffusion coefficient, and E^0 is the conditional potential of the electron transfer reaction. If we take $\varphi(\vartheta) = (\pi\vartheta)^{-1/2}\exp(-k\vartheta)$, then $\hat{\varphi}(s) = (s+k)^{-1/2}$. Consequently, $\hat{\varphi}_1(s) = (s+k)^{1/2}/s$ and $\varphi_1(\vartheta) = \exp(-k\vartheta)(\pi\vartheta)^{-1/2} + k^{1/2}\mathrm{erf}[(k\vartheta)^{1/2}]$. From Eq. (11.9), the solution is

$$\Psi(t) = \int_0^t \left\{\frac{\exp[-k(t-\tau)]}{[\pi(t-\tau)]^{1/2}} + k^{1/2}\mathrm{erf}\left[k^{1/2}(t-\tau)^{1/2}\right]\right\}\frac{\mathrm{d}f(\tau)}{\mathrm{d}\tau}\,\mathrm{d}\tau$$
$$+ f(0^+)\left[\frac{\exp(-k\,t)}{(\pi t)^{1/2}} + k^{1/2}\mathrm{erf}(k^{1/2}\,t^{1/2})\right]. \tag{11.27}$$

An application of the alternative formula (11.18) to the same example yields:

$$\Psi(t) = \exp(-k\,t)\int_0^t [\pi(t-\tau)]^{-1/2}\,\frac{\mathrm{d}}{\mathrm{d}\tau}[\exp(k\,\tau)\,f(\tau)]\,\mathrm{d}\tau + \exp(-k\,t)\,f(0^+)\,(\pi t)^{-1/2}\,. \tag{11.28}$$

Mirčeski and Tomovski [40] considered one more example, namely the IE for an electron transfer reaction accompanied by adsorption of the reactant and product. We omit the presentation of this example here. Obviously, many examples can be solved by using Eqs. (11.9), (11.12), and (11.18). The crucial factors for the success are: the availability of closed-form formulae or suitable numerical approximations for the functions $\varphi_k(\vartheta)$, and the application of a suitable numerical technique for the evaluation of the integrals.

11.1.2 Power Series Solutions

Any series expansion for the solution may be regarded "more analytical" or convenient than an integral expression, as the manipulation and analysis of the series terms is easy, at least in principle. The involvement of numerical computations is not entirely eliminated, because the series terms must ultimately be summed up to get the solution values, and the summation sometimes requires elaborate algorithms. A number of authors [2, 16, 17, 19, 20, 35, 51, 60, 63, 70, 71, 74, 75, 79, 82–84] have obtained analytical solutions of one-dimensional electroanalytical IEs or IDEs in the form of power series. It seems that the method has been particularly popular among researchers investigating adsorption at stationary and dropping mercury electrodes, but this fact is rather accidental. The idea of the method is to assume that the solution $\Psi(t)$ of an IE or IDE can be expressed as

$$\Psi(t) = \sum_{l=0}^{\infty} a_l \, t^{b_l} \,, \tag{11.29}$$

where a_l are expansion coefficients and b_l are power exponents, not necessarily integer. Substitution of Eq. (11.29) into an IE or IDE may yield a sequence of equations, possibly recursive, from which the unknown coefficients a_l and b_l can be determined. The series (11.29) can sometimes also be obtained by applying the successive approximation method (see Sect. 11.1.4). Either an ordinary series (with non-negative b_l) or asymptotic expansions (with negative b_l) can be searched. The IEs or IDEs may well be nonlinear.

As an example, let us solve one of the nonlinear IDEs derived by Reinmuth [60]:

$$\Psi(t) = [1 - \Psi(t)] \left[\lambda - 2 \int_0^t (t-\tau)^{-1/2} \frac{\mathrm{d}\Psi(\tau)}{\mathrm{d}\tau} \, \mathrm{d}\tau \right] . \tag{11.30}$$

The IDE describes an experiment in which an initially clean planar electrode (possibly maintained in this state by an appropriate polarisation, prior to the actual experiment beginning at the normalised time $t = 0$) is suddenly subject to a fast adsorption of a distributed species, controlled by diffusion from a semi-infinite domain. Equation (11.30) is a particular case of the IDE (9.27) derived in Sect. 9.1, corresponding to the Langmuir adsorption isotherm. Unknown function $\Psi(t)$ represents the fractional coverage of the electrode by the adsorbate. As the electrode is initially clean, the initial condition is $\Psi(0) = 0$. Parameter λ depends on the initial concentration and the adsorption equilibrium constant. One assumes

$$\Psi(t) = \sum_{l=0}^{\infty} a_l \, t^{l/2} \,, \tag{11.31}$$

with $a_0 = 0$, to satisfy the initial condition. Substitution of Eq. (11.31) into Eq. (11.30) gives

$$\sum_{i=0}^{\infty} a_i t^{i/2} = \left[1 - \sum_{i=0}^{\infty} a_i t^{i/2}\right]\left[\lambda - t^{-1/2} \sum_{i=1}^{\infty} i\, a_i t^{i/2} \mathrm{B}\left(\frac{i}{2}, \frac{1}{2}\right)\right] , \tag{11.32}$$

where

$$\mathrm{B}(z, w) = \int_0^1 t^{z-1}(1-t)^{w-1}\, dt \tag{11.33}$$

is the beta function [1]. By comparing the coefficients at the successive powers of $t^{1/2}$, at both sides of Eq. (11.32), one obtains:

$$a_1 = \lambda/\mathrm{B}\left(\frac{1}{2}, \frac{1}{2}\right) = \lambda/\pi , \tag{11.34}$$

$$a_2 = -a_1/\left[2\,\mathrm{B}\left(1, \frac{1}{2}\right)\right] , \tag{11.35}$$

and

$$a_i = \left[-a_{i-1} + \sum_{j=2}^{i-1} j\, a_j\, a_{i-j}\, \mathrm{B}\left(\frac{j}{2}, \frac{1}{2}\right)\right] / \left[i\, \mathrm{B}\left(\frac{i}{2}, \frac{1}{2}\right)\right] \tag{11.36}$$

for $i = 3, 4, \ldots$.

Assuming that the resulting equations for the coefficients a_i and b_i can be solved in a straightforward way, without any need to resort to numerical calculations, obtaining the solution should be easy. This is an attractive feature of the method. Unfortunately, the power series (11.31) may have limited convergence radii, which restricts the usefulness of this method.

11.1.3 Exponential Series Solutions

A different, and mathematically original analytical series expansion method was introduced by Reinmuth [61,62], for solving one-dimensional linear first and second kind Volterra IEs occurring in the theory of linear potential sweep voltammetry. The method makes use of the fact that, in the case of electrochemical reactions subject to Butler–Volmer kinetics, and also in the case of equilibrium electrochemical reactions, the coefficients of the relevant linear IEs are usually exponential functions of the potential. The electrode potential, in turn, is a linear function of time, in the case of the linear potential sweep voltammetry. Consequently, when presented in dimensionless form, the coefficients of the IEs involve terms dependent exponentially on time, such as $\exp[-(u-t)]$, where t is a dimensionless time (often denoted by at or bt in the literature), and u is a certain constant parameter. This suggests looking for the solution in the form of the exponential series:

$$\Psi(t) = \sum_{i=1}^{\infty} a_i \exp[-i\,(u-t)] \ , \tag{11.37}$$

where a_i are coefficients to be determined. To illustrate the principle of this method, we provide two examples below.

As a first example, let us take the equilibrium electron transfer reaction

$$\mathrm{X_1} + n\,\mathrm{e}^- \rightleftarrows \mathrm{X_2} \tag{11.38}$$

involving dynamic distributed species X_1 and X_2 with different diffusion coefficients $D_1 \neq D_2$. Assume planar diffusion and a semi-infinite spatial domain. The relevant dimensionless first kind Volterra IE can be written in the form [cf. the derivations in Sects. 5.1 and 5.5, leading to Eqs. (5.23) and (5.179)]:

$$\{1+\exp[-(u-t)]\} \int_0^t (t-\tau)^{-1/2}\, \Psi(\tau)\, \mathrm{d}\tau = \exp[-(u-t)] \ . \tag{11.39}$$

By setting Eq. (11.37) into Eq. (11.39), and integrating step-by-step the elements of the series, one obtains

$$\{1+\exp[-(u-t)]\} \sum_{i=1}^{\infty} a_i \left(\frac{\pi}{i}\right)^{1/2} \exp[-i\,(u-t)]\ \mathrm{erf}\left[(i\,t)^{1/2}\right] = \exp[-(u-t)] \ . \tag{11.40}$$

The coefficient u is proportional to the absolute value of the difference between the starting potential of the linear sweep, and the half-wave potential (5.180). If we assume that the difference is sufficiently large, then the time t after which the current function $\Psi(t)$ begins to differ significantly from zero is also large. Consequently, the physically interesting solution is characterised by t so large that $\mathrm{erf}\left[(i\,t)^{1/2}\right] \approx 1$ for all i. This simplifies Eq. (11.40). By comparing the coefficients standing at successive powers of $\exp[-(u-t)]$ at both sides of Eq. (11.40) one finds $a_i = \pi^{-1/2}(-1)^{i-1}\, i^{1/2}$, so that the solution is

$$\Psi(t) = \pi^{-1/2} \sum_{i=1}^{\infty} (-1)^{i-1}\, i^{1/2} \exp[-i\,(u-t)] \ . \tag{11.41}$$

As a second example, let us take the irreversible electron transfer reaction

$$X_1 + n\,e^- \rightarrow \text{products} \tag{11.42}$$

involving a dynamic distributed species X_1. Assume planar diffusion and a semi-infinite spatial domain. The relevant dimensionless second kind Volterra IE is obtainable as a particular version of Eq. (5.21) with $\mathscr{K}_1(t,\tau) = [\pi(t-\tau)]^{-1/2}$:

$$\int_0^t (t-\tau)^{-1/2}\,\Psi(\tau)\,d\tau + \exp(u-t)\,\Psi(t) - 1 = 0\,, \tag{11.43}$$

where the parameter u depends on the difference between the starting potential and the conditional potential of reaction (11.42), and also on the conditional rate constant. Substitution of Eq. (11.37) into Eq. (11.43) then gives, analogous to the first example:

$$\Psi(t) = \pi^{-1/2}\sum_{i=1}^{\infty}(-1)^{i-1}\left[\frac{\pi^i}{(i-1)!}\right]^{1/2}\exp\left[-i\,(u-t)\right]\,. \tag{11.44}$$

Exponential series solutions have been obtained for a number of IEs arising in the theory of linear potential sweep and cyclic voltammetry. In his original work [61], Reinmuth derived the series (11.41). He also derived the series (11.44) in a more general form, taking into account external or internal spherical diffusion, and a series for the EC reaction scheme involving distributed species, equilibrium electron transfer and irreversible homogeneous reaction, at the assumption of planar diffusion in a semi-infinite spatial domain. A collection of exponential series for standard EC and CE reaction schemes involving distributed species, assuming planar diffusion in a semi-infinite spatial domain, was reported by Nicholson and Shain [50]. Some of these series, including the ones for the catalytic EC reaction schemes, were obtained assuming equal diffusion coefficients of the reactants. The series for the catalytic reaction schemes with reactants having distinct diffusion coefficients were later supplemented by the present author [5]. Olmstead and Nicholson [54] derived the series for an EC reaction scheme assuming external spherical diffusion. The series for the EC catalytic reaction scheme involving adsorbed species $X_{1,ad}$ and $X_{2,ad}$ and distributed species X_3 and X_4:

$$X_{1,ad} + n\,e^- \rightleftarrows X_{2,ad}\,, \tag{11.45}$$

$$X_{2,ad} + X_3 \underset{k_b}{\overset{k_f}{\rightleftarrows}} X_{1,ad} + X_4\,, \tag{11.46}$$

was derived by Bieniasz et al. [6], assuming external spherical diffusion.

Readers interested in applying or developing this method should keep in mind that the method requires the existence of the exponential terms such as $\exp[-(u-t)]$, in the coefficients of the IEs. Hence, the method is not likely to work in the case of Marcus–Hush–Chidsey kinetics (see Sect. 2.7).

The main problem with using the above exponential series method is that, similarly to the power series method of Sect. 11.1.2, the series (11.37) may have limited convergence radii. Also, even if the series are convergent, the convergence may be slow. For example, the series (11.41) is convergent only for $t < u$, whereas the series (11.44) is convergent for all t. These difficulties have given rise to a number of studies where various rearrangements and reformulations of the exponential series have been proposed in order to improve the convergence. These studies have been thus far limited mostly to the few simplest (although otherwise important for electrochemistry) examples of electroanalytical experiments. In particular, the case of the series (11.41) has attracted much attention. Oldham [52, 53] applied the Weyl semi-integration to reformulate Eq. (11.41) into the following convergent series:

$$\Psi(t) = 2^{-1/2} \sum_{i=1}^{\infty} b_i^{-3} \left[b_i + (u-t)\right]^{1/2} \left[b_i - 2(u-t)\right], \tag{11.47}$$

with

$$b_i = \left[(2i-1)^2 \pi^2 + (u-t)^2\right]^{1/2}. \tag{11.48}$$

Myland and Oldham [46] subsequently presented generalisations of this series onto cyclic voltammetry and (together with Zoski [48]) onto voltammetry under mildly spherical diffusion conditions. Their works motivated Lether and Evans [26], Basha and Sangaranarayanan [4], Lether [25], and Ferreira et al. [13], to develop further modifications or new approximations to the series (11.47). A piecewise nonlinear approximation to $\Psi(t)$ was proposed [26], as well as a mini-max rational approximation [26], Padé approximation [4], and integral representation in terms of the Joncquière functions [25]. The latter was used to develop a new asymptotic and Taylor expansion for $\Psi(t)$ [13]. Acceleration of the convergence of the series (11.47) was achieved by adding and subtracting a specially designed series [26].

Although the series (11.44) for the irreversible electron transfer is convergent for all t, Arun Prasad and Sangaranarayanan [3] transformed the series into a Padé approximant, emphasising a simplicity of such a formulation. Exponential series similar to those of Reinmuth [61, 62] were proposed for the non-equilibrium reversible electron transfer reaction (11.38) by Myland and Oldham [47] and Cope [8]. All these developments, although mathematically elaborate and interesting, were focused on very particular examples of electroanalytical experiments. Their applicability to, and usefulness for other examples was not demonstrated. In contrast, Mocak [43], Mocak and Bond [44], and Sivakumar and Basha [69] investigated the merits of using several series summation algorithms for a number of different example exponential series. Series transformation techniques such as the *Epsilon* and *Eta* were tried [43], by using programs enabling arbitrary precision arithmetic computations, such as MATHEMATICA [32] and UBASIC [78]. Special series expansions implemented in MATHEMATICA, such as *PolyLog*, *LerchPhi*, and *Zeta* were utilised to express and sum the series of interest [44]. The *Epsilon* transformation was also used by Sivakumar and Basha [69], together with the replacement of the series by Padé approximants. Improvements in convergence and accuracy of the series were achieved.

11.1.4 The Method of Successive Approximations

The method of successive approximations is one of the classical methods for solving IEs or IDEs [29, 55, 56, 80]. In this method one assumes a starting analytical approximate expression for the unknown solution. The substitution of this expression into an IE or IDE, in an appropriate way, may yield a new analytical approximate expression that can be again substituted into the IE or IDE, to provide a next approximate expression, and so on. If the procedure is convergent, one can obtain a reasonably accurate approximate solution in a few iterations.

There have been very few applications of the successive approximation method to electrochemical IEs, for example by Matsuda and Ayabe [35] (in conjunction with the power series expansion), Senda and Tachi [67], Levich et al. [28], Holub [18], and Mas et al. [31]. In the latter case the procedure was limited to single iterations.

11.1.5 Application of Steady State Approximations

It sometimes happens that an approximate solution of the IEs or IDEs can be calculated analytically by making use of steady state assumptions. One obvious situation of this kind occurs when a particular integral concentration–production rate relationship turns into a non-integral relationship, in a limiting case of a model parameter value. One can then replace the integral(s) in the IE(s), by these limiting non-integral expressions, and solve the AE(s) obtained. In this way one obtains approximate analytical solutions, which are however valid only for the above model parameter values close to the limiting value. As an example, let us consider an IE for the irreversible electron transfer

$$\mathrm{X}_1 + n\,\mathrm{e}^- \rightarrow \text{products} \tag{11.49}$$

involving dynamic distributed species X_1 subject to spherical diffusion in a semi-infinite space domain. Assume reaction (11.49) obeys Butler–Volmer kinetics. The relevant IE describing a controlled potential experiment is [cf. Eq. (5.21)]:

$$k^0 \left[c_1^\star - \int_0^t \mathscr{K}_1^{\mathrm{s}}(t,\tau)\, \Psi(\tau)\, \mathrm{d}\tau \right] \exp\left\{ -\alpha^{\mathrm{f}} \frac{nF}{RT} \left[E(t) - E^0 \right] \right\} - \Psi(t) = 0 \tag{11.50}$$

where according to Eq. (5.74)

$$\mathscr{K}_1^{\mathrm{s}}(t,\tau) = D_1^{-1/2} \left\{ [\pi(t-\tau)]^{-1/2} - \varrho_1 \, \mathrm{erex}\left[\varrho_1 (t-\tau)^{1/2} \right] \right\} \tag{11.51}$$

and

$$\Psi(t) = -J_1^{\perp}(t) \,. \tag{11.52}$$

We have seen in Sect. 5.2.1 that the integral concentration–production rate relationship (5.87) for spherical diffusion in a semi-infinite spatial domain turns into the AE (5.90) when parameter $\varrho_1 \to \infty$, that is for microelectrodes with a sufficiently small radius. Consequently, the limiting steady state form of Eq. (11.50) is:

$$k^0 \left[c_1^\star - D_1^{-1/2} \varrho_1^{-1} \Psi(t) \right] \exp\left\{ -\alpha^{\mathrm{f}} \frac{nF}{RT} \left[E(t) - E^0 \right] \right\} - \Psi(t) = 0 \,. \tag{11.53}$$

Equation (11.53) is easily solved to obtain the analytical solution for the steady state:

$$\Psi(t) = \frac{k^0 c_1^\star \exp\left\{ -\alpha^{\mathrm{f}} \frac{nF}{RT} \left[E(t) - E^0 \right] \right\}}{1 + D_1^{-1/2} k^0 \varrho_1^{-1} \exp\left\{ -\alpha^{\mathrm{f}} \frac{nF}{RT} \left[E(t) - E^0 \right] \right\}} \,. \tag{11.54}$$

Other similar examples of steady state analytical solutions can be obtained by using Eqs. (5.127), (5.144), (8.52), (8.53), (8.54), etc.

A somewhat less obvious, but interesting possibility of obtaining approximate solutions occurs in the modelling of experiments characterised by periodic perturbations. In such cases, the exact solution very quickly (after one or a few periods after applying the perturbation) approaches closely an asymptotic solution, which oscillates with the same frequency, but is shifted by a certain phase shift with respect to the perturbation imposed. This sort of a steady state is likely to occur for all kinds of transport conditions and periodic perturbations. This is easiest to demonstrate by using the example of a controlled potential experiment described by a single IE with the Faradaic current $I(t)$ as an unknown function. To fix attention, assume the periodic electrode potential perturbation pertinent to AC polarography:

$$E(t) = E_{\mathrm{dc}} + E_{\mathrm{m}} \sin(\omega t) \,, \tag{11.55}$$

where E_{dc} is the DC component of the electrode potential (possibly slowly varying in time), ω is the perturbation frequency, and E_m is the AC perturbation amplitude. At the aforementioned steady state, the first harmonic of the AC response is expected to be:

$$I_{ac}(t) = I_m \sin(\omega t + \phi) , \tag{11.56}$$

where I_m is the Faradaic current amplitude, and ϕ is the phase shift relative to $E(t)$. Alternatively, one can write

$$I_{ac}(t) = a \sin(\omega t) + b \cos(\omega t) , \tag{11.57}$$

where a and b are suitable coefficients. The IE(s) for the first harmonic of the AC current usually contain one or more integrals of the form

$$Y(t) = \int_0^t \mathscr{K}(t,\tau)\, I(\tau)\, d\tau , \tag{11.58}$$

where $\mathscr{K}(t,\tau)$ is one of the kernels discussed in Chaps. 5, 6, or 8. Substitution of Eq. (11.57) into these IEs yields integrals

$$Y_{sin}(t) = \int_0^t \mathscr{K}(t,\tau) \sin(\omega\tau)\, d\tau , \tag{11.59}$$

$$Y_{cos}(t) = \int_0^t \mathscr{K}(t,\tau) \cos(\omega\tau)\, d\tau . \tag{11.60}$$

For sufficiently large t these integrals usually possess asymptotic approximations, by which they can be replaced in the IEs. This allows one to determine the unknown coefficients a and b at steady state. The asymptotic approximations have been obtained for a few typical kernels. In particular, for $\mathscr{K}(t,\tau) = [\pi(t-\tau)]^{-1/2}$ and $\mathscr{K}(t,\tau) = (\frac{3}{7}\pi)^{-1/2}\tau^{2/3}(t^{7/3} - \tau^{7/3})^{-1/2}$ one obtains for $t \to \infty$ [24, 38]:

$$Y_{sin}(t) \sim (2\omega)^{-1/2} [\sin(\omega t) - \cos(\omega t)] , \tag{11.61}$$

$$Y_{cos}(t) \sim (2\omega)^{-1/2} [\sin(\omega t) + \cos(\omega t)] , \tag{11.62}$$

whereas for the kernel $\mathscr{K}(t,\tau) = \exp[-k(t-\tau)](\frac{3}{7}\pi)^{-1/2}\tau^{2/3}(t^{7/3} - \tau^{7/3})^{-1/2}$ one gets [38]:

$$Y_{sin}(t) \sim (2\omega)^{-1/2} [p \sin(\omega t) - q \cos(\omega t)] , \tag{11.63}$$

$$Y_{cos}(t) \sim (2\omega)^{-1/2} [q \sin(\omega t) + p \cos(\omega t)] , \tag{11.64}$$

where

$$p = \frac{[1 + (k/\omega)^2]^{1/2} + k/\omega}{1 + (k/\omega)^2}, \tag{11.65}$$

$$q = \frac{[1 + (k/\omega)^2]^{1/2} - k/\omega}{1 + (k/\omega)^2}. \tag{11.66}$$

Examples of the application of such steady state solution technique are found in the works of Matsuda [33] on the modelling of controlled current oscillographic polarography; Delahay, Matsuda, and their co-workers [11, 37], and Smith [73], on the modelling of the Faradaic impedance; and Smith and co-workers [23, 24, 38, 39, 72], on the modelling of controlled potential AC polarography for various electrochemical systems. Approximate formulae for the fundamental harmonic of the current, valid for external cylindrical diffusion in a semi-infinite spatial domain, were also provided by Tokuda et al. [77]. Derivations of this sort can be facilitated by using symbolic algebra software, such as MATHEMATICA [32], as was suggested and described by Honeychurch (cf. chapter 7 in [21]).

11.2 Analytical Methods for Two-Dimensional IEs

Compared to one-dimensional IEs, analytical solution methods for two-dimensional IEs are scarce. We mention here only the method described by Daschbach [10], which makes use of the Neumann integral theorem and truncated Fourier expansion. The method was used to obtain IEs essentially equivalent to those of Mirkin and Bard [41, 42] (cf. Sects. 7.1 and 7.2) for the disk electrode, and their approximate solutions. Another analytical/numerical method employing series expansions, for solving the IEs resulting from the Cope and Tallman approach (cf. Sect. 7.1), will be described in Chap. 12.

References

1. Abramowitz M, Stegun IA (1972) Handbook of mathematical functions. Dover, New York
2. Aoki K, Tokuda K, Matsuda H (1986) Theory of stationary current–voltage curves of redox-electrode reactions in hydrodynamic voltammetry. Part XI. Wall-jet electrodes. J Electroanal Chem 206:37–46
3. Arun Prasad M, Sangaranarayanan MV (2004) Formulation of a simple analytical expression for irreversible electron transfer processes in linear sweep voltammetry and its experimental verification. Electrochim Acta 49:2569–2579
4. Basha CA, Sangaranarayanan MV (1989) On the evaluation of the current function in linear sweep voltammetry. J Electroanal Chem 261:431–436
5. Bieniasz LK (1985) Linear voltammetric current functions for a pseudo-first-order EC catalytic reaction scheme with $D_O \neq D_R$: series expansion algorithm. J Electroanal Chem 188:13–20

6. Bieniasz LK, González J, Molina Á, Laborda E (2010) Theory of linear sweep/cyclic voltammetry for the electrochemical reaction mechanism involving a redox catalyst couple attached to a spherical electrode. Electrochim Acta 56:543–552
7. Chryssoulakis Y, Iatrelli M, Kalogeropoulou S, Alexopoulou A (1985) Méthode simple de résolution des équations intégrals décrivant les courbes de la voltampérométrie cyclique ou triangulaire. J Appl Electrochem 15:307–311
8. Cope DK (1997) An expansion for quasi-reversible linear potential sweep voltammetry and the use of Euler's transformation of series. Anal Chem 69:1465–1469
9. Cottrell FG (1903) Der Reststrom bei galvanischer Polarisation, betrachtet als ein Diffusionsproblem. Z Phys Chem 42:385–431
10. Daschbach JL (1991) Time dependent responses at disk electrodes. In: Montenegro MI, Queirós MA, Daschbach JL (eds) Microelectrodes: theory and applications. Kluwer, Dordrecht, pp 67–82
11. Delahay P, Senda M, Weis CH (1961) Faradaic rectification and electrode processes. J Am Chem Soc 83:312–322
12. De Vries WT, van Dalen E (1963) The current–potential equation for linear-sweep voltammetry. J Electroanal Chem 6:490–493
13. Ferreira C, López JL, Miana P (2004) Two algorithms for computing the Randles-Sevcik function from electrochemistry. J Math Chem 35:131–137
14. Gokhshtein YaP (1959) General equations of oscillographic polarography, reversible processes at the cathodic and anodic polarisation. Dokl Akad Nauk SSSR 126:598–601 (in Russian)
15. Gorenflo R, Vessella S (1991) Abel integral equations, analysis and applications. Springer, Berlin
16. Holub K (1966) Surface reaction of an adsorbed substance transported by diffusion towards the dropping electrode. Collect Cecoslov Chem Commun 31:1461–1474
17. Holub K (1968) Current–time dependence for charge transfer processes with reactant adsorption at a dropping electrode. J Electroanal Chem 16:433–436
18. Holub K (1973) The reaction of an absorbed substance with a substance transported by diffusion to the electrode. Kinetics of Langmuir adsorption without desorption coupled with diffusion. J Electroanal Chem 48:253–264
19. Holub K, Van Leeuwen HP (1984) Influence of reactant adsorption on limiting currents in normal pulse polarography. Part II. Theory for the stationary, spherical electrode. J Electroanal Chem 162:55–65
20. Holub K, Van Leeuwen HP (1985) Influence of reactant adsorption on limiting currents in normal pulse polarography. Part IV. Theory for the expanding drop electrode. J Electroanal Chem 191:281–292
21. Honeychurch MJ (2006) Simulating electrochemical reactions with MATHEMATICA. IBNH, St Lucia
22. Honeychurch MJ, Rechnitz GA (1997) Cyclic voltammetry at monolayer covered electrodes: the effect of monolayers on the reduction of cytochrome c. J Phys Chem B 101:7472–7479
23. Hung HL, Smith DE (1966) Alternating current polarography with multi-step charge transfer. I. Theory for systems with reversible two-step charge transfer. J Electroanal Chem 11:237–254
24. Hung HL, Smith DE (1966) Alternating current polarography with multi-step charge transfer. II. Theory for systems with quasi-reversible two-step charge transfer. J Electroanal Chem 11:425–461
25. Lether FG (2001) A family of approximations for the numerical computation of the Randles-Sevcik function in electrochemistry. J Math Chem 30:289–297
26. Lether WG, Evans OM (1985) An algorithm for the computation of the reversible Randles-Sevcik function in electrochemistry. Appl Math Comput 16:253–264
27. Lether FG, Wenston PR (1987) An algorithm for the numerical evaluation of the reversible Randles-Sevcik function. Comput Chem 11:179–183
28. Levich VG, Podgaetskii EM, Filinovskii VYu (1970) A successive approximation method for solving a non-linear Volterra integral equation of the second kind. USSR Comput Math Math Phys 10:138–145

29. Linz P (1985) Analytical and numerical methods for Volterra equations. SIAM, Philadelphia
30. Mandal BN, Chakrabarti A (2011) Applied singular integral equations. CRC Press, Boca Raton; Science Publishers, Enfield
31. Mas F, Puy J, Sanz F, Virgili J (1983) Potentionstatic reversible reaction when both reactant and product are adsorbed at the dropping mercury electrode. Part I. Theoretical treatment. J Electroanal Chem 158:217–230
32. MATHEMATICA (2014) Wolfram Res. Inc., Champaigne, IL. http://www.wolfram.com. Accessed 10 June 2014
33. Matsuda H (1956) Zur Theorie der Heyrovský-Forejtschen oszillographischen Polarographie. Z Elektrochem 60:617–626
34. Matsuda H (1957) Beiträge zur Theorie der polarographischen Stromstärke. Allgemeine Formel der diffusionsbedingten Stromstärke und ihre Anwendung. Z Elektrochem 61:489–506
35. Matsuda H, Ayabe Y (1955) Theoretical analysis of polarographic waves. I. Reduction of simple metal ions. Bull Chem Soc Jpn 28:422–429
36. Matsuda H, Ayabe Y (1955) Zur Theorie der Randles-Sevčikschen Kathodenstrahl-Polarographie. Z Elektrochem 59:494–503
37. Matsuda H, Delahay P (1960) Faradaic rectification with control of alternating potential variations—application to electrode kinetics for fast processes. J Am Chem Soc 82:1547–1550
38. McCord TG, Smith DE (1968) Alternating current polarography: an extension of the general theory for systems with coupled first-order homogeneous chemical reactions. Anal Chem 40:1959–1966
39. McCord TG, Smith DE (1968) Second harmonic alternating current polarography: a general theory for systems with coupled first-order homogeneous chemical reactions. Anal Chem 40:1967–1970
40. Mirčeski V, Tomovski Ž (2008) Analytical solutions of integral equations for modelling of reversible electrode processes under voltammetric conditions. J Electroanal Chem 619–620:164–168
41. Mirkin MV, Bard AJ (1992) Multidimensional integral equations. Part 1. A new approach to solving microelectrode diffusion problems. J Electroanal Chem 323:1–27
42. Mirkin MV, Bard AJ (1992) Multidimensional integral equations: a new approach to solving microelectrode diffusion problems. Part 2. Applications to microband electrodes and the scanning electrochemical microscope. J Electroanal Chem 323:29–51
43. Mocak J (2002) Voltammetric current–potential calculations using infinite series solution. Electrochem Commun 4:803–807
44. Mocak J, Bond AM (2004) Use of MATHEMATICA software for theoretical analysis of linear sweep voltammograms. J Electroanal Chem 561:191–202
45. Moreno C, Campos A, Teixeira M, Le Gall J, Montenegro MI, Moura I, van Dijk C, Moura JGJ (1991) Simulation of the electrochemical behavior of multi-redox systems. Eur J Biochem 202:385–393
46. Myland JC, Oldham KB (1983) An analytical expression for the current–voltage relationship during reversible cyclic voltammetry. J Electroanal Chem 153:43–54
47. Myland JC, Oldham KB (1994) Quasireversible linear-potential-sweep voltammetry: an analytic solution for rational α. Anal Chem 66:1866–1872
48. Myland JC, Oldham KB, Zoski CG (1986) Calculation of reversible cyclic voltammograms at spherical electrodes. J Electroanal Chem 206:1–22
49. Natarajan A, Mohankumar N (1997) An algorithm for the numerical evaluation of the Randles-Sevcik function. Comput Chem 21:315–318
50. Nicholson RS, Shain I (1964) Theory of stationary electrode polarography. Single scan and cyclic methods applied to reversible, irreversible, and kinetic systems. Anal Chem 36:706–723.
51. Nicholson RS, Wilson JM, Olmstead ML (1966) Polarographic theory for an ECE mechanism. Application to reduction of p-nitrosophenol. Anal Chem 38:542–545
52. Oldham KB (1979) Analytical expressions for the reversible Randles-Sevcik function. J Electroanal Chem 105:373–375

53. Oldham KB (1983) The reformulation of an infinite sum via semiintegration. SIAM J Math Anal 14:974–981
54. Olmstead ML, Nicholson RS (1967) Theoretical evaluation of effects of electrode sphericity on stationary electrode polarography. J Electroanal Chem 14:133–141
55. Polyanin AD, Manzhirov AV (2008) Handbook of integral equations, 2nd edn. Chapman & Hall, Boca Raton
56. Rahman M (2007) Integral equations and their applications. WIT Press, Southampton
57. Ramamurthy AC, Rangarajan SK (1981) A Gaussian quadrature analysis of linear sweep voltammetry. Electrochim Acta 26:111–115
58. Rampazzo L (1966) A note on the theory of catalytic currents in linear-sweep polarography. A reversible chemical reaction (pseudo-first order) parallel to the electron-transfer process. Ric Sci 36:998–1007
59. Rampazzo L (1967) A note on the theory of catalytic currents in linear-sweep polarography. A reversible chemical reaction (pseudo-first order) parallel to the electron-transfer process. J Electroanal Chem 14:117–118
60. Reinmuth WH (1961) Diffusion to a plane with Langmuirian adsorption. J Phys Chem 65:473–476
61. Reinmuth WH (1961) Theory of stationary electrode polarography. Anal Chem 33:1793–1794
62. Reinmuth WH (1962) Theory of diffusion limited charge-transfer processes in electroanalytical techniques. Anal Chem 34:1446–1454
63. Reinmuth WH (1972) Theory of Ohmic loss in D.C. polarography. J Electroanal Chem 36:467–473
64. Savéant JM (1967) ECE mechanisms as studied by polarography and linear sweep voltammetry. Electrochim Acta 12:753–766
65. Savéant JM, Vianello E (1960) Recherches sur les courants catalytiques en polarographie-oscillographique à balayage linéaire de tension. Etude théorique. In: Advances in polarography. Pergamon Press, London, pp 367–374
66. Savéant JM, Vianello E (1965) Potential-sweep chronoamperometry: kinetic currents for first-order chemical reaction parallel to electron-transfer process (catalytic currents). Electrochim Acta 10:905–920
67. Senda M, Tachi I (1955) Studies on A.C. polarography. V. Theory of reversible wave. Bull Chem Soc Jpn 28:632–636
68. Ševčik A (1948) Oscillographic polarography with periodical triangular voltage. Collect Cecoslov Chem Commun 13:349–377
69. Sivakumar S, Basha CA (2005) Evaluation of the current function in linear sweep voltammetry by Pade approximation and epsilon convergence. Russ J Electrochem 41:421–438
70. Sluyters-Rehbach M, Sluyters JH (1977) A comparative study of approaches to the theory of reactant and/or product adsorption in D.C. polarography. J Electroanal Chem 75:371–385
71. Sluyters-Rehbach M, Sluyters JH (1977) A long-time series expansion as a solution of the diffusion problem in the case of Langmuirian adsorption at a dropping electrode. J Electroanal Chem 81:211–214
72. Smith DE (1963) Alternating current polarography of electrode processes with coupled homogeneous chemical reactions. I. Theory for systems with first-order preceding, following, and catalytic chemical reactions. Anal Chem 35:602–609
73. Smith DE (1964) Theory of the Faradaic impedance. Relationship between Faradaic impedances for various small amplitude alternating current techniques. Anal Chem 36:962–970
74. Smutek M (1955) Kinetics of electrode processes. XIII. A contribution to the theory of slow electrode processes. Collect Cecoslov Chem Commun 20:247–251
75. Sobel HR, Smith DE (1970) D.C. polarography: on the theory for the current–potential profile with an ECE mechanism. J Electroanal Chem 26:271–284
76. Suzuki T, Mori I, Nagamoto H, Ito MM, Inoue H (1992) A simple method for evaluating the current in the cyclic voltammogram for reversible systems. J Electroanal Chem 324:397–404

77. Tokuda K, Kitamura F, Kumagai A, Ohsaka T (1995) AC voltammetry at microcylinder electrodes. J Electroanal Chem 396:365–370
78. UBASIC (2014). http://en.wikipedia.org//wiki/UBASIC. Accessed 10 June 2014
79. Van Leeuwen HP, Sluyters-Rehbach M, Holub K (1982) The influence of reactant adsorption on limiting currents in normal pulse polarography. The ranges of low and high concentrations. J Electroanal Chem 135:13–24
80. Wazwaz AM (2011) Linear and nonlinear integral equations. Higher Education Press, Beijing; Springer, Berlin
81. Weidner JW, Fedkiw PS (1990) Reversible, linear-sweep voltammetry of a soluble redox couple: effect of initial concentrations. Anal Chem 62:875–877
82. Wein O (2010) Edge effects in voltage-step transient I. Ohmic losses in 1D approximation (revisited). Research Report ICPF No. 2010/3, Institute of Chemical Process Fundamentals ASCR, Prague
83. Wein O (2010) Voltage-step transient in redox systems II. 1D approximation (revisited). Research Report ICPF No. 2010/6, Institute of Chemical Process Fundamentals ASCR, Prague
84. Wein O, Tovchigrechko VV (2011) Voltage-step transient on circular electrodes. J Appl Electrochem 41:1065–1075

Chapter 12
Numerical Solution Methods

In this chapter we review numerical methods that have been used to obtain solutions of the IEs arising in electroanalytical chemistry. Numerical techniques applicable to one-dimensional Volterra IEs are discussed in Sect. 12.1. These techniques are in common use, and they are sufficiently powerful to solve almost any IE considered in electroanalytical chemistry. Numerical methods applicable to two- and higher-dimensional IEs are addressed much more briefly in Sect. 12.2.

12.1 Numerical Methods for One-Dimensional Volterra IEs

There exists an ample mathematical literature describing numerical methods designed for the solution of one-dimensional Volterra IEs, including weakly singular IEs of the type encountered in electrochemistry [7,8,34,35,41,56,65,68]. However, the arsenal of numerical techniques, used thus far to solve one-dimensional IEs of electroanalytical models, has been rather limited. It seems that the many techniques developed by mathematicians and numerical analysts over the past decades have not attracted attention of electrochemists, with a few exceptions.

The methods used in electrochemistry are based on the general idea of replacing the integrals

$$Y(t) = \int_0^t \mathscr{K}(t,\tau)\,\Psi(\tau)\,\mathrm{d}\tau\,, \tag{12.1}$$

occurring in the IEs, by finite sums. The sums involve unknown values of the function(s) $\Psi(t)$ at the nodes of discrete grids selected along the axis of the independent variable t. By discretisations of this kind, the IEs are converted into sets of AEs. By solving these AEs, one obtains approximate solutions at the grid nodes.

L.K. Bieniasz, *Modelling Electroanalytical Experiments by the Integral Equation Method*, Monographs in Electrochemistry, DOI 10.1007/978-3-662-44882-3_12

There is a large number of ways in which integrals (12.1) can be approximated by finite sums. The ways used in electrochemistry can be roughly divided into three groups: quadrature methods, product integration methods, and methods based on discrete differintegration. This division is to some extent artificial, because all these methods are similar, and some of the methods may actually belong to more than one group simultaneously.

By quadrature methods we understand here methods in which the entire integrands in Eq. (12.1) are discretised on a certain grid along the t axis, using a particular quadrature (for introduction to numerical quadratures see, for example, Kythe and Schäferkotter [66]). Such methods have been rarely used; examples are in [52, 54]. In general, we do not recommend them, unless the kernel $\mathscr{K}(t,\tau)$ is a very slowly varying and nonsingular function. In electrochemistry this condition is usually not satisfied, and the kernels are either weakly singular, or they vary much faster than the unknown function $\Psi(t)$. Therefore, a grid suitable for the discretisation of $\Psi(t)$ is usually too sparse for an accurate discretisation of the kernel.

The above problem can be avoided in the product integration methods (for a mathematical introduction to the product integration techniques, see [8,65]). In these methods the unknown functions $\Psi(t)$ are discretised on suitable grids, and certain approximations to $\Psi(t)$, based on these grids, are introduced. Following Nicholson and Olmstead [86], the most frequently used are two such methods. In the first of these methods, often called the step function method, the unknown function $\Psi(t)$ is approximated by a piecewise constant spline defined on the discrete grid (see Fig. 12.1a). In the second method, called the Huber method after its inventor [61], a piecewise linear spline is used (see Fig. 12.1b), which results in a better accuracy.

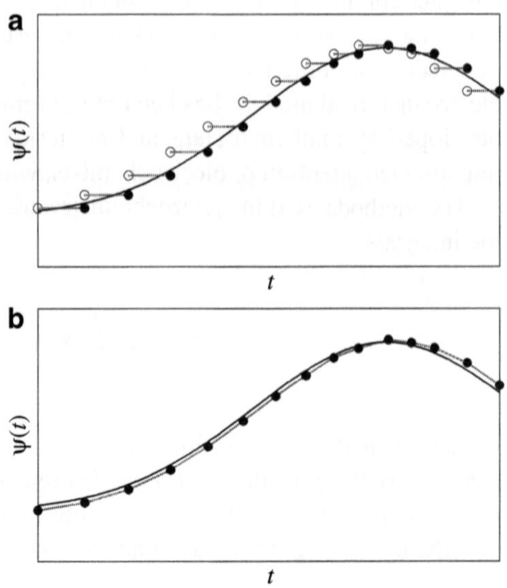

Fig. 12.1 Illustration of the concept of the step function method (**a**) and the Huber method (**b**). The exact solution $\Psi(t)$ (*solid lines*) is replaced by low-order spline functions (*dotted lines*), resulting in approximate solutions (*black circles*)

Product integration methods employing higher-order polynomial approximations to $\Psi(t)$ have also been used, but sporadically, so that they will not be discussed further. One of such methods is the Wagner method [107], which uses local parabolas in such a way that any two neighbouring local parabolas coincide at two consecutive nodes of the t-grid. A similar, but different method was developed and advocated by Olmstead and Nicholson [95]. Their method also uses local parabolas or higher-order polynomials, but any two neighbouring polynomials coincide at one grid node only.

In the product integration methods one obtains approximations to integrals (12.1) as linear combinations of the nodal $\Psi(t)$ values, with coefficients dependent on certain integrals (or moment integrals, see Sect. 12.1.1.1) of the kernel functions. It is most desirable to calculate these integrals analytically, or by means of highly accurate numerical approximations, such as truncated series expansions or polynomial approximants. However, less accurate numerical integration by means of quadratures is also acceptable, provided that the discretisation of the kernels is done on appropriately denser grids than the grids on which functions $\Psi(t)$ are discretised [78, 80]. The latter possibility indicates that the quadrature methods and product integration methods with analytical moment integral calculation present two extremes between which there can be intermediate methods, often also classified as product integration methods. For example, slowly varying multiplicative parts of the kernel functions can be combined with the functions $\Psi(t)$, and discretised jointly on the same grid. The resulting approximate expressions for the integrands are integrated analytically. Methods of this kind were used, for example, in [11, 89, 90], but they give generally worse results than the product integration methods with the analytical calculation of the moment integrals, and we do not recommend them.

The step function and Huber methods were originally published assuming uniform grids of the t variable, and in this version they have been predominantly used in electrochemistry. Relevant specialised algorithms have been elaborated for many electrochemical models, and occasionally published in the electrochemical literature. Apart from the aforementioned description in the review paper by Nicholson and Olmstead [86], algorithm descriptions exceeding just a presentation of the final formulae were given in [40, 45, 64, 69, 78–80, 87, 88, 99] for the step function method, and in [3–6, 9, 42–44, 49, 50, 59, 63, 67, 96, 99] for the Huber method. Particularly detailed descriptions of the uniform grid algorithms were given by Mirčeski [78], who discussed the problem of the numerical integration of the moment integrals in the step function method, and by Lovrić [69], who described the step function method algorithms for models of redox reactions complicated by the reactant and product adsorption at the surface of dropping and stationary mercury electrodes. A method similar to the step function method, but expressed by a somewhat different summation formula was described by Mirkin, Nilov, and Nauryzbaev [83]. A general purpose code based on the uniform grid versions of the step function and Huber methods was also developed and built into the problem solving environment ELSIM for electrochemical simulations [11]. Mirčeski et al. [79] developed a MATHCAD-based code for calculating square wave voltammograms by the step function method. Honeychurch [59] described codes for the uniform grid Huber method, written in MATHEMATICA [77], and applied them to a few simple electroanalytical models. In addition to the literature listed above, there have been electroanalytical modelling studies, in which a numerical technique equivalent to the uniform grid variant of the Huber method has been used, without referring to Huber, or even to the IEs. The technique originates from the numerical convolution procedure published by Oldham [92] and derived from one of the numerical algorithms for differintegration [94]. The procedure was initially designed and used for experimental data analysis, but later it was applied to the theoretical modelling as well, by Rudolph [102], Mahon and Oldham [70–75], Myland and Oldham [84], and Mahon et al. [76]. Those authors tend to avoid the use of the terminology typical for the IEs, but rather emphasise the role of the convolution, and fractional calculus [94] (especially semi-integration), in the modelling process. The Reader should be aware that such a change of terminology, and the related change of notation, does not modify the mathematical character of the equations considered, which remain IEs. For example, the expression $\int_0^t [\pi(t-\tau)]^{-1/2}\Psi(\tau)\mathrm{d}\tau$ can be called semi-integral and denoted as $\mathrm{d}^{-1/2}\Psi(t)/\mathrm{d}t^{-1/2}$, or it can be called a convolution product and denoted by $(\pi t)^{-1/2} \star \Psi(t)$. However, none of these alterations changes the fact that a certain integral is in mind, with all consequences regarding its mathematical properties and plausible numerical treatments.

Despite their popularity, uniform grids are generally not optimal, because the solutions $\Psi(t)$ vary with t. For t values at which $\Psi(t)$ exhibits more intense variations, a denser grid is appropriate than for other t values, to achieve a particular accuracy. This problem was realised by Mirkin and Nilov [82], who developed a nonuniform grid variant of the Huber method, and demonstrated its advantages in terms of the computing time and computer memory usage. Another deficiency of the fixed, uniform grid algorithms listed above, is the lack of error information. Until recently, the only exception was the algorithm described by Andrieux et al. [3], where elements of error control were included, based on the comparison of the solutions obtained on different (but uniform) grids.

In view of the popularity, and advantages, of the step function and Huber methods, Sect. 12.1.1 is devoted to a thorough description of these methods. In contrast to the descriptions listed above, the algorithms are formulated in a manner ensuring a large generality. Attention is also paid to the problem of computing the initial solution values, which has been largely overlooked in the literature. Following Mirkin and Nilov [82], nonuniform grids are assumed. In addition, Sect. 12.1.2 briefly presents an adaptive variant of the Huber method, recently developed by the present author [15–32]. The adaptive Huber method allows one to automatically solve the IEs with an accuracy of choice. The user of the method almost does not have to pay attention to the appropriate selection of computational grids or other details of the numerical procedure. The automatic operation of the method is achieved by a built-in algorithm that dynamically selects nonuniform grid steps in response to the estimates of the solution errors. The adaptive Huber method follows the contemporary trends in scientific computing, as adaptive numerical methods have become standard in many areas of computational science. It is therefore important to bring them to the attention of electrochemists. Development of such methods is also an indispensable element of the long-term programme of the automation of modelling procedures of electroanalytical chemistry, undertaken by the present author [14]. The programme aims at the creation of computational electrochemistry as a distinct subdiscipline of electrochemistry, employing computer experiments as a basic operational method of studying electrochemical phenomena. In order to achieve this goal, advanced computational algorithms and problem-solving software, dedicated to electrochemistry, are necessary [47, 60].

The problem of the calculation of the moment integrals of the kernel functions, needed by the product integration methods, is addressed in Sect. 12.1.1.4.

The third approach to discretise integrals (12.1), originating from the theory of differintegration, and proposed by Oldham and Myland [93], is discussed in Sect. 12.1.3.

A yet different method of calculating the integrals (12.1) was developed by the present author [12, 13]. The method is similar to the degenerate kernel approximation, frequently used (outside electrochemistry) to solve Fredholm IEs [7]. The method is described in Sect. 12.1.4.

12.1.1 The Step Function and the Huber Methods

Most frequently, one faces the problem of solving the following system of second kind Volterra IEs:

$$\vec{F}\left(t, \vec{\Psi}(t), \vec{Y}(t)\right) = \vec{0}\,, \tag{12.2}$$

where $\vec{\Psi}(t) = [\Psi_1(t), \ldots, \Psi_{N_\Psi}(t)]^{\mathrm{T}}$ is an N_Ψ-dimensional vector of unknown functions of the independent variable t, $\vec{Y}(t) = [Y_1(t), \ldots, Y_{N_Y}(t)]^{\mathrm{T}}$ is an N_Y-dimensional vector of integrals, $\vec{F}(\cdot) = [F_1(\cdot), \ldots, F_{N_\Psi}(\cdot)]^{\mathrm{T}}$ is an N_Ψ-dimensional vector of known functions representing the individual IEs, and $\vec{0}$ is a zero vector. Functions $\vec{F}(\cdot)$ can be nonlinear with respect to their variables. The integrals take the usual form:

$$Y_i(t) = \int\limits_0^t \mathscr{K}_\kappa(t,\tau)\,\Psi_\mu(\tau)\,\mathrm{d}\tau \tag{12.3}$$

for $i = 1, \ldots, N_Y$, where $\mathscr{K}_\kappa(t,\tau)$ for $\kappa = \kappa(i) = 1, \ldots, N_K$ are integral transformation kernels taken out of a set of N_K different kernels, and $\Psi_\mu(\tau)$ are unknown functions associated with the integrals, with $\mu = \mu(i) = 1, \ldots, N_\Psi$. Numbers N_Ψ, N_Y, and N_K need not be equal; the only constraint is $N_K \leq N_Y$. Therefore, Eq. (12.2) encompasses integro-algebraic equation systems (or even pure AE systems) as special cases. However, for the discussion of the product integration methods, the occurrence of at least one integral in Eq. (12.2) is assumed.

12.1.1.1 Discretisation

Let us assume a discrete grid of arbitrary nodes t_n (for $n = 0, 1, \ldots$) along the t axis, with $t_0 = 0$. Hence, the local step sizes $h_n = t_n - t_{n-1}$ (for $n = 1, 2, \ldots$) can be nonuniform. Let $\vec{\Psi}(t_n) = \vec{\Psi}_n = [\Psi_{1,n}, \ldots, \Psi_{N_\Psi,n}]^{\mathrm{T}}$ denote the true (exact) solution values at the grid nodes, and $\vec{\psi}_n = [\psi_{1,n}, \ldots, \psi_{N_\Psi,n}]^{\mathrm{T}}$ denote approximate nodal solution values, resulting from the solution of the discretised Eq. (12.2). In a similar way, let $\vec{Y}(t_n) = \vec{Y}_n = [Y_{1,n}, \ldots, Y_{N_Y,n}]^{\mathrm{T}}$ denote exact integral values at the grid nodes, and $\vec{y}_n = [y_{1,n}, \ldots, y_{N_Y,n}]^{\mathrm{T}}$ denote approximate nodal integral values. Our goal is to devise a procedure for calculating $\vec{\psi}_n$.

In accordance with Fig. 12.1, in the step function method $\vec{\Psi}(t)$ is approximated by a zero-order spline, that is:

$$\vec{\Psi}(t) \approx \vec{\psi}_k \tag{12.4}$$

for $t \in (t_{k-1}, t_k]$ with $k = 1, 2, \ldots$. In the Huber method, in turn, the first-order spline is used, that is:

$$\vec{\Psi}(t) \approx \vec{\psi}_{k-1} + \frac{\vec{\psi}_k - \vec{\psi}_{k-1}}{h_k}(t - t_{k-1}) \,. \tag{12.5}$$

Consequently, the approximate integrals are given by

$$y_{i,n} = \sum_{k=1}^{n} \int_{t_{k-1}}^{t_k} \mathcal{K}_\kappa(t_n, \tau)\, \psi_{\mu,k}\, \mathrm{d}\tau \tag{12.6}$$

in the step function method, and by

$$y_{i,n} = \sum_{k=1}^{n} \int_{t_{k-1}}^{t_k} \mathcal{K}_\kappa(t_n, \tau) \left[\psi_{\mu,k-1} + \frac{\psi_{\mu,k} - \psi_{\mu,k-1}}{h_k}(\tau - t_{k-1})\right] \mathrm{d}\tau \tag{12.7}$$

in the Huber method. By introducing the coefficients:

$$Q_{\kappa,m,n,l,k} = \int_{t_l}^{t_k} \mathcal{K}_\kappa(t_n, \tau)\, \tau^m \, \mathrm{d}\tau \ , \tag{12.8}$$

and

$$R_{\kappa,n,l,k} = 0 \ , \tag{12.9}$$

$$S_{\kappa,n,l,k} = Q_{\kappa,0,n,l,k} \tag{12.10}$$

for the step function method, or

$$R_{\kappa,n,l,k} = \frac{t_k \, Q_{\kappa,0,n,l,k} - Q_{\kappa,1,n,l,k}}{t_k - t_l} \ , \tag{12.11}$$

$$S_{\kappa,n,l,k} = \frac{Q_{\kappa,1,n,l,k} - t_l \, Q_{\kappa,0,n,l,k}}{t_k - t_l} \tag{12.12}$$

for the Huber method, Eqs. (12.6) and (12.7) can be written in the form common for both methods:

$$y_{i,n} = \sum_{k=1}^{n} \left(R_{\kappa,n,k-1,k} \, \psi_{\mu,k-1} + S_{\kappa,n,k-1,k} \, \psi_{\mu,k} \right) \ . \tag{12.13}$$

As can be seen, the integrals are replaced by linear combinations of the nodal solution values, with linear coefficients that depend on the moment integrals $Q_{\kappa,m,n,l,k}$ of the kernel functions. Zero-order moment integrals (for $m = 0$), i.e. ordinary integrals of the kernel functions, are needed by the step function method, whereas the Huber method requires zero- and first-order moment integrals. Higher-order moment integrals would be needed by product integration methods using higher-order polynomials or splines. They are also needed for error estimation (see Sect. 12.1.2). The availability of exact expressions or highly accurate approximations to $Q_{\kappa,m,n,l,k}$ is therefore necessary for the proper use of all these methods. This issue will be discussed in Sect. 12.1.1.4. For the moment, let us assume that the moment integrals are computable. It is convenient to introduce matrices $\overline{R}_{n,l,k}$ and $\overline{S}_{n,l,k}$, each of dimensions $N_Y \times N_\Psi$, such that there is only one nonzero element in every ith row of these matrices, equal to the coefficient $R_{\kappa,n,l,k}$ or $S_{\kappa,n,l,k}$, respectively, corresponding to the κth kernel associated with the ith integral. By using these matrices, Eq. (12.13) can be rewritten in vector notation:

$$\vec{y}_n = \sum_{k=1}^{n} \left(\overline{R}_{n,k-1,k} \, \vec{\psi}_{k-1} + \overline{S}_{n,k-1,k} \, \vec{\psi}_k \right) \ . \tag{12.14}$$

The solution of such discretised Eq. (12.2) proceeds step-wise: we determine the unknown vectors $\vec{\psi}_n$ at successive nodes t_n, all previous vectors $\vec{\psi}_k$ for $k = 0, 1, \ldots, n-1$ already having been determined. In Sect. 12.1.1.2 we describe the computation of $\vec{\psi}_n$ at such internal nodes. In Sect. 12.1.1.3 we discuss the problem of determining the initial solutions.

12.1.1.2 Solution at Internal Grid Nodes

At any internal node t_n, an approximate integral vector $\vec{y}_n$ given by Eq. (12.14) can be viewed as a function of only one unknown $\vec{\psi}_n$ (vector): $\vec{y}_n = \vec{y}_n(\vec{\psi}_n)$, and according to Eq. (12.14) this is a linear function. In order to determine $\vec{\psi}_n$, we substitute $\vec{\psi}_n$ and $\vec{y}_n$ into Eq. (12.2), in place of the exact solutions $\vec{\Psi}_n$ and exact integrals $\vec{Y}_n$ at the grid nodes. This gives

$$\vec{F}\left(t_n, \vec{\psi}_n, \vec{y}_n(\vec{\psi}_n)\right) = \vec{0}\,. \tag{12.15}$$

Equation (12.15) is a system of AEs for the unknowns $\vec{\psi}_n$, which can be solved by standard numerical algorithms for solving AEs. If the AEs are linear, the usual LU decomposition (see, for example, Press et al. [98]) is sufficient. If they are nonlinear, iterative methods such as the multidimensional Newton method (see, for example, Press et al. [98]) can be applied. In the Newton method, having a starting approximation for $\vec{\psi}_n$, one calculates a correction vector $\vec{\varsigma}_n$, by solving the linear system:

$$\overline{J}_n \vec{\varsigma}_n = -\vec{F}\left(t_n, \vec{\psi}_n, \vec{y}_n(\vec{\psi}_n)\right)\,, \tag{12.16}$$

where

$$\overline{J}_n = \overline{F}_\psi\left(t_n, \vec{\psi}_n, \vec{y}_n(\vec{\psi}_n)\right) + \overline{F}_Y\left(t_n, \vec{\psi}_n, \vec{y}_n(\vec{\psi}_n)\right)\overline{S}_{n,n-1,n} \tag{12.17}$$

is the Jacobian matrix for Eq. (12.15), with $\overline{F}_\psi(\cdot)$ and $\overline{F}_Y(\cdot)$ denoting matrices of partial derivatives of $\vec{F}(\cdot)$ with respect to the elements of $\vec{\Psi}(t)$ and $\vec{Y}(t)$, respectively. The correction vector $\vec{\varsigma}_n$ is added to $\vec{\psi}_n$, resulting in an improved approximation (if the method converges). The procedure is repeated iteratively until a satisfactory accuracy is obtained. As a starting approximation for $\vec{\psi}_n$ one can take $\vec{\psi}_{n-1}$.

12.1.1.3 Initial Solutions

The algorithm of Sect. 12.1.1.2 has to be completed with a procedure of determining $\vec{\psi}_0$. It is important to realise that, in contrast to ODEs, IEs do not require an independent specification of initial conditions. The solutions for $t = 0$ result from the IEs themselves. In the case of electrochemical IEs the determination of the initial solution values is a subtle and difficult matter, because of the frequent weak singularity of the kernels. The presence of equilibrium electrochemical reactions at the interface studied, in the case of spatially one-dimensional models, often leads to solutions singular at $t = 0$, meaning that for some $\Psi_i(t)$ there can be $\lim_{t\to 0+} \Psi_i(t) = \pm\infty$ [see, for example, Eqs. (11.23), (11.24), (11.27), and (11.28) in Sect. 11.1.1]. In such cases the application of the step function and Huber methods is actually disallowed, because the methods assume finite solutions when $t \to 0^+$ (see Fig. 12.1). Similar singularities may also occur at $t > 0$, if there are discontinuities in the coefficients of the IEs at $t > 0$. Such discontinuities typically arise when multiple potential step chronoamperometric transients are simulated. They can usually be transformed into initial singularities, by an appropriate change of variables and reformulation of the IEs.

Despite their limitation to finite initial solutions, the step function and Huber methods were often used when $\lim_{t\to 0+} \Psi_i(t) = \pm\infty$, by incorrectly assuming finite $\psi_{i,0}$. The fact that acceptable solutions were obtained results from the apparently fast error damping for typical kernels, such as $\mathscr{K}_\kappa(t,\tau) = (t-\tau)^{-1/2}$, which leads to large errors present at the few initial nodes not propagating further to subsequent nodes. Nevertheless, when the solution is singular at $t = 0$, a rigorous application of the step function and Huber methods requires a modification of the IEs, aimed at the removal of the singularity. This is usually a difficult problem, and very little has been done to develop suitable techniques for the elimination of the singularities. In the case of a simple first kind Volterra IE with a weakly singular kernel and singular solution $\Psi(t)$:

$$\int_0^t \mathscr{K}(t,\tau)\,\Psi(\tau)\,\mathrm{d}\tau = f(t)\,, \tag{12.18}$$

where $f(t)$ is a known function such that $f(0) \neq 0$, one can substitute $\Psi(t) = \Psi_\mathrm{s}(t) + \Psi_\mathrm{ns}(t)$, where $\Psi_\mathrm{s}(t)$ is a singular component, and $\Psi_\mathrm{ns}(t)$ is a nonsingular component. The singular component can be assumed to satisfy the IE:

$$\int_0^t \mathscr{K}(t,\tau)\,\Psi_\mathrm{s}(\tau)\,\mathrm{d}\tau = f(0)\,. \tag{12.19}$$

Equation (12.19) may be solvable analytically, by the methods outlined in Chap. 11. The nonsingular component then satisfies the IE:

$$\int_0^t \mathscr{K}(t,\tau)\,\Psi_{\rm ns}(\tau)\,\mathrm{d}\tau = f(t) - f(0)\,, \tag{12.20}$$

to which the product integration methods can be rigorously applied. Another option is to introduce some change of variables, so that the singular solution is transformed into a nonsingular one (see, for example, Amatore et al. [2]).

For the further discussion we assume that the initial singularities are removed, and $\vec{\Psi}(0^+)$ is finite in Eq. (12.2). In the step function method there is actually no need to determine $\vec{\psi}_0$, because this vector is not used by the algorithm [note that matrix $\overline{R}_{1,0,1}$ is zero in Eq. (12.14)]. Therefore, Eqs. (12.15)–(12.17) can be used to obtain solutions $\vec{\psi}_n$ for any $n \geq 1$. But, there is also no known general way to determine $\vec{\psi}_0$, which can be a problem, if the initial values are of interest. Things are more complicated in the Huber method, because $\overline{R}_{1,0,1}$ is not zero, so that $\vec{\psi}_0$ always occurs in the sum (12.14). Obviously, $\vec{\psi}_0$ is needed to determine $\vec{\psi}_1$, but how to determine $\vec{\psi}_0$? The simplest possible method is to always assume $\vec{\psi}_0 = 0$. This was done by some authors (see, for example, [4, 11, 42, 44, 99]), and the rationale is that prior to the transient experiments the electrochemical systems are usually in equilibrium (cf. Sect. 2.10). Consequently, all reaction rates and species production rates are zero [and $\Psi(t)$ usually represents these rates]. In this way, the discontinuity of $\Psi(t)$, theoretically present at $t_0 = 0$, is assumed to occur somewhere between t_0 and t_1. Of course, this assumption again disagrees with the expectation of regularity of the solution, required by the Huber method, and expressed by Fig. 12.1b, which may lead to large errors at the initial grid nodes. However, it should be noted that starting with $\vec{\psi}_0 = 0$ is consistent with the original proposal of Huber [61], who observed that solutions at the initial grid nodes (after t_0) can often be extrapolated back to t_0, giving improved initial value(s), if one begins with $\vec{\psi}_0 = 0$. But, in general, taking arbitrarily $\vec{\psi}_0 = 0$ is not rigorous, because $\Psi(0^-)$ is generally different from $\Psi(0^+)$, and by the initial solution we should understand here the value equal to $\Psi(0^+)$, not $\Psi(0^-)$. Therefore, a better method is to assume $\vec{\psi}_0 = \Psi(0^+)$, provided that $\Psi(0^+)$ is known or can be deduced from the IEs. This was done by a number of authors (see, for example, [5, 6, 63]). For example, a controlled potential experiment for an irreversible electron transfer reaction $X_1 + ne^- \rightarrow$ products is described by the second kind Volterra IE (5.21), which we rewrite in the form:

$$k^0\left[c_1^\star - \int_0^t \mathscr{K}_1(t,\tau)\,\Psi(\tau)\,\mathrm{d}\tau\right]\exp\left\{-\alpha^{\rm f}\frac{nF}{RT}\left[E(t)-E^0\right]\right\} - \Psi(t) = 0\,, \tag{12.21}$$

where $\Psi(t)$ is defined by Eq. (5.18), and $\mathscr{K}_1(t,\tau)$ is a suitable kernel, depending on the geometry and mode of transport. As the reaction has a finite rate, $\Psi(0^+)$ must be finite. Therefore, at $t = 0$ the integral vanishes, and we obtain from Eq. (12.21):

$$\Psi(0^+) = k^0\, c_1^\star\, \exp\left\{-\alpha^{\mathrm{f}}\frac{nF}{RT}\left[E(0^+) - E^0\right]\right\} \tag{12.22}$$

(the number n of electrons transferred, occurring in Eqs. (12.21) and (12.22) should not be confused with the index n of discrete time nodes, present in other equations in this chapter). Unfortunately, $\Psi(0^+)$ may not be equally easy to deduce in more general situations. Consequently, a desirable approach is to determine $\vec{\psi}_0$ numerically from the IEs. According to the proposal of the present author [16,19,21] this can be done in the following way.

In contrast to Eq. (12.15) for internal grid nodes, a discrete form of Eq. (12.2) (for the Huber method) at $t = t_1$ depends on two unknowns $\vec{\psi}_0$ and $\vec{\psi}_1$, so that it can be written:

$$\vec{F}\left(t_1, \vec{\psi}_1, \vec{y}_1(\vec{\psi}_0, \vec{\psi}_1)\right) = \vec{0}\,. \tag{12.23}$$

In order to determine both unknowns, an additional AE system connecting $\vec{\psi}_0$ and $\vec{\psi}_1$ is necessary. Such an additional equation can be:

$$\vec{F}\left(t_{1/2}, \vec{\psi}_{1/2}(\vec{\psi}_0, \vec{\psi}_1), \vec{y}_{1/2}(\vec{\psi}_0, \vec{\psi}_1)\right) = \vec{0}\,, \tag{12.24}$$

where

$$t_{1/2} = \frac{t_0 + t_1}{2} = \frac{t_1}{2} = \frac{h_1}{2}\,, \tag{12.25}$$

$$\vec{\psi}_{1/2}(\vec{\psi}_0, \vec{\psi}_1) = \frac{\vec{\psi}_0 + \vec{\psi}_1}{2}\,, \tag{12.26}$$

$$\vec{y}_{1/2}(\vec{\psi}_0, \vec{\psi}_1) = \overline{R}_{1/2,0,1/2}\,\vec{\psi}_0 + \overline{S}_{1/2,0,1/2}\,\vec{\psi}_{1/2}(\vec{\psi}_0, \vec{\psi}_1)\,. \tag{12.27}$$

Matrices $\overline{R}_{1/2,0,1/2}$ and $\overline{S}_{1/2,0,1/2}$ are calculated from Eqs. (12.9)–(12.12), analogous to $\overline{R}_{1,0,1}$ and $\overline{S}_{1,0,1}$, but by taking $h_1/2$ in place of h_1. Equations (12.23) and (12.24) are solved jointly as a system of $2N_\psi$ AEs, analogous to Eq. (12.15). The relevant linear equation system for the Newton corrections, analogous to Eq. (12.16), can be written using the block matrix/vector notation as

$$\begin{bmatrix} \overline{J}_{0,0} & \overline{J}_{0,1} \\ \overline{J}_{1,0} & \overline{J}_{1,1} \end{bmatrix} \begin{bmatrix} \vec{\varsigma}_0 \\ \vec{\varsigma}_1 \end{bmatrix} = -\begin{bmatrix} \vec{F}\left(t_{1/2}, \vec{\psi}_{1/2}(\vec{\psi}_0, \vec{\psi}_1), \vec{y}_{1/2}(\vec{\psi}_0, \vec{\psi}_1)\right) \\ \vec{F}\left(t_1, \vec{\psi}_1, \vec{y}_1(\vec{\psi}_0, \vec{\psi}_1)\right) \end{bmatrix}, \tag{12.28}$$

where

$$\overline{J}_{0,0} = \frac{1}{2}\overline{F}_\psi\left(t_{1/2}, \vec{\psi}_{1/2}(\vec{\psi}_0, \vec{\psi}_1), \vec{y}_{1/2}(\vec{\psi}_0, \vec{\psi}_1)\right)$$
$$+\overline{F}_Y\left(t_{1/2}, \vec{\psi}_{1/2}(\vec{\psi}_0, \vec{\psi}_1), \vec{y}_{1/2}(\vec{\psi}_0, \vec{\psi}_1)\right)\left(\overline{R}_{1/2,0,1/2} + \frac{1}{2}\overline{S}_{1/2,0,1/2}\right), \tag{12.29}$$

$$\overline{J}_{0,1} = \frac{1}{2}\overline{F}_\psi\left(t_{1/2}, \vec{\psi}_{1/2}(\vec{\psi}_0, \vec{\psi}_1), \vec{y}_{1/2}(\vec{\psi}_0, \vec{\psi}_1)\right)$$
$$+\frac{1}{2}\overline{F}_Y\left(t_{1/2}, \vec{\psi}_{1/2}(\vec{\psi}_0, \vec{\psi}_1), \vec{y}_{1/2}(\vec{\psi}_0, \vec{\psi}_1)\right)\overline{S}_{1/2,0,1/2}, \tag{12.30}$$

$$\overline{J}_{1,0} = \overline{F}_Y\left(t_1, \vec{\psi}_1, \vec{y}_1(\vec{\psi}_0, \vec{\psi}_1)\right)\overline{R}_{1,0,1}, \tag{12.31}$$

$$\overline{J}_{1,1} = \overline{F}_\psi\left(t_1, \vec{\psi}_1, \vec{y}_1(\vec{\psi}_0, \vec{\psi}_1)\right) + \overline{F}_Y\left(t_1, \vec{\psi}_1, \vec{y}_1(\vec{\psi}_0, \vec{\psi}_1)\right)\overline{S}_{1,0,1}. \tag{12.32}$$

12.1.1.4 Calculation of the Moment Integrals of the Kernels

An important aspect of the algorithm described in Sects. 12.1.1.1–12.1.1.3 is the computation of moment integrals (12.8) of the kernel functions. Ideally, these moment integrals should be calculated with the best accessible accuracy on a computer, that is with machine accuracy. If this is not the case, and the moment integrals are calculated by some kind of quadratures, the product integration methods may give wrong results. To illustrate the problem, Fig. 12.2 presents numerical solutions of the IE:

$$\int_0^t \mathscr{K}(t,\tau)\,\Psi(\tau)\,\mathrm{d}\tau = f(t) - f(0), \tag{12.33}$$

with

$$\mathscr{K}(t,\tau) = \exp[-k(t-\tau)](t-\tau)^{-1/2} \tag{12.34}$$

and

$$f(t) = [1 + \exp(u - t)]^{-1}. \tag{12.35}$$

Equation (12.33) describes a linear potential sweep voltammetric experiment for the catalytic EC reaction scheme:

$$X_1 + e^- \rightleftarrows X_2, \tag{12.36}$$

$$X_2 \rightarrow X_1 \tag{12.37}$$

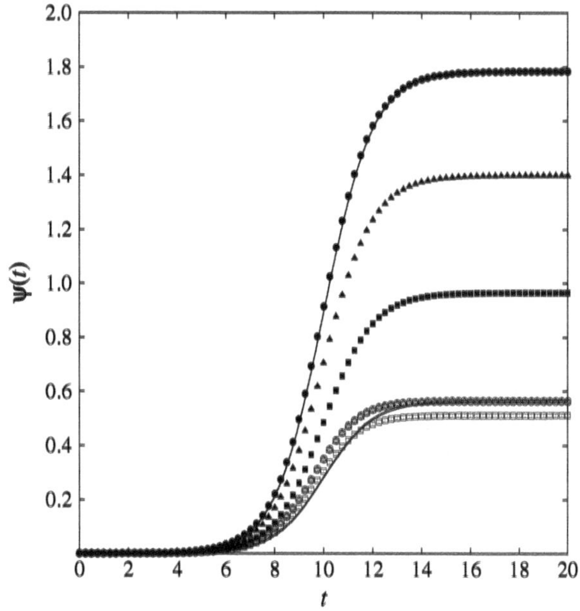

Fig. 12.2 Numerical solutions of Eq. (12.33) with $u = 10$ and $k = 1$ (*open symbols*) or $k = 10$ (*filled symbols*). Numerical methods: the Huber method with accurate moment integrals (*circles*); the Huber method with approximate moment integrals (*triangles*); the step function method with approximate moment integrals (*squares*). A uniform grid of 80 steps is assumed in every case. *Solid lines* denote steady state solutions (12.38)

involving an equilibrium electron transfer (12.36) and irreversible homogeneous reaction (12.37) between dynamic distributed species X_1 and X_2. Planar diffusion in a semi-infinite spatial domain, and equal diffusion coefficients are assumed. For derivations of Eq. (12.33) see [87, 103, 104] and the discussion of examples in Sect. 8.4. Function $\Psi(t)$ represents the dimensionless current function, defined in the standard way [87]. Parameter u is the dimensionless starting potential, and k is the dimensionless rate constant of reaction (12.37). The initial singularity is eliminated by subtracting $f(0)$, in the same manner as described in Sect. 12.1.1.3. When k is large, function $\Psi(t)$ approaches the steady state solution resulting from using Eq. (8.52):

$$\Psi(t) \approx (k/\pi)^{1/2}[f(t) - f(0)] . \tag{12.38}$$

Figure 12.2 compares solutions obtained using three methods: the Huber method with accurate moment integrals; the Huber method with approximate moment integrals; and the step function method with approximate moment integrals. The latter two methods are provided by the ELSIM program [11]. In ELSIM the step function and Huber discretisations are applied to the product $\exp[-k(t-\tau)]\Psi(\tau)$, rather than to $\Psi(\tau)$. This is advantageous from the point of view of the formal generality of the program, but such a discretisation means accurate moment integrals only when k is sufficiently small. For large k the moment integrals are inaccurate, unless one uses a very dense temporal grid. Figure 12.2 reveals that the Huber method with accurate moment integrals indeed gives the best results for large k. The other methods are then unreliable.

In view of the above arguments, accurate expressions for the various terms occurring in the kernel functions mentioned in this book, and for their moment integrals of order $m = 0$, 1 and 2, are of interest. Relatively easily obtainable exact analytical expressions exist only in the case of a few simple terms, such as

$$\mathscr{K}(t,\tau) = 1 , \tag{12.39}$$

$$\mathscr{K}(t,\tau) = \exp[-k(t-\tau)] , \tag{12.40}$$

$$\mathscr{K}(t,\tau) = (t-\tau)^{-1/2} , \tag{12.41}$$

$$\mathscr{K}(t,\tau) = (t-\tau)^{a-1} , \tag{12.42}$$

(where k and a are constants) occurring (for example) in kernels (4.7), (4.11), (5.72), and (5.178). The expressions were derived in [15, 19, 22, 25].

For the terms such as

$$\mathscr{K}(t,\tau) = \exp[-k(t-\tau)](t-\tau)^{-1/2} , \tag{12.43}$$

$$\mathscr{K}(t,\tau) = \exp[-a(t-\tau)]\,\mathrm{erex}[b(t-\tau)]^{1/2} , \tag{12.44}$$

$$\mathscr{K}(t,\tau) = \exp[-a(t-\tau)]\,\mathrm{daw}[b(t-\tau)]^{1/2} , \tag{12.45}$$

(where k, a, and b are constants) occurring in kernels (8.41), (5.74), (8.154), (8.156), and (9.20), analytical formulae for the moment integrals were derived in Bieniasz [15, 25]. The formulae involve error function erf(·) defined by Eq. (5.78) and/or related functions erfc(·) and erex(·) defined by Eqs. (5.76) and (5.77), and Dawson's integral daw(·) defined by Eq. (8.157). For a highly accurate calculation of functions erf(·), erfc(·) and erex(·), a number of computer codes are available, out of which the CALERF package of Cody [36], written in FORTRAN and freely available from netlib [85], is probably the best, providing results with machine accuracy for standard double precision variables (relative accuracy about 10^{-16} according to the IEEE 754 standard [51, 62]), or even better. For the daw(·) function an analogous FORTRAN subroutine of Cody et al. [37] is also available from netlib [85]. C++ translations of all these codes are obtainable from the present author upon request.

For the terms listed below, C++ procedures for calculating the terms and their moment integrals with an accuracy close to the machine accuracy for standard double precision variables were elaborated and published in the references indicated. In the references the Reader can also find further references to earlier known but less accurate approximations, if there were any. For the term

$$\mathscr{K}(t,\tau) = \mathrm{kcylw}[\varrho(t-\tau)^{1/2}] , \tag{12.46}$$

(where ϱ is a constant) occurring in kernel (5.85), see [23, 24]. For the term

$$\mathscr{K}(t,\tau) = \exp[-k(t-\tau)]\;\mathrm{kcylw}[\varrho(t-\tau)^{1/2}] , \tag{12.47}$$

(where ϱ and k are constants) see [27]. For the terms

$$\mathscr{K}(t,\tau) = \mathrm{kpli}[\mu(t-\tau)^{1/2}] \tag{12.48}$$

and

$$\mathscr{K}(t,\tau) = \mathrm{kplp}[\mu(t-\tau)^{1/2}]\,, \tag{12.49}$$

(where μ is a constant) occurring in kernels (5.120) and (5.121), see [28]. For the term

$$\mathscr{K}(t,\tau) = \mathrm{kins}[\varrho(t-\tau)^{1/2}]\,, \tag{12.50}$$

occurring in kernel (5.139), see [30]. For the term

$$\mathscr{K}(t,\tau) = \mathrm{kinc}[\varrho(t-\tau)^{1/2}]\,, \tag{12.51}$$

occurring in kernel (5.155), see [31]. For the term

$$\mathscr{K}(t,\tau) = \mathrm{kct}[\upsilon(t-\tau)^{1/2}]\,, \tag{12.52}$$

(where υ is a constant) occurring in kernel (6.70), see [29]. For the term

$$\mathscr{K}(t,\tau) = \tau^{2/3}\left(t^{7/3}-\tau^{7/3}\right)^{-1/2}\,, \tag{12.53}$$

occurring in kernel (6.31), see [32].

For the term

$$\mathscr{K}(t,\tau) = \mathrm{krd}[\upsilon(t-\tau)^{1/2}]\,, \tag{12.54}$$

occurring in kernel (6.47), an approximation for the term was reported by Seralathan and Rangarajan [105]. The approximation is probably more accurate than the replacement of function krd(·) by the function kct(·) (see the discussion in Sect. 6.2.3), but less accurate than machine accuracy.

Apart from accurately calculating the moment integrals of the kernel terms, one also needs accurate procedures for computing coefficients $R_{\kappa,n,n-1,n}$ and $S_{\kappa,n,n-1,n}$ defined by Eqs. (12.11) and (12.12), as well as some further similar coefficients occurring in the error estimators (see Sect. 12.1.2 below). The formulae for all these coefficients share the common property that they possess a $0/0$ limit when the step size h_n tends to zero. Consequently, they are very prone to machine errors when h_n is small. Therefore, the formulae such as Eqs. (12.11) and (12.12) cannot be directly used for numerically computing these coefficients. Instead, one should use series expansions into powers of h_n, or other well-conditioned representations of the formulae such as Eqs. (12.11) and (12.12). References [15–32] provide appropriate approximations for these coefficients.

12.1.2 The Adaptive Huber Method

An important question, accompanying the use of the step function and Huber methods, is how to construct the discrete grids in order to obtain solutions having a desired accuracy. In order to answer this question one needs two things. First, one needs to know how the true errors theoretically depend on the grid steps h_n. Second, one needs to have error estimators, that is special variables or expressions which tell approximately how big the errors are, in the situation when the true solutions $\vec{\Psi}_n$ are not known, but we only know the approximate solutions $\vec{\psi}_n$. By combining these two pieces of information it is possible to select the steps h_n in such a way that the estimated errors do not exceed a prescribed error tolerance. One hopes that the true errors are of comparable magnitude. Furthermore, such an adaptive grid selection can be automated by including a relevant decision algorithm into the numerical code, so that the method user does not have to pay attention to the details of the procedure.

12.1.2.1 Error Control

In principle, all solution errors are non-local, in the sense that the error at t_n depends not only on the step size h_n but also on all previous step sizes $h_1, \ldots, h_{n-1}$. However, any procedure of controlling the error contributions resulting from the previous step sizes would be very complicated and computationally expensive, because in order to reduce the error at t_n one would have to frequently modify the previous grid nodes and recalculate the solutions previously obtained at these nodes. Instead of controlling the global errors (errors involving contributions from all calculations at previous nodes) it is therefore easier to control local errors only, i.e. the errors resulting exclusively from the discretisation over the current interval $t \in [t_{n-1}, t_n]$. Such local errors present the major contribution to the (global) solution errors, in situations when the errors transferred from the past grid nodes are well damped while the calculations proceed to higher n. In practice, this error damping is usually satisfactory when the kernel functions $\mathscr{K}_\kappa(t, \tau)$ are non-decreasing functions of τ, at any fixed t. The past error damping is necessary for the numerical stability of the step function and Huber methods. The adaptive Huber method described in Bieniasz [15–32] is based on local error control. The essential formulae, enabling the grid adaptation by this method, are provided below.

Errors of $\vec{\psi}_n$ form a vector $\vec{\delta}_n$ such that

$$\vec{\psi}_n = \vec{\Psi}_n - \vec{\delta}_n \,. \tag{12.55}$$

According to the assumptions accepted, at $n > 1$ we determine local errors only. This means that for the purpose of the error control the discrete solutions and their integrals are considered to be exact for $t \in [0, t_{n-1}]$. The only source of the errors $\vec{\delta}_n$ are then the errors $\vec{\Delta}_n = [\Delta_{1,n}, \ldots, \Delta_{N_Y,n}]^{\mathrm{T}}$ of the Huber quadratures by which the integrals are discretised over the interval $[t_{n-1}, t_n]$. These quadrature errors are:

$$\Delta_{i,n} = \int_{t_{n-1}}^{t_n} \mathscr{K}_\kappa(t_n, \tau)\, \Psi_\mu(\tau)\, \mathrm{d}\tau$$

$$- \int_{t_{n-1}}^{t_n} \mathscr{K}_\kappa(t_n, \tau) \left[\Psi_{\mu,n-1} + \frac{\Psi_{\mu,n} - \Psi_{\mu,n-1}}{h_n} (\tau - t_{n-1}) \right] \mathrm{d}\tau \,, \tag{12.56}$$

or equivalently:

$$\Delta_{i,n} = \int_{t_{n-1}}^{t_n} \mathscr{K}_\kappa(t_n, \tau)\, \Psi_\mu(\tau)\, \mathrm{d}\tau - \left(R_{\kappa,n,n-1,n}\, \Psi_{\mu,n-1} + S_{\kappa,n,n-1,n}\, \Psi_{\mu,n} \right) . \tag{12.57}$$

We need to determine the connection between $\vec{\delta}_n$ and $\vec{\Delta}_n$. In view of Eq. (12.57) the exact integrals $Y_{i,n}$ are:

$$Y_{i,n} = \int_0^{t_{n-1}} \mathscr{K}_\kappa(t_n, \tau)\, \Psi_\mu(\tau)\, \mathrm{d}\tau + \left(R_{\kappa,n,n-1,n}\, \Psi_{\mu,n-1} + S_{\kappa,n,n-1,n}\, \Psi_{\mu,n} \right) + \Delta_{i,n} \,, \tag{12.58}$$

whereas the approximate integrals (in the local error model) are:

$$y_{i,n} = \int_0^{t_{n-1}} \mathscr{K}_\kappa(t_n, \tau)\, \Psi_\mu(\tau)\, \mathrm{d}\tau + \left(R_{\kappa,n,n-1,n}\, \Psi_{\mu,n-1} + S_{\kappa,n,n-1,n}\, \psi_{\mu,n} \right) . \tag{12.59}$$

By subtracting Eqs. (12.58) and (12.59) we obtain, in the vector notation:

$$\vec{y}_n = \vec{Y}_n - \overline{S}_{n,n-1,n}\, \vec{\delta}_n - \vec{\Delta}_n \,. \tag{12.60}$$

As the errors $\vec{\delta}_n$ and $\vec{\Delta}_n$ are expected ultimately to be rather small, the following linearisation of the functions $\vec{F}(\cdot)$ is acceptable:

$$\vec{F}\left(t_n, \vec{\psi}_n, \vec{y}_n\right) \approx \vec{F}\left(t_n, \vec{\Psi}_n, \vec{Y}_n\right) - \overline{F}_\Psi\left(t_n, \vec{\Psi}_n, \vec{Y}_n\right) \vec{\delta}_n$$

$$- \overline{F}_Y\left(t_n, \vec{\Psi}_n, \vec{Y}_n\right) \left(\vec{Y}_n - \vec{y}_n\right) , \tag{12.61}$$

where both $\vec{F}\left(t_n, \vec{\Psi}_n, \vec{Y}_n\right)$ and $\vec{F}\left(t_n, \vec{\psi}_n, \vec{y}_n\right)$ vanish, in accordance with Eqs. (12.2) and (12.15). By combining the remaining terms of Eq. (12.61) with Eq. (12.60) we obtain

$$\vec{0} \approx -\overline{F}_\Psi\left(t_n, \vec{\Psi}_n, \vec{Y}_n\right)\vec{\delta}_n - \overline{F}_Y\left(t_n, \vec{\Psi}_n, \vec{Y}_n\right)\left(\overline{S}_{n,n-1,n}\,\vec{\delta}_n + \vec{\Delta}_n\right) . \tag{12.62}$$

In practice the exact values $\vec{\Psi}_n$ and $\vec{Y}_n$ are not known, so that in Eq. (12.62) we have to replace them by the approximate values, which gives, after a rearrangement of Eq. (12.62):

$$\overline{J}_n\vec{\delta}_n \approx -\overline{F}_Y\left(t_n, \vec{\psi}_n, \vec{y}_n\right)\vec{\Delta}_n , \tag{12.63}$$

where $\overline{J}_n$ is given by Eq. (12.17). By solving the linear AE system (12.63) for $\vec{\delta}_n$, one obtains the local solution errors as functions of the quadrature errors $\vec{\Delta}_n$.

The calculation of the initial solution errors $\vec{\delta}_0$ and $\vec{\delta}_1$ proceeds in a similar way, but in the spirit of Eqs. (12.24) and (12.28) these solution errors have to be determined jointly. Actually, $\vec{\delta}_0$ and $\vec{\delta}_1$ have the meaning of global errors, because at t_0 and t_1 there are no previous errors to be neglected. The following relationships define the quadrature errors $\vec{\Delta}_1$ and $\vec{\Delta}_{1/2}$:

$$\vec{Y}_1 = \left(\overline{R}_{1,0,1}\,\vec{\Psi}_0 + \overline{S}_{1,0,1}\,\vec{\Psi}_1\right) + \vec{\Delta}_1 , \tag{12.64}$$

$$\vec{Y}_{1/2} = \left(\overline{R}_{1/2,0,1/2}\,\vec{\Psi}_0 + \overline{S}_{1/2,0,1/2}\,\frac{\vec{\Psi}_0 + \vec{\Psi}_1}{2}\right) + \vec{\Delta}_{1/2} . \tag{12.65}$$

Subtraction of the approximate integrals $\vec{y}_1$ and $\vec{y}_{1/2}$, defined by Eqs. (12.14) and (12.27), gives:

$$\vec{Y}_1 - \vec{y}_1 = \left(\overline{R}_{1,0,1}\,\vec{\delta}_0 + \overline{S}_{1,0,1}\,\vec{\delta}_1\right) + \vec{\Delta}_1 , \tag{12.66}$$

$$\vec{Y}_{1/2} - \vec{y}_{1/2} = \left(\overline{R}_{1/2,0,1/2}\,\vec{\delta}_0 + \overline{S}_{1/2,0,1/2}\,\frac{\vec{\delta}_0 + \vec{\delta}_1}{2}\right) + \vec{\Delta}_{1/2} . \tag{12.67}$$

In addition,

$$\vec{\Psi}_{1/2} = \vec{\Psi}\left(t_{1/2}\right) = \frac{\vec{\Psi}_0 + \vec{\Psi}_1}{2} + \vec{\sigma}_{1/2} , \tag{12.68}$$

where $\vec{\sigma}_{1/2} = \left[\sigma_{1,1/2}, \ldots, \sigma_{N_\Psi,1/2}\right]^{\mathrm{T}}$ is the vector of the errors of the linear interpolation of $\vec{\Psi}(t)$ at $t_{1/2}$. Hence, in view of Eq. (12.26):

$$\vec{\Psi}_{1/2} - \vec{\psi}_{1/2} = \frac{\vec{\delta}_0 + \vec{\delta}_1}{2} + \vec{\sigma}_{1/2}\,. \tag{12.69}$$

After setting Eqs. (12.66), (12.67), and (12.69) into functions $\vec{F}(\cdot)$, linearising and replacing exact solutions and their integrals by approximate ones, analogous to Eqs. (12.61)–(12.63), we obtain:

$$\begin{bmatrix} \overline{J}_{0,0} & \overline{J}_{0,1} \\ \overline{J}_{1,0} & \overline{J}_{1,1} \end{bmatrix} \begin{bmatrix} \vec{\delta}_0 \\ \vec{\delta}_1 \end{bmatrix}$$
$$\approx - \begin{bmatrix} \overline{F}_\Psi\left(t_{1/2}, \vec{\psi}_{1/2}, \vec{y}_{1/2}\right)\vec{\sigma}_{1/2} + \overline{F}_Y\left(t_{1/2}, \vec{\psi}_{1/2}, \vec{y}_{1/2}\right)\vec{\Delta}_{1/2} \\ \overline{F}_Y\left(t_1, \vec{\psi}_1, \vec{y}_1\right)\vec{\Delta}_1 \end{bmatrix}. \tag{12.70}$$

By solving Eq. (12.70) the solution errors $\vec{\delta}_0$ and $\vec{\delta}_1$ are obtained as functions of the quadrature errors $\vec{\Delta}_1$ and $\vec{\Delta}_{1/2}$, and interpolation errors $\vec{\sigma}_{1/2}$.

In practice, we do not know exactly the errors $\vec{\Delta}_n$, $\vec{\Delta}_{1/2}$, and $\vec{\sigma}_{1/2}$ in Eqs. (12.63) and (12.70), so that we have to replace them by suitable a posteriori error estimators $\mathrm{est}\left\langle\vec{\Delta}_n\right\rangle$, $\mathrm{est}\left\langle\vec{\Delta}_{1/2}\right\rangle$, and $\mathrm{est}\left\langle\vec{\sigma}_{1/2}\right\rangle$. Subsequently, by solving Eqs. (12.63) and (12.70) we obtain estimators $\mathrm{est}\left\langle\vec{\delta}_n\right\rangle$ of $\vec{\delta}_n$. Let us pass to the problem of obtaining the estimators $\mathrm{est}\left\langle\vec{\Delta}_n\right\rangle$, $\mathrm{est}\left\langle\vec{\Delta}_{1/2}\right\rangle$, and $\mathrm{est}\left\langle\vec{\sigma}_{1/2}\right\rangle$. By expanding $\vec{\Psi}(t)$ in a Taylor series (possibly truncated) around t_{n-1}, where $n > 1$, one obtains for $t \in (t_{n-1}, t_n]$:

$$\Psi_\mu(t) = \Psi_{\mu,n-1} + \left.\frac{\mathrm{d}\Psi_\mu(t)}{\mathrm{d}t}\right|_{t=t_{n-1}}(t - t_{n-1}) + \frac{1}{2}\left.\frac{\mathrm{d}^2\Psi_\mu(t)}{\mathrm{d}t^2}\right|_{t=t_{n-1}}(t - t_{n-1})^2$$
$$+\frac{1}{6}\left.\frac{\mathrm{d}^3\Psi_\mu(t)}{\mathrm{d}t^3}\right|_{t=t_{n-1}}(t - t_{n-1})^3 + \cdots\,. \tag{12.71}$$

Substitution of Eq. (12.71) into Eq. (12.57) and integration gives

$$\Delta_{i,n} = \frac{1}{2}\left.\frac{\mathrm{d}^2\Psi_\mu(t)}{\mathrm{d}t^2}\right|_{t=t_{n-1}}\left[(t_{n-1}^2 + h_n t_{n-1})Q_{\kappa,0,n,n-1,n}\right.$$
$$\left.-(2t_{n-1} + h_n)Q_{\kappa,1,n,n-1,n} + Q_{\kappa,2,n,n-1,n}\right]$$

$$+\frac{1}{6}\left.\frac{d^3\Psi_\mu(t)}{dt^3}\right|_{t=t_{n-1}} \left[(h_n^2 t_{n-1} - t_{n-1}^3) Q_{\kappa,0,n,n-1,n} + (3t_{n-1}^2 - h_n^2) Q_{\kappa,1,n,n-1,n} \right.$$
$$\left. -3t_{n-1} Q_{\kappa,2,n,n-1,n} + Q_{\kappa,3,n,n-1,n}\right] + \dots \qquad (12.72)$$

In order to obtain computable error estimators est $\langle \Delta_{i,n} \rangle$ we retain only the first expansion term in Eq. (12.72), and approximate the second solution derivative by the standard three-point backward difference approximation for nonuniform grids (see, for example, Roos et al. [101, p. 84]). The following estimator is obtained:

$$\text{est}\,\langle \Delta_{i,n} \rangle = \left[\frac{2}{h_{n-1}(h_{n-1}+h_n)} \Psi_{\mu,n-2} - \frac{2}{h_{n-1}h_n} \Psi_{\mu,n-1} + \frac{2}{h_n(h_{n-1}+h_n)} \Psi_{\mu,n}\right]$$
$$\times \frac{1}{2}\left[(t_{n-1}^2 + h_n t_{n-1}) Q_{\kappa,0,n,n-1,n} - (2t_{n-1} + h_n) Q_{\kappa,1,n,n-1,n} + Q_{\kappa,2,n,n-1,n}\right] . \qquad (12.73)$$

As can be seen, in addition to the zero- and first-order moment integrals $Q_{\kappa,0,n,l,k}$ and $Q_{\kappa,1,n,l,k}$, already shown to be necessary for computing the solutions by the Huber method, the error estimation additionally requires some values of the second-order moment integrals $Q_{\kappa,2,n,l,k}$.

Somewhat different formulae have to be used to determine est $\langle \Delta_{i,1} \rangle$ and est $\langle \Delta_{i,1/2} \rangle$. In this case it is not a good idea to expand the solution around $t_0 = 0$, because the solution may not possess the required derivatives at this point. We therefore expand around t_1, and use the expansion for $t \in [0, t_1)$:

$$\Psi_\mu(t) = \Psi_{\mu,1} + \left.\frac{d\Psi_\mu(t)}{dt}\right|_{t=t_1} (t - t_1) + \frac{1}{2}\left.\frac{d^2\Psi_\mu(t)}{dt^2}\right|_{t=t_1} (t-t_1)^2$$
$$+\frac{1}{6}\left.\frac{d^3\Psi_\mu(t)}{dt^3}\right|_{t=t_1} (t-t_1)^3 + \dots . \qquad (12.74)$$

Substitution of Eq. (12.74) into Eq. (12.57), assuming integration limits $t_n = t_1$, $t_{n-1} = 0$, and integration give

$$\Delta_{i,1} = \frac{1}{2}\left.\frac{d^2\Psi_\mu(t)}{dt^2}\right|_{t=t_1} \left[Q_{\kappa,2,1,0,1} - h_1 Q_{\kappa,1,1,0,1}\right]$$
$$+\frac{1}{6}\left.\frac{d^3\Psi_\mu(t)}{dt^3}\right|_{t=t_1} \left[Q_{\kappa,3,1,0,1} - 3h_1 Q_{\kappa,2,1,0,1} + 2h_1^2 Q_{\kappa,1,1,0,1}\right]$$
$$+ \dots , \qquad (12.75)$$

whereas substitution of Eq. (12.74) into Eq. (12.57), assuming integration limits $t_n = t_{1/2}$, $t_{n-1} = 0$, and integration give

$$\Delta_{l,1/2} = \frac{1}{2}\left.\frac{d^2\Psi_\mu(t)}{dt^2}\right|_{t=t_1}\left[\mathcal{Q}_{\kappa,2,1/2,0,1/2} - h_1\mathcal{Q}_{\kappa,1,1/2,0,1/2}\right]$$
$$+\frac{1}{6}\left.\frac{d^3\Psi_\mu(t)}{dt^3}\right|_{t=t_1}\left[\mathcal{Q}_{\kappa,3,1/2,0,1/2} - 3h_1\mathcal{Q}_{\kappa,2,1/2,0,1/2} + 2h_1^2\mathcal{Q}_{\kappa,1,1/2,0,1/2}\right]$$
$$+\ldots. \tag{12.76}$$

Consequently, the computable error estimators are:

$$\text{est}\,\langle\Delta_{l,1}\rangle = \frac{1}{2}\frac{\psi_{\mu,0} - 2\psi_{\mu,1} + \tilde{\psi}_{\mu,2}}{h_1^2}\left[\mathcal{Q}_{\kappa,2,1,0,1} - h_1\mathcal{Q}_{\kappa,1,1,0,1}\right] \tag{12.77}$$

and

$$\text{est}\left\langle\Delta_{l,1/2}\right\rangle = \frac{1}{2}\frac{\psi_{\mu,0} - 2\psi_{\mu,1} + \tilde{\psi}_{\mu,2}}{h_1^2}\left[\mathcal{Q}_{\kappa,2,1/2,0,1/2} - h_1\mathcal{Q}_{\kappa,1,1/2,0,1/2}\right], \tag{12.78}$$

where $\tilde{\psi}_{\mu,2}$ is a provisional solution at $t = t_2$, calculated assuming $h_2 = h_1$. The provisional solution serves only for obtaining the estimate of the second derivative, and is discarded after the solutions $\psi_{\mu,0}$ and $\psi_{\mu,1}$ are obtained with a desired accuracy.

It remains to determine the estimator of the interpolation error. By expanding $\Psi_{\mu,0}$ and $\Psi_{\mu,1/2}$ in a Taylor series around t_1, and setting the expansions into Eq. (12.68) one obtains

$$\sigma_{\mu,1/2} = -\frac{1}{8}\left.\frac{d^2\Psi_\mu(t)}{dt^2}\right|_{t=t_1} h_1^2 + \frac{3}{48}\left.\frac{d^3\Psi_\mu(t)}{dt^3}\right|_{t=t_1} h_1^3 + \cdots, \tag{12.79}$$

so that for the computable error estimator we take

$$\text{est}\left\langle\sigma_{\mu,1/2}\right\rangle = -\frac{1}{8}\left(\psi_{\mu,0} - 2\psi_{\mu,1} + \tilde{\psi}_{\mu,2}\right). \tag{12.80}$$

The estimates $\text{est}\left\langle\overrightarrow{\delta}_n\right\rangle$ of the solution errors can be used for the error control and adaptive step selection in the following way. At every grid node one first calculates the solutions assuming a certain tentative value of the grid step h_n. At $n = 1$ an arbitrary step size h_{start} is initially attempted, and at $n > 1$ the control-theoretic predictive algorithm of Gustafsson [55] is used to predict the most adequate h_n value, based on the recently obtained error estimates. The algorithm is described in detail in [55], so that we do not present it here. In agreement with the findings of Sect. 12.1.2.2 below, the algorithm is supplemented with the information that errors are proportional to h_n^2. After obtaining the solutions, error estimates are calculated, and one checks whether the accuracy conditions are satisfied:

$$\max\left\{\left\|\mathrm{est}\left\langle\vec{\delta}_0\right\rangle\right\|, \left\|\mathrm{est}\left\langle\vec{\delta}_1\right\rangle\right\|\right\} \le tol/\theta\ , \tag{12.81}$$

$$\left\|\mathrm{est}\left\langle\vec{\delta}_n\right\rangle\right\| \le tol \quad \text{for } n > 1\ . \tag{12.82}$$

In inequalities (12.81) and (12.82) $\|\cdot\|$ is a suitable vector norm. The infinity norm is adequate in cases when the various elements of $\vec{\Psi}(t)$ are comparable in magnitude. If this is not the case, alternative norms, involving individual scaling of the various elements of $\mathrm{est}\left\langle\vec{\delta}_n\right\rangle$, may be necessary. Parameter "*tol*" is a pre-defined absolute error tolerance, that is the target absolute accuracy one wants to achieve, and θ is a heuristic safety factor. The safety factor is introduced to effectively diminish the error tolerance at $n = 1$, where the error estimates may be somewhat unreliable in cases when the solution is not differentiable at $t = 0$. Values of θ between 10 and 100 are usually appropriate. Inequalities (12.81) and (12.82) imply absolute error control; it is not recommended to control relative errors, because this usually leads to a considerable inefficiency of the adaptive Huber method, when a significant effort is spent on resolving uninteresting parts of the solutions, with values close to zero. If the conditions (12.81) and (12.82) are satisfied, the solution obtained is accepted as sufficiently accurate. The adaptive Huber method then proceeds with analogous calculations for the next grid node, using a new step h_{n+1} predicted by the Gustafsson algorithm [55]. In the opposite case, if the accuracy condition (12.81) or (12.82) is not satisfied, the solution obtained at t_n is discarded, and the predictive algorithm of Gustafsson yields a smaller tentative value of the step h_n to be tried. The calculations of the solutions and error estimates are then repeated, possibly several times, until criteria (12.81) and (12.82) are met. An upper limit $h_{\max}$ on the step sizes is also imposed to avoid a situation where too few grid nodes are selected within the entire t interval of interest.

12.1.2.2 Performance

The performance of the adaptive Huber method is best tested on IEs that possess easily and accurately computable analytical solutions. This allows a reliable determination of the true errors of the numerical solutions. The following two example IEs satisfy these requirements:

$$\int_0^t (t-\tau)^{-1/2}\,\Psi(\tau)\,\mathrm{d}\tau = 1 - \exp(-t)\ , \tag{12.83}$$

$$1-\int_{0}^{t}\left\{[\pi(t-\tau)]^{-1/2}-\varrho\,\mathrm{erex}\left[\varrho(t-\tau)^{1/2}\right]\right\}\Psi(\tau)\,\mathrm{d}\tau-k^{-1}\Psi(t)=0\,. \tag{12.84}$$

Equation (12.83) is a linear first kind Volterra IE (or the Abel IE) for the dimensionless linear potential sweep voltammetric current function corresponding to the electron transfer

$$\mathrm{X}_1+n\,\mathrm{e}^-\rightleftarrows\mathrm{X}_{2,\mathrm{s}}\,, \tag{12.85}$$

where X_1 is a dynamic distributed species and $X_{2,s}$ is a static localised species (possibly electrode metal). Planar diffusion in a semi-infinite spatial domain is assumed. The analytical solution of this example was obtained by Berzins and Delahay [10], without using IEs. The solution is:

$$\Psi(t)=\frac{2}{\pi}\mathrm{daw}(t^{1/2})\,, \tag{12.86}$$

where daw(·) is the Dawson integral defined by Eq. (8.157). Equation (12.83) was later derived by White and Lawson [111]. Equation (12.84) is a linear second kind Volterra IE (5.21) for the dimensionless potential step chronoamperometric current corresponding to the irreversible electron transfer

$$\mathrm{X}_1+n\,\mathrm{e}^-\rightarrow\mathrm{X}_2\,, \tag{12.87}$$

where X_1 is a dynamic distributed species diffusing to a spherical electrode in a semi-infinite spatial domain. Function erex(·) is defined by Eq. (5.76). Parameter k is a dimensionless rate constant of reaction (12.87) at the applied potential, and parameter ϱ characterises the electrode sphericity [cf. Eqs. (5.74) and (5.75) in Sect. 5.2.1]. The analytical solution of this example was obtained by Shain et al. [106] without using IEs. The solution is:

$$\Psi(t)=k\,\frac{\varrho+k\,\mathrm{erex}\left[(\varrho+k)\,t^{1/2}\right]}{\varrho+k}\,. \tag{12.88}$$

Essential performance characteristics of the adaptive Huber method, obtained by solving Eqs. (12.83) and (12.84), are depicted in Figs. 12.3, 12.4, and 12.5. These characteristics are representative of the large collection of similar results obtained in Bieniasz [15–32] for a variety of electroanalytical IEs.

Figure 12.3 reveals that the true errors obtained for Eq. (12.83) are close to the error tolerance "*tol*". This result is representative of first kind Volterra IEs, and it demonstrates that the error control is successful. In the case of second kind Volterra IEs, exemplified by Eq. (12.84), the true errors tend to be somewhat larger than the tolerance "*tol*", and this discrepancy increases with decreasing "*tol*".

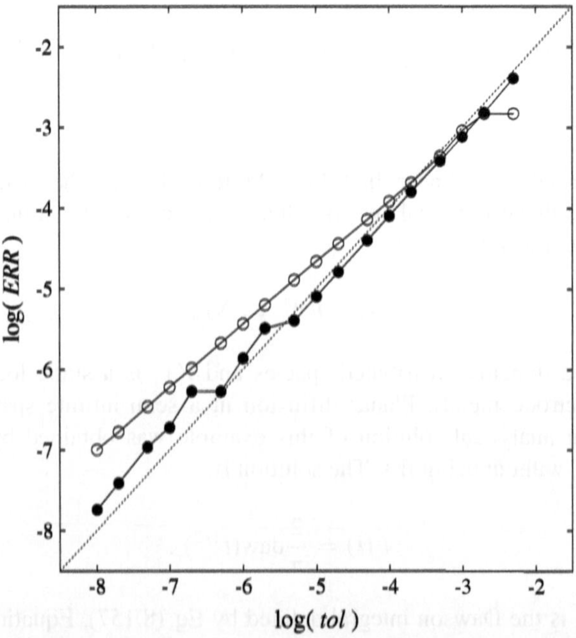

Fig. 12.3 The dependence on the error tolerance parameter "*tol*", of the largest modulus *ERR* of the true absolute error obtained [The calculations were performed on a personal computer having an Intel Pentium D processor operating at 3 GHz, and a 2 GB operational memory. The numerical code was written in C++ using extended precision (*long double* variables having 80 bits or 18 digit precision), and compiled as a 32 bit application using the MinGW compiler. The code was run under MS Windows XP Professional.] by the adaptive Huber method for Eq. (12.83) with $t \in [0, 2]$ and method parameters $h_{\text{start}} = 0.1$, $h_{\max} = 0.1$, $\theta = 10$ (*solid line with filled circles*), and for Eq. (12.84) with $k = 1$, $\varrho = 1$, $t \in [0, 1]$, and method parameters $h_{\text{start}} = 10^{-5}$, $h_{\max} = 0.1$, $\theta = 10$ (*solid line with open circles*). The *dotted line* is the plot of the function $ERR = tol$

However, it is clear that by selecting an appropriate "*tol*" value one is able to modify and ensure a desired accuracy of the numerical solution. The range of practically achievable errors is usually between 10^{-2} and 10^{-8} of the maximum solution value. Smaller errors are difficult to obtain, due to the increasing computational cost and interference of machine errors. Figure 12.4 shows that in the Huber method the true errors are approximately proportional to N^{-2}, where N is the total number of integration steps in the time interval where the solution is sought. Since the average time step is inversely proportional to N, this implies that the practical accuracy order is close to two. Finally, Fig. 12.5 reveals that the true errors are inversely proportional to the computational time *ct*. By combining this result with the proportionality of the errors to N^{-2}, we obtain that the computational time grows as N^2. The latter result holds true for kernels dependent on two variables (t and τ). For IEs involving only the constant kernel $\mathscr{K}(t, \tau) = 1$ one can obtain a proportionality of the *ct* to N, because there is no need to recalculate the integrals at past time grid nodes, when the solution at a new node is to be derived.

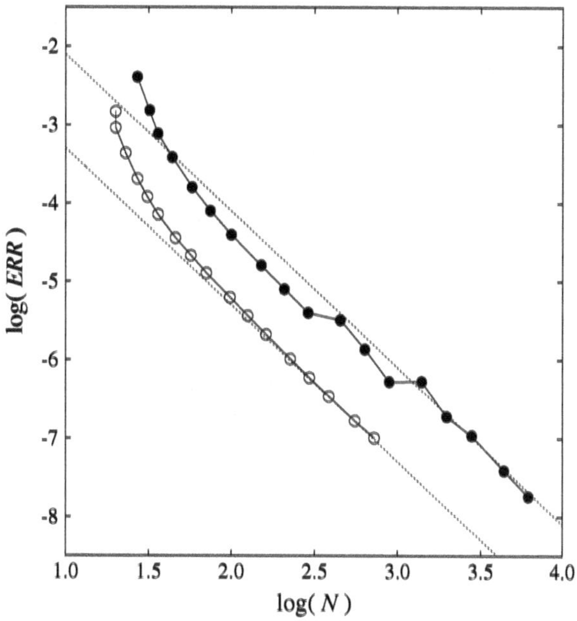

Fig. 12.4 The dependence on the number of integration steps N, of the largest modulus *ERR* of the true absolute error obtained by the adaptive Huber method for Eqs. (12.83) and (12.84). Parameters and other details are as in Fig. 12.3. The *dotted lines* are plots of functions $\log(ERR) = -0.1 - 2\log(N)$ and $\log(ERR) = -1.3 - 2\log(N)$

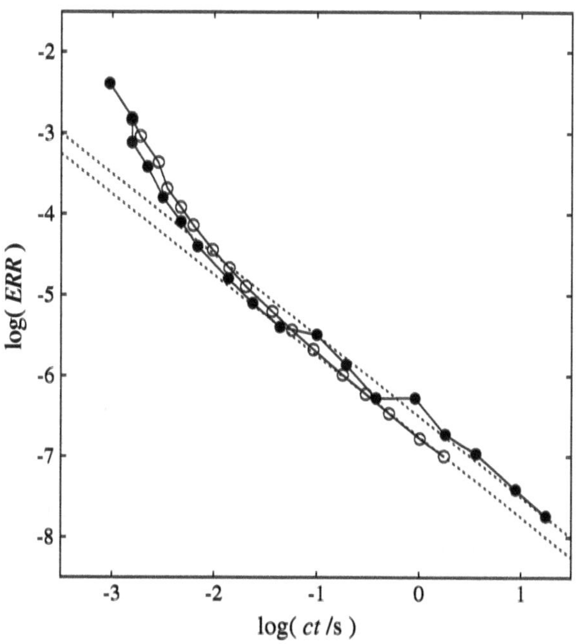

Fig. 12.5 The dependence on the computational time ct, of the largest modulus *ERR* of the true absolute error obtained by the adaptive Huber method for Eqs. (12.83) and (12.84). Parameters and other details are as in Fig. 12.3. The *dotted lines* are plots of functions $\log(ERR) = -6.5 - \log(ct)$ and $\log(ERR) = -6.75 - \log(ct)$. Computational time values are averaged over 50 program runs

12.1.3 Methods Based on Discrete Differintegration

At the beginning of Sect. 12.1 we have mentioned the connections between electroanalytical IEs and the fractional calculus. Such connections have been explored for many years, mostly by Oldham and co-workers. In particular, in a paper by Oldham and Myland [93], a numerical technique for solving IEs was suggested, which originated from the Grünwald definition of differintegration [53,91,94]. The technique seems to differ conceptually from the methods based on quadratures, and discussed in Sects. 12.1.1 and 12.1.2, although from the algorithmic point of view it also consists in replacing the integrals, occurring in the IEs, by finite sums.

According to Grünwald, the generalised differintegration operator of (possibly fractional) order ν is defined by the limiting operation

$$\frac{\mathrm{d}^{\nu}\Psi(t)}{\mathrm{d}t^{\nu}} = \lim_{h\to 0}\left\{\frac{1}{h^{\nu}\Gamma(-\nu)}\sum_{n=0}^{t/h-1}\frac{\Gamma(n-\nu)}{\Gamma(n+1)}\Psi(t-n\,h)\right\}, \tag{12.89}$$

where $\Gamma(\cdot)$ is the Euler gamma function [1] and n is a summation index. By taking $\nu = -1/2$ one obtains the semi-integral, which frequently occurs in the electroanalytical IEs:

$$\frac{\mathrm{d}^{-1/2}\Psi(t)}{\mathrm{d}t^{-1/2}} = \int_0^t [\pi(t-\tau)]^{-1/2}\Psi(\tau)\mathrm{d}\tau = \lim_{h\to 0}\left\{h^{1/2}\sum_{n=0}^{t/h-1}\frac{(2n-1)!!}{(2n)!!}\Psi(t-nh)\right\}, \tag{12.90}$$

where the twin exclamation mark denotes the double factorial; for example, $5!! = 1\cdot 3\cdot 5$ and $6!! = 2\cdot 4\cdot 6$. Oldham and Myland [93] suggested approximating the semi-integral by taking a finite number N of terms instead of the infinite number t/h in Eq. (12.90):

$$\int_0^t [\pi(t-\tau)]^{-1/2}\,\Psi(\tau)\,\mathrm{d}\tau \approx h^{1/2}\sum_{n=0}^{N-1}\frac{(2n-1)!!}{(2n)!!}\,\Psi(t-n\,h)\,. \tag{12.91}$$

Parameter $h = t/N$ thus acquired the meaning of a uniform grid step along the t axis. By denoting

$$w_i = \begin{cases} h^{1/2} & \text{for } i = N \\ \frac{2(N-i)-1}{2(N-i)}w_{i+1} & \text{for } i = N-1,\ldots,1 \\ 0 & \text{for } i = 0 \end{cases}, \tag{12.92}$$

Eq. (12.91) can be written, using a reversed summation order, as

$$\int_0^t [\pi(t-\tau)]^{-1/2}\, \Psi(\tau)\,\mathrm{d}\tau \approx \sum_{i=0}^{N} w_i\, \Psi(i\,h)\,. \tag{12.93}$$

Hence, similarly to the step function and Huber methods, formula (12.93) approximates the semi-integral by a linear combination of the nodal values of the function $\Psi(t)$, albeit with different linear coefficients w_i. Based on the equations of Sect. 12.1.1, it is easy to show that in the uniform grid variant of the step function method, Eq. (12.93) holds with coefficients

$$w_i = \begin{cases} 0 & \text{for } i = 0 \\ 2\left(\frac{h}{\pi}\right)^{1/2}\left[(N-i+1)^{1/2} - (N-i)^{1/2}\right] & \text{for } i = 1, \ldots, N \end{cases} . \tag{12.94}$$

Similarly, in the Huber method, Eq. (12.93) holds with coefficients

$$w_i = \begin{cases} 2\left(\frac{h}{\pi}\right)^{1/2}\left\{N^{1/2} + \frac{2}{3}\left[(N-1)^{3/2} - N^{3/2}\right]\right\} & \text{for } i = 0 \\ \frac{4}{3}\left(\frac{h}{\pi}\right)^{1/2}\left[(N-i-1)^{3/2} - 2(N-i)^{3/2}\right. & \\ \quad \left. +(N-i+1)^{3/2}\right] & \text{for } i = 1, \ldots, N-1\,. \\ \frac{4}{3}\left(\frac{h}{\pi}\right)^{1/2} & \text{for } i = N \end{cases} \tag{12.95}$$

It is interesting to check which of the three methods gives the most accurate approximation to the semi-integral, because this determines the accuracy of the IE solutions. Since many functions $\Psi(t)$ can be expanded in power series, an informative test is to compare the errors resulting from the formulae (12.92)–(12.95), applied to functions $\Psi(t) = t^m$, with $m = 0, 1, 2, \ldots$. All three methods prove to be exact for $m = 0$, and the piecewise linear approximation of the Huber method is exact also for $m = 1$. Therefore, in Fig. 12.6 we show errors obtained for $m = 2$. Results for greater m are similar. As can be seen, formula (12.92) gives a slightly more accurate result than the semi-integral obtained by the step function method, and it has a similar order of accuracy, close to 1 [meaning that the error is $\mathcal{O}(h)$]. The piecewise linear approximation of the Huber method gives a much more accurate semi-integral, with an accuracy order close to 2 [meaning that the error is $\mathcal{O}(h^2)$]. This result is consistent with the previous findings of Sect. 12.1.2.2. As the computational cost is comparable in all three cases, we conclude that the piecewise linear approximation outperforms the remaining approximations. Hence, there is little reason for using Eq. (12.91), the more so because it is not clear how to generalise Eq. (12.91) to other kernels occurring in the electrochemical IEs, different from $\mathscr{K}(t,\tau) = [\pi(t-\tau)]^{-1/2}$. Obviously there is no such problem with the Huber method. Oldham and Myland [93] used Eq. (12.91) also for the kernel $\mathscr{K}(t,\tau) = \exp[-k(t-\tau)]\,[\pi(t-\tau)]^{-1/2}$, by discretising the product $\exp(kt)\,\Psi(t)$. However, we

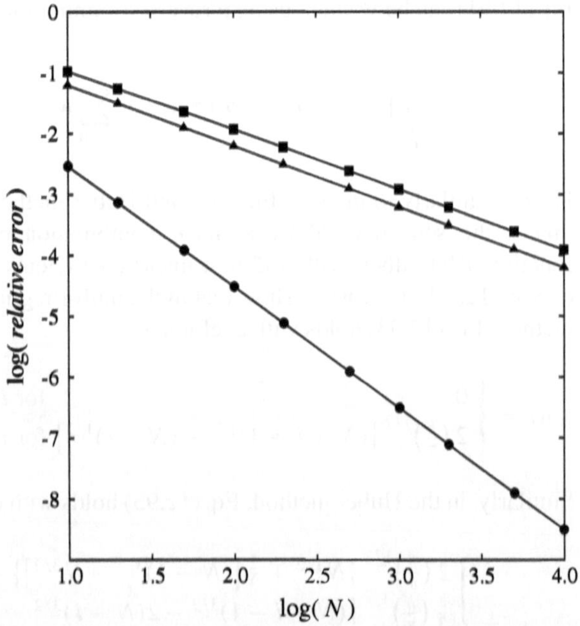

Fig. 12.6 Relative errors (at $t = 1$) of the uniform grid discretisation (12.93) of the semi-integral of the function $\Psi(t) = t^2$, by means of the step function approximation to $\Psi(t)$ (*squares*), piecewise linear approximation to $\Psi(t)$ (*circles*), and discrete Grünwald differintegration (*triangles*). N denotes the number of uniform grid steps h into which the interval $[0, 1]$ of t is divided ($N = 1/h$). The errors are calculated with reference to the exact semi-integral $\int_0^t [\pi(t-\tau)]^{-1/2}\, \tau^2\, d\tau = (16/15)\pi^{-1/2}\, t^{5/2}$

have already indicated at the beginning of Sect. 12.1, and in Sect. 12.1.1.4 that this is not a satisfactory approach, because it requires the discrete grid to be extremely dense when parameter k is large. Concluding, formula (12.91) seems to have no advantages compared to the Huber method with accurate moment integrals.

The above observations do not mean that there do not exist alternative numerical semi-integral approximations that might be more attractive. For example, Wein [108, 109], and Wein and Tovchigrechko [110] used an approximation that is almost identical to the Huber method, and probably has a comparable accuracy. Wein [108] described also another approximation, related to the Simpson quadrature.

12.1.4 Approximate (Degenerate) Kernel Method

We have seen in Sect. 12.1.2.2 that in the product integration methods the amount of computations, and the related computational time, increase quadratically with the number N of integration steps used along the t axis (the same property has the method discussed in Sect. 12.1.3). Although modern personal computers are fast, so that the computational time needed to solve a typical IE occurring in electroanalytical chemistry is usually of the order of a fraction of a second, the above feature is somewhat unsatisfactory. In the alternative simulation approach based on the direct numerical solution of PDEs or ODEs, the computational time usually increases linearly with N, provided that spatial grids do not change with time. If the IE approach is to be competitive, it would be nice to have IE solution methods in which the computational time also increases linearly with N. One such method has been proposed by the present author [12, 13], and it is briefly described below.

The idea of the method is borrowed from the so-called degenerate kernel methods often used to solve Fredholm IEs [7]. In the degenerate kernel methods one looks for a set of $M + 1$ functions $\mathscr{U}_i(t)$ and $\mathscr{W}_i(t)$, enabling the following approximate representation of the kernel function $\mathscr{K}(t, \tau)$ of interest:

$$\mathscr{K}(t,\tau) \approx \sum_{i=0}^{M} \mathscr{U}_i(t)\, \mathscr{W}_i(\tau)\,. \tag{12.96}$$

Substitution of Eq. (12.96) into the usual integral (12.1) gives

$$Y(t) \approx \sum_{i=0}^{M} \mathscr{U}_i(t) \int_0^t \mathscr{W}_i(\tau)\, \Psi(\tau)\, \mathrm{d}\tau\,. \tag{12.97}$$

We observe that whereas in Eq. (12.1) the integrand depends on t through the kernel $\mathscr{K}(t, \tau)$, there is no such dependence any more in the integrands in Eq. (12.97). Consequently, when it comes to evaluating the integrals in Eq. (12.97) by finite sums for any particular grid node t_n of a discrete grid, there is no need to recalculate the integrals over all previous grid nodes. One only adds contributions from the interval $[t_{n-1}, t_n]$, to the integrals previously evaluated at t_{n-1}. As a result, the computational time is proportional to $(M + 1)\, N$, i.e. it increases linearly with the number N of integration steps used.

The main problem is how to devise the approximations (12.96). Bieniasz [12] obtained an approximation of this sort for the convolution kernel $\mathscr{K}(t, \tau) = (t - \tau)^{-1/2}$, by employing exponential functions:

$$(t-\tau)^{-1/2} \approx \sum_{i=0}^{M} a_i \exp[-b_i(t-\tau)] = \sum_{i=0}^{M} a_i \exp(-b_i t)\exp(b_i \tau)\,. \tag{12.98}$$

By taking $M = 15$, coefficients a_i and b_i were determined, ensuring an absolute error of the approximation (12.96) not exceeding 0.005, for $0.00155 \le t - \tau \le 30$. This allowed a solution of a number of typical IEs (single IEs and systems of two IEs) involving the above kernel, and also kernels in the form $\mathscr{K}(t, \tau) = \mathscr{H}(t, \tau)\,(t-\tau)^{-1/2}$, where $\mathscr{H}(t, \tau)$ is a slowly varying function with separated variables, i.e. $\mathscr{H}(t, \tau) = \mathscr{P}(t)\,\mathscr{Q}(\tau)$. In particular, the kernel $\mathscr{K}(t, \tau) = \exp[-k\,(t - \tau)](t - \tau)^{-1/2}$ characteristic of homogeneous reaction-diffusion problems in one-dimensional semi-infinite spatial domain at planar interfaces [see Eq. (8.41) in Sect. 8.2] was considered. Functions $\mathscr{P}(t)$ and $\mathscr{Q}(t)$ were discretised using the same uniform grid on which function $\Psi(t)$ was discretised, so that the values of k could not be too large.

The accuracy of the above method depends on three factors: the accuracy of the approximation (12.96), the quadrature used to numerically evaluate the integrals in Eq. (12.97), and the integration step sizes. The first factor is limiting, because the error of the approximation (12.96) cannot be reduced by diminishing step sizes. The method proved to be substantially faster than the product integration methods employing the same numbers of (uniform) grid steps. However, due to the non-reducible error of the approximation (12.96), the accuracy of the numerical solution cannot be fully controlled. It seems that in order to make this method competitive relative to the contemporary product integration methods of Sects. 12.1.1 and 12.1.2, new and robust algorithms have to be developed, for obtaining kernel approximations (12.96) for arbitrary kernels, arbitrary intervals of t and τ, and with a prescribed accuracy.

12.1.5 The Analog Method

Although the contemporary world of scientific computing is dominated by digital computers, it should be mentioned, for completeness, that in the past electrochemical IEs were also solved by analog computers. A relevant method was described by Holub and Němec [57,58].

12.2 Numerical Methods for Multidimensional Volterra IEs

Compared to spatially one-dimensional models, much less has been published about numerical algorithms appropriate for solving electrochemical Volterra IEs arising for spatially two- and three-dimensional models. We therefore provide only a brief overview of these methods and suggest that substantial work has yet to be done to develop appropriate methods.

For solving the IEs resulting from the Cope and Tallman approach (cf. Sect. 7.1), Coen, Cope, and Tallman [38] introduced a method earlier described by Gladwell and Coen [48]. In this approach kernels with a logarithmic spatial singularity typically arise. The method of Gladwell and Coen [48] assumes that the solution can be represented as a series of Chebyshev polynomials of a spatial variable. The series involves a spatially singular factor selected in such a way so that the kernel singularity is properly resolved. Substitution of the series into the IEs gives systems of AEs for the expansion coefficients. Inherent in this method is a numerical algorithm for calculating inverse Laplace transforms. The algorithm of Piessens [97], tested by Cope [39] was used for this purpose.

With regard to the Mirkin and Bard approach (cf. Sects. 7.1 and 7.2), the only numerical method was provided in [81]. It can be classified as a quadrature method, which implies (by analogy with the discussion in Sect. 12.1.1.4) that it is not likely to be effective in dealing with rapidly varying kernels. The integrands present in integrals (7.10) were discretised on a nonuniform time grid and on a spatial grid which was almost uniform, except at the end(s) of spatial interval(s), where a denser grid spacing was used. For the temporal integration a quadrature termed "trapezoid rule" was used, but the formula reported seems to resemble a rectangle quadrature. In the last discrete temporal interval, where singularities of the kernels occur, a different formula was found more appropriate, which partially resembled the trapezium quadrature (although it was stated that the trapezoid rule could not be used). For spatial integration an alternative quadrature, termed "modified midpoint rule" was applied, because of some unspecified "symmetry problems". From the above discretisations AE systems were obtained, for the nodal values of the approximate solutions at successive temporal grid nodes. A "surprising peculiarity of multidimensional IEs" was discovered, namely that the best stability and accuracy of the solution is achieved by choosing an extremely small initial time step, and a very large coefficient of the temporal grid expansion. This result may indicate some numerical weakness of the method.

With regard to the solution of the time-dependent IEs resulting from the electrochemical applications of the BIM (cf. Sects. 7.1 and 7.3), the literature is not very specific about the numerical methods used. Judging, for example, from the review of Fan et al. [46] it seems that the techniques typical of the BEM [33] were mostly used. This means that spatio-temporal domains of integration were divided into finite elements, spatial and temporal basis functions and collocation points were chosen, and resulting AE systems were solved. Alternatively, the so-called dual reciprocity method was used [100]. The users and advocators of the BIM did not provide detailed discussions of how the expected singularities of the IEs were handled. There is generally a lack of comprehensive and lucid descriptions of the BEM in the context of electrochemical IEs.

References

1. Abramowitz M, Stegun IA (1972) Handbook of mathematical functions. Dover Publications, New York
2. Amatore C, Garreau D, Hammi M, Pinson J, Savéant JM (1985) Kinetic analysis of reversible electrodimerization reactions by the combined use of double potential step chronoamperometry and linear sweep voltammetry. Application to the reduction of 9-cyanoanthracene. J Electroanal Chem 184:1–24
3. Andrieux CP, Limoges B, Marchal D, Savéant JM (2006) Redox enzymes immobilized on electrodes with solution cosubstrates. General procedure for simulation of time-resolved catalytic responses. Anal Chem 78:3138–3143
4. Aoki K, Kato N (1988) Analysis of the cyclic voltammograms associated with deposition or precipitation of the electrochemical product. J Electroanal Chem 245:51–60
5. Aoki K, Tokuda K, Matsuda H (1984) Theory of differential pulse voltammetry at stationary planar electrodes. J Electroanal Chem 175:1–13
6. Aoki K, Tokuda K, Matsuda H (1986) Linear sweep and cyclic voltammetry for electrocatalysis at modified electrodes with very thin films. J Electroanal Chem 199:69–79
7. Atkinson KE (1997) The numerical solution of integral equations of the second kind. Cambridge University Press, Cambridge

8. Baker CTH (1978) The numerical treatment of integral equations. Clarendon Press, Oxford
9. Balducci G, Costa G (1993) The four-member square scheme in cyclic voltammetry: general solution for Nernstian electron transfers. J Electroanal Chem 348:355–365
10. Berzins T, Delahay P (1953) Oscillographic polarographic waves for the reversible deposition of metals on solid electrodes. J Am Chem Soc 75:555–559
11. Bieniasz LK (1992) ELSIM—a user-friendly PC program for electrochemical kinetic simulations. Version 1.0—solution of integral equations for linear scan and cyclic voltammetry. Comput Chem 16:11–14
12. Bieniasz LK (1992) An efficient numerical method of solving the Abel integral equation for cyclic voltammetry. Comput Chem 16:311–317
13. Bieniasz LK (1993) An efficient numerical method of solving integral equations for cyclic voltammetry. J Electroanal Chem 347:15–30
14. Bieniasz LK (2002) Towards computational electrochemistry—a kineticist's perspective. Mod Asp Electrochem 35:135–195
15. Bieniasz LK (2008) An adaptive Huber method with local error control, for the numerical solution of the first kind Abel integral equations. Computing 83:25–39
16. Bieniasz LK (2008) Initialisation of the adaptive Huber method for solving the first kind Abel integral equation. Computing 83:163–174
17. Bieniasz LK (2008) Cyclic voltammetric current functions determined with a prescribed accuracy by the adaptive Huber method for Abel integral equations. Anal Chem 80:9659–9665
18. Bieniasz LK (2010) Automatic simulation of cyclic voltammograms by the adaptive Huber method for weakly singular second kind Volterra integral equations. Electrochim Acta 55:721–728
19. Bieniasz LK (2010) An adaptive Huber method for weakly singular second kind Volterra integral equations with nonlinear dependencies between unknowns and their integrals. Computing 87:35–54
20. Bieniasz LK (2010) Automatic simulation of cyclic voltammograms by the adaptive Huber method for systems of weakly singular Volterra integral equations. J Electroanal Chem 642:127–134
21. Bieniasz LK (2011) An adaptive Huber method for non-linear systems of weakly singular second kind Volterra integral equations. Appl Math Comput 217:5622–5631
22. Bieniasz LK (2011) Extension of the adaptive Huber method for solving integral equations occurring in electro-analysis onto kernel function representing fractional diffusion. Electroanalysis 23:1506–1511
23. Bieniasz LK (2011) A highly accurate, inexpensive procedure for computing integral transformation kernel and its moment integrals for cylindrical wire electrodes. J Electroanal Chem 661:280–286
24. Bieniasz LK (2011) Automatic simulation of electrochemical transients at cylindrical wire electrodes, by the adaptive Huber method for Volterra integral equations. J Electroanal Chem 662:371–378
25. Bieniasz LK (2011) Extension of the adaptive Huber method for Volterra integral equations arising in electroanalytical chemistry, to convolution kernels $\exp[-\alpha(t-\tau)]\,\mathrm{erex}[\beta(t-\tau)]^{1/2}$ and $\exp[-\alpha(t-\tau)]\,\mathrm{daw}[\beta(t-\tau)]^{1/2}$. J Comput Methods Sci Eng 11:323–338
26. Bieniasz LK (2012) Automatic simulation of electrochemical transients by the adaptive Huber method for Volterra integral equations involving kernel terms $\exp[-\alpha(t-\tau)]\,\mathrm{erex}[\beta(t-\tau)]^{1/2}$ and $\exp[-\alpha(t-\tau)]\,\mathrm{daw}[\beta(t-\tau)]^{1/2}$. J Math Chem 50:765–781
27. Bieniasz LK (2012) Automatic solution of integral equations pertinent to diffusion with first order homogeneous reactions at cylindrical wire electrodes. J Electroanal Chem 674:38–47
28. Bieniasz LK (2012) Automatic simulation of electrochemical transients, assuming finite diffusion space at planar interfaces, by the adaptive Huber method for Volterra integral equations. J Electroanal Chem 684:20–31
29. Bieniasz LK (2013) Automatic solution of the Singh and Dutt integral equations for channel or tubular electrodes, by the adaptive Huber method. J Electroanal Chem 693:95–104

30. Bieniasz LK (2013) Automatic solution of integral equations describing electrochemical transients under conditions of internal spherical diffusion. J Electroanal Chem 694:104–113
31. Bieniasz LK (2013) Automatic solution of integral equations describing electrochemical transients under conditions of internal cylindrical diffusion. J Electroanal Chem 700:30–39
32. Bieniasz LK (2013) Automatic solution of integral equations describing electrochemical transients at dropping mercury electrodes. J Electroanal Chem 705:44–51
33. Brebbia CA, Telles JCF, Wrobel LC (1984) Boundary element techniques. Springer, Berlin
34. Brunner H (2004) Collocation methods for Volterra integral and related functional differential equations. Cambridge University Press, Cambridge
35. Brunner H, Van der Houwen PJ (1986) The numerical solution of Volterra equations. North-Holland, Amsterdam
36. Cody WJ (1969) Rational Chebyshev approximations for the error function. Math Comput 23:631–637
37. Cody WJ, Paciorek KA, Thacher HC Jr (1970) Chebyshev approximations for Dawson's integral. Math Comput 24:171–178
38. Coen S, Cope DK, Tallman DE (1986) Diffusion current at a band electrode by an integral equation method. J Electroanal Chem 215:29–48
39. Cope DK (1990) Convergence of Piessens' method for numerical inversion of the Laplace transform on the real line. SIAM J Numer Anal 27:1345–1354
40. Delmastro JR, Booman GL (1969) Polarographic kinetic currents for first-order preceding and regeneration reactions at spherical electrodes. Anal Chem 41:1409–1420
41. Delves LM, Mohamed JL (1985) Computational methods for integral equations. Cambridge University Press, Cambridge
42. De Vries WT (1965) Exact treatment of anodic stripping voltammetry with a plane mercury-film electrode. J Electroanal Chem 9:448–456
43. De Vries WT (1968) Double layer charging in constant-current chronopotentiometry at a mercury-film electrode. J Electroanal Chem 19:41–53
44. Diao G, Zhang Z (1996) Theory and application of cyclic voltammetry at a hemispherical microelectrode for a quasi-reversible reaction. J Electroanal Chem 410:155–162
45. Dračka O (1969) Study of the kinetics of electrode processes by means of electrolysis with constant current. XIII. Correction for the influence of double layer charging in measurement of transition times. Collect Cecoslov Chem Commun 34:2627–2644
46. Fan TH, Mayle EJ, Kottke PA, Fedorov AG (2006) Simulation of electroanalysis using the boundary integral method. Trends Anal Chem 25:52–65
47. Gallopoulos E, Houstis E, Rice JR (1992) Future research directions in problem solving environments for computational science. Technical Report CSD TR-92-032, Computer Science Department, Purdue University
48. Gladwell GML, Coen S (1975) A Chebyshev approximation method for microstrip problems. IEEE Trans Microw Theor Tech MTT-23:865–870
49. Gokhshtein YaP, Gokhshtein AYa (1959) Consecutive electrochemical reactions in oscillographic polarography. Dokl Akad Nauk SSSR 128:985–987 (in Russian)
50. Gokhshtein YaP, Gokhshtein AYa (1960) Multistage electrochemical reactions in oscillographic polarography. In: Longmuir IS (ed) Advances in polarography, vol 2. Pergamon Press, Oxford, pp 465–481
51. Goldberg D (1991) What every computer scientist should know about floating-point arithmetic. ACM Comput Surv 23:5–48
52. Gonzalez J, Hapiot P, Konovalov V, Savéant JM (1999) Investigating the reduction characteristics of transient free radicals by laser-pulse electron photo-injection—mechanism diagnostic criteria and determination of reactivity parameters from time-resolved experiments. J Electroanal Chem 463:157–189
53. Grünwald AK (1867) Ueber "begrenzte" Derivationen und deren Anwendung. Z Math Phys Leipzig 12:441–480
54. Gueshi T, Tokuda K, Matsuda H (1979) Voltammetry at partially covered electrodes. Part II. Linear potential sweep and cyclic voltammetry. J Electroanal Chem 101:29–38

55. Gustafsson K (1994) Control–theoretic techniques for stepsize selection in implicit Runge-Kutta methods. ACM Trans Math Softw 20:496–517
56. Hackbusch W (1995) Integral equations, theory and numerical treatment. Birkhäuser, Basel
57. Holub K, Němec L (1966) The analog method for solution of problems involving diffusion to the electrode. I. Diffusion to a sphere with Langmuirian adsorption solved by analog method. J Electroanal Chem 11:1–11
58. Holub K, Němec L (1968) The analog method for solution of problems involving diffusion to the electrode. II. Kinetics of adsorption controlled by diffusion to the plane, spherical, and dropping electrodes for an arbitrary adsorption isotherm. J Electroanal Chem 18:209–214
59. Honeychurch MJ (2006) Simulating electrochemical reactions with MATHEMATICA. IBNH, St Lucia
60. Houstis EN, Rice JR (2000) Future problem solving environments for computational science. Math Comput Simul 54:243–257
61. Huber A (1939) Eine Näherungsmethode zur Auflösung Volterrascher Integralgleichungen. Monatschr Math Phys 47:240–246
62. IEEE 754 standard (2014). http://grouper.ieee.org/groups/754. Accessed 10 June 2014
63. Kohler H, Piron DL, Bélanger G (1987) A linear sweep voltammetry theory for irreversible electrode reactions with an order of one or higher. I. Mathematical formulation. J Electrochem Soc 134:120–125
64. Kumar VT, Birke RL (1993) Evaluation of electrochemical parameters for an EC mechanism from a global analysis of current–potential–time data: application to reductive cleavage of methylcobalamin. Anal Chem 65:2428–2436
65. Kythe PK, Puri P (2002) Computational methods for linear integral equations. Birkhäuser, Boston
66. Kythe PK, Schäferkotter MR (2005) Handbook of computational methods for integration. Chapman and Hall/CRC, Boca Raton
67. Lexa D, Rentien P, Savéant JM, Xu F (1985) Methods for investigating the mechanistic and kinetic role of ligand exchange reactions in coordination electrochemistry. Cyclic voltammetry of chloroiron(III)tetraphenylporphyrin in dimethylformamide. J Electroanal Chem 191:253–279
68. Linz P (1985) Analytical and numerical methods for Volterra equations. SIAM, Philadelphia
69. Lovrić M (1996) Simulation of electrochemical problems by numerical integration. Russ J Electrochem 32:988–995
70. Mahon PJ, Oldham KB (1998) Voltammetric modelling via extended semiintegrals. J Electroanal Chem 445:179–195
71. Mahon PJ, Oldham KB (1999) Convolutive modelling: a third route to predict voltammograms. ECS Proc 99-5:90–97
72. Mahon PJ, Oldham KB (1999) Convolutive modelling of electrochemical processes based on the relationship between the current and the surface concentration. J Electroanal Chem 464:1–13
73. Mahon PJ, Oldham KB (2001) Incorporating electrode kinetics into the convolutive modeling of reactions at planar, cylindrical and spherical electrodes. Electrochim Acta 46:953–965
74. Mahon PJ, Oldham KB (2004) Convolutive modelling of the disk electrode geometry under reversible conditions. Electrochim Acta 49:5049–5054
75. Mahon PJ, Oldham KB (2006) An examination of the conditions resulting in a temporal convolution related to electrochemical processes. J Solid State Electrochem 10:785–791
76. Mahon PJ, Myland JC, Oldham KB (2002) A fresh approach to voltammetric modelling. J Electroanal Chem 537:1–5
77. MATHEMATICA (2014) Wolfram Res. Inc., Champaigne, IL. http://www.wolfram.com. Accessed 10 June 2014
78. Mirčeski V (2003) Modification of the step-function method for solving linear integral equations and application in modelling of a voltammetric experiment. J Electroanal Chem 545:29–37

79. Mirčeski V, Gulaboski R, Kuzmanovski I (1999) MATHCAD—a tool for numerical calculation of square-wave voltammograms. Bull Chem Technol Maced 18:57–64
80. Mirčeski V, Komorsky-Lovrić Š, Lovrić M (2007) Square-wave voltammetry, theory and application. Springer, Berlin (Appendix A)
81. Mirkin MV, Bard AJ (1992) Multidimensional integral equations. Part 1. A new approach to solving microelectrode diffusion problems. J Electroanal Chem 323:1–27
82. Mirkin MV, Nilov AP (1991) Modification of the Huber method for solving integral equations on a non-uniform grid. Comput Chem 15:55–58
83. Mirkin MV, Nilov AP, Nauryzbaev MK (1990) A general method of calculating polarization curves for various analytical techniques. J Electroanal Chem 281:41–60
84. Myland JC, Oldham KB (2002) Convolutive modelling in the absence of supporting electrolyte: coping with migration and changing resistance in predicting voltammetry. J Electroanal Chem 529:66–74
85. Netlib software repository (2014). http://www.netlib.org. Accessed 10 June 2014
86. Nicholson RS, Olmstead ML (1972) Numerical solution of integral equations. In: Mattson JS, Mark HB Jr, MacDonald HC Jr (eds) Electrochemistry, calculations, simulation, and instrumentation. Computers in chemistry and instrumentation, vol 2. Marcel Dekker, New York, pp 119–138
87. Nicholson RS, Shain I (1964) Theory of stationary electrode polarography. Single scan and cyclic methods applied to reversible, irreversible, and kinetic systems. Anal Chem 36:706–723
88. Nicholson RS, Shain I (1965) Theory of stationary electrode polarography for a chemical reaction coupled between two charge transfers. Anal Chem 37:178–190
89. Nishihara C, Matsuda H (1974) Contributions to the theory of kinetic and catalytic currents in polarography. J Electroanal Chem 51:287–293
90. Ohsaka T, Sotomura T, Matsuda H, Oyama N (1983) Double potential step chronoamperometry for reversible follow-up chemical reactions. Application to the equation kinetics of bis(ethylenediaminemonoacetato)cobalt(II). Bull Chem Soc Jpn 56:3065–3072
91. Oldham KB (1981) An algorithm for semiintegration, semidifferentiation and other instances of differintegration. J Electroanal Chem 121:341–342
92. Oldham KB (1986) Convolution: a general electrochemical procedure implemented by a universal algorithm. Anal Chem 58:2296–2300
93. Oldham KB, Myland JC (2011) Modelling cyclic voltammetry without digital simulation. Electrochim Acta 56:10612–10625
94. Oldham KB, Spanier J (1974) The fractional calculus. Theory and applications of differentiation and integration to arbitrary order. Academic, New York
95. Olmstead ML, Nicholson RS (1968) A method based on polynomial approximations for numerical solution of Volterra integral equations. J Electroanal Chem 16:145–151
96. Olmstead ML, Hamilton RG, Nicholson RS (1969) Theory of cyclic voltammetry for a dimerization reaction initiated electrochemically. Anal Chem 41:260–267
97. Piessens R (1972) A new numerical method for the inversion of the Laplace transform. J Inst Math Appl 10:185–192
98. Press WH, Teukolsky SA, Vetterling WT, Flannery BP (1992) Numerical recipes in C. The art of scientific computing, 2nd edn. Cambridge University Press, Cambridge
99. Puy J, Mas F, Sanz F, Virgili J (1983) Potentiostatic reversible reaction when both reactant and product are adsorbed at the dropping mercury electrode. Part II. Langmuirian and Frumkinian adsorption. J Electroanal Chem 158:231–252
100. Qiu FL, Fisher AC (2000) The dual reciprocity method: simulation of potential step voltammetry at microelectrodes. Electrochem Commun 2:738–742
101. Roos HG, Stynes M, Tobiska L (1996) Numerical methods for singularly perturbed differential equations. Convection-diffusion and flow problems. Springer, Berlin
102. Rudolph M (1990) An algorithm of general application for the digital simulation of electrochemical processes. J Electroanal Chem 292:1–7

103. Savéant JM, Vianello E (1960) Recherches sur les courants catalytiques en polarographie-oscillographique á balayage linéaire de tension. Etude théorique. In: Advances in polarography. Pergamon Press, London, pp 367–374
104. Savéant JM, Vianello E (1965) Potential-sweep chronoamperometry: kinetic currents for first-order chemical reaction parallel to electron-transfer process (catalytic currents). Electrochim Acta 10:905–920
105. Seralathan M, Rangarajan SK (1985) Scheme of squares. Part I. Systems formalism for potentiostatic studies. J Electroanal Chem 191:209–228
106. Shain I, Martin KJ, Ross JW (1961) Electrolysis with constant potential: irreversible reactions at a hanging mercury drop electrode. J Phys Chem 65:259–261
107. Wagner C (1954) On the numerical solution of Volterra integral equations. J Math Phys 32:289–301
108. Wein O (2010) Edge effects in voltage-step transient I. Ohmic losses in 1D approximation (revisited). Research Report ICPF No. 2010/3, Institute of Chemical Process Fundamentals ASCR, Prague
109. Wein O (2010) Voltage-step transient in redox systems II. 1D approximation (revisited). Research Report ICPF No. 2010/6, Institute of Chemical Process Fundamentals ASCR, Prague
110. Wein O, Tovchigrechko VV (2011) Voltage-step transient on circular electrodes. J Appl Electrochem 41:1065–1075
111. White N, Lawson F (1970) Potential sweep voltammetry of metal deposition and dissolution. Part I. Theoretical analysis. J Electroanal Chem 25:409–419

Appendices

Appendix A
Solution–Flux Relationships for One-Dimensional Diffusion Equations

We derive here analytical solution–flux relationships for the archetypal linear diffusion PDE

$$\frac{\partial u}{\partial t} = D \triangle u \,, \tag{A.1}$$

defined over a single one-dimensional spatial domain (interval) that can be either semi-infinite or finite with length l. In Eq. (A.1) D is a (positive) diffusion coefficient, and u is an unknown function of spatial coordinates and time t. Solution u can be positive or negative; it does not have to represent a concentration that is always non-negative. The trivial solution $u \equiv 0$ is not of interest. We assume that only initially $u = 0$. We determine solution–flux relationships at one boundary (corresponding to an electrochemical interface studied), which must hold irrespective of the actual boundary conditions (not explicitly formulated here) imposed at this boundary. This is accomplished for three particular boundary conditions imposed at the second boundary: $u = 0$ in the case of the semi-infinite spatial domain, and $u = 0$ (for a permeable second boundary) or $\mathbf{grad}\, u = 0$ (for an impermeable second boundary) in the case of finite spatial domains. Three geometries of the diffusion field: planar, spherical, and cylindrical are considered' respectively, in Sects. A.1–A.3. The derivations for the transient Eq. (A.1) are performed by employing the Laplace transform method. Whenever appropriate, we also determine the limiting form of the solution–flux relationship under the steady state condition $\partial u / \partial t \approx 0$.

As we shall see, the solution–flux relationships corresponding to the various cases considered in this appendix may depend on certain parameters, defined identically for all the cases. These parameters are:

$$\varrho = D^{1/2} / r_0 \,, \tag{A.2}$$

L.K. Bieniasz, *Modelling Electroanalytical Experiments by the Integral Equation Method*, Monographs in Electrochemistry, DOI 10.1007/978-3-662-44882-3

which is applicable in the cases of spherical and cylindrical diffusion to interfaces with radius r_0, and

$$\mu = D^{1/2}/l \, , \tag{A.3}$$

which is applicable in the cases of finite spatial domains with thickness l.

A.1 Planar Diffusion

In the case of planar diffusion Eq. (A.1) can be written as

$$\frac{\partial u(x,t)}{\partial t} = D\frac{\partial^2 u(x,t)}{\partial x^2} \, . \tag{A.4}$$

The initial condition is

$$u(x,0) = 0 \, , \tag{A.5}$$

and the boundary conditions at the second boundary are

$$u(\infty,t) = 0 \tag{A.6}$$

or

$$u(l,t) = 0 \tag{A.7}$$

or

$$\left.\frac{\partial u(x,t)}{\partial x}\right|_{x=l} = 0 \, . \tag{A.8}$$

By applying the Laplace transformation (3.12) to Eq. (A.4), and by taking into account the initial condition (A.5) we obtain an equation for $\hat{u}(x,s)$:

$$\mathscr{L}\left\{\frac{\partial u(x,t)}{\partial t}\right\} = s\hat{u}(x,s) - 0 = \mathscr{L}\left\{D\frac{\partial^2 u(x,t)}{\partial x^2}\right\} = D\frac{\partial^2 \hat{u}(x,s)}{\partial x^2} \, . \tag{A.9}$$

Equation (A.9) is a second-order ODE in the independent variable x, and it is equivalent to

$$\frac{\partial^2 \hat{u}(x,s)}{\partial x^2} - \frac{s}{D}\hat{u}(x,s) = 0 \, . \tag{A.10}$$

Application of the Laplace transform to the boundary conditions (A.6)–(A.8) gives in turn

$$\hat{u}(\infty, s) = 0 , \tag{A.11}$$

or

$$\hat{u}(l, s) = 0 , \tag{A.12}$$

or

$$\left. \frac{\partial \hat{u}(x, s)}{\partial x} \right|_{x=l} = 0 . \tag{A.13}$$

Equation (A.10) is a linear homogeneous ODE with constant coefficients, so that its general solution is [3]

$$\hat{u}(x, s) = a(s) \exp\left[-\left(\frac{s}{D}\right)^{1/2} x \right] + b(s) \exp\left[\left(\frac{s}{D}\right)^{1/2} x \right] , \tag{A.14}$$

where $a(s)$ and $b(s)$ are integration constants independent of x. Differentiation with respect to x gives the Laplace transform of the flux:

$$\hat{J}(x, s) = -D \frac{\partial \hat{u}(x, s)}{\partial x}$$

$$= (D\, s)^{1/2} \left\{ a(s) \exp\left[-\left(\frac{s}{D}\right)^{1/2} x \right] - b(s) \exp\left[\left(\frac{s}{D}\right)^{1/2} x \right] \right\} . \tag{A.15}$$

Further discussion depends on the choice of the boundary conditions (A.6)–(A.8).

In the case of steady state planar diffusion, Eq. (A.4) becomes a second-order ODE

$$\frac{\partial^2 u(x, t)}{\partial x^2} = 0 . \tag{A.16}$$

The general solution of Eq. (A.16) is

$$u(x, t) = a(t)\, x + b(t) , \tag{A.17}$$

where $a(t)$ and $b(t)$ are integration constants independent of x. Consequently, the flux is

$$J(x, t) = -D \frac{\partial u(x, t)}{\partial x} = -D\, a(t) . \tag{A.18}$$

Further discussion depends on the choice of the boundary conditions (A.6)–(A.8).

A.1.1 Semi-infinite Spatial Domain

In the case of the boundary condition (A.6), $b(s) = 0$ is the only option for the transient solution transform (A.14). Hence, Eqs. (A.14) and (A.15) become

$$\hat{u}(x,s) = a(s)\,\exp\left[-\left(\frac{s}{D}\right)^{1/2} x\right] , \tag{A.19}$$

$$\hat{J}(x,s) = (D\,s)^{1/2}\,a(s)\,\exp\left[-\left(\frac{s}{D}\right)^{1/2} x\right] . \tag{A.20}$$

By eliminating $a(s)$ from Eqs. (A.19) and (A.20) we obtain

$$\hat{u}(x,s) = (D\,s)^{-1/2}\,\hat{J}(x,s) . \tag{A.21}$$

The convolution theorem (3.17) applied to Eq. (A.21) gives

$$u(x,t) = \int_0^t \mathscr{K}(t,\tau)\;J(x,\tau)\;\mathrm{d}\tau , \tag{A.22}$$

where the integral transformation kernel $\mathscr{K}(t,\tau)$ is of the convolution type, and in view of the analytical inverse transform

$$\mathscr{L}^{-1}\left\{s^{-1/2}\right\} = (\pi t)^{-1/2} , \tag{A.23}$$

it does not depend on the spatial coordinate x:

$$\mathscr{K}(t,\tau) = D^{-1/2}\,[\pi(t-\tau)]^{-1/2} . \tag{A.24}$$

Hence, in particular, at the boundary at $x = 0$, corresponding to the interface studied:

$$u(0,t) = \int_0^t \mathscr{K}(t,\tau)\;J(0,\tau)\;\mathrm{d}\tau . \tag{A.25}$$

In the special case of the steady state solution (A.17), boundary condition (A.6) requires that both $a(t)$ and $b(t)$ are identically zero. Therefore, a non-trivial steady state solution does not exist in this case.

A.1.2 Finite Spatial Domain, Permeable Second Boundary

By applying the boundary condition (A.7) to Eq. (A.14) we find that

$$a(s)\ \exp\left[-\left(\frac{s}{D}\right)^{1/2} l\right] + b(s)\ \exp\left[\left(\frac{s}{D}\right)^{1/2} l\right] = 0\ , \tag{A.26}$$

so that one of the coefficients $a(s)$ or $b(s)$ can be eliminated, for example $b(s)$:

$$b(s) = -a(s)\ \exp\left[-2\left(\frac{s}{D}\right)^{1/2} l\right]\ . \tag{A.27}$$

Substituting this result into Eqs. (A.14) and (A.15) yields

$$\hat{u}(x,s) = a(s)\ \left\{\exp\left[-\left(\frac{s}{D}\right)^{1/2} x\right] - \exp\left[\left(\frac{s}{D}\right)^{1/2} (x-2l)\right]\right\}\ , \tag{A.28}$$

$$\hat{J}(x,s) = (D\,s)^{1/2}\,a(s)\ \left\{\exp\left[-\left(\frac{s}{D}\right)^{1/2} x\right] + \exp\left[\left(\frac{s}{D}\right)^{1/2} (x-2l)\right]\right\}\ . \tag{A.29}$$

By eliminating $a(s)$ from Eqs. (A.28) and (A.29) we obtain

$$\hat{u}(x,s) = (D\,s)^{-1/2}\ \tanh\left(\frac{l-x}{D^{1/2}}\,s^{1/2}\right)\ \hat{J}(x,s)\ . \tag{A.30}$$

The convolution theorem (3.17) applied to Eq. (A.30) gives

$$u(x,t) = \int_0^t \mathscr{K}(t,\tau)\ J(x,\tau)\ \mathrm{d}\tau\ , \tag{A.31}$$

where the integral transformation kernel $\mathscr{K}(t,\tau)$ is of the convolution type, and it also generally depends on the spatial coordinate x. The formula for the kernel is:

$$\mathscr{K}(t,\tau) = \varphi(t-\tau) \tag{A.32}$$

where

$$\mathscr{K}(\vartheta) = D^{-1/2}\,\mathscr{L}^{-1}\left\{s^{-1/2}\tanh\left(\frac{l-x}{D^{1/2}}\,s^{1/2}\right)\right\}\ . \tag{A.33}$$

In Eq. (A.33) the Laplace transform is between the ϑ and s domains. The inverse Laplace transform can be expressed (see, for example, Roberts and Kaufman [5,

p. 283]) by means of the exponential theta function $\Theta_2(\nu \,|\, z)$ (see, for example, Oldham et al. [4, p. 261]) with parameter $\nu = 0$, which gives

$$\mathscr{K}(t,\tau) = (l-x)^{-1}\,\Theta_2\left[0\,\middle|\,\frac{D}{(l-x)^2}\,(t-\tau)\right] . \tag{A.34}$$

At the boundary at $x = 0$, corresponding to the interface studied, Eqs. (A.31) and (A.34) simplify to

$$u(0,t) = \int_0^t \mathscr{K}(t,\tau)\, J(0,\tau)\, \mathrm{d}\tau , \tag{A.35}$$

$$\mathscr{K}(t,\tau) = l^{-1}\,\Theta_2\left[0\,\middle|\,\mu^2\,(t-\tau)\right] , \tag{A.36}$$

with μ defined by Eq. (A.3). It is this form of the kernel that was originally discussed in the literature. However, by using the function kplp(·) defined by Eq. (5.122), and taking into account basic rules of the change of variables in Laplace transforms [cf. Eq. (3.15)], it is more convenient to express the kernel as

$$\mathscr{K}(t,\tau) = D^{-1/2}\left\{[\pi(t-\tau)]^{-1/2} - \frac{D^{1/2}}{l-x}\,\mathrm{kplp}\left[\frac{D^{1/2}}{l-x}(t-\tau)^{1/2}\right]\right\} \tag{A.37}$$

when $x \neq 0$, or

$$\mathscr{K}(t,\tau) = D^{-1/2}\left\{[\pi(t-\tau)]^{-1/2} - \mu\,\mathrm{kplp}\left[\mu(t-\tau)^{1/2}\right]\right\} \tag{A.38}$$

when $x = 0$.

In the special case of the steady state solution (A.17), by applying the boundary condition (A.7) we obtain

$$a(t)\,l + b(t) = 0 , \tag{A.39}$$

so that

$$b(t) = -a(t)\,l . \tag{A.40}$$

This gives

$$u(x,t) = a(t)\,(x-l) . \tag{A.41}$$

By combining Eqs. (A.18) and (A.41) the unknown $a(t)$ can be eliminated, which gives the solution–flux relationship for the steady state:

$$u(x,t) = D^{-1/2}\left[\frac{D^{1/2}}{l-x}\right]^{-1} J(x,t) . \tag{A.42}$$

In particular, at the boundary at $x = 0$, corresponding to the interface studied,

$$u(0,t) = D^{-1/2}\mu^{-1}\, J(0,t)\ . \tag{A.43}$$

A.1.3 Finite Spatial Domain, Impermeable Second Boundary

By applying the boundary condition (A.8) to Eq. (A.14) we find that

$$-\,a(s)\,\exp\left[-\left(\frac{s}{D}\right)^{1/2} l\right] + b(s)\,\exp\left[\left(\frac{s}{D}\right)^{1/2} l\right] = 0\ , \tag{A.44}$$

so that one of the coefficients $a(s)$ or $b(s)$ can be eliminated, for example $b(s)$:

$$b(s) = a(s)\,\exp\left[-2\left(\frac{s}{D}\right)^{1/2} l\right]\ . \tag{A.45}$$

Substituting this result into Eqs. (A.14) and (A.15) yields

$$\hat{u}(x,s) = a(s)\left\{\exp\left[-\left(\frac{s}{D}\right)^{1/2} x\right] + \exp\left[\left(\frac{s}{D}\right)^{1/2} (x-2l)\right]\right\}\ , \tag{A.46}$$

$$\hat{J}(x,s) = (D\,s)^{1/2}\,a(s)\left\{\exp\left[-\left(\frac{s}{D}\right)^{1/2} x\right] - \exp\left[\left(\frac{s}{D}\right)^{1/2} (x-2l)\right]\right\}\ . \tag{A.47}$$

By eliminating $a(s)$ from Eqs. (A.46) and (A.47) we obtain

$$\hat{u}(x,s) = (D\,s)^{-1/2}\coth\left(\frac{l-x}{D^{1/2}}\,s^{1/2}\right)\hat{J}(x,s)\ . \tag{A.48}$$

The convolution theorem (3.17) applied to Eq. (A.48) gives

$$u(x,t) = \int_0^t \mathscr{K}(t,\tau)\, J(x,\tau)\, d\tau\ , \tag{A.49}$$

where the integral transformation kernel $\mathscr{K}(t,\tau)$ is of the convolution type, and it also generally depends on the spatial coordinate x. The formula for the kernel is:

$$\mathscr{K}(t,\tau) = \varphi(t-\tau) \tag{A.50}$$

where

$$\varphi(\vartheta) = D^{-1/2}\,\mathscr{L}^{-1}\left\{s^{-1/2}\coth\left(\frac{l-x}{D^{1/2}}\,s^{1/2}\right)\right\} . \quad \text{(A.51)}$$

In Eq. (A.51) the Laplace transform is between the ϑ and s domains. The inverse Laplace transform can be expressed (see, for example, Roberts and Kaufman [5, p. 283]) by means of the exponential theta function $\Theta_3(\nu\,|\,z)$ (see, for example, Oldham et al. [4, p. 261]) with parameter $\nu = 0$, which gives

$$\mathscr{K}(t,\tau) = (l-x)^{-1}\,\Theta_3\left[0\left|\frac{D}{(l-x)^2}\,(t-\tau)\right.\right] . \quad \text{(A.52)}$$

At the boundary at $x = 0$, corresponding to the interface studied, Eqs. (A.49) and (A.52) simplify to

$$u(0,t) = \int_0^t \mathscr{K}(t,\tau)\,J(0,\tau)\,\mathrm{d}\tau , \quad \text{(A.53)}$$

$$\mathscr{K}(t,\tau) = l^{-1}\,\Theta_3\left[0\,|\mu^2\,(t-\tau)\right] , \quad \text{(A.54)}$$

with μ defined by Eq. (A.3). This form of the kernel was originally discussed in the literature. However, by using the function kpli(·) defined by Eq. (5.123), and taking into account basic rules of the change of variables in Laplace transforms [cf. Eq. (3.15)], it is more convenient to express the kernel as

$$\mathscr{K}(t,\tau) = D^{-1/2}\left\{[\pi(t-\tau)]^{-1/2} + \frac{D^{1/2}}{l-x}\,\mathrm{kpli}\left[\frac{D^{1/2}}{l-x}(t-\tau)^{1/2}\right]\right\} \quad \text{(A.55)}$$

when $x \neq 0$, or

$$\mathscr{K}(t,\tau) = D^{-1/2}\left\{[\pi(t-\tau)]^{-1/2} + \mu\,\mathrm{kpli}\left[\mu(t-\tau)^{1/2}\right]\right\} , \quad \text{(A.56)}$$

when $x = 0$.

In the special case of the steady state solution (A.17), the boundary condition (A.8) implies that $a(t) \equiv 0$, so that only a spatially uniform solution is possible, which is not interesting:

$$u(x,t) = b(t) , \quad \text{(A.57)}$$

$$J(x,t) = 0 . \quad \text{(A.58)}$$

No definite solution–flux relationship exists in this case.

A.2 Spherical Diffusion

In the case of spherical diffusion Eq. (A.1) can be written as

$$\frac{\partial u(r,t)}{\partial t} = D\left[\frac{\partial^2 u(r,t)}{\partial r^2} + \frac{2}{r}\frac{\partial u(r,t)}{\partial r}\right] . \tag{A.59}$$

The initial condition is

$$u(r,0) = 0 , \tag{A.60}$$

and the boundary conditions at the second boundary are

$$u(\infty,t) = 0 , \tag{A.61}$$

or

$$u(r_0 \pm l, t) = 0 , \tag{A.62}$$

or

$$\left.\frac{\partial u(r,t)}{\partial r}\right|_{r=r_0\pm l} = 0 . \tag{A.63}$$

By applying the Laplace transformation (3.12) to both sides of Eq. (A.59), and by taking into account the initial condition (A.60) we obtain the following equation for $\hat{u}(r,s)$:

$$\mathscr{L}\left\{\frac{\partial u(r,t)}{\partial t}\right\} = s\hat{u}(r,s) - 0$$

$$= \mathscr{L}\left\{D\left[\frac{\partial^2 u(r,t)}{\partial r^2} + \frac{2}{r}\frac{\partial u(r,t)}{\partial r}\right]\right\} = D\left[\frac{\partial^2 \hat{u}(r,s)}{\partial r^2} + \frac{2}{r}\frac{\partial \hat{u}(r,s)}{\partial r}\right] , \tag{A.64}$$

which is a second-order ODE in independent variable r. Equation (A.64) is equivalent to

$$\frac{\partial^2 \hat{u}(r,s)}{\partial r^2} + \frac{2}{r}\frac{\partial \hat{u}(r,s)}{\partial r} - \frac{s}{D}\hat{u}(r,s) = 0 . \tag{A.65}$$

Application of the Laplace transformation to the boundary conditions (A.61)–(A.63) gives

$$\hat{u}(\infty,s) = 0 , \tag{A.66}$$

or

$$\hat{u}(r_0 \pm l, s) = 0 \,, \tag{A.67}$$

or

$$\left.\frac{\partial \hat{u}(r,s)}{\partial r}\right|_{r=r_0 \pm l} = 0 \,. \tag{A.68}$$

The change of variables:

$$w(r,t) = u(r,t)\, r \tag{A.69}$$

turns Eq. (A.65) into

$$\frac{\partial^2 \hat{w}(r,s)}{\partial r^2} - \frac{s}{D}\hat{w}(r,s) = 0 \,. \tag{A.70}$$

Equation (A.70) is analogous to Eq. (A.10) for the planar diffusion case, so that by proceeding in the way identical to that in Sect. A.1, we obtain the result analogous to Eq. (A.14):

$$\hat{w}(r,s) = a(s)\, \exp\left[-\left(\frac{s}{D}\right)^{1/2} r\right] + b(s)\, \exp\left[\left(\frac{s}{D}\right)^{1/2} r\right] , \tag{A.71}$$

where $a(s)$ and $b(s)$ are integration constants independent of r. Hence, by returning to the original solution variable we obtain

$$\hat{u}(r,s) = \frac{a(s)}{r}\, \exp\left[-\left(\frac{s}{D}\right)^{1/2} r\right] + \frac{b(s)}{r}\, \exp\left[\left(\frac{s}{D}\right)^{1/2} r\right] . \tag{A.72}$$

Differentiation with respect to r gives the Laplace transform of the flux:

$$\begin{aligned} \hat{J}(r,s) &= -D\frac{\partial \hat{u}(r,s)}{\partial r} \\ &= \frac{D}{r^2}\left\{a(s)\left[\left(\frac{s}{D}\right)^{1/2} r + 1\right] \exp\left[-\left(\frac{s}{D}\right)^{1/2} r\right]\right. \\ &\quad \left. - b(s)\left[\left(\frac{s}{D}\right)^{1/2} r - 1\right] \exp\left[\left(\frac{s}{D}\right)^{1/2} r\right]\right\} . \end{aligned} \tag{A.73}$$

Further discussion depends on the choice of the boundary conditions (A.61)–(A.63).

In the case of steady state spherical diffusion, Eq. (A.59) becomes a second-order ODE

$$\frac{\partial^2 u(r,t)}{\partial r^2} + \frac{2}{r}\frac{\partial u(r,t)}{\partial r} = 0 \,. \tag{A.74}$$

The change of variables (A.69) transforms Eq. (A.74) into

$$\frac{\partial^2 w(r,t)}{\partial r^2} = 0 \,, \tag{A.75}$$

which has the general solution

$$w(r,t) = a(t) + b(t)\, r \,, \tag{A.76}$$

where $a(t)$ and $b(t)$ are integration constants independent of r. Therefore, from Eqs. (A.69) and (A.76)

$$u(r,t) = r^{-1} a(t) + b(t) \,. \tag{A.77}$$

Consequently, the flux is

$$J(r,t) = -D \frac{\partial u(r,t)}{\partial r} = D\, r^{-2} a(t) \,. \tag{A.78}$$

Further discussion depends on the choice of the boundary conditions (A.61)–(A.63).

A.2.1 Semi-infinite Spatial Domain

In the case of the boundary condition (A.61), $b(s) = 0$ is the only option. Hence, Eqs. (A.72) and (A.73) become:

$$\hat{u}(r,s) = \frac{a(s)}{r} \exp\left[-\left(\frac{s}{D}\right)^{1/2} r\right] , \tag{A.79}$$

$$\hat{J}(r,s) = \frac{D}{r^2}\, a(s) \left[\left(\frac{s}{D}\right)^{1/2} r + 1\right] \exp\left[-\left(\frac{s}{D}\right)^{1/2} r\right] . \tag{A.80}$$

By combining Eqs. (A.79) and (A.80) the constant $a(s)$ can be eliminated, which gives

$$\hat{u}(r,s) = D^{-1/2} \left(s^{1/2} + \frac{D^{1/2}}{r}\right)^{-1} \hat{J}(r,s) \,. \tag{A.81}$$

The convolution theorem (3.17) applied to Eq. (A.81) gives:

$$u(r,t) = \int_0^t \mathscr{K}(t,\tau)\; J(r,\tau)\, \mathrm{d}\tau \,, \tag{A.82}$$

where the integral transformation kernel is of the convolution type, but in contrast to planar semi-infinite diffusion, it depends on the spatial coordinate r. In view of the analytical inverse transform

$$\mathscr{L}^{-1}\left\{\left(s^{1/2}+\frac{D^{1/2}}{r}\right)^{-1}\right\}=(\pi t)^{-1/2}-\frac{D^{1/2}}{r}\,\mathrm{erex}\left(\frac{D^{1/2}}{r}t^{1/2}\right)\,, \tag{A.83}$$

with function erex $(\cdot)$ defined by Eq. (5.76), the formula for the kernel is

$$\mathscr{K}(t,\tau)=D^{-1/2}\left\{[\pi(t-\tau)]^{-1/2}-\frac{D^{1/2}}{r}\,\mathrm{erex}\left[\frac{D^{1/2}}{r}(t-\tau)^{1/2}\right]\right\}\,. \tag{A.84}$$

At the boundary at $r = r_0$, corresponding to the interface studied, Eqs. (A.82) and (A.84) simplify to

$$u(r_0,t)=\int\limits_0^t \mathscr{K}(t,\tau)\;J(r_0,\tau)\,\mathrm{d}\tau\,, \tag{A.85}$$

$$\mathscr{K}(t,\tau)=D^{-1/2}\left\{[\pi(t-\tau)]^{-1/2}-\varrho\,\mathrm{erex}\left[\varrho(t-\tau)^{1/2}\right]\right\}\,, \tag{A.86}$$

with ϱ defined by Eq. (A.2).

In the special case of the steady state solution (A.77), the application of the boundary condition (A.61) implies that $b(t)$ must be zero, so that

$$u(r,t)=r^{-1}a(t)\,. \tag{A.87}$$

By combining Eqs. (A.78) and (A.87) the unknown $a(t)$ can be eliminated, which gives the solution–flux relationship for the steady state:

$$u(r,t)=D^{-1/2}\left(\frac{D^{1/2}}{r}\right)^{-1}J(r,t)\,. \tag{A.88}$$

Specifically, at the boundary at $r = r_0$, corresponding to the interface studied,

$$u(r_0,t)=D^{-1/2}\,\varrho^{-1}\,J(r_0,t)\,. \tag{A.89}$$

Equation (A.89) can also be deduced directly from Eq. (A.85), in the following way [2]. The time needed for a steady state to be established must decrease with decreasing radius r_0, i.e. with increasing ϱ; obviously, this time would be infinite in the planar case, when $\varrho = 0$. Assume therefore that $t \gg \varrho^{-2}$ so that the steady state is achieved. The expression erex $\left[\varrho(t-\tau)^{1/2}\right]$ in Eq. (A.85) is then practically equal to the first term of its asymptotic expansion (see, for example, Eq. (7.1.23) in Abramowitz and Stegun [1]), that is

$$\operatorname{erex}\left[\varrho(t-\tau)^{1/2}\right] \approx \varrho^{-1}\pi^{-1/2}(t-\tau)^{-1/2}\,, \tag{A.90}$$

for all τ with the exception of a narrow interval of τ close to $\tau = t$. By setting expression (A.90) into Eq. (A.86) we see that $\mathscr{K}(t,\tau) \approx 0$ everywhere except for this narrow interval, and only the contribution from this interval determines the integral in Eq. (A.85). If we want the steady state to be approximately present in a time-dependent electroanalytical experiment, any changes of $J(r_0, t)$ resulting from time-dependent boundary conditions at $r = r_0$ must be slow compared to the time needed for the steady state to establish. We can therefore assume that $J(r_0,\tau) \approx J(r_0,t)$ in the above narrow interval of τ. As a consequence, Eq. (A.85) is replaced by

$$u(r_0,t) = J(r_0,t)\int_0^t \mathscr{K}(t,\tau)\,\mathrm{d}\tau\,. \tag{A.91}$$

The integral in Eq. (A.91) can be calculated analytically:

$$\int_0^t \mathscr{K}(t,\tau)\,\mathrm{d}\tau = D^{-1/2}\varrho^{-1}\left[1-\operatorname{erex}\left(\varrho t^{1/2}\right)\right] \approx D^{-1/2}\varrho^{-1}\,, \tag{A.92}$$

where the term $\operatorname{erex}\left(\varrho t^{1/2}\right)$ has been neglected, owing to the assumed $t \gg \varrho^{-2}$. Hence, by combining Eqs. (A.91) and (A.92) we obtain Eq. (A.89).

A.2.2 Finite Spatial Domain, Permeable Second Boundary

Assume first that the second boundary is located at $r = r_0 \pm l > 0$. Boundary condition (A.62) applied to Eq. (A.72) then gives

$$\frac{a(s)}{r_0 \pm l}\exp\left[-\left(\frac{s}{D}\right)^{1/2}(r_0 \pm l)\right] + \frac{b(s)}{r_0 \pm l}\exp\left[\left(\frac{s}{D}\right)^{1/2}(r_0 \pm l)\right] = 0\,, \tag{A.93}$$

which (in view of $r_0 \pm l > 0$) implies a relationship between $a(s)$ and $b(s)$:

$$b(s) = -a(s)\exp\left[-2\left(\frac{s}{D}\right)^{1/2}(r_0 \pm l)\right]. \tag{A.94}$$

Therefore, Eqs. (A.72) and (A.73) become

$$\hat{u}(r,s) = \frac{a(s)}{r}\left\langle \exp\left[-\left(\frac{s}{D}\right)^{1/2} r\right] - \exp\left\{\left(\frac{s}{D}\right)^{1/2}[r - 2(r_0 \pm l)]\right\}\right\rangle , \tag{A.95}$$

$$\begin{aligned}\hat{J}(r,s) = D\frac{a(s)}{r^2}\Bigg\langle &\left[\left(\frac{s}{D}\right)^{1/2} r + 1\right]\exp\left[-\left(\frac{s}{D}\right)^{1/2} r\right] \\ &+ \left[\left(\frac{s}{D}\right)^{1/2} r - 1\right]\exp\left\{\left(\frac{s}{D}\right)^{1/2}[r - 2(r_0 \pm l)]\right\}\Bigg\rangle .\end{aligned} \tag{A.96}$$

By combining Eqs. (A.95) and (A.96) the constant $a(s)$ can be eliminated, which gives

$$\hat{u}(r,s) = \pm D^{-1/2}\frac{\tanh\left[\left(\frac{s}{D}\right)^{1/2}(r_0 \pm l - r)\right]}{s^{1/2} + \frac{D^{1/2}}{r}\tanh\left[\left(\frac{s}{D}\right)^{1/2}(r_0 \pm l - r)\right]}\left[\pm\hat{J}(r,s)\right] . \tag{A.97}$$

The convolution theorem (3.17) applied to Eq. (A.97) gives:

$$u(r,t) = \int_0^t \mathscr{K}(t,\tau)\ [\pm J(r,\tau)]\ \mathrm{d}\tau , \tag{A.98}$$

where the integral transformation kernel is of the convolution type, and it depends on the spatial coordinate r. The formula for the kernel is:

$$\mathscr{K}(t,\tau) = \varphi(t-\tau) , \tag{A.99}$$

where

$$\varphi(\vartheta) = D^{-1/2}\mathscr{L}^{-1}\left\{\frac{\pm\tanh\left[\left(\frac{s}{D}\right)^{1/2}(r_0 \pm l - r)\right]}{s^{1/2} + \frac{D^{1/2}}{r}\tanh\left[\left(\frac{s}{D}\right)^{1/2}(r_0 \pm l - r)\right]}\right\} . \tag{A.100}$$

The Laplace transform in Eq. (A.100) is between the ϑ and s domains. At the boundary at $r = r_0$, corresponding to the interface studied, Eqs. (A.98) and (A.100) simplify to

$$u(r_0,t) = \int_0^t \mathscr{K}(t,\tau)\ [\pm J(r_0,\tau)]\ \mathrm{d}\tau , \tag{A.101}$$

$$\varphi(\vartheta) = D^{-1/2}\,\mathscr{L}^{-1}\left\{\frac{\pm\tanh\left(\pm\frac{s^{1/2}}{\mu}\right)}{s^{1/2}+\varrho\tanh\left(\pm\frac{s^{1/2}}{\mu}\right)}\right\}, \tag{A.102}$$

with parameters ϱ and μ defined by Eqs. (A.2) and (A.3).

The above results cannot be reasonably extended to the case of the second boundary located at $r \equiv r_0 - l = 0$. From Taylor expansions of the exponential functions, at low r Eq. (A.72) takes the form

$$\hat{u}(r,s) = \frac{a(s)+b(s)}{r} + \left(\frac{s}{D}\right)^{1/2}[b(s)-a(s)] + \mathscr{O}(r)\,. \tag{A.103}$$

Therefore, $\hat{u}(r,s)$ can be finite at $r \to 0$ only when $a(s) \equiv -b(s)$. Furthermore, in order to satisfy the boundary condition (A.62), there must be $a(s) \equiv b(s) \equiv 0$ in Eq. (A.103), which means an uninteresting solution.

In the special case of the steady state solution (A.77), the application of the boundary condition (A.62) implies that

$$\frac{a(t)}{r_0 \pm l} + b(t) = 0\,. \tag{A.104}$$

Therefore,

$$b(t) = -\frac{a(t)}{r_0 \pm l}, \tag{A.105}$$

and we obtain from Eq. (A.77):

$$u(r,t) = a(t)\left(\frac{1}{r} - \frac{1}{r_0 \pm l}\right). \tag{A.106}$$

By combining Eqs. (A.78) and (A.106) the unknown $a(t)$ can be eliminated, which gives the solution–flux relationship for the steady state:

$$u(r,t) = \pm\frac{r}{D}\left(1 - \frac{r}{r_0 \pm l}\right)[\pm J(r,t)]\,. \tag{A.107}$$

In particular, at the boundary at $r = r_0$, corresponding to the interface studied, Eq. (A.107) becomes

$$u(r_0,t) \equiv D^{-1/2}\frac{\varrho^{-1}\mu^{-1}}{\varrho^{-1} \pm \mu^{-1}}[\pm J(r_0,t)]\,, \tag{A.108}$$

with parameters ϱ and μ defined by Eqs. (A.2) and (A.3).

A.2.3 Finite Spatial Domain, Impermeable Second Boundary

Assume first that the second boundary is located at $r = r_0 \pm l > 0$. Boundary condition (A.63) applied to Eq. (A.72) then gives

$$\frac{a(s)}{(r_0 \pm l)^2}\left[\left(\frac{s}{D}\right)^{1/2}(r_0 \pm l) + 1\right]\exp\left[-\left(\frac{s}{D}\right)^{1/2}(r_0 \pm l)\right]$$
$$-\frac{b(s)}{(r_0 \pm l)^2}\left[\left(\frac{s}{D}\right)^{1/2}(r_0 \pm l) - 1\right]\exp\left[\left(\frac{s}{D}\right)^{1/2}(r_0 \pm l)\right] = 0\,, \qquad \text{(A.109)}$$

which [in view of $r_0 \pm l > 0$, and by taking s with sufficiently large real part so that $(s/D)^{1/2}(r_0 \pm l) - 1 \neq 0$] implies a relationship between $a(s)$ and $b(s)$:

$$b(s) = a(s)\,\frac{\left(\frac{s}{D}\right)^{1/2}(r_0 \pm l) + 1}{\left(\frac{s}{D}\right)^{1/2}(r_0 \pm l) - 1}\exp\left[-2\left(\frac{s}{D}\right)^{1/2}(r_0 \pm l)\right]. \qquad \text{(A.110)}$$

Therefore, Eqs. (A.72) and (A.73) become

$$\hat{u}(r,s) = \frac{a(s)}{r}\left\langle \exp\left[-\left(\frac{s}{D}\right)^{1/2} r\right]\right.$$
$$\left. + \frac{\left(\frac{s}{D}\right)^{1/2}(r_0 \pm l) + 1}{\left(\frac{s}{D}\right)^{1/2}(r_0 \pm l) - 1}\exp\left\{\left(\frac{s}{D}\right)^{1/2}[r - 2(r_0 \pm l)]\right\}\right\rangle, \qquad \text{(A.111)}$$

$$\hat{J}(r,s) = D\,\frac{a(s)}{r^2}\left\langle\left[\left(\frac{s}{D}\right)^{1/2} r + 1\right]\exp\left[-\left(\frac{s}{D}\right)^{1/2} r\right]\right.$$
$$\left. - \frac{\left(\frac{s}{D}\right)^{1/2}(r_0 \pm l) + 1}{\left(\frac{s}{D}\right)^{1/2}(r_0 \pm l) - 1}\left[\left(\frac{s}{D}\right)^{1/2} r - 1\right]\exp\left\{\left(\frac{s}{D}\right)^{1/2}[r - 2(r_0 \pm l)]\right\}\right\rangle. \qquad \text{(A.112)}$$

By combining Eqs. (A.111) and (A.112) the constant $a(s)$ can be eliminated, which gives

$$\hat{u}(r,s) = \frac{\pm D^{-1/2}\left\{\left(\frac{s}{D}\right)^{1/2}(r_0 \pm l) - \tanh\left[\left(\frac{s}{D}\right)^{1/2}(r_0 \pm l - r)\right]\right\}}{s^{1/2}\frac{(r_0 \pm l - r)}{r} + \left[\frac{s}{D^{1/2}}(r_0 \pm l) - \frac{D^{1/2}}{r}\right]\tanh\left[\left(\frac{s}{D}\right)^{1/2}(r_0 \pm l - r)\right]}$$
$$\times\left[\pm \hat{J}(r,s)\right]. \qquad \text{(A.113)}$$

The convolution theorem (3.17) applied to Eq. (A.113) gives:

$$u(r,t) = \int_0^t \mathcal{K}(t,\tau)\;[\pm J(r,\tau)]\;\mathrm{d}\tau\,, \qquad \text{(A.114)}$$

where the integral transformation kernel is of the convolution type, and it depends on the spatial coordinate r. The formula for the kernel is:

$$\mathscr{K}(t,\tau) = \varphi(t-\tau)\ , \tag{A.115}$$

where

$$\varphi(\vartheta) = D^{-1/2}\,\mathscr{L}^{-1}\left\{\frac{\pm\left\{\left(\frac{s}{D}\right)^{1/2}(r_0 \pm l) - \tanh\left[\left(\frac{s}{D}\right)^{1/2}(r_0 \pm l - r)\right]\right\}}{s^{1/2}\frac{(r_0 \pm l - r)}{r} + \left[s\,\frac{r_0 \pm l}{D^{1/2}} - \frac{D^{1/2}}{r}\right]\tanh\left[\left(\frac{s}{D}\right)^{1/2}(r_0 \pm l - r)\right]}\right\}\ . \tag{A.116}$$

The Laplace transform in Eq. (A.116) is between the ϑ and s domains. At the boundary at $r = r_0$, corresponding to the interface studied, Eqs. (A.114) and (A.116) simplify to

$$u(r_0,t) = \int_0^t \mathscr{K}(t,\tau)\ [\pm J(r_0,\tau)]\ \mathrm{d}\tau\ , \tag{A.117}$$

$$\varphi(\vartheta) = D^{-1/2}\,\mathscr{L}^{-1}\left\{\frac{s^{1/2}\left(\frac{1}{\varrho} \pm \frac{1}{\mu}\right) - \tanh\left(\pm\frac{s^{1/2}}{\mu}\right)}{s^{1/2}\frac{\varrho}{\mu} \pm \left[s\left(\frac{1}{\varrho} \pm \frac{1}{\mu}\right) - \varrho\right]\tanh\left(\pm\frac{s^{1/2}}{\mu}\right)}\right\}\ , \tag{A.118}$$

with parameters ϱ and μ defined by Eqs. (A.2) and (A.3).

The above results can be extended to the case of the second boundary located at $r = r_0 - l = 0$. From Taylor expansions of the exponential functions, at low r Eq. (A.73) takes the form

$$\hat{J}(r,s) = D\left\{\frac{a(s)+b(s)}{r^2} - \frac{s}{2D}\ [a(s)+b(s)]\ +\ \mathscr{O}(r)\right\}\ . \tag{A.119}$$

We see that $\hat{J}(r,s)$ is finite at $r \to 0$ when $a(s) = -b(s)$, similarly to $\hat{u}(r,s)$ [cf. Eq. (A.103)]. In addition, the second term in braces in Eq. (A.119) then vanishes, so that $\hat{J}(r,s)$ obeys the boundary condition (A.63). Therefore, Eqs. (A.117) and (A.118) are applicable also when $r_0 - l \to 0$. Equation (A.118) then simplifies to

$$\varphi(\vartheta) = D^{-1/2}\,\mathscr{L}^{-1}\left\{\frac{\tanh\left(\frac{s^{1/2}}{\varrho}\right)}{s^{1/2} - \varrho\tanh\left(\frac{s^{1/2}}{\varrho}\right)}\right\}\ . \tag{A.120}$$

By using the function kins(·) defined by Eq. (5.138), the corresponding kernel function can be expressed as

$$\mathscr{K}(t,\tau) = D^{-1/2}\left\{[\pi(t-\tau)]^{-1/2} + \varrho\,\mathrm{kins}\left[\varrho(t-\tau)^{1/2}\right]\right\}\ . \tag{A.121}$$

In the special case of the steady state solution (A.77), the boundary condition (A.63) implies that $a(t) \equiv 0$, so that only a spatially uniform solution is possible, which is not interesting:

$$u(r,t) = b(t) \,, \tag{A.122}$$

$$J(r,t) = 0 \,. \tag{A.123}$$

No definite solution–flux relationship exists in this case.

A.3 Cylindrical Diffusion

In the case of cylindrical diffusion Eq. (A.1) can be written as

$$\frac{\partial u(r,t)}{\partial t} = D\left[\frac{\partial^2 u(r,t)}{\partial r^2} + \frac{1}{r}\,\frac{\partial u(r,t)}{\partial r}\right] . \tag{A.124}$$

The initial condition is

$$u(r,0) = 0 \,, \tag{A.125}$$

and the boundary conditions at the second boundary are:

$$u(\infty,t) = 0 \,, \tag{A.126}$$

or

$$u(r_0 \pm l, t) = 0 \,, \tag{A.127}$$

or

$$\left.\frac{\partial u(r,t)}{\partial r}\right|_{r=r_0\pm l} = 0 \,. \tag{A.128}$$

By applying the Laplace transformation (3.12) to both sides of Eq. (A.124), and by taking into account the initial condition (A.125) we obtain the following equation for $\hat{u}(r,s)$:

$$\begin{aligned} &\mathscr{L}\left\{\frac{\partial u(r,t)}{\partial t}\right\} = s\hat{u}(r,s) - 0 \\ &= \mathscr{L}\left\{D\left[\frac{\partial^2 u(r,t)}{\partial r^2} + \frac{1}{r}\,\frac{\partial u(r,t)}{\partial r}\right]\right\} = D\left[\frac{\partial^2 \hat{u}(r,s)}{\partial r^2} + \frac{1}{r}\,\frac{\partial \hat{u}(r,s)}{\partial r}\right] . \end{aligned} \tag{A.129}$$

Equation (A.129) is a second-order ODE in independent variable r, equivalent to

$$\frac{\partial^2 \hat{u}(r,s)}{\partial r^2} + \frac{1}{r}\frac{\partial \hat{u}(r,s)}{\partial r} - \frac{s}{D}\hat{u}(r,s) = 0 . \tag{A.130}$$

Application of the Laplace transformation to the boundary conditions (A.126)–(A.128) gives

$$\hat{u}(\infty, s) = 0 , \tag{A.131}$$

or

$$\hat{u}(r_0 \pm l, s) = 0 , \tag{A.132}$$

or

$$\left.\frac{\partial \hat{u}(r,s)}{\partial r}\right|_{r=r_0 \pm l} = 0 . \tag{A.133}$$

We replace variable r by a new independent variable ξ defined by

$$\xi = (s/D)^{1/2}\, r , \tag{A.134}$$

owing to which Eq. (A.130) becomes

$$\xi^2 \frac{\partial^2 \hat{u}(\xi,s)}{\partial \xi^2} + \xi \frac{\partial \hat{u}(\xi,s)}{\partial \xi} - \xi^2 \hat{u}(\xi,s) = 0 . \tag{A.135}$$

Equation (A.135) is the homogeneous modified Bessel equation of order zero [1], so that it possesses the general solution

$$\hat{u}(\xi,s) = a(s)\,\mathrm{K}_0(\xi) + b(s)\,\mathrm{I}_0(\xi) , \tag{A.136}$$

where $\mathrm{I}_0(\cdot)$ and $\mathrm{K}_0(\cdot)$ are the modified Bessel functions of the first and second kind, respectively, and of order zero [1], and $a(s)$ and $b(s)$ are integration constants independent of ξ. By returning to the variable r, Eq. (A.136) becomes:

$$\hat{u}(r,s) = a(s)\,\mathrm{K}_0\left[\left(\frac{s}{D}\right)^{1/2} r\right] + b(s)\,\mathrm{I}_0\left[\left(\frac{s}{D}\right)^{1/2} r\right] . \tag{A.137}$$

Differentiation with respect to r gives the Laplace transform of the flux:

$$\hat{J}(r,s) = -D\frac{\partial \hat{u}(r,s)}{\partial r}$$

$$= (D\,s)^{1/2}\left\{a(s)\,\mathrm{K}_1\left[\left(\frac{s}{D}\right)^{1/2} r\right] - b(s)\,\mathrm{I}_1\left[\left(\frac{s}{D}\right)^{1/2} r\right]\right\} , \tag{A.138}$$

where $\mathrm{I}_1(\cdot)$ and $\mathrm{K}_1(\cdot)$ are the modified Bessel functions of the first and second kind, respectively, and of order one. They appear in Eq. (A.138) owing to the known relationships (cf. Eq. (9.6.27) in Abramowitz and Stegun [1]):

$$\frac{\mathrm{dK}_0(\xi)}{\mathrm{d}\xi} = -\mathrm{K}_1(\xi)\,, \tag{A.139}$$

$$\frac{\mathrm{dI}_0(\xi)}{\mathrm{d}\xi} = \mathrm{I}_1(\xi)\,. \tag{A.140}$$

Further discussion depends on the choice of the boundary conditions (A.126)–(A.128).

In the case of steady state cylindrical diffusion, Eq. (A.124) becomes a second-order ODE

$$\frac{\partial^2 u(r,t)}{\partial r^2} + \frac{1}{r}\,\frac{\partial u(r,t)}{\partial r} = 0\,. \tag{A.141}$$

After multiplying both sides by r, an equivalent equation (assuming $r > 0$) is

$$\frac{\partial}{\partial r}\left[r\,\frac{\partial u(r,t)}{\partial r}\right] = 0\,, \tag{A.142}$$

so that

$$r\,\frac{\partial u(r,t)}{\partial r} = a(t) \tag{A.143}$$

and

$$u(r,t) = a(t)\,\ln(r) + b(t)\,. \tag{A.144}$$

where $a(t)$ and $b(t)$ are integration constants independent of r. Consequently, the flux is

$$J(r,t) = -D\frac{\partial u(r,t)}{\partial r} = -D\,r^{-1}a(t)\,. \tag{A.145}$$

Further discussion depends on the choice of the boundary conditions (A.126)–(A.128).

A.3.1 Semi-infinite Spatial Domain

In the case of the boundary condition (A.126), in view of the fact that $\lim_{\xi\to\infty} \mathrm{I}_0(\xi) = \infty$, $b(s)$ must be zero, so that Eqs. (A.137) and (A.138) reduce to

$$\hat{u}(r,s) = a(s)\,\mathrm{K}_0\left[\left(\frac{s}{D}\right)^{1/2} r\right] , \tag{A.146}$$

$$\hat{J}(r,s) = (D\,s)^{1/2}\,a(s)\,\mathrm{K}_1\left[\left(\frac{s}{D}\right)^{1/2} r\right] . \tag{A.147}$$

Combination of Eqs. (A.146) and (A.147) gives, after elimination of $a(s)$:

$$\hat{u}(r,s) = D^{-1/2}\,\frac{\left(\frac{r}{D^{1/2}}\right)\mathrm{K}_0\left(\frac{r}{D^{1/2}}\,s^{1/2}\right)}{\left(\frac{r}{D^{1/2}}\,s^{1/2}\right)\mathrm{K}_1\left(\frac{r}{D^{1/2}}\,s^{1/2}\right)}\,\hat{J}(r,s) . \tag{A.148}$$

The convolution theorem (3.17) applied to Eq. (A.148) gives

$$u(r,t) = \int_0^t \mathscr{K}(t,\tau)\; J(r,\tau)\;\mathrm{d}\tau , \tag{A.149}$$

where the integral transformation kernel $\mathscr{K}(t,\tau)$ is of the convolution type, and similarly to the spherical semi-infinite diffusion case [cf. Eq. (A.84)] it depends on r. In view of Eq. (A.148), by making use of the function $\varphi(\vartheta)$ defined by Eq. (5.81), and taking into account basic rules of the change of variables in Laplace transforms [cf. Eq. (3.15)], the kernel can be expressed as

$$\mathscr{K}(t,\tau) = r^{-1}\varphi\left[\frac{D}{r^2}(t-\tau)\right] . \tag{A.150}$$

However, by using the function kcylw(·) defined by Eq. (5.84), it is more convenient to express the kernel as

$$\mathscr{K}(t,\tau) = D^{-1/2}\left\{[\pi(t-\tau)]^{-1/2} - \frac{D^{1/2}}{r}\,\mathrm{kcylw}\left[\frac{D^{1/2}}{r}(t-\tau)^{1/2}\right]\right\} . \tag{A.151}$$

At the boundary at $r = r_0$, corresponding to the interface studied, Eqs. (A.149), (A.150), and (A.151) simplify to

$$u(r_0,t) = \int_0^t \mathscr{K}(t,\tau)\; J(r_0,\tau)\;\mathrm{d}\tau , \tag{A.152}$$

with

$$\mathcal{K}(t,\tau) = r_0^{-1}\varphi\left[\varrho^2(t-\tau)\right] \tag{A.153}$$

or

$$\mathcal{K}(t,\tau) = D^{-1/2}\left\{[\pi(t-\tau)]^{-1/2} - \varrho\,\mathrm{kcylw}\left[\varrho(t-\tau)^{1/2}\right]\right\}\,, \tag{A.154}$$

where ϱ is defined by Eq. (A.2).

In the special case of the steady state solution (A.144), the application of the boundary condition (A.126) implies that both $a(t)$ and $b(t)$ must be simultaneously zero. Therefore, a non-trivial steady state solution does not exist in this case.

A.3.2 *Finite Spatial Domain, Permeable Second Boundary*

Assume first that the second boundary is located at $r = r_0 \pm l > 0$. Boundary condition (A.132) applied to Eq. (A.137) then gives

$$a(s)\,\mathrm{K}_0\left[\left(\frac{s}{D}\right)^{1/2}(r_0 \pm l)\right] + b(s)\,\mathrm{I}_0\left[\left(\frac{s}{D}\right)^{1/2}(r_0 \pm l)\right] = 0\,, \tag{A.155}$$

which [in view of $r_0 \pm l > 0$, and hence of finite values of $\mathrm{K}_0(\cdot)$] implies a relationship between $a(s)$ and $b(s)$:

$$b(s) = -a(s)\,\frac{\mathrm{K}_0\left[\left(\frac{s}{D}\right)^{1/2}(r_0 \pm l)\right]}{\mathrm{I}_0\left[\left(\frac{s}{D}\right)^{1/2}(r_0 \pm l)\right]}\,. \tag{A.156}$$

Therefore, Eqs. (A.137) and (A.138) become

$$\hat{u}(r,s) = a(s)\left\{\mathrm{K}_0\left[\left(\frac{s}{D}\right)^{1/2} r\right] - \frac{\mathrm{K}_0\left[\left(\frac{s}{D}\right)^{1/2}(r_0 \pm l)\right]}{\mathrm{I}_0\left[\left(\frac{s}{D}\right)^{1/2}(r_0 \pm l)\right]}\,\mathrm{I}_0\left[\left(\frac{s}{D}\right)^{1/2} r\right]\right\}, \tag{A.157}$$

$$\hat{J}(r,s) = D\left(\frac{s}{D}\right)^{1/2} a(s)\left\{\mathrm{K}_1\left[\left(\frac{s}{D}\right)^{1/2} r\right] + \frac{\mathrm{K}_0\left[\left(\frac{s}{D}\right)^{1/2}(r_0 \pm l)\right]}{\mathrm{I}_0\left[\left(\frac{s}{D}\right)^{1/2}(r_0 \pm l)\right]} \times \mathrm{I}_1\left[\left(\frac{s}{D}\right)^{1/2} r\right]\right\}. \tag{A.158}$$

By combining Eqs. (A.157) and (A.158) the constant $a(s)$ can be eliminated, which gives

$$\hat{u}(r,s) = \pm D^{-1/2} \frac{\mathrm{K}_0\left[\left(\frac{s}{D}\right)^{1/2} r\right] - \frac{\mathrm{K}_0\left[\left(\frac{s}{D}\right)^{1/2}(r_0 \pm l)\right]}{\mathrm{I}_0\left[\left(\frac{s}{D}\right)^{1/2}(r_0 \pm l)\right]} \mathrm{I}_0\left[\left(\frac{s}{D}\right)^{1/2} r\right]}{s^{1/2}\left\{\mathrm{K}_1\left[\left(\frac{s}{D}\right)^{1/2} r\right] + \frac{\mathrm{K}_0\left[\left(\frac{s}{D}\right)^{1/2}(r_0 \pm l)\right]}{\mathrm{I}_0\left[\left(\frac{s}{D}\right)^{1/2}(r_0 \pm l)\right]} \mathrm{I}_1\left[\left(\frac{s}{D}\right)^{1/2} r\right]\right\}} \times \left[\pm \hat{J}(r,s)\right] . \tag{A.159}$$

The convolution theorem (3.17) applied to Eq. (A.159) gives:

$$u(r,t) = \int_0^t \mathscr{K}(t,\tau) \; [\pm J(r,\tau)] \; \mathrm{d}\tau , \tag{A.160}$$

where the integral transformation kernel is of the convolution type, and it depends on the spatial coordinate r. The formula for the kernel is

$$\mathscr{K}(t,\tau) = \varphi(t-\tau) , \tag{A.161}$$

where

$$\varphi(\vartheta) = \pm D^{-1/2} \mathscr{L}^{-1} \left\{ \frac{\mathrm{K}_0\left[\left(\frac{s}{D}\right)^{1/2} r\right] - \frac{\mathrm{K}_0\left[\left(\frac{s}{D}\right)^{1/2}(r_0 \pm l)\right]}{\mathrm{I}_0\left[\left(\frac{s}{D}\right)^{1/2}(r_0 \pm l)\right]} \mathrm{I}_0\left[\left(\frac{s}{D}\right)^{1/2} r\right]}{s^{1/2}\left\{\mathrm{K}_1\left[\left(\frac{s}{D}\right)^{1/2} r\right] + \frac{\mathrm{K}_0\left[\left(\frac{s}{D}\right)^{1/2}(r_0 \pm l)\right]}{\mathrm{I}_0\left[\left(\frac{s}{D}\right)^{1/2}(r_0 \pm l)\right]} \mathrm{I}_1\left[\left(\frac{s}{D}\right)^{1/2} r\right]\right\}} \right\} . \tag{A.162}$$

The Laplace transformation in Eq. (A.162) is between the ϑ and s domains. At the boundary at $r = r_0$, corresponding to the interface studied, Eqs. (A.160) and (A.162) simplify to

$$u(r_0,t) = \int_0^t \mathscr{K}(t,\tau) \; [\pm J(r_0,\tau)] \; \mathrm{d}\tau , \tag{A.163}$$

$$\varphi(\vartheta) = \pm D^{-1/2} \mathscr{L}^{-1} \left\{ \frac{\mathrm{K}_0\left(\frac{s^{1/2}}{\varrho}\right) - \frac{\mathrm{K}_0\left[s^{1/2}\left(\frac{1}{\varrho} \pm \frac{1}{\mu}\right)\right]}{\mathrm{I}_0\left[s^{1/2}\left(\frac{1}{\varrho} \pm \frac{1}{\mu}\right)\right]} \mathrm{I}_0\left(\frac{s^{1/2}}{\varrho}\right)}{s^{1/2}\left\{\mathrm{K}_1\left(\frac{s^{1/2}}{\varrho}\right) + \frac{\mathrm{K}_0\left[s^{1/2}\left(\frac{1}{\varrho} \pm \frac{1}{\mu}\right)\right]}{\mathrm{I}_0\left[s^{1/2}\left(\frac{1}{\varrho} \pm \frac{1}{\mu}\right)\right]} \mathrm{I}_1\left(\frac{s^{1/2}}{\varrho}\right)\right\}} \right\} , \tag{A.164}$$

with parameters ϱ and μ defined by Eqs. (A.2) and (A.3).

The above results cannot be reasonably extended to the case of the second boundary located at $r = r_0 - l = 0$. From the asymptotic expansion of the function $\mathrm{K}_0(z)$ for $z \to 0$, and Taylor expansion of the function $\mathrm{I}_0(z)$ we know that at small z

$$\mathrm{K}_0(z) = \ln(2) - \gamma_\mathrm{E} - \ln(z) + \mathcal{O}(z^2)\,, \tag{A.165}$$

$$\mathrm{I}_0(z) = 1 + \mathcal{O}(z^2)\,, \tag{A.166}$$

where γ_E is the Euler constant [1]. Consequently, at low r Eq. (A.137) takes the form

$$\hat{u}(r,s) = -a(s)\ln\left[\left(\frac{s}{D}\right)^{1/2} r\right] + \{a(s)\,[\ln(2) - \gamma_\mathrm{E}] + b(s)\} + \mathcal{O}(r^2)\,. \tag{A.167}$$

Therefore, $\hat{u}(r,s)$ can be finite at $r \to 0$ only when $a(s) \equiv 0$. If we then want the boundary condition (A.132) to be satisfied, then $b(s)$ must be zero, too, which means an uninteresting solution.

In the special case of the steady state solution (A.144), the application of the boundary condition (A.127) implies that

$$a(t)\ln(r_0 \pm l) + b(t) = 0\,. \tag{A.168}$$

Therefore,

$$b(t) = -a(t)\ln(r_0 \pm l)\,, \tag{A.169}$$

and we obtain from Eq. (A.144):

$$u(r,t) = a(t)\ln\left(\frac{r}{r_0 \pm l}\right)\,. \tag{A.170}$$

By combining Eqs. (A.145) and (A.170) the unknown $a(t)$ can be eliminated, which gives the solution–flux relationship for the steady state:

$$u(r,t) = \mp\frac{r}{D}\ln\left(\frac{r}{r_0 \pm l}\right)[\pm J(r,t)]\,. \tag{A.171}$$

In particular, at the boundary at $r = r_0$, corresponding to the interface studied, Eq. (A.171) becomes

$$u(r_0,t) = \mp D^{-1/2}\varrho^{-1}\ln\left(\frac{\varrho^{-1}}{\varrho^{-1} \pm \mu^{-1}}\right)[\pm J(r_0,t)]\,, \tag{A.172}$$

with parameters ϱ and μ defined by Eqs. (A.2) and (A.3).

A.3.3 Finite Spatial Domain, Impermeable Second Boundary

Assume first that the second boundary is located at $r = r_0 \pm l > 0$. Boundary condition (A.133) applied to Eq. (A.137) then gives

$$a(s)\,\mathrm{K}_1\left[\left(\frac{s}{D}\right)^{1/2}(r_0 \pm l)\right] - b(s)\,\mathrm{I}_1\left[\left(\frac{s}{D}\right)^{1/2}(r_0 \pm l)\right] = 0\,, \tag{A.173}$$

which [in view of $r_0 \pm l > 0$, for which the values of $\mathrm{K}_1(\cdot)$ and $\mathrm{I}_1(\cdot)$ are finite and nonzero] implies a relationship between $a(s)$ and $b(s)$:

$$b(s) = a(s)\,\frac{\mathrm{K}_1\left[\left(\frac{s}{D}\right)^{1/2}(r_0 \pm l)\right]}{\mathrm{I}_1\left[\left(\frac{s}{D}\right)^{1/2}(r_0 \pm l)\right]}\,. \tag{A.174}$$

Therefore, Eqs. (A.137) and (A.138) become

$$\begin{aligned}\hat{u}(r,s) &= a(s)\left\{\mathrm{K}_0\left[\left(\frac{s}{D}\right)^{1/2} r\right] + \frac{\mathrm{K}_1\left[\left(\frac{s}{D}\right)^{1/2}(r_0 \pm l)\right]}{\mathrm{I}_1\left[\left(\frac{s}{D}\right)^{1/2}(r_0 \pm l)\right]}\right.\\ &\quad \left.\times\, \mathrm{I}_0\left[\left(\frac{s}{D}\right)^{1/2} r\right]\right\}\,,\end{aligned} \tag{A.175}$$

$$\begin{aligned}\hat{J}(r,s) &= D\left(\frac{s}{D}\right)^{1/2} a(s)\left\{\mathrm{K}_1\left[\left(\frac{s}{D}\right)^{1/2} r\right] - \frac{\mathrm{K}_1\left[\left(\frac{s}{D}\right)^{1/2}(r_0 \pm l)\right]}{\mathrm{I}_1\left[\left(\frac{s}{D}\right)^{1/2}(r_0 \pm l)\right]}\right.\\ &\quad \left.\times\, \mathrm{I}_1\left[\left(\frac{s}{D}\right)^{1/2} r\right]\right\}\,.\end{aligned} \tag{A.176}$$

By combining Eqs. (A.175) and (A.176) the constant $a(s)$ can be eliminated, which gives

$$\begin{aligned}\hat{u}(r,s) &= \pm D^{-1/2}\,\frac{\mathrm{K}_0\left[\left(\frac{s}{D}\right)^{1/2} r\right] + \frac{\mathrm{K}_1\left[\left(\frac{s}{D}\right)^{1/2}(r_0 \pm l)\right]}{\mathrm{I}_1\left[\left(\frac{s}{D}\right)^{1/2}(r_0 \pm l)\right]}\,\mathrm{I}_0\left[\left(\frac{s}{D}\right)^{1/2} r\right]}{s^{1/2}\left\{\mathrm{K}_1\left[\left(\frac{s}{D}\right)^{1/2} r\right] - \frac{\mathrm{K}_1\left[\left(\frac{s}{D}\right)^{1/2}(r_0 \pm l)\right]}{\mathrm{I}_1\left[\left(\frac{s}{D}\right)^{1/2}(r_0 \pm l)\right]}\,\mathrm{I}_1\left[\left(\frac{s}{D}\right)^{1/2} r\right]\right\}}\\ &\quad \times\left[\pm\hat{J}(r,s)\right]\,.\end{aligned} \tag{A.177}$$

The convolution theorem (3.17) applied to Eq. (A.177) gives:

$$u(r,t) = \int_0^t \mathcal{K}(t,\tau)\ [\pm J(r,\tau)]\ \mathrm{d}\tau\ , \tag{A.178}$$

where the integral transformation kernel is of the convolution type, and it depends on the spatial coordinate r. The formula for the kernel is:

$$\mathcal{K}(t,\tau) = \varphi(t-\tau)\ , \tag{A.179}$$

where

$$\varphi(\vartheta) = \pm D^{-1/2}\,\mathcal{L}^{-1}\left\{\frac{\mathrm{K}_0\left[\left(\frac{s}{D}\right)^{1/2} r\right] + \frac{\mathrm{K}_1\left[\left(\frac{s}{D}\right)^{1/2}(r_0 \pm l)\right]}{\mathrm{I}_1\left[\left(\frac{s}{D}\right)^{1/2}(r_0 \pm l)\right]}\,\mathrm{I}_0\left[\left(\frac{s}{D}\right)^{1/2} r\right]}{s^{1/2}\left\{\mathrm{K}_1\left[\left(\frac{s}{D}\right)^{1/2} r\right] - \frac{\mathrm{K}_1\left[\left(\frac{s}{D}\right)^{1/2}(r_0 \pm l)\right]}{\mathrm{I}_1\left[\left(\frac{s}{D}\right)^{1/2}(r_0 \pm l)\right]}\,\mathrm{I}_1\left[\left(\frac{s}{D}\right)^{1/2} r\right]\right\}}\right\}. \tag{A.180}$$

The Laplace transformation in Eq. (A.180) is between the ϑ and s domains. At the boundary at $r = r_0$, corresponding to the interface studied, Eqs. (A.178) and (A.180) simplify to

$$u(r_0,t) = \int_0^t \mathcal{K}(t,\tau)\ [\pm J(r_0,\tau)]\ \mathrm{d}\tau\ , \tag{A.181}$$

$$\varphi(\vartheta) = \pm D^{-1/2}\,\mathcal{L}^{-1}\left\{\frac{\mathrm{K}_0\left(\frac{s^{1/2}}{\varrho}\right) + \frac{\mathrm{K}_1\left[s^{1/2}\left(\frac{1}{\varrho} \pm \frac{1}{\mu}\right)\right]}{\mathrm{I}_1\left[s^{1/2}\left(\frac{1}{\varrho} \pm \frac{1}{\mu}\right)\right]}\,\mathrm{I}_0\left(\frac{s^{1/2}}{\varrho}\right)}{s^{1/2}\left\{\mathrm{K}_1\left(\frac{s^{1/2}}{\varrho}\right) - \frac{\mathrm{K}_1\left[s^{1/2}\left(\frac{1}{\varrho} \pm \frac{1}{\mu}\right)\right]}{\mathrm{I}_1\left[s^{1/2}\left(\frac{1}{\varrho} \pm \frac{1}{\mu}\right)\right]}\,\mathrm{I}_1\left(\frac{s^{1/2}}{\varrho}\right)\right\}}\right\}, \tag{A.182}$$

with parameters ϱ and μ defined by Eqs. (A.2) and (A.3).

The above results can be extended to the case of the second boundary located at $r = r_0 - l = 0$. From the asymptotic expansion of the function $\mathrm{K}_1(z)$ for $z \to 0$, and Taylor expansion of the function $\mathrm{I}_1(z)$ we know that at small z

$$\mathrm{K}_1(z) = \frac{1}{z} + \mathcal{O}(z)\ , \tag{A.183}$$

$$\mathrm{I}_1(z) = \mathcal{O}(z)\ . \tag{A.184}$$

Consequently, at low r Eq. (A.138) takes the form

$$\hat{J}(r,s) = (D\,s)^{1/2}\left\{a(s)\left[\left(\frac{s}{D}\right)^{-1/2} r^{-1} + \mathcal{O}(r)\right] - b(s)\,\mathcal{O}(r)\right\} . \tag{A.185}$$

The boundary condition (A.133) is satisfied by Eq. (A.185) in the limit $r_0 - l \to 0$ only when $a(s) \equiv 0$, but for arbitrary $b(s)$. As for $a(s) \equiv 0$ the boundary solution $\hat{u}(0,s)$ is finite [cf. Eq. (A.167)], the resulting solution is physically acceptable. In this case Eq. (A.182) simplifies to

$$\varphi(\vartheta) = D^{-1/2}\,\mathscr{L}^{-1}\left\{\frac{\mathrm{I}_0\left(\frac{s^{1/2}}{\varrho}\right)}{s^{1/2}\,\mathrm{I}_1\left(\frac{s^{1/2}}{\varrho}\right)}\right\} . \tag{A.186}$$

By using the function kinc(·) defined by Eq. (5.154), the corresponding kernel function can be expressed as

$$\mathscr{K}(t,\tau) = D^{-1/2}\left\{[\pi(t-\tau)]^{-1/2} + \varrho\,\mathrm{kinc}\left[\varrho(t-\tau)^{1/2}\right]\right\} . \tag{A.187}$$

In the special case of the steady state solution (A.144), the boundary condition (A.128) implies that $a(t) \equiv 0$, so that only a spatially uniform solution is possible, which is not interesting:

$$u(r,t) = b(t)\,, \tag{A.188}$$

$$J(r,t) = 0\,. \tag{A.189}$$

No definite solution–flux relationship exists in this case.

References

1. Abramowitz M, Stegun IA (1972) Handbook of mathematical functions. Dover Publications, New York
2. Bieniasz LK, González J, Molina Á, Laborda E (2010) Theory of linear sweep/cyclic voltammetry for the electrochemical reaction mechanism involving a redox catalyst couple attached to a spherical electrode. Electrochim Acta 56:543–552
3. Kamke E (1962) Differentialgleichungen. I. Gewöhnliche Differentialgleichungen, Geest & Portig, Leipzig
4. Oldham K, Myland J, Spanier J (2009) An atlas of functions. Springer, New York
5. Roberts GE, Kaufman H (1966) Table of Laplace transforms. Saunders, Philadelphia

Appendix B
Solution–Flux Relationships for One-Dimensional Convection-Diffusion Equations

In this appendix we derive analytical solution–flux relationships for a number of convection-diffusion PDEs defined over a single one-dimensional and semi-infinite spatial domain (interval), and occurring in models of electroanalytical experiments. The boundary value at infinity is assumed equal to the initial value. The solution–flux relationships are determined at the boundary corresponding to an electrochemical interface studied. They must hold irrespective of the actual boundary conditions (not explicitly formulated here) imposed at this boundary. The convection-diffusion fields considered include: the case of a constant convection velocity (discussed in Sect. B.1), convection to a DME (using the expanding plane model, discussed in Sect. B.2), convection to an RDE (discussed in Sect. B.3), and convection in channel or tubular electrodes (using the Singh and Dutt approximation, discussed in Sect. B.4).

B.1 Constant Convection Velocity

In the case of the constant convection velocity, the incomplete convection-diffusion IBVP describing the transport of distributed species [cf. Eqs. (6.7)–(6.9) in Chap. 6] includes the PDE:

$$\frac{\partial u(x,t)}{\partial t} = D\frac{\partial^2 u(x,t)}{\partial x^2} + v_0\,\frac{\partial u(x,t)}{\partial x}\,, \tag{B.1}$$

where the convection velocity $v_0 \geq 0$. The initial condition is

$$u(x,0) = 0\,, \tag{B.2}$$

L.K. Bieniasz, *Modelling Electroanalytical Experiments by the Integral Equation Method*, Monographs in Electrochemistry, DOI 10.1007/978-3-662-44882-3

and the boundary condition at $x \to \infty$ is

$$u(\infty, t) = 0 \,. \tag{B.3}$$

The flux of variable $u(x,t)$ is given by

$$J(x,t) = -D\frac{\partial u(x,t)}{\partial x} - v_0\, u(x,t) \,. \tag{B.4}$$

By applying the Laplace transformation (3.12) to Eqs. (B.1) and (B.3), and by taking into account the initial condition (B.2) we obtain the equation for $\hat{u}(x,s)$:

$$\mathcal{L}\left\{\frac{\partial u(x,t)}{\partial t}\right\} = s\hat{u}(x,s) - 0$$

$$= \mathcal{L}\left\{D\frac{\partial^2 u(x,t)}{\partial x^2} + v_0\frac{\partial u(x,t)}{\partial x}\right\} = D\frac{\partial^2 \hat{u}(x,s)}{\partial x^2} + v_0\frac{\partial \hat{u}(x,s)}{\partial x} \tag{B.5}$$

with the boundary condition

$$\hat{u}(\infty, s) = 0 \,. \tag{B.6}$$

Equation (B.5) is a second-order ODE in the independent variable x, equivalent to

$$\frac{\partial^2 \hat{u}(x,s)}{\partial x^2} + \frac{v_0}{D}\frac{\partial \hat{u}(x,s)}{\partial x} - \frac{s}{D}\hat{u}(x,s) = 0 \,. \tag{B.7}$$

By introducing an additional unknown

$$\hat{w}(x,s) = \frac{\partial \hat{u}(x,s)}{\partial x} \,, \tag{B.8}$$

Eqs. (B.7) and (B.8) are cast into the form of a linear ODE system with constant coefficients:

$$\frac{\partial}{\partial x}\begin{bmatrix}\hat{u}(x,s)\\ \hat{w}(x,s)\end{bmatrix} = \begin{bmatrix}0 & 1\\ \frac{s}{D} & -\frac{v_0}{D}\end{bmatrix}\begin{bmatrix}\hat{u}(x,s)\\ \hat{w}(x,s)\end{bmatrix} \,. \tag{B.9}$$

The matrix of the system (B.9) has the eigenvalues $\lambda_1(s)$ and $\lambda_2(s)$:

$$\lambda_{1/2}(s) = -\frac{1}{2}\left\{\frac{v_0}{D} \pm \left[\left(\frac{v_0}{D}\right)^2 + 4\frac{s}{D}\right]^{1/2}\right\} \,, \tag{B.10}$$

which are distinct when $\Re\{ s \} > 0$, so that the general solution of Eq. (B.9) is [3]:

$$\begin{bmatrix} \hat{u}(x,s) \\ \hat{w}(x,s) \end{bmatrix} = a(s) \begin{bmatrix} 1 \\ \lambda_1(s) \end{bmatrix} \exp[\lambda_1(s)\, x] + b(s) \begin{bmatrix} 1 \\ \lambda_2(s) \end{bmatrix} \exp[\lambda_2(s)\, x] \ , \tag{B.11}$$

where $a(s)$ and $b(s)$ are integration constants independent of x. In view of the boundary condition (B.6), $b(s)$ must be zero, so that from Eq. (B.11) and Laplace transformed Eq. (B.4) we obtain

$$\hat{u}(x,s) = a(s) \exp[\lambda_1(s)\, x] \ , \tag{B.12}$$

$$\hat{J}(x,s) = -a(s)\ [D\, \lambda_1(s) + v_0] \exp[\lambda_1(s)\, x] \ . \tag{B.13}$$

By combining Eqs. (B.12) and (B.13), the coefficient $a(s)$ can be eliminated, resulting in the relationship:

$$\hat{u}(x,s) = -\,[D\, \lambda_1(s) + v_0]^{-1}\ \hat{J}(x,s) \ , \tag{B.14}$$

or equivalently, by taking into account Eq. (B.10):

$$\hat{u}(x,s) = D^{-1/2} \left[(s + \upsilon^2)^{1/2} - \upsilon\right]^{-1}\ \hat{J}(x,s) \ , \tag{B.15}$$

where

$$\upsilon = \frac{v_0}{2D^{1/2}} \ . \tag{B.16}$$

The convolution theorem (3.17) applied to Eq. (B.15) gives

$$u(x,t) = \int_0^t \mathcal{K}(t,\tau)\ J(x,\tau)\ \mathrm{d}\tau \ , \tag{B.17}$$

where the integral transformation kernel is of the convolution type. It can be easily obtained by finding the inverse Laplace transform of the multiplicative factor at $\hat{J}(x,s)$ in Eq. (B.15) in standard tables [1]:

$$\mathcal{K}(t,\tau) = D^{-1/2} \left\langle \frac{\exp\left[-\upsilon^2(t-\tau)\right]}{[\pi(t-\tau)]^{1/2}} + \upsilon\left\{1 + \mathrm{erf}\left[\upsilon(t-\tau)^{1/2}\right]\right\} \right\rangle \ . \tag{B.18}$$

Equation (B.17) holds at any x, in particular at the boundary at $x = 0$, corresponding to the interface studied.

A steady state characterised by the condition $\partial u(x,t)/\partial t \approx 0$ is possible. At steady state Eq. (B.1) becomes a second-order ODE:

$$D\frac{\partial^2 u(x,t)}{\partial x^2} + v_0 \frac{\partial u(x,t)}{\partial x} = 0\,. \tag{B.19}$$

By denoting

$$w(x,t) = \frac{\partial u(x,t)}{\partial x}\,, \tag{B.20}$$

Eq. (B.19) can be rewritten as a first-order ODE in variable x:

$$\frac{\partial w(x,t)}{\partial x} + \frac{v_0}{D}\, w(x,t) = 0\,. \tag{B.21}$$

The general solution of Eq. (B.21) is

$$w(x,t) = a(t)\,\exp\left(-\frac{v_0}{D}\,x\right)\,, \tag{B.22}$$

where $a(t)$ is an integration constant. Consequently, by combining Eqs. (B.20) and (B.22), and integrating between x and infinity, we obtain:

$$u(\infty,t) - u(x,t) = a(t)\int_x^\infty \exp\left(-\frac{v_0}{D}\,\xi\right)\,\mathrm{d}\xi = a(t)\,\frac{D}{v_0}\,\exp\left(-\frac{v_0}{D}\,x\right)\,. \tag{B.23}$$

In view of the boundary condition (B.3), Eq. (B.23) simplifies to

$$u(x,t) = -a(t)\,\frac{D}{v_0}\,\exp\left(-\frac{v_0}{D}\,x\right)\,. \tag{B.24}$$

Hence, a non-trivial steady state solution exists when $a(t) \neq 0$. However, from Eqs. (B.4) and (B.24) we find that at steady state the flux of variable u is identically zero:

$$J(x,t) = -D\frac{\partial u(x,t)}{\partial x} - v_0\, u(x,t)$$

$$= -a(t)\left[D\,\exp\left(-\frac{v_0}{D}\,x\right) - v_0\,\frac{D}{v_0}\,\exp\left(-\frac{v_0}{D}\,x\right)\right] = 0\,. \tag{B.25}$$

Therefore, there is no definite relationship between $u(x,t)$ and $J(x,t)$ at steady state.

B.2 Expanding Plane

In the case of the expanding plane model of the DME, the incomplete convection-diffusion IBVP describing the transport of distributed species [cf. Eqs. (6.23)–(6.25) in Sect. 6.2.2] includes the PDE:

$$\frac{\partial u(x,t)}{\partial t} = D\frac{\partial^2 u(x,t)}{\partial x^2} + \frac{2x}{3t}\frac{\partial u(x,t)}{\partial x} . \tag{B.26}$$

The initial condition is

$$u(x,0) = 0 , \tag{B.27}$$

and the boundary condition at $x \to \infty$ is

$$u(\infty,t) = 0 . \tag{B.28}$$

The flux of variable $u(x,t)$ is given by

$$J(x,t) = -D\frac{\partial u(x,t)}{\partial x} - \frac{2x}{3t}\, u(x,t) . \tag{B.29}$$

By introducing new independent variables ξ and θ:

$$\xi = x\, t^{2/3} , \tag{B.30}$$

$$\theta = \frac{3}{7} t^{7/3} , \tag{B.31}$$

Eq. (B.26) is transformed into the planar diffusion PDE:

$$\frac{\partial u(\xi,\theta)}{\partial \theta} = D\frac{\partial^2 u(\xi,\theta)}{\partial \xi^2} . \tag{B.32}$$

Therefore, a relationship analogous to Eq. (A.25) with the kernel given by Eq. (A.24) holds in the (ξ,θ) domain:

$$u(\xi,\theta) = \int_0^\theta D^{-1/2}\,[\pi(\theta-\vartheta)]^{-1/2}\left[-D\frac{\partial u(\xi,\vartheta)}{\partial \xi}\right] \mathrm{d}\vartheta . \tag{B.33}$$

By returning to the original variables x and t, Eq. (B.33) becomes

$$u(x,t) = \int_0^t \mathscr{K}(t,\tau)\left[-D\frac{\partial u(x,\tau)}{\partial x}\right] \mathrm{d}\tau , \tag{B.34}$$

where the integral transformation kernel $\mathcal{K}(t, \tau)$ is not of the convolution type:

$$\mathcal{K}(t,\tau) = D^{-1/2}\left(\frac{3}{7}\pi\right)^{-1/2} \tau^{2/3}\left(t^{7/3} - \tau^{7/3}\right)^{-1/2} . \tag{B.35}$$

Although the derivative expression in brackets in Eq. (B.34) is not identical with the flux (B.29) when $x > 0$, it coincides with the flux when $x = 0$, i.e. exactly at the boundary corresponding to the interface studied. Therefore, for $x = 0$ Eq. (B.34) becomes

$$u(0,t) = \int_0^t \mathcal{K}(t,\tau)\, J(0,\tau)\, \mathrm{d}\tau . \tag{B.36}$$

Equation (B.36) thus represents the solution–flux relationship at the expanding planar interface.

B.3 Rotating Disk Electrode

In the case of the RDE, the incomplete convection-diffusion IBVP describing the transport of distributed species [cf. Eqs. (6.37)–(6.39) in Sect. 6.2.3] includes the PDE:

$$\frac{\partial u(x,t)}{\partial t} = D\frac{\partial^2 u(x,t)}{\partial x^2} + \beta\, x^2\, \frac{\partial u(x,t)}{\partial x} , \tag{B.37}$$

where coefficient $\beta = 0.51023\,\omega^{3/2}\nu^{-1/2}$. The initial condition is

$$u(x,0) = 0 . \tag{B.38}$$

Assuming a semi-infinite spatial domain, the boundary condition at $x \to \infty$ is

$$u(\infty,t) = 0 . \tag{B.39}$$

The flux of variable $u(x,t)$ is given by:

$$J(x,t) = -D\frac{\partial u(x,t)}{\partial x} - \beta\, x^2 u(x,t) . \tag{B.40}$$

We introduce new independent variables ξ and ϑ:

$$\xi = D^{-1/2}\,\upsilon\, x , \tag{B.41}$$

$$\vartheta = \upsilon^2\, t , \tag{B.42}$$

where

$$\upsilon = \left(D^{1/2}\beta\right)^{1/3} . \tag{B.43}$$

Equations (B.37) and (B.38) become:

$$\frac{\partial u(\xi,\vartheta)}{\partial \vartheta} = \frac{\partial^2 u(\xi,\vartheta)}{\partial \xi^2} + \xi^2 \frac{\partial u(\xi,\vartheta)}{\partial \xi} , \tag{B.44}$$

$$u(\xi,0) = 0 , \tag{B.45}$$

whereas Eq. (B.39) turns into the boundary condition for $\xi \to \infty$:

$$u(\infty,\vartheta) = 0 . \tag{B.46}$$

By applying the Laplace transformation (3.12) (between the domains of ϑ and s) to Eqs. (B.44) and (B.46), and by taking into account the initial condition (B.45), we obtain the equation for $\hat{u}(\xi,s)$:

$$\mathcal{L}\left\{\frac{\partial u(\xi,\vartheta)}{\partial \vartheta}\right\} = s\hat{u}(\xi,s) - 0$$

$$= \mathcal{L}\left\{\frac{\partial^2 u(\xi,\vartheta)}{\partial \xi^2} + \xi^2 \frac{\partial u(\xi,\vartheta)}{\partial \xi}\right\} = \frac{\partial^2 \hat{u}(\xi,s)}{\partial \xi^2} + \xi^2 \frac{\partial \hat{u}(\xi,s)}{\partial \xi} . \tag{B.47}$$

with the boundary condition

$$\hat{u}(\infty,s) = 0 . \tag{B.48}$$

Equation (B.47) is a second-order ODE in the independent variable x, and it is equivalent to

$$\frac{\partial^2 \hat{u}(\xi,s)}{\partial \xi^2} + \xi^2 \frac{\partial \hat{u}(\xi,s)}{\partial \xi} - s\,\hat{u}(\xi,s) = 0 . \tag{B.49}$$

We introduce a new dependent variable $w(\xi,\vartheta)$ such that

$$u(\xi,\vartheta) = w(\xi,\vartheta)\exp\left(-\xi^3/6\right) . \tag{B.50}$$

Consequently, Eqs. (B.48) and (B.49) are transformed into

$$\frac{\partial^2 \hat{w}(\xi,s)}{\partial \xi^2} - \left(s + \xi + \frac{1}{4}\xi^4\right)\hat{w}(\xi,s) = 0 , \tag{B.51}$$

$$\hat{w}(\infty,s) = 0 . \tag{B.52}$$

Solution of Eqs. (B.51) and (B.52) cannot be expressed by known analytical functions. However, as the ODE (B.51) is linear, we expect, by analogy with other similar ODEs discussed in this book, that there exists a relationship between $\hat{w}(0,s)$ and $-\partial\hat{w}(\xi,s)/\partial\xi|_{\xi=0}$, having the form:

$$\hat{w}(0,s) = \hat{\varphi}(s)\left[-\left.\frac{\partial\hat{w}(\xi,s)}{\partial\xi}\right|_{\xi=0}\right], \tag{B.53}$$

where $\hat{\varphi}(s)$ is a suitable function. By applying the convolution theorem (3.17) to Eq. (B.53), we then obtain

$$w(0,\vartheta) = \int_0^\vartheta \varphi(\vartheta-\theta)\left[-\left.\frac{\partial w(\xi,\theta)}{\partial\xi}\right|_{\xi=0}\right] d\theta\,. \tag{B.54}$$

Function $\varphi(\vartheta)$ was determined approximately by Filinovskii and Kiryanov [2], by assuming that significant variations of $w(\xi,\vartheta)$ are limited to a small neighbourhood of $\xi = 0$. Under this assumption the term $\xi^4/4$ in Eq. (B.51) has a small effect on the solution, and can be neglected, so that Eq. (B.51) simplifies to

$$\frac{\partial^2\hat{w}(\xi,s)}{\partial\xi^2} - (s+\xi)\,\hat{w}(\xi,s) = 0\,. \tag{B.55}$$

Replacing ξ by a new independent variable ζ:

$$\zeta = s + \xi\,, \tag{B.56}$$

transforms Eq. (B.55) into

$$\frac{\partial^2\hat{w}(\zeta,s)}{\partial\zeta^2} - \zeta\,\hat{w}(\zeta,s) = 0\,. \tag{B.57}$$

Equation (B.57) is the Airy equation [1], so that it posesses the general solution

$$\hat{w}(\zeta,s) = a(s)\,\mathrm{Ai}(\zeta) + b(s)\,\mathrm{Bi}(\zeta)\,, \tag{B.58}$$

where $\mathrm{Ai}(\cdot)$ and $\mathrm{Bi}(\cdot)$ are the Airy functions, and $a(s)$ and $b(s)$ are integration constants independent of ζ. By returning to the variable ξ, Eq. (B.58) becomes

$$\hat{w}(\xi,s) = a(s)\,\mathrm{Ai}(s+\xi) + b(s)\,\mathrm{Bi}(s+\xi)\,. \tag{B.59}$$

In view of the boundary condition (B.52) $b(s)$ must be zero. Consequently,

$$\hat{w}(\xi,s) = a(s)\,\mathrm{Ai}(s+\xi) \tag{B.60}$$

and

$$\frac{\partial \hat{w}(\xi, s)}{\partial \xi} = a(s)\,\mathrm{Ai}'(s+\xi)\,, \tag{B.61}$$

where $\mathrm{Ai}'(\cdot)$ is the first derivative of the Airy function $\mathrm{Ai}(\cdot)$. By combining Eqs. (B.60) and (B.61), the coefficient $a(s)$ can be eliminated, resulting in the relationship

$$\hat{w}(\xi, s) = \left[-\frac{\mathrm{Ai}\,(s+\xi)}{\mathrm{Ai}'\,(s+\xi)}\right]\left[-\frac{\partial \hat{w}(\xi, s)}{\partial \xi}\right]. \tag{B.62}$$

In particular, at $\xi = 0$,

$$\hat{w}(0, s) = \left[-\frac{\mathrm{Ai}\,(s)}{\mathrm{Ai}'\,(s)}\right]\left[-\left.\frac{\partial \hat{w}(\xi, s)}{\partial \xi}\right|_{\xi=0}\right]. \tag{B.63}$$

Equation (B.63) has the form of Eq. (B.53). Hence, in the approximation of Filinovskii and Kiryanov [2],

$$\varphi(\vartheta) \approx \mathscr{L}^{-1}\left\{-\frac{\mathrm{Ai}\,(s)}{\mathrm{Ai}'\,(s)}\right\}. \tag{B.64}$$

The Airy functions can be expressed [1] in terms of the modified Bessel functions $\mathrm{K}_{1/3}\,(\cdot)$ and $\mathrm{K}_{2/3}\,(\cdot)$ of the second kind and fractional orders:

$$\mathrm{Ai}(s) = \frac{1}{\pi}\left(\frac{s}{3}\right)^{1/2}\mathrm{K}_{1/3}\left(\frac{2}{3}\,s^{3/2}\right), \tag{B.65}$$

$$\mathrm{Ai}'(s) = -\frac{1}{\pi}\,\frac{s}{3^{1/2}}\,\mathrm{K}_{2/3}\left(\frac{2}{3}\,s^{3/2}\right). \tag{B.66}$$

Therefore,

$$\varphi(\vartheta) \approx \mathscr{L}^{-1}\left\{\frac{\mathrm{K}_{1/3}\left(\frac{2}{3}s^{3/2}\right)}{s^{1/2}\,\mathrm{K}_{2/3}\left(\frac{2}{3}s^{3/2}\right)}\right\}, \tag{B.67}$$

or, more conveniently,

$$\varphi(\vartheta) = (\pi\vartheta)^{-1/2} - \mathrm{krd}(\vartheta^{1/2})\,, \tag{B.68}$$

where the function $\mathrm{krd}(\cdot)$ is approximated by formula (6.49).

Equation (B.68) remains valid even if we abandon the approximation from Filinovskii and Kiryanov [2] and do not neglect the term $\xi^4/4$ in Eq. (B.51). The function krd(·) can then be evaluated by solving the ODE (B.51) numerically with the boundary conditions (B.52) and

$$\left.\frac{\partial \hat{w}(\xi, s)}{\partial \xi}\right|_{\xi=0} = -g(s) , \tag{B.69}$$

where $g(s)$ is a prescribed function. From Eqs. (B.53) and (B.68) we then obtain:

$$\mathrm{krd}(z) = \mathrm{krd}(\vartheta^{1/2}) = (\pi\vartheta)^{-1/2} - \mathscr{L}^{-1}\left\{\frac{\hat{w}(0, s)}{g(s)}\right\} , \tag{B.70}$$

where $\hat{w}(0, s)$ is a boundary value of the numerical solution. Function $g(s)$ in Eqs. (B.69) and (B.70) can be arbitrary, since the function $\hat{\varphi}(s)$ does not depend on the boundary conditions at $\xi = 0$. The choice $g(s) = -1/s$ corresponds to a model of current step chronopotentiometry, but numerical calculations are somewhat easier if we take $g(s) = -1$. We note that numerical inversion of the Laplace transform is needed in Eq. (B.70), which presents high demands on the accuracy of the numerical solution of Eq. (B.51), and makes such determinations of krd(·) difficult.

By combining Eqs. (B.54) and (B.68), returning to the original variables x, t, and $u(x, t)$, and taking into account Eqs. (B.40) and (B.50) for $x = 0$, we finally obtain:

$$u(0, t) = \int_0^t \mathscr{K}(t, \tau)\, J(0, \tau)\, d\tau , \tag{B.71}$$

where the integral transformation kernel is of the convolution type:

$$\begin{aligned}\mathscr{K}(t, \tau) &= D^{-1/2}\, \upsilon\, \varphi[\upsilon^2(t - \tau)] \\ &= D^{-1/2} \left\{[\pi(t - \tau)]^{-1/2} - \upsilon\, \mathrm{krd}\left[\upsilon(t - \tau)^{1/2}\right]\right\} .\end{aligned} \tag{B.72}$$

Equation (B.71) is the solution–flux relationship at the boundary at $x = 0$, corresponding to the interface studied.

A special case of Eq. (B.37) corresponds to the steady state characterised by the condition $\partial u(x, t)/\partial t \approx 0$. Equation (B.37) then becomes a second-order ODE:

$$D\frac{\partial^2 u(x, t)}{\partial x^2} + \beta\, x^2\, \frac{\partial u(x, t)}{\partial x} = 0 . \tag{B.73}$$

By denoting

$$w(x, t) = \frac{\partial u(x, t)}{\partial x} , \tag{B.74}$$

Eq. (B.73) can be rewritten as a first-order ODE:

$$\frac{\partial w(x,t)}{\partial x} + \frac{\beta}{D} x^2 w(x,t) = 0 . \tag{B.75}$$

The general solution of Eq. (B.75) is

$$w(x,t) = a(t) \exp\left(-\frac{\beta}{3D} x^3\right) , \tag{B.76}$$

where $a(t)$ is an integration constant. Consequently, by combining Eqs. (B.74) and (B.76), and integrating between x and infinity, we obtain:

$$u(\infty,t) - u(x,t) = a(t) \int_x^\infty \exp\left(-\frac{\beta}{3D} \xi^3\right) \mathrm{d}\xi , \tag{B.77}$$

where ξ is an integration variable. In view of the boundary condition (B.39), Eq. (B.77) simplifies to

$$u(x,t) = -a(t) \int_x^\infty \exp\left(-\frac{\beta}{3D} \xi^3\right) \mathrm{d}\xi . \tag{B.78}$$

From Eqs. (B.40) and (B.78) the general solution for the flux of variable u is

$$J(x,t) = -D\frac{\partial u(x,t)}{\partial x} - \beta x^2 u(x,t)$$

$$= -a(t) \left[D \exp\left(-\frac{\beta}{3D} x^3\right) - \beta x^2 \int_x^\infty \exp\left(-\frac{\beta}{3D} \xi^3\right) \mathrm{d}\xi \right] . \tag{B.79}$$

By eliminating $a(t)$ from Eqs. (B.78) and (B.79) we get the solution–flux relationship for the steady state:

$$u(x,t) = \left[D \frac{\exp\left(-\frac{\beta}{3D} x^3\right)}{\int_x^\infty \exp\left(-\frac{\beta}{3D} \xi^3\right) \mathrm{d}\xi} - \beta x^2 \right]^{-1} J(x,t) . \tag{B.80}$$

In particular, at the boundary at $x = 0$, corresponding to the interface studied:

$$u(0,t) = \left[D^{-1} \int_0^\infty \exp\left(-\frac{\beta}{3D} \xi^3\right) \mathrm{d}\xi \right] J(0,t) . \tag{B.81}$$

As we know that $\int_0^\infty \exp\left(-z^3\right) \mathrm{d}z = \Gamma(1/3)/3$, where $\Gamma(z)$ is the Euler gamma function [1], after simple substitutions of variables Eq. (B.81) can be rewritten as

$$u(0,t) = D^{-1/2}\,\upsilon^{-1}\,3^{-2/3}\,\Gamma(1/3)\,J(0,t)\,. \tag{B.82}$$

B.4 Channel and Tubular Electrodes

Assuming the Singh and Dutt approximation discussed in Sect. 6.2.4, the incomplete convection-diffusion IBVP describing the transport of distributed species [cf. Eqs. (6.62)–(6.64)] includes the PDE:

$$\frac{\partial u(x,t)}{\partial t} = D\frac{\partial^2 u(x,t)}{\partial x^2} - \gamma\, x\, u(x,t)\,, \tag{B.83}$$

where according to Eqs. (6.56) and (6.58), coefficient $\gamma = 2v_0(hl)^{-1}$ for channel, or $\gamma = 2v_0(r_0 l)^{-1}$ for tubular electrodes. The initial condition is

$$u(x,0) = 0\,, \tag{B.84}$$

and the boundary condition in the middle of the channel or tube (assumed infinitely distant from the electrode surface) is

$$u(\infty,t) = 0\,. \tag{B.85}$$

As Eq. (B.83) is formally of the reaction-diffusion type, the flux $J(x,t)$ of variable $u(x,t)$ does not contain a convective term:

$$J(x,t) = -D\frac{\partial u(x,t)}{\partial x}\,. \tag{B.86}$$

We introduce new independent variables:

$$\xi = D^{-1/2}\upsilon\, x\,, \tag{B.87}$$

$$\vartheta = \upsilon^2\, t\,, \tag{B.88}$$

where

$$\upsilon = \left(D^{1/2}\gamma\right)^{1/3}\,. \tag{B.89}$$

Equations (B.83) and (B.84) become:

$$\frac{\partial u(\xi,\vartheta)}{\partial \vartheta} = \frac{\partial^2 u(\xi,\vartheta)}{\partial \xi^2} - \xi\, u(\xi,\vartheta)\ , \tag{B.90}$$

$$u(\xi,0) = 0\ , \tag{B.91}$$

and Eq. (B.85) turns into the boundary condition for $\xi \to \infty$:

$$u(\infty,\vartheta) = 0\ . \tag{B.92}$$

By applying the Laplace transformation (3.12) (between the ϑ and s domains) to Eqs. (B.90) and (B.92), and by taking into account the initial condition (B.91), we obtain the equation for $\hat{u}(x,s)$:

$$\mathscr{L}\left\{\frac{\partial u(\xi,\vartheta)}{\partial \vartheta}\right\} = s\hat{u}(\xi,s) - 0$$

$$= \mathscr{L}\left\{\frac{\partial^2 u(\xi,\vartheta)}{\partial \xi^2} - \xi\, u(\xi,\vartheta)\right\} = \frac{\partial^2 \hat{u}(\xi,s)}{\partial \xi^2} - \xi\, \hat{u}(\xi,s)\ . \tag{B.93}$$

with the boundary condition

$$\hat{u}(\infty,s) = 0\ . \tag{B.94}$$

Equation (B.93) is a second-order ODE in the independent variable ξ, and it is equivalent to

$$\frac{\partial^2 \hat{u}(\xi,s)}{\partial \xi^2} - (s+\xi)\ \hat{u}(\xi,s) = 0\ . \tag{B.95}$$

Despite the different origin, Eqs. (B.94) and (B.95) are formally identical to Eqs. (B.52) and (B.55), discussed in connection with convection-diffusion at the RDEs. Therefore, by proceeding in the same manner as described in Sect. B.3, one obtains [cf. Eqs. (B.54), (B.64), and (B.67)]:

$$u(0,\vartheta) = \int_0^{\vartheta} \varphi(\vartheta-\theta) \left[-\left.\frac{\partial u(\xi,\theta)}{\partial \xi}\right|_{\xi=0}\right] \mathrm{d}\theta\ . \tag{B.96}$$

where

$$\varphi(\vartheta) = \mathscr{L}^{-1}\left\{-\frac{\mathrm{Ai}\,(s)}{\mathrm{Ai}'\,(s)}\right\} = \mathscr{L}^{-1}\left\{\frac{\mathrm{K}_{1/3}\left(\frac{2}{3}s^{3/2}\right)}{s^{1/2}\,\mathrm{K}_{2/3}\left(\frac{2}{3}s^{3/2}\right)}\right\}\ , \tag{B.97}$$

with Ai(·) and Ai′(·) denoting the Airy function and its derivative, and $K_{1/3}(\cdot)$ and $K_{2/3}(\cdot)$ denoting the modified Bessel functions of the second kind and fractional orders [1].

By returning to the variables x and t, Eqs. (B.96) and (B.97) yield:

$$u(0,t) = \int_0^t \mathscr{K}(t,\tau)\; J(0,\tau)\, \mathrm{d}\tau\,, \tag{B.98}$$

where the integral transformation kernel is of the convolution type:

$$\begin{aligned}\mathscr{K}(t,\tau) &= D^{-1/2}\, \upsilon\, \varphi[\upsilon^2(t-\tau)] \\ &= D^{-1/2}\left\{[\pi(t-\tau)]^{-1/2} - \upsilon\, \mathrm{kct}\left[\upsilon(t-\tau)^{1/2}\right]\right\},\end{aligned} \tag{B.99}$$

where the function kct(·) is defined by Eq. (6.71). Equation (B.98) is the solution–flux relationship at the boundary at $x = 0$, corresponding to the interface studied.

A special case of Eq. (B.83) corresponds to the steady state characterised by the condition $\partial u(x,t)/\partial t \approx 0$. Equation (B.83) then becomes a second-order ODE:

$$D\frac{\partial^2 u(x,t)}{\partial x^2} - \gamma\, x\, u(x,t) = 0\,. \tag{B.100}$$

By introducing a new spatial variable ξ defined by Eq. (B.87), Eq. (B.100) becomes the Airy equation [1]:

$$\frac{\partial^2 u(\xi,t)}{\partial \xi^2} - \xi\, u(\xi,t) = 0\,, \tag{B.101}$$

whereas the boundary condition (B.85) for $x \to \infty$ turns into the condition for $\xi \to \infty$:

$$u(\infty,t) = 0\,. \tag{B.102}$$

The general solution of Eq. (B.101) is

$$u(\xi,t) = a(t)\,\mathrm{Ai}(\xi) + b(t)\,\mathrm{Bi}(\xi)\,, \tag{B.103}$$

where $a(t)$ and $b(t)$ are integration constants independent of ξ. In view of the boundary condition (B.102) $b(t)$ must be zero. By returning to the variable x, Eq. (B.103) becomes

$$u(x,t) = a(t)\,\mathrm{Ai}\left(D^{-1/2}\upsilon\, x\right)\,. \tag{B.104}$$

Consequently, the flux (B.86) is

$$J(x,t) = -D^{1/2}\,\upsilon\, a(t)\ \mathrm{Ai}'\left(D^{-1/2}\upsilon\, x\right)\ . \tag{B.105}$$

By eliminating $a(t)$ from Eqs. (B.104) and (B.105) we get the solution–flux relationship for the steady state:

$$u(x,t) = D^{-1/2}\,\upsilon^{-1}\left[-\frac{\mathrm{Ai}(D^{-1/2}\upsilon\, x)}{\mathrm{Ai}'(D^{-1/2}\upsilon\, x)}\right] J(x,t)\ . \tag{B.106}$$

In particular, at the boundary at $x = 0$, corresponding to the interface studied,

$$u(0,t) = D^{-1/2}\,\upsilon^{-1}\, 3^{-1}\frac{\Gamma(1/3)}{\Gamma(2/3)}\, J(0,t)\ , \tag{B.107}$$

since $\mathrm{Ai}(0) = 3^{-2/3}/\,\Gamma(2/3)$ and $\mathrm{Ai}'(0) = -3^{-1/3}/\,\Gamma(1/3)$ (cf. Eqs. (10.4.4) and (10.4.5) in Abramowitz and Stegun [1]), with $\Gamma(\cdot)$ denoting the Euler gamma function [1].

References

1. Abramowitz M, Stegun IA (1972) Handbook of mathematical functions. Dover Publications, New York
2. Filinovskii VYu, Kiryanov VA (1964) On the theory of non-stationary convective diffusion at the rotating disk electrode. Dokl Akad Nauk SSSR 156:1412–1415 (in Russian)
3. Kamke E (1962) Differentialgleichungen. I. Gewöhnliche Differentialgleichungen, Geest & Portig, Leipzig

Appendix C
Solution–Flux Relationships for Multidimensional Diffusion Equations (Mirkin and Bard Approach)

We derive here analytical solution–flux relationships for the archetypal transient linear diffusion PDE

$$\frac{\partial u}{\partial t} = D \bigtriangleup u \,, \tag{C.1}$$

defined over example two-dimensional spatial domains that are infinite or semi-infinite in one spatial direction, and semi-infinite in the second spatial direction. In Eq. (C.1) D is a (positive) diffusion coefficient, and u is an unknown function of spatial coordinates and time t. Solution u can be positive or negative; it does not have to represent a concentration that is always non-negative. The trivial solution $u \equiv 0$ is not of interest. We assume that only initially $u = 0$. We determine solution–flux relationships at one boundary (corresponding to an electrochemical interface studied), which must hold irrespective of the actual boundary conditions (not explicitly formulated here) imposed at this boundary. We consider two geometries of the diffusion fields: diffusion to infinite parallel band(s) on a planar interface studied, and diffusion to a disk/ring(s) system on the planar interface. As the bands are assumed to be infinitely long, edge effects that might result from a finite length of the bands are not taken into account, and two spatial coordinates are sufficient for the description. Relevant derivations are contained in Sects. C.1 and C.2, respectively. The derivations are performed by employing the Mirkin and Bard approach [3, 4]. This implies that the Laplace transformation is used with respect to the time variable, and the Fourier or Hankel transformations are used with respect to the selected spatial variables.

L.K. Bieniasz, *Modelling Electroanalytical Experiments by the Integral Equation Method*, Monographs in Electrochemistry, DOI 10.1007/978-3-662-44882-3

C.1 Diffusion to Infinite Parallel Band(s) on a Plane

In the case of transient diffusion to a number of infinite bands on a plane, Eq. (C.1) can be written using the Cartesian coordinate system (cf. Table 2.1 and Fig. 2.3):

$$\frac{\partial u(x,z,t)}{\partial t} = D\left[\frac{\partial^2 u(x,z,t)}{\partial x^2} + \frac{\partial^2 u(x,z,t)}{\partial z^2}\right] . \tag{C.2}$$

The initial condition is

$$u(x,z,0) = 0 , \tag{C.3}$$

and the boundary conditions at other boundaries are

$$u(\pm\infty,z,t) = 0 , \tag{C.4}$$

$$u(x,\infty,t) = 0 . \tag{C.5}$$

By applying the Fourier transformation (3.20), with respect to the variable x, to Eq. (C.2), and by taking into account the boundary condition (C.4), we obtain an equation for $\breve{u}(k,z,t)$:

$$\mathscr{F}\left\{\frac{\partial u(x,z,t)}{\partial t}\right\} = \frac{\partial \breve{u}(k,z,t)}{\partial t}$$

$$= \mathscr{F}\left\{D\left[\frac{\partial^2 u(x,z,t)}{\partial x^2} + \frac{\partial^2 u(x,z,t)}{\partial z^2}\right]\right\} = -D\,k^2\,\breve{u}(k,z,t) + D\frac{\partial^2 \breve{u}(k,z,t)}{\partial z^2} , \tag{C.6}$$

where k is the Fourier variable, and where formula (3.24) with $n = 2$ is utilised. We note that the Fourier transform of the initial condition (C.3) is

$$\breve{u}(k,z,0) = \mathscr{F}\{0\} = 0 . \tag{C.7}$$

Therefore, application of the Laplace transformation (3.12), with respect to the variable t, to Eq. (C.6) gives:

$$\mathscr{L}\left\{\frac{\partial \breve{u}(k,z,t)}{\partial t}\right\} = s\hat{\breve{u}}(k,z,s) - 0$$

$$= \mathscr{L}\left\{-D\,k^2\,\breve{u}(k,z,t) + D\frac{\partial^2 \breve{u}(k,z,t)}{\partial z^2}\right\} = -D\,k^2\,\hat{\breve{u}}(k,z,s) + D\frac{\partial^2 \hat{\breve{u}}(k,z,s)}{\partial z^2} . \tag{C.8}$$

Equation (C.8) is a second-order linear homogeneous ODE in the independent variable z, and with constant coefficients. It is equivalent to

$$\frac{\partial^2 \hat{\tilde{u}}(k,z,s)}{\partial z^2} - \left(k^2 + \frac{s}{D}\right) \hat{\tilde{u}}(k,z,s) = 0 \,. \tag{C.9}$$

Application of the Fourier and Laplace transformations to the boundary condition (C.5) gives in turn

$$\hat{\tilde{u}}(k,\infty,s) = \mathscr{L}\{\mathscr{F}\{0\}\} = 0 \,. \tag{C.10}$$

The general solution of Eq. (C.9) is [2]:

$$\hat{\tilde{u}}(k,z,s) = a(s) \exp\left[-\left(k^2 + \frac{s}{D}\right)^{1/2} z\right] + b(s) \exp\left[\left(k^2 + \frac{s}{D}\right)^{1/2} z\right], \tag{C.11}$$

where $a(s)$ and $b(s)$ are integration constants independent of z. Differentiation with respect to z gives the Fourier–Laplace transform of the flux component in the direction of z:

$$\begin{aligned}\hat{\tilde{J}}_z(k,z,s) &= -D\frac{\partial \hat{\tilde{u}}(k,z,s)}{\partial z} \\ &= D^{1/2}\left(D\,k^2 + s\right)^{1/2}\left\{a(s) \exp\left[-\left(k^2 + \frac{s}{D}\right)^{1/2} z\right]\right. \\ &\quad \left. -b(s) \exp\left[\left(k^2 + \frac{s}{D}\right)^{1/2} z\right]\right\} \,. \end{aligned} \tag{C.12}$$

In view of the boundary condition (C.10), $b(s) = 0$ is the only option, so that Eqs. (C.11) and (C.12) become

$$\hat{\tilde{u}}(k,z,s) = a(s) \exp\left[-\left(k^2 + \frac{s}{D}\right)^{1/2} z\right], \tag{C.13}$$

$$\hat{\tilde{J}}_z(k,z,s) = D^{1/2}\left(D\,k^2 + s\right)^{1/2} a(s) \exp\left[-\left(k^2 + \frac{s}{D}\right)^{1/2} z\right]. \tag{C.14}$$

By eliminating $a(s)$ from Eqs. (C.13) and (C.14) we obtain

$$\hat{\tilde{u}}(k,z,s) = D^{-1/2}\left(D\,k^2 + s\right)^{-1/2} \hat{\tilde{J}}_z(k,z,s) \,. \tag{C.15}$$

The inverse Laplace transformation (3.14) of $\left(D\,k^2 + s\right)^{-1/2}$ is

$$\mathscr{L}^{-1}\left\{\left(D\,k^2 + s\right)^{-1/2}\right\} = (\pi t)^{-1/2} \exp\left(-D\,k^2 t\right) \,. \tag{C.16}$$

Hence, from the convolution theorem (3.17):

$$\breve{u}(k,z,t) = D^{-1/2} \int_0^t \frac{\exp\left[-D\,k^2\,(t-\tau)\right]}{\left[\pi(t-\tau)\right]^{1/2}}\;\breve{J}_z(k,z,\tau)\;\mathrm{d}\tau\;. \tag{C.17}$$

We now apply the inverse Fourier transformation (3.22) to Eq. (C.17), and substitute the following direct Fourier transform for $\breve{J}_z(k,z,\tau)$:

$$\breve{J}_z(k,z,\tau) = \int_{-\infty}^{\infty} \exp\left(-\mathrm{i}k\zeta\right)\;J_z(\zeta,z,\tau)\;\mathrm{d}\zeta\;. \tag{C.18}$$

This gives

$$u(x,z,t)$$
$$= \frac{1}{2\pi}\int_{-\infty}^{\infty} \mathrm{e}^{\mathrm{i}kx}\left\{D^{-1/2}\int_0^t \frac{\mathrm{e}^{-Dk^2(t-\tau)}}{\left[\pi(t-\tau)\right]^{1/2}}\left[\int_{-\infty}^{\infty}\mathrm{e}^{-\mathrm{i}k\zeta}\,J_z(\zeta,z,\tau)\;\mathrm{d}\zeta\right]\mathrm{d}\tau\right\}\mathrm{d}k\;. \tag{C.19}$$

By changing the integration order in Eq. (C.19), we arrive at the integral which can be calculated analytically with the following result (see, for example, formula (7.42) in Abramowitz and Stegun [1]):

$$\int_{-\infty}^{\infty} \exp\left[-D\,k^2\,(t-\tau)\right]\exp\left[\mathrm{i}k\,(x-\zeta)\right]\mathrm{d}k = \left[\frac{\pi}{D(t-\tau)}\right]^{1/2}\exp\left[-\frac{(x-\zeta)^2}{4D(t-\tau)}\right]. \tag{C.20}$$

Consequently, Eq. (C.19) becomes

$$u(x,z,t) = \int_{\mathbb{D}} \mathscr{K}\left(\begin{bmatrix} x \\ t \end{bmatrix},\begin{bmatrix} \zeta \\ \tau \end{bmatrix}\right) J_z(\zeta,z,\tau)\,\mathrm{d}\zeta\,\mathrm{d}\tau\;, \tag{C.21}$$

where the integral transformation kernel is a function of two vector variables $[x,\,t]^{\mathrm{T}}$ and $[\zeta,\,\tau]^{\mathrm{T}}$:

$$\mathscr{K}\left(\begin{bmatrix} x \\ t \end{bmatrix},\begin{bmatrix} \zeta \\ \tau \end{bmatrix}\right) = D^{-1}\left[2\pi(t-\tau)\right]^{-1}\exp\left[-\frac{(x-\zeta)^2}{4D(t-\tau)}\right]. \tag{C.22}$$

The integration domain in Eq. (C.21) is $\mathbb{D}=\mathbb{D}_\zeta\times\mathbb{D}_\tau$, where $\zeta\in\mathbb{D}_\zeta=(-\infty,\infty)$ and $\tau\in\mathbb{D}_\tau=[0,t)$.

In particular, at the boundary at $z = 0$, corresponding to the interface studied, Eq. (C.21) predicts:

$$u(x,0,t) = \int\limits_{\mathbb{D}} \mathcal{K}\left(\begin{bmatrix} x \\ t \end{bmatrix}, \begin{bmatrix} \zeta \\ \tau \end{bmatrix}\right) J_z(\zeta,0,\tau)\,\mathrm{d}\zeta\,\mathrm{d}\tau\,. \tag{C.23}$$

C.2 Diffusion to Disk/Ring(s) on a Plane

In the case of transient diffusion to a disk and a finite number of rings on a plane, Eq. (C.1) can be written using the cylindrical coordinate system (cf. Table 2.1 and Fig. 2.4):

$$\frac{\partial u(r,z,t)}{\partial t} = D\left[\frac{\partial^2 u(r,z,t)}{\partial r^2} + \frac{1}{r}\frac{\partial u(r,z,t)}{\partial r} + \frac{\partial^2 u(r,z,t)}{\partial z^2}\right]. \tag{C.24}$$

The initial condition is

$$u(r,z,0) = 0\,, \tag{C.25}$$

and the boundary conditions at other boundaries are

$$u(\infty,z,t) = 0\,, \tag{C.26}$$

$$u(r,\infty,t) = 0\,. \tag{C.27}$$

By applying the Hankel transformation (3.25) of order zero, with respect to the variable r, to Eq. (C.24), and by taking into account the boundary condition (C.26), we obtain an equation for $\breve{u}(k,z,t)$:

$$\begin{aligned} \mathcal{H}\left\{\frac{\partial u(r,z,t)}{\partial t}\right\} &= \frac{\partial \breve{u}(p,z,t)}{\partial t} \\ = \mathcal{H}\left\{D\left[\frac{\partial^2 u(r,z,t)}{\partial r^2} + \frac{1}{r}\frac{\partial u(r,z,t)}{\partial r} + \frac{\partial^2 u(r,z,t)}{\partial z^2}\right]\right\} & \\ = -D\,p^2\,\breve{u}(p,z,t) + D\frac{\partial^2 \breve{u}(p,z,t)}{\partial z^2}\,, & \end{aligned} \tag{C.28}$$

where p is the Hankel variable, and where formula (3.28) is utilised. We note that the Hankel transform of the initial condition (C.25) is

$$\breve{u}(p,z,0) = \mathcal{H}\{0\} = 0\,. \tag{C.29}$$

Therefore, application of the Laplace transformation (3.12), with respect to the variable t, to Eq. (C.28), gives:

$$\mathscr{L}\left\{\frac{\partial \breve{u}(p,z,t)}{\partial t}\right\} = s\hat{\breve{u}}(p,z,s) - 0$$

$$= \mathscr{L}\left\{-D\, p^2\, \breve{u}(p,z,t) + D\frac{\partial^2 \breve{u}(p,z,t)}{\partial z^2}\right\} = -D\, p^2\, \hat{\breve{u}}(p,z,s) + D\frac{\partial^2 \hat{\breve{u}}(p,z,s)}{\partial z^2}\,. \tag{C.30}$$

Equation (C.30) is a second-order linear homogeneous ODE in the independent variable z, and with constant coefficients. It is equivalent to

$$\frac{\partial^2 \hat{\breve{u}}(p,z,s)}{\partial z^2} - \left(p^2 + \frac{s}{D}\right)\hat{\breve{u}}(p,z,s) = 0\,. \tag{C.31}$$

Application of the Hankel and Laplace transformations to the boundary condition (C.27) gives in turn

$$\hat{\breve{u}}(p,\infty,s) = \mathscr{L}\{\mathscr{H}\{0\}\} = 0\,. \tag{C.32}$$

The general solution of Eq. (C.31) is [2]:

$$\hat{\breve{u}}(p,z,s) = a(s)\,\exp\left[-\left(p^2+\frac{s}{D}\right)^{1/2} z\right] + b(s)\,\exp\left[\left(p^2+\frac{s}{D}\right)^{1/2} z\right], \tag{C.33}$$

where $a(s)$ and $b(s)$ are integration constants independent of z. Differentiation with respect to z gives the Hankel–Laplace transform of the flux component in the direction of z:

$$\hat{\breve{J}}_z(p,z,s) = -D\frac{\partial \hat{\breve{u}}(p,z,s)}{\partial z}$$

$$= D^{1/2}\left(D\,p^2+s\right)^{1/2}\left\{a(s)\,\exp\left[-\left(p^2+\frac{s}{D}\right)^{1/2} z\right] - b(s)\,\exp\left[\left(p^2+\frac{s}{D}\right)^{1/2} z\right]\right\}. \tag{C.34}$$

In view of the boundary condition (C.32), $b(s) = 0$ is the only option, so that Eqs. (C.33) and (C.34) become

$$\hat{\breve{u}}(p,z,s) = a(s)\,\exp\left[-\left(p^2+\frac{s}{D}\right)^{1/2} z\right] \tag{C.35}$$

$$\hat{\breve{J}}_z(p,z,s) = D^{1/2}\left(D\,p^2+s\right)^{1/2} a(s)\,\exp\left[-\left(p^2+\frac{s}{D}\right)^{1/2} z\right]. \tag{C.36}$$

By eliminating $a(s)$ from Eqs. (C.35) and (C.36) we obtain

$$\mathring{\check{u}}(p,z,s) = D^{-1/2}\left(D\,p^2+s\right)^{-1/2}\,\mathring{\check{J}}_z(p,z,s)\,. \tag{C.37}$$

The inverse Laplace transform (3.14) of $\left(D\,p^2+s\right)^{-1/2}$ is

$$\mathscr{L}^{-1}\left\{\left(D\,p^2+s\right)^{-1/2}\right\} = (\pi t)^{-1/2}\exp\left(-D\,p^2 t\right)\,. \tag{C.38}$$

Hence, from the convolution theorem (3.17):

$$\check{u}(p,z,t) = D^{-1/2}\int_0^t \frac{\exp\left[-D\,p^2\,(t-\tau)\right]}{[\pi(t-\tau)]^{1/2}}\,\check{J}_z(p,z,\tau)\,\mathrm{d}\tau\,. \tag{C.39}$$

We now apply the inverse Hankel transformation (3.27) to Eq. (C.39), and substitute the following direct Hankel transform for $\check{J}_z(p,z,\tau)$:

$$\check{J}_z(p,z,\tau) = \int_0^\infty \mathrm{J}_0\,(p\zeta)\,\zeta\;J_z(\zeta,z,\tau)\,\mathrm{d}\zeta\,. \tag{C.40}$$

This gives

$$u(r,z,t)$$
$$= \int_0^\infty \mathrm{J}_0\,(pr)\,p\left\{D^{-1/2}\int_0^t \frac{\mathrm{e}^{-Dp^2(t-\tau)}}{[\pi(t-\tau)]^{1/2}}\left[\int_0^\infty \mathrm{J}_0\,(p\zeta)\,\zeta\;J_z(\zeta,z,\tau)\,\mathrm{d}\zeta\right]\mathrm{d}\tau\right\}\mathrm{d}p\,. \tag{C.41}$$

By changing the integration order in Eq. (C.41), we arrive at the integral which can be calculated analytically with the following result (cf. Watson [5, p. 395]):

$$\int_0^\infty \exp\left[-D\,p^2\,(t-\tau)\right]\mathrm{J}_0\,(p\zeta)\,\mathrm{J}_0\,(pr)\,p\,\mathrm{d}p$$
$$= \frac{1}{2D(t-\tau)}\exp\left[-\frac{\zeta^2+r^2}{4D(t-\tau)}\right]\mathrm{I}_0\left[\frac{r\zeta}{2D(t-\tau)}\right], \tag{C.42}$$

where $\mathrm{I}_0(\cdot)$ denotes the modified Bessel function of the first kind and order zero [1]. Consequently, Eq. (C.41) becomes

$$u(r,z,t) = \int_{\mathbb{D}} \mathscr{K}\left(\begin{bmatrix} r \\ t \end{bmatrix}, \begin{bmatrix} \zeta \\ \tau \end{bmatrix}\right) J_z(\zeta,z,\tau)\,\mathrm{d}\zeta\,\mathrm{d}\tau\,, \tag{C.43}$$

where the integral transformation kernel is a function of two vector variables $[r,\, t]^{\mathrm{T}}$ and $[\zeta,\, \tau]^{\mathrm{T}}$:

$$\mathcal{K}\left(\begin{bmatrix} r \\ t \end{bmatrix}, \begin{bmatrix} \zeta \\ \tau \end{bmatrix}\right) = D^{-3/2} \frac{\zeta \exp\left[-\frac{\zeta^2+r^2}{4D(t-\tau)}\right] \mathrm{I}_0\left[\frac{r\zeta}{2D(t-\tau)}\right]}{2\pi^{1/2}(t-\tau)^{3/2}} \,. \tag{C.44}$$

The integration domain in Eq. (C.43) is $\mathbb{D} = \mathbb{D}_\zeta \times \mathbb{D}_\tau$, where $\zeta \in \mathbb{D}_\zeta = [0, \infty)$ and $\tau \in \mathbb{D}_\tau = [0, t)$.

In particular, at the boundary at $z = 0$, corresponding to the interface studied, Eq. (C.43) predicts:

$$u(r, 0, t) = \int_{\mathbb{D}} \mathcal{K}\left(\begin{bmatrix} r \\ t \end{bmatrix}, \begin{bmatrix} \zeta \\ \tau \end{bmatrix}\right) J_z(\zeta, 0, \tau)\, \mathrm{d}\zeta\, \mathrm{d}\tau \,. \tag{C.45}$$

References

1. Abramowitz M, Stegun IA (1972) Handbook of mathematical functions. Dover Publications, New York
2. Kamke E (1962) Differentialgleichungen. I. Gewöhnliche Differentialgleichungen, Geest & Portig, Leipzig
3. Mirkin MV, Bard AJ (1992) Multidimensional integral equations. Part 1. A new approach to solving microelectrode diffusion problems. J Electroanal Chem 323:1–27
4. Mirkin MV, Bard AJ (1992) Multidimensional integral equations: a new approach to solving microelectrode diffusion problems. Part 2. Applications to microband electrodes and the scanning electrochemical microscope. J Electroanal Chem 323:29–51
5. Watson GN (1944) A treatise on the theory of Bessel functions. Cambridge University Press, Cambridge

Appendix D
Solution–Flux Relationships for Multidimensional Diffusion Equations (the BIM)

We derive here analytical solution–flux relationships for the archetypal linear diffusion PDE

$$\frac{\partial u}{\partial t} = D \bigtriangleup u \,, \tag{D.1}$$

defined over a generally three-dimensional spatial domain Ω surrounded by a closed surface S. In Eq. (D.1) D is a (positive) diffusion coefficient, and u is an unknown function of spatial coordinates and time t. Solution u is assumed to be positive or zero; it may represent a concentration that is always non-negative. The trivial solution $u \equiv 0$ is not of interest. We assume that initially $u = u^{\star}$. We determine a solution–flux relationship at the boundary S by employing the theory of the BIM [3]. A special case of a steady state Eq. (D.1) is also considered. The solution–flux relationship obtained must hold irrespective of the actual boundary conditions (not explicitly formulated here) imposed at the boundary S.

Let us use Cartesian coordinates, in which case $u = u(\boldsymbol{x}, t)$, $\boldsymbol{x} = [x,\, y,\, z]^{\mathrm{T}}$. The initial condition thus can be written as

$$u(\boldsymbol{x}, 0) = u^{\star} \,. \tag{D.2}$$

By $\boldsymbol{n} = \left[n_x,\, n_y,\, n_z\right]^{\mathrm{T}}$ we shall denote the unit vector normal to the surface S and directed outwards.

If $u(\boldsymbol{x}, t)$ is to be a solution of Eq. (D.1), then the residual must be zero, i.e.

$$D \bigtriangleup u - \frac{\partial u}{\partial t} = 0 \,. \tag{D.3}$$

L.K. Bieniasz, *Modelling Electroanalytical Experiments by the Integral Equation Method*, Monographs in Electrochemistry, DOI 10.1007/978-3-662-44882-3

If we multiply Eq. (D.3) by an arbitrary function $\mathcal{G} = \mathcal{G}\left(\begin{bmatrix} x \\ t \end{bmatrix}, \begin{bmatrix} \chi \\ \tau \end{bmatrix}\right)$ and integrate over Ω and over time, we shall get zero, too:

$$\int_0^t \int_\Omega \left(D \triangle u - \frac{\partial u}{\partial \tau}\right) \mathcal{G} \, \mathrm{d}\Omega \, \mathrm{d}\tau = 0 \,. \tag{D.4}$$

In Eq. (D.4) we assume tacitly that $u = u(\chi, \tau)$, $\chi = [\xi, \eta, \zeta]^\mathrm{T}$ and τ are integration variables, and $\mathrm{d}\Omega = \mathrm{d}\xi \, \mathrm{d}\eta \, \mathrm{d}\zeta$.

From the Green–Gauss–Ostrogradski theorem (also known as the divergence theorem [1, 2]) we know that for any continuous and differentiable vector field $\boldsymbol{w}$,

$$\int_\Omega \operatorname{div} \boldsymbol{w} \, \mathrm{d}\Omega = \oint_S \boldsymbol{w} \cdot \boldsymbol{n} \, \mathrm{d}s \,. \tag{D.5}$$

In particular, if we take $\boldsymbol{w} = \mathcal{G} \, \mathbf{grad} \, u$, assuming that $\mathcal{G}$ and u possess continuous first and second spatial derivatives, then

$$\operatorname{div} \boldsymbol{w} = \nabla \cdot (\mathcal{G} \, \mathbf{grad} \, u) = \nabla \mathcal{G} \cdot \nabla u + \mathcal{G} \, \nabla^2 u \,, \tag{D.6}$$

$$\boldsymbol{w} \cdot \boldsymbol{n} = (\mathcal{G} \, \mathbf{grad} \, u) \cdot \boldsymbol{n} = \mathcal{G} \frac{\partial u}{\partial n} \,, \tag{D.7}$$

where $\partial u / \partial n$ is the directional spatial derivative in the direction of $\boldsymbol{n}$. Consequently, we obtain from Eq. (D.5):

$$\int_\Omega (\nabla \mathcal{G} \cdot \nabla u + \mathcal{G} \, \nabla^2 u) \, \mathrm{d}\Omega = \oint_S \mathcal{G} \frac{\partial u}{\partial n} \, \mathrm{d}s \,. \tag{D.8}$$

Similarly for $\boldsymbol{w} = u \, \mathbf{grad} \, \mathcal{G}$ we obtain

$$\int_\Omega (\nabla u \cdot \nabla \mathcal{G} + u \, \nabla^2 \mathcal{G}) \, \mathrm{d}\Omega = \oint_S u \frac{\partial \mathcal{G}}{\partial n} \, \mathrm{d}s \,. \tag{D.9}$$

Subtraction of Eqs. (D.8) and (D.9) gives

$$\int_\Omega (\mathcal{G} \triangle u - u \triangle \mathcal{G}) \, \mathrm{d}\Omega = \oint_S \left(\mathcal{G} \frac{\partial u}{\partial n} - u \frac{\partial \mathcal{G}}{\partial n}\right) \mathrm{d}s \,, \tag{D.10}$$

which is known as the Green second identity [3]. From Eq. (D.10) we obtain

$$\int_{\Omega} \mathcal{G} \triangle u \, d\Omega = \int_{\Omega} u \triangle G \, d\Omega + \oint_{S} \left(\mathcal{G} \frac{\partial u}{\partial n} - u \frac{\partial \mathcal{G}}{\partial n} \right) ds \,, \tag{D.11}$$

which we substitute into Eq. (D.4) together with another identity:

$$\int_{0}^{t} \frac{\partial u}{\partial \tau} \mathcal{G} \, d\tau = (u\mathcal{G})|_{\tau=0}^{\tau=t} - \int_{0}^{t} u \frac{\partial \mathcal{G}}{\partial \tau} \, d\tau \,. \tag{D.12}$$

This gives

$$\begin{aligned} &\int_{0}^{t} \int_{\Omega} u \left(D \triangle \mathcal{G} + \frac{\partial \mathcal{G}}{\partial \tau} \right) d\Omega \, d\tau \\ &= \int_{\Omega} u \mathcal{G} \, d\Omega \Bigg|_{\tau=0}^{\tau=t} - D \int_{0}^{t} \oint_{S} \left(\mathcal{G} \frac{\partial u}{\partial n} - u \frac{\partial \mathcal{G}}{\partial n} \right) ds \, d\tau \,. \end{aligned} \tag{D.13}$$

In the above equations $\mathcal{G}$ is an arbitrary function, but we can be more specific. Let us select for $\mathcal{G}$ the solution of the equation

$$D \triangle \mathcal{G} + \frac{\partial \mathcal{G}}{\partial \tau} = -\delta(\boldsymbol{\chi} - \boldsymbol{x}) \, \delta(\tau - t) \,, \tag{D.14}$$

where $\delta(\cdot)$ denotes the Dirac delta. Equation (D.14) has an analytical solution, which in the three-dimensional case is [3]:

$$\mathcal{G}\left(\begin{bmatrix} \boldsymbol{x} \\ t \end{bmatrix}, \begin{bmatrix} \boldsymbol{\chi} \\ \tau \end{bmatrix} \right) = \begin{cases} [4\pi D(t-\tau)]^{-3/2} \exp\left[-\frac{(x-\xi)^2 + (y-\eta)^2 + (z-\zeta)^2}{4D(t-\tau)} \right] & \text{for } t > \tau \\ 0 & \text{for } t < \tau \end{cases} . \tag{D.15}$$

As this function is singular at $\tau = t$, the integrals in Eq. (D.13) are improper integrals. In particular,

$$\int_{0}^{t} \int_{\Omega} u \left(D \triangle \mathcal{G} + \frac{\partial \mathcal{G}}{\partial \tau} \right) d\Omega \, d\tau = -\lim_{\varepsilon \to 0} \int_{0}^{t-\varepsilon} \int_{\Omega} u \, \delta(\boldsymbol{\chi} - \boldsymbol{x}) \, \delta(\tau - t) \, d\Omega \, d\tau = 0 \,, \tag{D.16}$$

$$\int_{\Omega} u \mathcal{G} \, d\Omega \Bigg|_{\tau=t} = \lim_{\varepsilon \to 0} \int_{\Omega} u \mathcal{G} \, d\Omega \Bigg|_{\tau=t-\varepsilon} = \int_{\Omega} u \, \delta(\boldsymbol{\chi} - \boldsymbol{x}) \, d\Omega = \lambda u \,, \tag{D.17}$$

where $\lambda = 1$ for $\boldsymbol{x} \in \Omega \setminus S$ and $\lambda = 1/2$ for $\boldsymbol{x} \in S$ (provided that the surface S is locally smooth). The integral $D \int\limits_0^t \oint\limits_S u\ (\partial \mathscr{G} / \partial n)\ \mathrm{d}s\, \mathrm{d}\tau$ is in general to be evaluated in the sense of the Cauchy principal value, as its kernel $\partial \mathscr{G}/\partial n$ may be strongly singular. Thus, with this choice of $\mathscr{G}$ Eq. (D.13) becomes

$$\lambda u = u^{\star} \int\limits_{\Omega} \mathscr{G}\, \mathrm{d}\Omega \Bigg|_{\tau=0} + D \int\limits_0^t \oint\limits_S \left(\mathscr{G} \frac{\partial u}{\partial n} - u \frac{\partial \mathscr{G}}{\partial n} \right) \mathrm{d}s\, \mathrm{d}\tau \,. \tag{D.18}$$

When $\boldsymbol{x} \in S$, Eq. (D.18) presents an integral relationship between the solution u at the boundary S, and its flux $-D\ \partial u/\partial n$ normal to the surface S.

In three-dimensional spatial domains nonzero steady state solutions of the diffusion equation are often possible, as limiting cases. In such cases we have to solve the Laplace equation

$$\triangle\ u = 0 \,. \tag{D.19}$$

If we multiply Eq. (D.19) by an arbitrary function $\mathscr{G} = \mathscr{G}(\boldsymbol{x}, \boldsymbol{\chi})$ and integrate over Ω, we obtain

$$\int\limits_{\Omega} \mathscr{G} \triangle\ u\, \mathrm{d}\Omega = 0 \,. \tag{D.20}$$

By employing Eq. (D.11) this is equivalent to

$$\int\limits_{\Omega} u \triangle\ \mathscr{G}\, \mathrm{d}\Omega + \oint\limits_S \left(\mathscr{G} \frac{\partial u}{\partial n} - u \frac{\partial \mathscr{G}}{\partial n} \right) \mathrm{d}s = 0 \,. \tag{D.21}$$

Let us select for $\mathscr{G}$ the solution of the equation

$$\triangle\ \mathscr{G} = -\delta(\boldsymbol{\chi} - \boldsymbol{x}) \,. \tag{D.22}$$

Equation (D.22) possesses an analytical solution, which in the three-dimensional case is [3]:

$$\mathscr{G}(\boldsymbol{x}, \boldsymbol{\chi}) = \left\{ 4\pi \left[(x-\xi)^2 + (y-\eta)^2 + (z-\zeta)^2 \right]^{1/2} \right\}^{-1} \,. \tag{D.23}$$

Setting Eq. (D.22) into Eq. (D.21) gives

$$\lambda u = \oint\limits_S \left(\mathscr{G} \frac{\partial u}{\partial n} - u \frac{\partial \mathscr{G}}{\partial n} \right) \mathrm{d}s \,. \tag{D.24}$$

Equation (D.24) presents the solution–flux relationship for the steady state.

References

1. Colley SJ (2012) Vector calculus. Pearson, Boston
2. Encyclopedia of Mathematics (2014). http://www.encyclopediaofmath.org. Accessed 10 June 2014
3. Kythe PK (1996) Fundamental solutions for differential operators and applications. Birkhäuser, Boston

Appendix E
Solution–Flux Relationships for One-Dimensional Reaction-Diffusion Equations

In this appendix we present example analytical derivations of the solution–flux relationships for a number of selected reaction-diffusion IBVPs or BVPs defined over a single one-dimensional and semi-infinite spatial domain. The solution–flux relationships are determined at one boundary (corresponding to an electrochemical interface studied). They must hold irrespective of the actual boundary conditions (not explicitly formulated here) imposed at this boundary. We employ the Laplace transformation method, and the classical theory of BVPs in second-order ODEs.

E.1 A Single Transient Linear Reaction-Diffusion PDE

We begin with analytical solution–flux relationships for the archetypal single transient linear reaction-diffusion PDE

$$\frac{\partial u}{\partial t} = D \, \triangle \, u - k \, u \, , \tag{E.1}$$

where D is a (positive) diffusion coefficient, k is a coefficient (positive, negative or zero), and u is an unknown function of spatial coordinates and time t. Solution u can be positive or negative; it does not have to represent a concentration that is always non-negative. The trivial solution $u \equiv 0$ is not of interest. We assume that only initially $u = 0$. We determine solution–flux relationships at one boundary (corresponding to an electrochemical interface studied), which must hold irrespective of the actual boundary conditions (not explicitly formulated here) imposed at this boundary. This is accomplished for the boundary condition $u = 0$ imposed at infinity. Three geometries of the diffusion field: planar, spherical, and cylindrical are considered.

L.K. Bieniasz, *Modelling Electroanalytical Experiments by the Integral Equation Method*, Monographs in Electrochemistry, DOI 10.1007/978-3-662-44882-3

In the case of planar diffusion Eq. (E.1) can be written as

$$\frac{\partial u(x,t)}{\partial t} = D\frac{\partial^2 u(x,t)}{\partial x^2} - k\,u(x,t)\,, \tag{E.2}$$

and the initial and boundary conditions are

$$u(x,0) = 0\,, \tag{E.3}$$

$$u(\infty,t) = 0\,. \tag{E.4}$$

In the case of spherical diffusion, Eq. (E.1) can be written as

$$\frac{\partial u(r,t)}{\partial t} = D\left[\frac{\partial^2 u(r,t)}{\partial r^2} + \frac{2}{r}\frac{\partial u(r,t)}{\partial r}\right] - k\,u(r,t)\,, \tag{E.5}$$

and the initial and boundary conditions are

$$u(r,0) = 0\,, \tag{E.6}$$

$$u(\infty,t) = 0\,. \tag{E.7}$$

Finally, in the case of cylindrical diffusion, Eq. (E.1) can be written as

$$\frac{\partial u(r,t)}{\partial t} = D\left[\frac{\partial^2 u(r,t)}{\partial r^2} + \frac{1}{r}\frac{\partial u(r,t)}{\partial r}\right] - k\,u(r,t)\,, \tag{E.8}$$

and the initial and boundary conditions are

$$u(r,0) = 0\,, \tag{E.9}$$

$$u(\infty,t) = 0\,. \tag{E.10}$$

By applying the Laplace transformation (3.12) to Eqs. (E.2) and (E.4), (E.5) and (E.7), or (E.8) and (E.10), and by taking into account the initial conditions (E.3), (E.6), or (E.9), we obtain the following ODEs for the Laplace transforms of the solutions. For planar diffusion

$$\mathscr{L}\left\{\frac{\partial u(x,t)}{\partial t}\right\} = s\hat{u}(x,s) - 0$$

$$= \mathscr{L}\left\{D\frac{\partial^2 u(x,t)}{\partial x^2} - k\,u(x,t)\right\} = D\frac{\partial^2 \hat{u}(x,s)}{\partial x^2} - k\,\hat{u}(x,s)\,, \tag{E.11}$$

or, equivalently,

$$\frac{\partial^2 \hat{u}(x,s)}{\partial x^2} - \frac{s+k}{D}\,\hat{u}(x,s) = 0\,. \tag{E.12}$$

For spherical diffusion

$$\begin{aligned}\mathscr{L}\left\{\frac{\partial u(r,t)}{\partial t}\right\} &= s\hat{u}(r,s) - 0\\ = \mathscr{L}\left\{D\left[\frac{\partial^2 u(r,t)}{\partial r^2} + \frac{2}{r}\frac{\partial u(r,t)}{\partial r}\right] - k\, u(r,t)\right\}\\ = D\left[\frac{\partial^2 \hat{u}(r,s)}{\partial r^2} + \frac{2}{r}\frac{\partial \hat{u}(r,s)}{\partial r}\right] - k\,\hat{u}(r,s)\,,\end{aligned} \tag{E.13}$$

or, equivalently,

$$\frac{\partial^2 \hat{u}(r,s)}{\partial r^2} + \frac{2}{r}\frac{\partial \hat{u}(r,s)}{\partial r} - \frac{s+k}{D}\,\hat{u}(r,s) = 0\,. \tag{E.14}$$

For cylindrical diffusion

$$\begin{aligned}\mathscr{L}\left\{\frac{\partial u(r,t)}{\partial t}\right\} &= s\hat{u}(r,s) - 0\\ = \mathscr{L}\left\{D\left[\frac{\partial^2 u(r,t)}{\partial r^2} + \frac{1}{r}\frac{\partial u(r,t)}{\partial r}\right] - k\, u(r,t)\right\}\\ = D\left[\frac{\partial^2 \hat{u}(r,s)}{\partial r^2} + \frac{1}{r}\frac{\partial \hat{u}(r,s)}{\partial r}\right] - k\,\hat{u}(r,s)\,,\end{aligned} \tag{E.15}$$

or, equivalently,

$$\frac{\partial^2 \hat{u}(r,s)}{\partial r^2} + \frac{1}{r}\frac{\partial \hat{u}(r,s)}{\partial r} - \frac{s+k}{D}\,\hat{u}(r,s) = 0\,. \tag{E.16}$$

Equations (E.12), (E.14), or (E.16) are analogous to Eqs. (A.10), (A.65), or (A.130), respectively, previously obtained for the pure diffusion case, with the only difference that every explicit occurrence of the Laplace variable s is now replaced by the sum $s + k$. Therefore, the derivations are analogous as for pure diffusion, if only we put $s + k$ in place of s. We thus obtain the following analogues of Eqs. (A.21), (A.81), or (A.148). For planar diffusion:

$$\hat{u}(x,s) = D^{-1/2}\,(s+k)^{-1/2}\,\hat{J}(x,s)\,, \tag{E.17}$$

for spherical diffusion

$$\hat{u}(r,s) = D^{-1/2}\left[(s+k)^{1/2} + \frac{D^{1/2}}{r}\right]^{-1}\hat{J}(r,s)\,, \tag{E.18}$$

and for cylindrical diffusion

$$\hat{u}(r,s) = D^{-1/2} \frac{\left(\frac{r}{D^{1/2}}\right) \mathrm{K}_0\left[\frac{r}{D^{1/2}}(s+k)^{1/2}\right]}{\left[\frac{r}{D^{1/2}}(s+k)^{1/2}\right] \mathrm{K}_1\left[\frac{r}{D^{1/2}}(s+k)^{1/2}\right]} \hat{J}(r,s) . \tag{E.19}$$

The addition of a constant to s in a Laplace transform is equivalent to multiplying the original function by an exponential function [cf. Eq. (3.16)]. Therefore, by applying the convolution theorem (3.17) to Eqs. (E.17)–(E.19) we obtain the following solution–flux relationships. For planar diffusion

$$u(x,t) = \int_0^t \mathscr{K}(t,\tau)\, J(x,\tau)\, d\tau , \tag{E.20}$$

where in view of Eq. (A.24)

$$\mathscr{K}(t,\tau) = D^{-1/2} \exp[-k(t-\tau)]\, [\pi(t-\tau)]^{-1/2} . \tag{E.21}$$

For spherical diffusion

$$u(r,t) = \int_0^t \mathscr{K}(t,\tau)\, J(r,\tau)\, d\tau , \tag{E.22}$$

where in view of Eq. (A.84)

$$\mathscr{K}(t,\tau) = D^{-1/2} \exp[-k(t-\tau)] \left\{ [\pi(t-\tau)]^{-1/2} - \frac{D^{1/2}}{r} \operatorname{erex}\left[\frac{D^{1/2}}{r}(t-\tau)^{1/2}\right] \right\} . \tag{E.23}$$

For cylindrical diffusion

$$u(r,t) = \int_0^t \mathscr{K}(t,\tau)\, J(r,\tau)\, d\tau , \tag{E.24}$$

where in view of Eq. (A.151)

$$\mathscr{K}(t,\tau) = D^{-1/2} \exp[-k(t-\tau)] \left\{ [\pi(t-\tau)]^{-1/2} - \frac{D^{1/2}}{r} \operatorname{kcylw}\left[\frac{D^{1/2}}{r}(t-\tau)^{1/2}\right] \right\} . \tag{E.25}$$

In the absence of the reaction term in Eq. (E.1), that is for $k = 0$, Eqs. (E.21), (E.23), or (E.25) reduce to Eqs. (A.24), (A.84), or (A.151), respectively.

At the boundary corresponding to the interface studied, Eqs. (E.20)–(E.25) reduce to the following formulae. For planar diffusion

$$u(0,t) = \int_0^t \mathscr{K}(t,\tau)\; J(0,\tau)\; d\tau\,, \tag{E.26}$$

where

$$\mathscr{K}(t,\tau) = D^{-1/2} \exp\left[-k(t-\tau)\right] \left[\pi(t-\tau)\right]^{-1/2} . \tag{E.27}$$

For spherical diffusion

$$u(r_0,t) = \int_0^t \mathscr{K}(t,\tau)\; J(r_0,\tau)\; d\tau\,, \tag{E.28}$$

where

$$\mathscr{K}(t,\tau) = D^{-1/2} \exp\left[-k(t-\tau)\right] \left\{\left[\pi(t-\tau)\right]^{-1/2} - \varrho\, \mathrm{erex}\left[\varrho(t-\tau)^{1/2}\right]\right\} \tag{E.29}$$

and ϱ is defined by Eq. (A.2). For cylindrical diffusion

$$u(r_0,t) = \int_0^t \mathscr{K}(t,\tau)\; J(r_0,\tau)\; d\tau\,, \tag{E.30}$$

where

$$\mathscr{K}(t,\tau) = D^{-1/2} \exp\left[-k(t-\tau)\right] \left\{\left[\pi(t-\tau)\right]^{-1/2} - \varrho\, \mathrm{kcylw}\left[\varrho(t-\tau)^{1/2}\right]\right\} . \tag{E.31}$$

E.2 A Single Steady State Linear Reaction-Diffusion ODE

We pass to the special case of Eq. (E.1), corresponding to the steady state condition $\partial u/\partial t \approx 0$.

In the case of planar diffusion, Eq. (E.2) becomes a second-order, reaction-diffusion ODE with constant coefficients:

$$\frac{\partial^2 u(x,t)}{\partial x^2} - \frac{k}{D}\, u(x,t) = 0\,. \tag{E.32}$$

The general integral of Eq. (E.32) is [4]:

$$u(x,t) = a(t)\,\exp[-(k/D)^{1/2}x] + b(t)\,\exp[(k/D)^{1/2}x]\,, \tag{E.33}$$

where $a(t)$ and $b(t)$ are integration constants independent of x. In view of the boundary condition (E.4), $b(t) = 0$ is the only possibility. Hence,

$$u(x,t) = a(t)\,\exp[-(k/D)^{1/2}x]\,, \tag{E.34}$$

and

$$J(x,t) = -D\frac{\partial u(x,t)}{\partial x} = (k\,D)^{1/2}\,a(t)\,\exp[-(k/D)^{1/2}x]\,. \tag{E.35}$$

By combining Eqs. (E.34) and (E.35) the constant $a(t)$ can be eliminated, which gives

$$u(x,t) = D^{-1/2}\,k^{-1/2}\,J(x,t)\,. \tag{E.36}$$

Specifically at the boundary at $x = 0$, corresponding to the interface studied,

$$u(0,t) = D^{-1/2}\,k^{-1/2}\,J(0,t)\,. \tag{E.37}$$

Equation (E.37) is a particular subcase of Eq. (E.26), valid for the steady state.

In the case of spherical diffusion, Eq. (E.5) becomes a second-order ODE

$$\left[\frac{\partial^2 u(r,t)}{\partial r^2} + \frac{2}{r}\frac{\partial u(r,t)}{\partial r}\right] - \frac{k}{D}\,u(r,t) = 0\,. \tag{E.38}$$

The change of variables (A.69) transforms Eq. (E.38) into

$$\frac{\partial^2 w(r,t)}{\partial r^2} - \frac{k}{D}\,w(r,t) = 0\,, \tag{E.39}$$

which is analogous to Eq. (E.32) for planar diffusion. Therefore, the general integral of Eq. (E.38) is

$$u(r,t) = a(t)\,r^{-1}\,\exp[-(k/D)^{1/2}r] + b(t)\,r^{-1}\,\exp[(k/D)^{1/2}r]\,, \tag{E.40}$$

where $a(t)$ and $b(t)$ are integration constants independent of r. In view of the boundary condition (E.7), $b(t) = 0$ is the only possibility. Hence,

$$u(r,t) = a(t)\,r^{-1}\,\exp[-(k/D)^{1/2}r] \tag{E.41}$$

and

$$J(r,t) = -D\frac{\partial u(r,t)}{\partial r} = D\left[r^{-1} + (k/D)^{1/2}\right] a(t)\, r^{-1}\, \exp[-(k/D)^{1/2} r] \,. \tag{E.42}$$

By combining Eqs. (E.41) and (E.42) the constant $a(t)$ can be eliminated, which gives

$$u(r,t) = D^{-1/2}\left[\frac{D^{1/2}}{r} + k^{1/2}\right]^{-1} J(r,t) \,. \tag{E.43}$$

Specifically at the boundary at $r = r_0$, corresponding to the interface studied,

$$u(r_0,t) = D^{-1/2}\left(\varrho + k^{1/2}\right)^{-1} J(r_0,t) \,. \tag{E.44}$$

Equation (E.44) is a particular subcase of Eq. (E.28), valid for the steady state.

In the case of cylindrical diffusion, Eq. (E.8) becomes a second-order ODE

$$\left[\frac{\partial^2 u(r,t)}{\partial r^2} + \frac{1}{r}\frac{\partial u(r,t)}{\partial r}\right] - \frac{k}{D}\, u(r,t) = 0 \,. \tag{E.45}$$

The change of variables

$$\xi = r(k/D)^{1/2} \tag{E.46}$$

transforms Eq. (E.45) into

$$\xi^2\,\frac{\partial^2 u(\xi,t)}{\partial \xi^2} + \xi\,\frac{\partial u(\xi,t)}{\partial \xi} - \xi^2\, u(\xi,t) = 0 \,. \tag{E.47}$$

Equation (E.47) is the homogeneous modified Bessel equation of order zero [1], so that it possesses the general solution

$$u(\xi,t) = a(t)\,\mathrm{K}_0(\xi) + b(t)\,\mathrm{I}_0(\xi) \,, \tag{E.48}$$

where $a(t)$ and $b(t)$ are integration constants independent of r.

In view of the boundary condition (E.10), $b(t) = 0$ is the only possibility. Hence, Eq. (E.48) reduces to

$$u(r,t) = a(t)\,\mathrm{K}_0\left[r(k/D)^{1/2}\right] \,. \tag{E.49}$$

Consequently,

$$J(r,t) = -D\frac{\partial u(r,t)}{\partial r} = (k\,D)^{1/2}\, a(t)\,\mathrm{K}_1\left[r(k/D)^{1/2}\right] \,, \tag{E.50}$$

where the equality (A.139) has been taken into account.

By combining Eqs. (E.49) and (E.50) the constant $a(t)$ can be eliminated, which gives

$$u(r,t) = D^{-1/2}\,k^{-1/2}\,\frac{\mathrm{K}_0\left[k^{1/2}/(D^{1/2}/r)\right]}{\mathrm{K}_1\left[k^{1/2}/(D^{1/2}/r)\right]}\,J(r,t)\;. \tag{E.51}$$

Specifically at the boundary at $r = r_0$, corresponding to the interface studied

$$u(r_0,t) = D^{-1/2}\,k^{-1/2}\,\frac{\mathrm{K}_0\left(k^{1/2}/\varrho\right)}{\mathrm{K}_1\left(k^{1/2}/\varrho\right)}\,J(r_0,t)\;. \tag{E.52}$$

Equation (E.52) is a particular subcase of Eq. (E.30), valid for the steady state.

E.3 A System of Coupled Transient Linear Reaction-Diffusion PDEs

In this section we consider a system of N_s^{dd} (further denoted by N, for brevity) coupled reaction-diffusion PDEs for the concentrations of dynamic distributed species. We assume planar diffusion in a semi-infinite one-dimensional spatial domain $x \in [0,\infty)$, and (pseudo) first-order homogeneous reactions. The PDEs are:

$$\frac{\partial \vec{u}(x,t)}{\partial t} = \overline{D}\,\frac{\partial^2 \vec{u}(x,t)}{\partial x^2} - \overline{K}\,\vec{u}(x,t)\;, \tag{E.53}$$

and the initial and boundary conditions are:

$$\vec{u}(x,0) = \vec{0}\;, \tag{E.54}$$

$$\vec{u}(\infty,t) = \vec{0}\;. \tag{E.55}$$

In Eqs. (E.53)–(E.55) $\vec{u}(x,t) = [u_1(x,t),\ \ldots,\ u_N(x,t)]^{\mathrm{T}}$ is the solution vector containing N elements, $\overline{D} = \mathrm{diag}(D_1,\ \ldots,\ D_N)$ is an $N \times N$ diagonal matrix of positive diffusion coefficients, $\overline{K}$ is a (generally full) $N \times N$ matrix of coefficients dependent on the homogeneous rate constants, and $\vec{0}$ denotes zero vectors. The task is to derive analytical relationships between $\vec{u}(x,t)$ and the vector $\vec{J}(x,t) = [-D_1\partial u_1(x,t)/\partial x,\ \ldots,\ -D_N\partial u_N(x,t)/\partial x]^{\mathrm{T}}$ of its fluxes.

By applying the Laplace transformation (3.12) to Eqs. (E.53) and (E.55), and by taking into account the initial condition (E.54), we obtain the following second-order ODE system for the Laplace transform $\hat{\vec{u}}(x,s)$ of $\vec{u}(x,t)$:

$$\mathscr{L}\left\{\frac{\partial \vec{u}(x,t)}{\partial t}\right\} = s\,\hat{\vec{u}}(x,s) - \vec{0}$$

$$= \mathscr{L}\left\{\overline{D}\,\frac{\partial^2 \vec{u}(x,t)}{\partial x^2} - \overline{K}\,\vec{u}(x,t)\right\} = \overline{D}\,\frac{\partial^2 \hat{\vec{u}}(x,s)}{\partial x^2} - \overline{K}\,\hat{\vec{u}}(x,s)\,. \tag{E.56}$$

After introducing matrix $\overline{B}(s)$ defined by [cf. Eq. (8.127)]:

$$\overline{B}(s) = \overline{D}^{-1}\,(s\,\overline{1} + \overline{K})\,, \tag{E.57}$$

where $\overline{1}$ denotes a unit matrix, Eq. (E.56) can be written in a more compact way:

$$\frac{\partial^2 \hat{\vec{u}}(x,s)}{\partial x^2} - \overline{B}(s)\,\hat{\vec{u}}(x,s) = \vec{0}\,. \tag{E.58}$$

By defining a block vector $\vec{y}(x,t) = \left[\vec{u}(x,t),\, \vec{J}(x,t)\right]^{\mathrm{T}}$ and its Laplace transform $\hat{\vec{y}}(x,s)$, Eq. (E.58) can be further rewritten as a first-order ODE system with constant coefficients, for the $2N$ unknowns in $\vec{y}(x,t)$:

$$\frac{\partial \hat{\vec{y}}(x,s)}{\partial x} - \overline{A}(s)\,\hat{\vec{y}}(x,s) = \vec{0}\,, \tag{E.59}$$

where

$$\overline{A}(s) = -\begin{bmatrix} \overline{0} & \overline{D}^{-1} \\ \overline{D}\,\overline{B}(s) & \overline{0} \end{bmatrix}, \tag{E.60}$$

with $\overline{0}$ denoting zero submatrices. According to the theory of first-order linear ODEs with constant coefficients [4], the general solution of Eq. (E.59) depends on the numbers of distinct eigenvalues and linearly independent eigenvectors of matrix $\overline{A}(s)$. The general solution is obtained from an appropriate linear combination of ordinary or generalised eigenvectors of $\overline{A}(s)$. We therefore face the problem of determining the eigenvalues and generalised eigenvectors of $\overline{A}(s)$. By definition, a vector $\vec{v}^{\,(n)}(s)$ (with $n = 1, 2, \ldots$), satisfying the equation

$$\left[\overline{A}(s) - \Lambda(s)\,\overline{1}\right]^n \vec{v}^{\,(n)}(s) = \vec{0}\,, \tag{E.61}$$

and, in the case of $n > 1$ also

$$\left[\overline{A}(s) - \Lambda(s)\,\overline{1}\right]^k \vec{v}^{\,(n)}(s) \neq \vec{0}\,, \tag{E.62}$$

for $k = 1, 2, \ldots, n-1$, is called the nth-order generalised eigenvector associated with any eigenvalue $\Lambda(s)$ of $\overline{A}(s)$. Ordinary eigenvectors are generalised eigenvectors of order $n = 1$. By splitting $\vec{v}^{\,(n)}(s)$ into two blocks $\vec{u}^{\,(n)}(s)$ and $\vec{w}^{\,(n)}(s)$ of N elements each: $\vec{v}^{\,(n)}(s) = \left[\vec{u}^{\,(n)}(s), \vec{w}^{\,(n)}(s)\right]^{\mathrm{T}}$, it can be shown [2] that Eq. (E.61) is equivalent to the equation pair

$$\left[\overline{B}(s) - \Lambda(s)^2\,\overline{1}\right]^n \vec{u}^{\,(n)}(s) = \vec{0}\,, \tag{E.63}$$

$$\vec{w}^{\,(n)}(s) = \overline{T}^{\,(n)}(s)\,\vec{u}^{\,(n)}(s)\,, \tag{E.64}$$

where $\overline{T}^{\,(n)}(s)$ is an $N \times N$ matrix defined by

$$\overline{T}^{\,(n)}(s) = \begin{cases} -\Lambda(s)\,\overline{D} & \text{for } n = 1 \\ \overline{D}\left\{\sum\limits_{m=1}^{n-1} p_m\left[\overline{B}(s) - \Lambda(s)^2\,\overline{1}\right]^m - \Lambda(s)\,\overline{1}\right\} & \text{for } n = 2, 3, \ldots \end{cases} . \tag{E.65}$$

Coefficients p_m in Eq. (E.65) obey the equation

$$2\Lambda(s)\,p_m = p_1\,p_{m-1} + p_2\,p_{m-2} + \cdots + p_{m-1}\,p_1\,, \tag{E.66}$$

so that $p_1 = -[2\Lambda(s)]^{-1}$, $p_2 = [8\Lambda^3(s)]^{-1}$, $p_3 = -[16\Lambda^5(s)]^{-1}$, $p_4 = 5[128\Lambda^7(s)]^{-1}$, etc. In view of Eqs. (E.63) and (E.64), the problem of determining the generalised eigenvectors $\vec{v}^{\,(n)}(s)$ of $\overline{A}(s)$ is equivalent to determining the generalised eigenvectors $\vec{u}^{\,(n)}(s)$ of $\overline{B}(s)$, and the remaining blocks $\vec{w}^{\,(n)}(s)$ of $\vec{v}^{\,(n)}(s)$ are simply obtained by multiplying $\vec{u}^{\,(n)}(s)$ by $\overline{T}^{\,(n)}(s)$. We also see that every nonzero eigenvalue $\lambda(s)$ of $\overline{B}(s)$ contributes two eigenvalues

$$\Lambda^{\pm}(s) = \pm\,[\lambda(s)]^{1/2} \tag{E.67}$$

of $\overline{A}(s)$.

Depending on the number of distinct eigenvalues and eigenvectors of $\overline{B}(s)$, the following three cases of the general solution of Eq. (E.59) can be distinguished. The first, simplest case occurs when the matrix $\overline{A}(s)$ possesses $2N$ distinct eigenvalues $\Lambda_1(s), \ldots, \Lambda_{2N}(s)$. The existence of corresponding $2N$ linearly independent ordinary eigenvectors $\vec{v}_1^{\,(1)}(s), \ldots, \vec{v}_{2N}^{\,(1)}(s)$ of $\overline{A}(s)$ is then guaranteed, and the general solution of Eq. (E.59) can be expressed as

$$\hat{\vec{y}}(x, s) = \sum_{i=1}^{2N} a_i(s)\,\exp\left[\Lambda_i(s)\,x\right]\,\vec{v}_i^{\,(1)}(s)\,, \tag{E.68}$$

where $a_i(s)$ are integration constants.

The second, somewhat more complicated case occurs when the matrix $\overline{A}(s)$ does not have $2N$ distinct eigenvalues, but possesses $2N$ linearly independent ordinary eigenvectors. In such a case Eq. (E.68) remains valid, but some of the $\Lambda_i(s)$ are repeated.

The third, and most complicated case occurs when the matrix $\overline{A}(s)$ is deficient, i.e. it does not have $2N$ independent ordinary eigenvectors. In such a case, in order to obtain the general solution of Eq. (E.59), one has to determine the so-called Jordan normal form of the matrix $\overline{A}(s)$. The Jordan normal form of a matrix is a block-diagonal matrix, in which every block corresponds to one of the available linearly independent ordinary eigenvectors. One eigenvalue may be associated with more than one Jordan blocks. The general solution of Eq. (E.59) is then a linear combination of linearly independent specific solutions corresponding to the various Jordan blocks. Every Jordan block contributes the number of independent specific solutions equal to the block size. Consequently,

$$\vec{\hat{y}}(x,s) = \sum_{i=1}^{M} \exp\left[\Lambda_i(s)\,x\right] \sum_{j=1}^{m_i^g} \sum_{k=1}^{n_{i,j}} a_{i,j,k}(s)\,\vec{z}_{i,j,k}(x,s)\ , \tag{E.69}$$

where

$$\vec{z}_{i,j,k}(x,s) = \sum_{\mu=0}^{k-1} \frac{x^{\mu}}{\mu!}\,\vec{v}_{i,j}^{\,(k-\mu)}(s)\ . \tag{E.70}$$

In Eq. (E.69) i counts the distinct eigenvalues, M ($<2N$) is the number of the distinct eigenvalues, j counts the Jordan blocks corresponding to any ith eigenvalue. The number of such blocks is denoted by m_i^g. This number is equal to the number of linearly independent ordinary eigenvectors associated with the eigenvalue, called the geometric multiplicity of the eigenvalue. Index k counts the linearly independent vectors $\vec{z}_{i,j,k}(x,s)$, in accordance with Eq. (E.70) constructed as polynomials in x. The polynomial coefficients involve generalised eigenvectors $\vec{v}_{i,j}^{\,(k-\mu)}(s)$ of various orders, which form the so-called Jordan chains associated with a given Jordan block. The sizes of the blocks are denoted by $n_{i,j}$. The sum of these sizes, for all Jordan blocks associated with an ith eigenvalue, gives the multiplicity of $\Lambda_i(s)$ as a root of the characteristic equation, that is the so-called algebraic multiplicity m_i^a of this eigenvalue: $\sum_{j=1}^{m_i^g} n_{i,j} = m_i^a$. The sum of the algebraic multiplicities of all eigenvalues gives, in turn, the matrix size: $\sum_{i=1}^{M} m_i^a = 2N$. As a result, there are exactly $2N$ linearly independent vectors $\vec{z}_{i,j,k}(x,s)$, and $2N$ integration constants $a_{i,j,k}(s)$ in Eq. (E.69).

Equation (E.69) represents the most general situation, and will be used for further discussion. From Eq. (E.67) we know that in Eq. (E.69) there are exactly $M/2$

eigenvalues of $\overline{A}(s)$ with positive real parts, and $M/2$ eigenvalues with negative real parts. Let us denote them by $\Lambda_1^+(s), \ldots, \Lambda_{M/2}^+(s)$ and $\Lambda_1^-(s), \ldots, \Lambda_{M/2}^-(s)$, respectively. Zero eigenvalues are avoided, if only $\Re\{s\}$ is sufficiently large (and we are always free to select such a large s, see Sect. 8.4). This is because Eq. (E.57) predicts that for large $\Re\{s\}$ matrix $\overline{B}(s)$ is diagonally dominant, which guarantees that $\lambda(s) \neq 0$. In order to satisfy the boundary condition (E.55), all terms corresponding to eigenvalues of $\overline{A}(s)$ with positive real parts have to be excluded from the solution (E.69). Thus, Eq. (E.69) becomes

$$\vec{y}(x,s) = \sum_{l=1}^{M/2} \exp\left[\Lambda_l^-(s)\,x\right] \sum_{j=1}^{m_l^g} \sum_{k=1}^{n_{l,j}} a_{l,j,k}(s)\,\vec{z}_{\,l,j,k}(x,s)\,, \tag{E.71}$$

where the numbering of the terms by index l is now synchronised with the numbering of $\Lambda_l^-(s)$. As every eigenvector of $\overline{B}(s)$ contributes two linearly independent eigenvectors of $\overline{A}(s)$, corresponding to $\Lambda_l^-(s)$ and $\Lambda_l^+(s)$, there are exactly N linearly independent vectors $\vec{z}_{\,l,j,k}(x,s)$ in Eq. (E.71), and N coefficients $a_{l,j,k}(s)$.

In order to determine which of the above three cases actually occurs, any particular matrix $\overline{B}(s)$ has to be analysed individually. However, the analysis can be facilitated by the following two general observations (a) and (b).

(a) In the case of equal diffusion coefficients we have $\overline{D} = D\,\overline{1}$, where D is the common diffusion coefficient. Matrix $\overline{B}(s)$ then becomes

$$\overline{B}(s) = (s\,\overline{1} + \overline{K})/D\,. \tag{E.72}$$

As D is a scalar, eigenvectors $\vec{u}^{\,(n)}(s)$ of $\overline{B}(s)$ are identical with the eigenvectors of $\overline{K}$, and any eigenvalue of $\overline{B}(s)$ is related to one eigenvalue κ of $\overline{K}$ by the formula that depends linearly on s:

$$\lambda(s) = (s + \kappa)/D\,. \tag{E.73}$$

Thus whether or not matrix $\overline{B}(s)$ is deficient, depends exclusively on $\overline{K}$. If $\overline{K}$ has N distinct eigenvalues, then matrix $\overline{K}$ is similar to a diagonal matrix, and we obtain the case discussed in Sect. 8.2.

(b) If all diffusion coefficients are different, then for sufficiently large s all eigenvalues of $\overline{B}(s)$ become distinct. This can be shown by using the Gershgorin theorems (see, for example, Bronson [3], or a didactically lucid presentation in Smith [5]). According to the Gershgorin circle theorem, each eigenvalue of $\overline{B}(s)$ lies inside or on the boundary of at least one of the circles defined by

$$|\,\lambda(s) - B_{i,i}(s)\,| = \sum_{\substack{j=1 \\ j \neq i}}^{N} |\,B_{i,j}(s)\,|\,, \tag{E.74}$$

where $i = 1, \ldots, N$. In view of Eq. (E.57), after dividing by (nonzero) $|s|$, Eq. (E.74) becomes

$$\left| z - \left(\frac{1}{D_i} + \frac{K_{i,i}}{D_i\, s} \right) \right| = \sum_{\substack{j=1 \\ j \neq i}}^{N} \left| \frac{K_{i,j}}{D_i\, s} \right| , \tag{E.75}$$

where $z = \lambda(s)/s$ is a new complex variable. By taking s with sufficiently large $\Re\{s\}$ we can force all circles defined by Eq. (E.75) to be located arbitrarily close to $z = 1/D_i$, and to have arbitrarily small radii. Since D_i are now distinct, this implies that we can make all circles disjoint. The Gershgorin third theorem then ensures that there are N distinct eigenvalues, each one in the corresponding isolated circle.

By eliminating the unknown integration constants $a_{i,j,k}(s)$ in Eq. (E.71), one can obtain the desired relationships between the block subvectors $\vec{u}(x,t)$ and $\vec{J}(x,t)$ of $\vec{y}(x,t)$. In principle, this can be done for any x, and later the resulting formulae can be taken for $x = 0$, which corresponds to the interface studied. However, derivations are simpler, if one assumes $x = 0$ from the beginning. In such a case, every exponential factor in Eq. (E.71) reduces to unity, and also in accordance with Eq. (E.70), any vector $\vec{z}_{i,j,k}(x,s)$ reduces to $\vec{v}^{\,(k)}_{i,j}(s)$. We therefore obtain

$$\hat{\vec{y}}(0,s) = \sum_{i=1}^{M/2} \sum_{j=1}^{m_i^g} \sum_{k=1}^{n_{i,j}} a_{i,j,k}(s)\, \vec{v}^{\,(k)}_{i,j}(s) , \tag{E.76}$$

or, after taking into account Eq. (E.64) with appropriate lower indices added to the quantities occurring in it:

$$\hat{\vec{u}}(0,s) = \sum_{i=1}^{M/2} \sum_{j=1}^{m_i^g} \sum_{k=1}^{n_{i,j}} a_{i,j,k}(s)\, \vec{u}^{\,(k)}_{i,j}(s) , \tag{E.77}$$

$$\hat{\vec{J}}(0,s) = \sum_{i=1}^{M/2} \sum_{j=1}^{m_i^g} \sum_{k=1}^{n_{i,j}} a_{i,j,k}(s)\, \overline{T}^{\,(k)}_{i,j}(s)\, \vec{u}^{\,(k)}_{i,j}(s) . \tag{E.78}$$

As there are N linearly independent vectors $\vec{u}^{\,(k)}_{i,j}(s)$ and $\overline{T}^{\,(k)}_{i,j}(s)\,\vec{u}^{\,(k)}_{i,j}(s)$ in Eqs. (E.77) and (E.78), we can arrange them as columns of two square and invertible $N \times N$ matrices, denoted here by $\overline{U}(s)$ and $\overline{W}(s)$, respectively. The N coefficients

$a_{i,j,k}(s)$ can, in turn, be arranged to form a vector denoted by $\vec{a}(s)$. This allows one to re-write Eqs. (E.77) and (E.78) in compact form:

$$\hat{\vec{u}}(0,s) = \overline{U}(s)\,\vec{a}(s)\;, \tag{E.79}$$

$$\hat{\vec{J}}(0,s) = \overline{W}(s)\,\vec{a}(s)\;. \tag{E.80}$$

Owing to the invertibility of $\overline{W}$, vector $\vec{a}(s)$ can be calculated from Eq. (E.80):

$$\vec{a}(s) = \overline{W}^{-1}(s)\,\hat{\vec{J}}(0,s) \tag{E.81}$$

and substituted to Eq. (E.79), which gives

$$\hat{\vec{u}}(0,s) = \hat{\overline{\varphi}}(s)\,\hat{\vec{J}}(0,s)\;, \tag{E.82}$$

where the new square matrix $\overline{\varphi}(s)$ is defined by

$$\hat{\overline{\varphi}}(s) = \overline{U}(s)\,\overline{W}^{-1}(s)\;. \tag{E.83}$$

Application of the convolution theorem (3.17) to Eq. (E.82) gives

$$\vec{u}(0,t) = \int_0^t \overline{\mathcal{K}}(t,\tau)\,\vec{J}(0,\tau)\,\mathrm{d}\tau\;, \tag{E.84}$$

where

$$\overline{\mathcal{K}}(t,\tau) = \overline{\varphi}(t-\tau) \tag{E.85}$$

is the $N \times N$ matrix of convolution kernels resulting from diffusion and coupling of the PDEs (E.53) by homogeneous reactions.

E.4 A Single Steady State Reaction-Diffusion ODE for Planar Diffusion and an *m*th-Order Reaction

Finally, we consider the steady state for the single reaction-diffusion PDE

$$\frac{\partial u(x,t)}{\partial t} = D\,\frac{\partial^2 u(x,t)}{\partial x^2} - k\,u(x,t)^m \tag{E.86}$$

involving planar diffusion and nonlinear reaction terms, defined over a single semi-infinite one-dimensional spatial domain. In Eq. (E.86) D is a (positive) diffusion coefficient, k is a positive reaction rate constant, $m = 2, 3, \ldots,$ and u is an unknown real function of spatial coordinate x and time t. Solution u is assumed to be positive. The trivial solution $u \equiv 0$ is not of interest. We assume that only initially $u = 0$. We determine solution–flux relationships at one boundary (corresponding to an electrochemical interface studied), which must hold irrespective of the actual boundary conditions (not explicitly formulated here) imposed at this boundary. This is accomplished for the boundary condition $u = 0$ imposed at infinity.

Under the steady state condition $\partial u/\partial t \approx 0$ Eq. (E.86) becomes a second-order ODE in the independent variable x:

$$\frac{\partial^2 u(x,t)}{\partial x^2} - \frac{k}{D}\, u(x,t)^m = 0\,, \tag{E.87}$$

with the boundary condition

$$u(\infty,t) = 0\,. \tag{E.88}$$

As the ODE (E.87) does not depend explicitly on x, the standard change of variables can be used:

$$w = \frac{\partial u}{\partial x} \tag{E.89}$$

(to simplify the formulae we omit here and below the lists of independent variables). This gives

$$\frac{\partial^2 u}{\partial x^2} = \frac{\partial w}{\partial x} = \frac{\partial w}{\partial u}\frac{\partial u}{\partial x} = w\frac{\partial w}{\partial u}\,, \tag{E.90}$$

so that Eq. (E.87) can be rewritten as

$$w\frac{\partial w}{\partial u} - \frac{k}{D}\, u^m = 0\,. \tag{E.91}$$

Integration of both sides of Eq. (E.91) with respect to the variable u gives

$$\frac{1}{2} w^2 = \frac{k}{D}\frac{1}{m+1}\, u^{m+1} + b\,, \tag{E.92}$$

where b is an integration constant. In view of the condition (E.88) we expect that asymptotically also $w \to 0$ when $x \to \infty$. Therefore, both u and w vanish in Eq. (E.92) when $x \to \infty$, so that there must be $b = 0$. Thus,

$$w = \pm\left(\frac{2}{m+1}\frac{k}{D}\right)^{1/2} u^{(m+1)/2}\,. \tag{E.93}$$

As we are looking for positive u, the solution with "+" is non-physical, because with increasing x the variable u must approach zero from the side of positive values, hence w must be negative. Therefore, from Eqs. (E.89) and (E.93)

$$\frac{\partial u}{\partial x} = -\left(\frac{2}{m+1}\frac{k}{D}\right)^{1/2} u^{(m+1)/2} . \tag{E.94}$$

After dividing Eq. (E.94) by the (positive) $u^{(m+1)/2}$, and integrating both sides with respect to x, we finally identify the dependence of u on x and t:

$$\frac{2}{1-m} u(x,t)^{(1-m)/2} = -\left(\frac{2}{m+1}\frac{k}{D}\right)^{1/2} x + a(t) , \tag{E.95}$$

where $a(t)$ is an integration constant independent of x. Equation (E.95) can be rewritten:

$$u(x,t) = \left\{\frac{m-1}{2}\left[\left(\frac{2}{m+1}\frac{k}{D}\right)^{1/2} x - a(t)\right]\right\}^{-2/(m-1)} , \tag{E.96}$$

so that

$$J(x,t) = -D\frac{\partial u(x,t)}{\partial x}$$

$$= D\left(\frac{2}{m+1}\frac{k}{D}\right)^{1/2} \left\{\frac{m-1}{2}\left[\left(\frac{2}{m+1}\frac{k}{D}\right)^{1/2} x - a(t)\right]\right\}^{-(m+1)/(m-1)} . \tag{E.97}$$

Combination of Eqs. (E.96) and (E.97) allows us to eliminate the unknown $a(t)$, giving

$$u(x,t) = \left[\frac{m+1}{2}(k\,D)^{-1}\right]^{1/(m+1)} J(x,t)^{2/(m+1)} . \tag{E.98}$$

Specifically, at the boundary at $x = 0$, corresponding to the interface studied,

$$u(0,t) = \left[\frac{m+1}{2}(k\,D)^{-1}\right]^{1/(m+1)} J(0,t)^{2/(m+1)} . \tag{E.99}$$

References

1. Abramowitz M, Stegun IA (1972) Handbook of mathematical functions. Dover Publications, New York
2. Bieniasz LK (2011) Analysis of the applicability of the integral equation method in the theory of transient electroanalytical experiments for homogeneous reaction-diffusion systems: the case of planar electrodes. J Electroanal Chem 657:91–97
3. Bronson R (1989) Theory and problems of matrix operations. McGraw-Hill, New York
4. Kamke E (1962) Differentialgleichungen. I. Gewöhnliche Differentialgleichungen, Geest & Portig, Leipzig
5. Smith GD (1985) Numerical solution of partial differential equations. Finite difference methods. Clarendon Press, Oxford

Glossary

This glossary explains the meaning of several important terms that are often used in the present book, but that may not be widely known or adopted.

Distributed species A chemical species present throughout spatially extended phases.

Dynamic species A chemical species, the concentration of which varies during transient electroanalytical experiments.

Equilibrium reaction Any reversible reaction that is assumed, in a mathematical model, to be in a permanent equilibrium. Electrochemical equilibrium reactions are usually called "reversible" (or Nernstian), but this terminology is inconsistent with the general meaning of reversible reactions (see **Reversible reaction** below).

External (spherical or cylindrical) diffusion Diffusion in a symmetrical spatial domain surrounding a sphere or cylinder.

Impermeable boundary A spatial boundary that cannot be crossed by distributed species, so that the flux normal to the boundary is zero.

Incomplete IBVP (or BVP) An IBVP (or BVP) without boundary conditions at the interface studied.

Internal (spherical or cylindrical) diffusion Diffusion in a symmetrical spatial domain inside a sphere or cylinder.

Irreversible reaction Any reaction that is assumed, in a mathematical model, to proceed in one direction (forward or backward) only.

Localised species A chemical species present only at interface(s) between spatially extended phases, for example adsorbed or deposited.

Non-equilibrium reaction Any reversible or irreversible reaction having finite forward and backward reaction rates. In the literature, electrochemical non-equilibrium

L.K. Bieniasz, *Modelling Electroanalytical Experiments by the Integral Equation Method*, Monographs in Electrochemistry, DOI 10.1007/978-3-662-44882-3

(but not irreversible) reactions are usually called "quasi-reversible". The latter terminology is not used in the present book.

Permeable boundary A spatial boundary ensuring continuity of concentration profiles of distributed species, across the boundary.

Positive equilibrium A state of a reaction, in which the forward and backward reaction rates are positive and equal.

Static species A chemical species, the concentration of which remains constant during transient electroanalytical experiments.

Reversible reaction Any reaction that is assumed, in a mathematical model, to proceed in both directions (forward and backward). In the literature, this notion is not consistently used, because in addition to the present general meaning, by "reversible" electrochemical reactions one often means reactions that are in a permanent equilibrium. To avoid this inconsistency, in the present book the latter reactions are called **Equilibrium reactions**.

Zero equilibrium A state of a reaction, in which the forward and backward reaction rates are both equal to zero. This happens, for example, if some reactants are absent. In the case of electrochemical reactions this may also happen when the electrode potential is set to a value at which the reaction does not practically proceed.

List of Symbols

This list contains the explanation of the most important symbols used in the present book. In the text, some of the symbols may have additional indices, superscripts, vector arrows or matrix bars, and/or lists of arguments, which are not provided here. In the case of symbols that represent special functions, the list of arguments is usually omitted here, for brevity, and replaced by the dot $(\cdot)$.

I. Roman Symbols

a, b	Various coefficients, depending on the context
$\boldsymbol{a}, \boldsymbol{b}$, etc.	Vectors (in physical space), see the convention in p. 13
$\vec{a}, \vec{b}$, etc.	Vectors (in spaces corresponding to species, reactions, or unknowns), see the convention in p. 13
$\overline{A}, \overline{B}$, etc.	Matrices (acting on vectors in spaces corresponding to species, reactions, or unknowns, see the convention in p. 13)
A	Area (macroscopic) of an interface studied, m^2
A_{micr}	Area (microscopic) of an interface studied, m^2
$\mathrm{Ai}(\cdot)$, $\mathrm{Ai}'(\cdot)$	Airy function and its derivative
$\mathfrak{A}$	Integral operator
$\mathrm{B}(\cdot)$	Beta function, defined by Eq. (11.33)
$\mathrm{Bi}(\cdot)$	Airy function
c	Concentration of a distributed species, $\mathrm{mol\,m^{-3}}$
$c^{\star}$	Initial/bulk concentration of a distributed species, $\mathrm{mol\,m^{-3}}$
$c^{\dagger}$	Concentration of a distributed species at an interface studied, $\mathrm{mol\,m^{-3}}$
C_{DL}	Differential double layer capacitance, $\mathrm{C\,V^{-1}}$
ct	Computational time, s
d_{F}	Fractal dimension of an irregular interface
div	Divergence operator
D	Diffusion coefficient, $\mathrm{m^2\,s^{-1}}$
$\mathbb{D}$	Integration domain

L.K. Bieniasz, *Modelling Electroanalytical Experiments by the Integral Equation Method*, Monographs in Electrochemistry, DOI 10.1007/978-3-662-44882-3

$\overline{D}$	Matrix of diffusion coefficients
$\mathrm{daw}(\cdot)$	Dawson integral, defined by Eq. (8.157)
e^-	Electron
$\mathrm{erf}(\cdot)$	Error function, defined by Eq. (5.78)
$\mathrm{erex}(\cdot)$	Special function, defined by Eq. (5.77)
$\mathrm{erfc}(\cdot)$	Complementary error function, defined by Eq. (5.76)
$\mathrm{est}\,\langle\cdot\rangle$	A posteriori error estimator (in the adaptive Huber method)
E	Electrode potential, V
E_{rest}	Rest electrode potential, V
$E_{1/2}$	Half-wave potential, V
E^0	Conditional electrode potential, V
E'	True electrode potential, V
F	Faraday constant, $\mathrm{C\,mol^{-1}}$
$F(\cdot)$, $\vec{F}(\cdot)$	Function(s) defining an IE or an IE system
$\mathscr{F}\{f\}$, $\breve{f}$	Fourier transform, defined by Eq. (3.20)
$\mathscr{F}^{-1}\left\{\breve{f}\right\}$	Inverse Fourier transform, defined by Eq. (3.22)
grad	Gradient operator
$\mathrm{G}(\cdot)$	Carslaw and Jaeger integral, defined by Eq. (5.83)
$\mathscr{G}(\boldsymbol{x}, \boldsymbol{\chi})$	Kernel for steady state diffusion in an arbitrary spatial domain
$\mathscr{G}\left(\begin{bmatrix}\boldsymbol{x}\\ t\end{bmatrix}, \begin{bmatrix}\boldsymbol{\chi}\\ \tau\end{bmatrix}\right)$	Kernel for transient diffusion in an arbitrary spatial domain
h	Discrete integration step size
h_{start}	Starting integration step size (in the adaptive Huber method)
h_{max}	Maximum integration step size (in the adaptive Huber method)
$\mathscr{H}_\nu\{f\}$, $\check{f}$	Hankel transform of order ν, defined by Eq. (3.25)
$\mathscr{H}_\nu^{-1}\left\{\check{f}\right\}$	Inverse Hankel transform of order ν, defined by Eq. (3.27)
i, j, k, n, m, etc.	Integer indices
$\boldsymbol{i}_{\mathrm{F}}$	Faradaic current density (vector in ordinary space)
I	Electric current, A
I_{F}	Faradaic current, A
I_{C}	Capacitive current, A
$\mathrm{I}_n(\cdot)$	Modified Bessel function of first kind and order n
J	Flux vector component(s), $\mathrm{mol\,m^{-2}\,s^{-1}}$
$J^{\perp}$	Flux vector component perpendicular to the interface studied, $\mathrm{mol\,m^{-2}\,s^{-1}}$
$\boldsymbol{J}$	Flux vector (in ordinary space), $\mathrm{mol\,m^{-2}\,s^{-1}}$
$\overline{J}$	Jacobian matrix
$\mathrm{J}_n(\cdot)$	Bessel function of the first kind and order n
k	Fourier variable
$k^{\mathrm{f}}, k^{\mathrm{b}}, k^0, k$	Reaction rate constants (forward, backward, conditional), or expressions formally playing the role of such constants)
K	Equilibrium constant

$\overline{K}$	Matrix of rate constants for (pseudo) first order homogeneous reactions
$\mathrm{K}_n(\cdot)$	Modified Bessel function of second kind and order n
$\mathscr{K}(\cdot)$	Kernel of an integral operator
$\mathscr{K}^{\mathrm{inv}}(\cdot)$	Inverse kernel of an integral operator
$\mathscr{K}^{\mathrm{ap}}(t,\tau)$	Kernel for anomalous planar diffusion in a one-dimensional semi-infinite spatial domain, defined by Eq. (5.178)
$\mathscr{K}^{\mathrm{b}}\left(\begin{bmatrix} x \\ t \end{bmatrix}, \begin{bmatrix} \zeta \\ \tau \end{bmatrix}\right)$	Kernel for diffusion to infinite band(s) on a plane (assuming a two-dimensional, infinite spatial domain), defined by Eq. (7.28)
$\mathscr{K}^{\mathrm{c}}(t,\tau)$	Kernel for external cylindrical diffusion in a one-dimensional semi-infinite spatial domain, defined by Eq. (5.85)
$\mathscr{K}^{\mathrm{conv}}(t,\tau)$	Kernel for constant-velocity convection-diffusion in a one-dimensional semi-infinite spatial domain, defined by Eq. (6.15)
$\mathscr{K}^{\mathrm{ct}}(t,\tau)$	Kernel for convection-diffusion in channel or tubular electrodes (assuming a one-dimensional, semi-infinite spatial domain), defined by Eq. (6.70)
$\mathscr{K}^{\mathrm{dr}}\left(\begin{bmatrix} r \\ t \end{bmatrix}, \begin{bmatrix} \zeta \\ \tau \end{bmatrix}\right)$	Kernel for diffusion to disk/ring(s) on a plane (assuming a two-dimensional, infinite spatial domain), defined by Eq. (7.30)
$\mathscr{K}^{\mathrm{ep}}(t,\tau)$	Kernel for convection-diffusion to an expanding plane (assuming a one-dimensional, semi-infinite spatial domain), defined by Eq. (6.31)
$\mathscr{K}^{\mathrm{exci}}(t,\tau)$	Kernel for external cylindrical diffusion in a one-dimensional finite spatial domain with an impermeable second boundary, defined by Eq. (5.150)
$\mathscr{K}^{\mathrm{excp}}(t,\tau)$	Kernel for external cylindrical diffusion in a one-dimensional finite spatial domain with a permeable second boundary, defined by Eq. (5.148)
$\mathscr{K}^{\mathrm{exsi}}(t,\tau)$	Kernel for external spherical diffusion in a one-dimensional finite spatial domain with an impermeable second boundary, defined by Eq. (5.134)
$\mathscr{K}^{\mathrm{exsp}}(t,\tau)$	Kernel for external spherical diffusion in a one-dimensional finite spatial domain with a permeable second boundary, defined by Eq. (5.131)
$\mathscr{K}^{\mathrm{inc}}(t,\tau)$	Kernel for internal diffusion in a one-dimensional finite spatial domain inside an entire cylinder, defined by Eq. (5.155)
$\mathscr{K}^{\mathrm{inci}}(t,\tau)$	Kernel for internal cylindrical diffusion in a one-dimensional finite spatial domain with an impermeable second boundary, defined by Eq. (5.150)

$\mathscr{K}^{\mathrm{incp}}(t,\tau)$	Kernel for internal cylindrical diffusion in a one-dimensional finite spatial domain with a permeable second boundary, defined by Eq. (5.148)
$\mathscr{K}^{\mathrm{ins}}(t,\tau)$	Kernel for internal diffusion in a one-dimensional finite spatial domain inside an entire sphere, defined by Eq. (5.139)
$\mathscr{K}^{\mathrm{insi}}(t,\tau)$	Kernel for internal spherical diffusion in a one-dimensional finite spatial domain with an impermeable second boundary, defined by Eq. (5.134)
$\mathscr{K}^{\mathrm{insp}}(t,\tau)$	Kernel for internal spherical diffusion in a one-dimensional finite spatial domain with a permeable second boundary, defined by Eq. (5.131)
$\mathscr{K}^{\mathrm{p}}(t,\tau)$	Kernel for planar diffusion in a one-dimensional semi-infinite spatial domain, defined by Eq. (5.72)
$\mathscr{K}^{\mathrm{pli}}(t,\tau)$	Kernel for planar diffusion in a one-dimensional finite spatial domain with an impermeable second boundary, defined by Eq. (5.121)
$\mathscr{K}^{\mathrm{plp}}(t,\tau)$	Kernel for planar diffusion in a one-dimensional finite spatial domain with a permeable second boundary, defined by Eq. (5.120)
$\mathscr{K}^{\mathrm{rd}}(t,\tau)$	Kernel for convection-diffusion to an RDE (assuming a one-dimensional semi-infinite spatial domain), defined by Eq. (6.47)
$\mathscr{K}^{\mathrm{s}}(t,\tau)$	Kernel for external spherical diffusion in a one-dimensional semi-infinite spatial domain, defined by Eq. (5.74)
$\overline{\overline{\mathscr{K}}}(t,\tau)$	Matrix of kernels for one-dimensional transport
$\mathrm{kct}(\cdot)$	Special function, defined by Eq. (6.71)
$\mathrm{kcylw}(\cdot)$	Special function, defined by Eq. (5.84)
$\mathrm{kinc}(\cdot)$	Special function, defined by Eq. (5.154)
$\mathrm{kins}(\cdot)$	Special function, defined by Eq. (5.138)
$\mathrm{kpli}(\cdot)$	Special function, defined by Eq. (5.123)
$\mathrm{kplp}(\cdot)$	Special function, defined by Eq. (5.122)
$\mathrm{krd}(\cdot)$	Special function, defined in Sect. 6.2.3
l	Length of a finite spatial interval, m
$\mathscr{L}\{f\}, \hat{f}$	Laplace transform, defined by Eq. (3.12)
$\mathscr{L}^{-1}\left\{\hat{f}\right\}$	Inverse Laplace transform, defined by Eq. (3.14)
n	Number of electrons transferred in an electron transfer reaction
$\boldsymbol{n}$	Vector normal to a closed surface or curve, and directed outwards
N	Number of integration steps (in Chap. 12)
N	Number of coupled reaction-diffusion PDEs (in Appendix E)
N_{r}	Number of reactions
$N_{\mathrm{r}}^{\mathrm{het}}$	Number of heterogeneous reactions
$N_{\mathrm{r}}^{\mathrm{hom}}$	Number of homogeneous reactions

N_s	Number of species
N_s^{dd}	Number of dynamic distributed species
N_s^{distr}	Number of distributed species
N_s^{dl}	Number of dynamic localised species
N_s^{dyn}	Number of dynamic species
N_s^{loc}	Number of localised species
N_s^{sd}	Number of static distributed species
N_s^{stat}	Number of static species
N_K	Number of distinct kernels in an IE or an IDE system
N_Y	Number of integrals in an IE or an IDE system
N_Ψ	Number of unknown functions, solutions of an IE or IDE system
$\overline{N}$	Stoichiometric matrix, defined in Sect. 2.7
$\overline{N}^{dd,hom}$	Stoichiometric submatrix, corresponding to dynamic distributed species and homogeneous reactions
$\overline{N}^{dl,het}$	Stoichiometric submatrix, corresponding to dynamic localised species and heterogeneous reactions
$\overline{N}^{dyn}$	Stoichiometric submatrix, corresponding to dynamic species
$\overline{N}^{dyn,het}$	Stoichiometric submatrix, corresponding to dynamic species and heterogeneous reactions
$\mathcal{O}(\cdot)$	A small quantity, tending to zero according to the formula provided as an argument
p	Hankel variable
$\boldsymbol{p}$	Species production rates
$\boldsymbol{p}^{het}$	Species production rates in heterogeneous reactions, mol m^{-2} s^{-1}
$\boldsymbol{p}^{dd,het}$	Production rates of dynamic distributed species in heterogeneous reactions, mol m^{-2} s^{-1}
$\boldsymbol{p}^{dl,het}$	Production rates of dynamic localised species in heterogeneous reactions, mol m^{-2} s^{-1}
$\boldsymbol{p}^{het,\star}$	Species production rates in heterogeneous reactions, in the initial state of an electrochemical system, mol m^{-2} s^{-1}
$\boldsymbol{p}^{hom}$	Species production rates in homogeneous reactions, mol m^{-3} s^{-1}
Q_{DL}	Double layer charge, C
$Q_{\kappa,m,n,l,k}$	Moment integrals of the kernels, defined by Eq. (12.8)
r, r^f, r^b	Reaction rates (total, forward, backward)
r^{het}	Rate of a heterogeneous reaction, mol m^{-2} s^{-1}
r^{hom}	Rate of a homogeneous reaction, mol m^{-3} s^{-1}
r	Radial coordinate, m
r_0	Radius of a (spherical or cylindrical) interface studied, m
R	Gas constant, J K^{-1} mol^{-1}
$R_{\kappa,n,l,k}$, $\overline{R}_{n,l,k}$	Coefficients of product integration methods, and related matrices

R_U, R_{U1}, R_{U2}	Uncompensated resistance(s), Ω
$\Re\{\cdot\}$	Real part of a complex number
s	Laplace variable
S	Closed surface in a three-dimensional space, or closed curve in a two-dimensional space
$S_{\kappa,n,l,k}$, $\overline{S}_{n,l,k}$	Coefficients of product integration methods, and related matrices
$\mathrm{S}(\cdot)$	Special function defined by Eq. (2.40)
$\overline{S}$	Similarity transformation matrix
t	Time, s
tol	Error tolerance parameter (in the adaptive Huber method)
T	Absolute temperature, K
$[\cdot]^{\mathrm{T}}$	Transposed vector
$\mathfrak{u}$	Ionic mobility, $\mathrm{m^2\,mol\,s^{-1}\,J^{-1}}$
u, w, etc.	Auxiliary dependent variables
$\overrightarrow{u}^{(n)}$, $\overrightarrow{v}^{(n)}$	generalised eigenvector(s) of nth order (in spaces corresponding to species, reactions, or unknowns)
v	Convection velocity vector component, $\mathrm{m\,s^{-1}}$
$\boldsymbol{v}$	Convection velocity vector, $\mathrm{m\,s^{-1}}$
x	Cartesian coordinate, m
$\mathrm{X_1}$, $\mathrm{X_2}$, etc.	Abstract names of chemical species
y	Cartesian coordinate, m
y	Integrals, approximate values obtained by numerical methods
Y	Integrals, exact values
$\mathrm{Y}_n(\cdot)$	Bessel function of the second kind and order n
z	Cartesian coordinate, m
$\mathfrak{z}$	Number of proton charges carried by a species
$\overrightarrow{0}$	Zero vector
$\overline{0}$	Zero matrix
$\overline{1}$	Unit matrix

II. Greek Symbols

α	Parameter characterising anomalous diffusion, defined by Eq. (2.64)
α^{f}, α^{b}	Charge transfer coefficients (for forward and backward electrode reactions)
β	Parameter characterising convection to an RDE, defined by Eq. (6.33)
γ_{E}	Euler's constant
γ	Parameter characterising convection in tubular or channel electrodes, defined by Eqs. (6.56) or (6.58), respectively
$\varGamma$	Concentration of a localised species, $\mathrm{mol\,m^{-2}}$
$\varGamma^{\star}$	Initial concentration of a localised species, $\mathrm{mol\,m^{-2}}$
$\Gamma(\cdot)$	Euler Gamma function
δ	Error(s) of the approximate solution(s) ψ

$\delta(\cdot)$	Dirac delta function
$\mathit{\Delta}$	Error(s) of the approximate integral(s) y
$\triangle$	Laplacian operator
$\Delta \Phi$	Galvani potential difference, V
$\Delta \Phi_{\mathrm{rest}}$	Rest value of the Galvani potential difference, V
$\Delta \Phi'$	True Galvani potential difference, V
ε	Dimensionless electrode potential
ζ, η, ξ	Auxiliary independent variables (usually spatial coordinates, possibly dimensionless)
λ	Reorganisation energy, J mol^{-1}
λ, Λ	Eigenvalue(s)
$\lambda^{\mathrm{f}}, \lambda^{\mathrm{b}}$	Reaction orders (for forward and backward reactions)
λ_{inn}	Inner cut-off, defined in Sect. 2.12
λ_{out}	Outer cut-off, defined in Sect. 2.12
μ	Parameter characterising diffusion in finite, one-dimensional spatial domains (intervals), defined by Eq. (A.3), s$^{-1/2}$
ν	Kinematic viscosity, m^2 s^{-1}
ν	Stoichiometric number of a rate determining step
$\nu^{\mathrm{f}}, \nu^{\mathrm{b}}$	Stoichiometric coefficients (of forward or backward reactions)
ω	Rotation frequency of an RDE, rad s^{-1}
ω	Current or potential perturbation frequency (in AC electroanalytical techniques), rad s^{-1}
Ω	Three- or two-dimensional spatial domain
ϱ	Parameter characterising spherical or cylindrical diffusion, defined by Eq. (A.2), s$^{-1/2}$
$\sigma_{j,\mathrm{F}}$	Parameter characterising chronoamperometric limiting current under conditions of anomalous diffusion, defined by Eqs. (2.65) and (2.66)
$\vec{\sigma}$	Vector of homogeneous reaction rate terms dependent exclusively on the concentrations of static species, defined by Eq. (2.52)
$\vec{\sigma}_{1/2}$	Vector of interpolation errors (in Chap. 12)
τ	Dummy integration variable (time), s
υ	Parameter characterising some kernels for convection-diffusion, defined by Eqs. (B.16) or (B.43) or (B.89), depending on the convection field
ϑ, θ	Auxiliary independent variables (usually time-like coordinates, possibly dimensionless). Often $\vartheta = t - \tau$
$\Theta_n(\cdot)$	Exponential theta-n function
$\varphi(\cdot)$	Function(s) representing some of the integral transformation kernel(s) of the convolution type
ϕ	Phase shift (in AC electroanalytical techniques)
Φ	Inner electric potential (Galvani potential), V
ψ	Solution(s) of IEs or IDEs, approximate values
$\Psi, \Psi(\cdot)$	Solution(s) of IEs or IDEs, exact values

About the Author

Lesław K. Bieniasz (M.Sc., Ph.D., Dr.Scient.) is currently assistant professor at the Faculty of Physics, Mathematics, and Computer Science, of the Cracow University of Technology, in Cracow, Poland. Formerly he was associated with the Institute of Physical Chemistry of the Polish Academy of Sciences. He received his M.Sc. degree with honours in technical physics at the University of Mining and Metallurgy, Cracow; Ph.D. in electrochemistry at the Institute of Physical Chemistry of the Polish Academy of Sciences, Warsaw, Poland; and DrScient degree at the University of Aarhus, in Denmark. He was a visiting researcher at the University of Aarhus, Denmark (Prof. D. Britz, in 1992–1993);Tübingen University, Germany (Prof. B. Speiser, in 1997 and 1998); Institute of C.E.A. in Saclay, France (Dr. C. Bureau, in 1999); Saitama University, Japan (Prof. S. Nakabayashi, in 2002); Princeton University, USA (Prof. H. Rabitz, in 2005); and University of Murcia, Spain (Prof. Á. Molina, in 2010). His scientific interests include theoretical and computational electrochemistry, as well as computational science and computer science, with the focus on the automation of scientific research. In particular, he is interested in the development of automatic numerical methods and computer-aided approaches serving for the theoretical modelling, computer simulation, and data analysis in electroanalytical chemistry. He is the author of over 100 scientific papers and conference publications corresponding to these research areas.

L.K. Bieniasz, *Modelling Electroanalytical Experiments by the Integral Equation Method*, Monographs in Electrochemistry, DOI 10.1007/978-3-662-44882-3

About the Editor

Fritz Scholz is professor at the University of Greifswald, Germany. Following studies of chemistry at Humboldt University, Berlin, he obtained a Dr. rer. nat. and a Dr. sc. nat. (habilitation) from that University. In 1987 and 1989 he worked with Alan Bond in Australia. His main interest is in electrochemistry and electroanalysis. He has published more than 300 scientific papers, and he is editor and co-author of the book "Electroanalytical Methods" (Springer, 2002, 2005, 2010, and Russian Edition: BINOM, 2006), co-author of the book "Electrochemistry of Immobilized Particles and Droplets"(Springer 2005), co-editor of the "Electrochemical Dictionary" (Springer, 2008; 2nd edn. 2012), and co-editor of volumes 7a and 7b of the "Encyclopedia of Electrochemistry" (Wiley-VCH 2006). In 1997 he has founded the *Journal of Solid State Electrochemistry* (Springer) and serves as Editor-in-Chief since that time. In 2014 he has founded the journal *ChemTexts—The Textbook Journal* (Springer). He is the editor of the series "Monographs in Electrochemistry" (Springer) in which modern topics of electrochemistry are presented. Scholz introduced the technique "Voltammetry of Immobilized Microparticles" for studying the electrochemistry of solid compounds and materials, he introduced three-phase electrodes to determine the Gibbs energies of ion transfer between immiscible liquids, and currently he is studying the interaction of free oxygen radicals with metal surfaces, as well as the interaction of liposomes with the surface of mercury electrodes in order to assess membrane properties. In the last years, his research also comprises the interaction of electrode surfaces with free oxygen radicals.

L.K. Bieniasz, *Modelling Electroanalytical Experiments by the Integral Equation Method*, Monographs in Electrochemistry, DOI 10.1007/978-3-662-44882-3

Index

L.K. Bieniasz, *Modelling Electroanalytical Experiments by the Integral Equation Method*, Monographs in Electrochemistry, DOI 10.1007/978-3-662-44882-3

J

K

L

M

N

O

P

Q

R

S

MIX
Papier aus verantwortungsvollen Quellen
Paper from responsible sources
FSC® C105338

If you have any concerns about our products,
you can contact us on
ProductSafety@springernature.com

In case Publisher is established outside the EU,
the EU authorized representative is:
Springer Nature Customer Service Center GmbH
Europaplatz 3, 69115 Heidelberg, Germany

Printed by Libri Plureos GmbH
in Hamburg, Germany